THE SCIENCE OF GENETICS
An introduction to heredity

THE SCIENCE OF GENETICS
An introduction to heredity
FIFTH EDITION

George W. Burns
Ohio Wesleyan University

Macmillan Publishing Co., Inc.
New York

Collier Macmillan Publishers
London

Cover photograph from R. S. Verma et al.: A rapid method of constitutive heterochromatin of secondary constriction regions of human chromosomes 1, 9, and 16. *Journal of Heredity*, **73**: 74–76, 1982. Copyright © 1982 by the American Genetic Association.

Macmillan Publishing Co., Inc.
866 Third Avenue, New York, New York 10022

Collier Macmillan Canada, Inc.

Library of Congress Cataloging in Publication Data

Burns, George W. Date:
   The science of genetics.

Includes index.
   1. Genetics.   I. Title.   [DNLM: 1. Genetics.
QH 430 B967S]
QH430.B87   1983   575.1   82-8963
ISBN 0-02-317120-0   AACR2

Printing: 1 2 3 4 5 6 7 8   Year: 3 4 5 6 7 8 9 0

ISBN   0-02-317120-0

Preface

The rate of doubling of genetic knowledge has now dropped to just over one year. Not only does this fact lend an air of excitement to any textbook writer's effort, but it is also a source of frustration. It becomes increasingly difficult to determine what must be omitted to keep within reasonable bounds as answers to "old" questions become better defined and new questions can be asked. The many thoughtful and useful suggestions made by users of previous editions have been carefully considered. If all suggested additional topics had been incorporated, the length of the book would have been unreasonably extended. Therefore, the final responsibility for inclusions and exclusions rests with the author. Of course excluded topics give the instructor the opportunity of adding material in class.

As in previous editions, human genetic patterns have been emphasized, but not at the expense of plant and lower animal illustrations. Treatments of sex anomalies, chromosome aberrations, and genetic polymorphisms of human beings have been retained. The human karyotype, both normal and abnormal, has been treated at some length. A list of selected chromosome assignments for human genetic loci is included. The major histocompatibility complex, and its human leukocyte antigens (HLA), with its relationship to disease has been expanded.

The exciting, continuing story of recombinant DNA, and its promises for the human race, has been thoroughly updated. Bacterial plasmids are, of course, included. Problems have been revised and augmented, and key references to the literature have been included.

Again the problem approach has been followed, using work of men and women who have contributed most to the development of the science. Topics are developed inductively, from observation to explanation to principle. The historical theme—starting with "classical" genetics and progressing through molecular genetics, wherein the story of our increasing knowledge about the field unfolds for the student as it has over the years for humanity—has been employed again in this edition. Some past users have found it feasible and desirable to begin with the subject matter of Chapters 16 through 19, then return to Mendelian genetics. Continuity of material lends itself to either that approach or one of following chapters in sequence. I have used the latter pathway with classes more or less evenly divided between majors in the sciences and in the arts and humanities. Although intended primarily for upperclass students, the content and sequence in this text have succeeded as well with freshmen in my classes.

Salient facts of cell structure and behavior are reviewed where needed. Appendices contain answers to problems, life cycles of animals and plants used in genetic research, structural formulas of the biologically important amino acids, a summary of mathematical formulas, ratios and statistical tests that have been developed in the text, a table of metric values and a list of journals. The glossary has been slightly expanded.

The continuing help of many persons who have made this book possible is gratefully acknowledged. Although a number of my former students, as well as professional geneticists around the world, have been kind enough to provide photos of interesting and useful material, special thanks must go to my former student, Ms. Dawn DeLozier (who is know at the Institut Universitaire de Génétique Médicale in Geneva) for her tireless work in providing me with photographs of outstanding quality and relevance. Another former student, Ms. Janet Raup, now a legislative intern with the Ohio Legislature, has given invaluable aid by furnishing timely information on and copy of new state legislation regarding paternity testing. Persons graciously providing photographs and electron micrographs from their own research have been made available by Drs. Barbara J. Bachmann, G. F. Bahr, Theodore A. Baramki, Mihaly Bartalos, E. J. DuPraw, Richard C. Juberg, C. C. Lin, K. Brooks Low, Herbert A. Lubs, A. L. Olins, and Wayne Wray. In addition, Ms. Barbara Bruce, another of my former students, provided several reprints of papers covering research in which she participated. The National Foundation-March of Dimes generously gave permission to reproduce drawings and tables from *Paris Conference (1971): Standardization in Human Cytogenetics*, and also from *Paris Conference (1971), Supplement (1975), Standardization in Human Cytogenetics.* I am also indebted to the literary executor of the late Sir Ronald A. Fisher, F.R.S., and to Oliver and Boyd, Ltd., Edinburgh, for their permission to reprint Table 3 from *Statistical Methods for Research Workers.* Finally, and above all, to my students, who have always made the teaching of genetics an exciting and rewarding experience, this book is dedicated with real affection.

G. W. B.

Contents

CHAPTER 19   MOLECULAR STRUCTURE OF THE GENE   367

CHAPTER 20   REGULATION OF GENE ACTION   392

CHAPTER 21   CYTOPLASMIC GENETIC SYSTEMS   406

CHAPTER 22   THE NEW GENETICS, RECOMBINANT DNA, AND THE FUTURE   421

CHAPTER 1

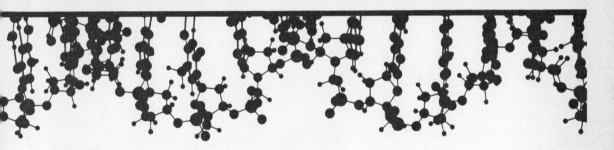

Introduction

One of the most exciting fields of biological science, if not of all science, is genetics, the study of the mechanisms of heredity, by which traits or characteristics are passed from generation to generation. Essentially a product of the twentieth century, modern genetics has made almost explosive progress from the rediscovery in 1900 of Mendel's basic observations of the 1860s to a fairly comprehensive understanding of underlying principles at the molecular level. As our knowledge of these operating mechanisms developed, it became apparent that they are remarkably similar in their fundamental behavior for all kinds of organisms, whether man or mouse, bacterium or corn. But geneticists' quest for truth and understanding is far from completed; as in other sciences, the answer to one question raises new ones and opens entirely new avenues of inquiry.

Genetics is personally relevant to everyone. Humans are genetic animals; each is the product of a long series of matings. People differ with regard to the expression of many traits; one has certain inherited characteristics of both the father and the mother, but often, as well, some not exhibited by either parent. Familiar examples abound—hair or eye color, curly or straight hair, height, intelligence, and baldness, to list but a few. Less obvious genetic traits include such diverse ones as form of ear lobes (Fig. 1-1), ability to taste the chemical phenylthiocarbamide (PTC), red-green color blindness, hemophilia or "bleeder's disease," extra fingers or toes, astigmatism, blood group (A, B, AB, O), production of rhesus antigen (Rh+ or Rh−), handedness, drooping of eyelids (ptosis), sickle-cell disorder, or ability to produce insulin (lack of which results in diabetes). Note that some of these characteristics seem purely morphological, being concerned primarily with form and structure, whereas others are clearly physiological. Look about you at your friends and family for points of difference or

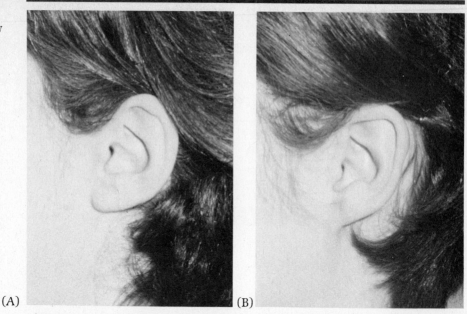

(A)   (B)

Figure 1-1. Inherited difference in form of ear lobe. (A) Free ear lobe, the result of a dominant gene. (B) Attached ear lobe, caused by a recessive allele.

similarity. Most of these characteristics have genetic bases. You are undoubtedly aware also of mentally retarded persons in our population and of infants so grossly deformed that they are born dead or die within a very short time. Many of these unfortunate cases have underlying genetic and/or cytological causes. The determination of one's sex and the occurrence of sex intermediates (hermaphrodites and pseudohermaphrodites) also have genetic and cytological causes.

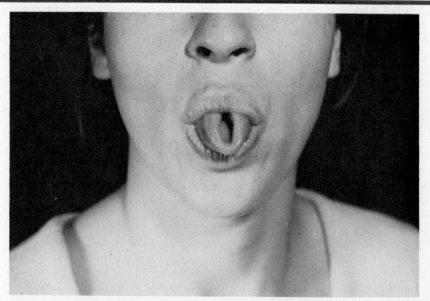

Figure 1-2. Ability to roll the tongue in this fashion was formerly believed to be an inherited trait, but more recent evidence indicates it is a learned ability.

Figure 1-3. **Homunculus,** *"little man in a sperm cell."* (Drawing
by N. Hartsoecker, *Journal des Scavans,* Feb. 7, 1695.)

A survey of our long interest in heredity is outside the scope of this book, but
it is well established that as far back as 6,000 years ago, records of pedigrees of
such domestic animals as the horse and of such crop plants as rice were kept.
Because certain animals and plants were necessary for survival and culture,
attempts have been made since the beginning of recorded history at least to
develop improved varieties. But the story of human concern with heredity during
our species' lifetime on this planet has been, until recently, one of interest largely
in results rather than in fundamental understanding of the mechanisms involved.

The development of ideas relating to these mechanisms has been found to be
replete with misconceptions, many of which are naïve in the light of modern
knowledge. Some, however, have been surprisingly close to the truth, especially
in view of a lack of sophisticated equipment. These theories may be divided
roughly into three categories: (1) *"vapors and fluids,"* (2) *preformation,* and (3)
particulate inheritance. Such early Greek philosophers as Pythagoras (500 B.C.)
proposed that "vapors" derived from various organs unite to form a new individual. Then Aristotle assigned a "vitalizing" effect to semen, which, he suggested, was highly purified blood, a notion that was to influence thinking for
almost 2,000 years.

By the seventeenth century sperm and egg had been discovered, and the Dutch
scientist Swammerdam theorized that sex cells contained miniatures of the adult.
Literature of that time contains drawings of models or manikins within sperm
heads, which imaginative workers reported seeing (Fig. 1-3). Such theories of
preformation persisted well into the eighteenth century, by which time the German investigator Wolff offered experimental evidence that no preformed embryo
existed in the egg of the chicken.

But Maupertuis in France, recognizing that preformation could not easily account for transmission of traits to the offspring from both parents had proposed

in the early 1800s that minute particles, one from each body part, united in sexual reproduction to form a new individual. In some instances, he reasoned, particles from the male parent might dominate those from the female, and in other cases the reverse might be true. Thus the notion of particulate inheritance came into consideration. Maupertuis was actually closer to the truth, in general terms, than anyone realized for more than a century.

Charles Darwin postulated in the nineteenth century, essentially the same basic mechanism in his theory of pangenesis, the central idea of which had first been suggested by Hippocrates (400 B.C.). Under this concept, each part of the body produced minute particles ("gemmules"), which were contained in the blood of the entire body but eventually concentrated in the reproductive organs. Thus, an individual would represent a "blending" of both parents. Moreover, acquired characters would be inherited because as parts of the body changed, so did the pangenes they produced. A champion weight lifter, therefore, should produce children with strong arm muscles. Such transmission of acquired traits, however, does not occur.

Pangenesis was disproved later in the same century by the German biologist Weismann. In a well-known experiment he cut off the tails of mice for 22 generations, yet each new lot of offspring consisted only of animals with tails. If the source of pangenes was removed, how, he reasoned, could the next generation have tails? In spite of these early misunderstandings concerning the idea of particulate inheritance, however, this basic concept is central to our modern understanding.

Most attempts to explain observed breeding results failed because investigators generally tried to encompass simultaneously *all* variations, whether heritable or not. Nor was the progress of scientific thought or the development of suitable equipment and techniques ready to help point the way. The Augustinian monk Gregor Mendel laid the groundwork for the modern concept of the particulate theory. By attacking the problem in logical fashion, concentrating on one or a few observable, contrasting traits in a controlled breeding program, and suggesting causal "factors" (which are now called *genes*), Mendel opened the door for modern investigations. Mendel came closer to a real understanding of heredity than anyone had in the preceding 5,000 years or more. Still to be elucidated, however, was an understanding of the cellular mechanisms involved.

Characteristics of useful experimental organisms

Even though Mendel's approach was probably as much the result of luck as of thoughtful planning, it was superb in simplicity and logic. Mendel was fortunate to have in the garden pea (*Pisum sativum*) an organism well suited to genetic study. There are six important considerations in choosing a plant or animal for genetic experiments:

1. Variation.   The organism chosen should show a number of detectable differences. Nothing could be learned of the inheritance of skin color in man, for example, if all human beings were alike in this respect. In general, the larger the number of discontinuously differing traits and the more clearly marked they are, the greater the usefulness of the species for genetic study.

2. Recombination.   Genetic analysis of a species is greatly expedited if there exists some effective means of combining, in one individual, traits of two parents. Such *recombination* permits comparison of one expression of a character with another expression of the same trait (e.g., tall versus dwarf *size*, brown versus blue *eye color*) through several generations. In many organisms recombination occurs as the result of *sexual reproduction*, in which two sex cells (**gametes**), generally from two different parents, combine as a fertilized egg (**zygote**). The sexual process is characteristic of higher animals and plants, although it occurs in many lower forms as well. In bacteria and viruses there occur such processes as conjugation and transduction, which also bring about recombination. Life cycles of several genetically important organisms are reviewed in Appendix B.

On the other hand, many organisms reproduce *asexually* or *vegetatively* and cannot furnish recombinational information. Basically, asexual reproduction may involve specialized cells (often called **spores**), daughter cells (in unicellular forms), parts of a single parent (cuttings, grafts, fragmentation, etc.), or **partheno-genesis,** in which an individual develops from an unfertilized egg, as in the male honeybee. By and large, a means of recombination is required for genetic study.

3. Controlled Matings.   Systematic study of an organism's genetics is far easier if it is possible to control mating, choose parental lines with particular purposes in mind, and keep careful records of offspring through several generations. The mouse (*Mus*), fruit fly (*Drosophila*, Fig. 1-4), corn (*Zea*), and the orange-pink bread mold (*Neurospora*), for instance, make better genetic subjects in this respect than do humans. The study of human genetics has been largely depend-ent on pedigree analysis, or studies of traits in a given family line for several past generations. Such analyses are, of course, useful, but are somewhat limiting in that geneticists must rely on lines of progeny that have been established and cannot devise crosses of their own. However, it should be noted that geneticists now have powerful new investigative tools, e.g., biochemical studies (such as the biochemistry of inherited disorders and the sequencing of proteins and of the genetic material—deoxyribonucleic acid, or DNA—of most organisms); cy-tological studies, including electron microscopy and fluorescence microscopy; hybrid cells created in the laboratory; and recombinant DNA. Most of these techniques and their fruits will be examined in subsequent chapters.

4. Short life cycle.   Acquisition of genetic knowledge is facilitated if the or-ganism chosen requires only a short time between generations. Mice, which are sexually mature at five or six weeks of age and have a gestation period of about 19 to 31 days, are much more useful, for instance, than elephants, which mature in 8 to 16 years and have a gestation period of nearly two years. Likewise, the fruit fly, *Drosophila*, is much used in investigations because it can provide a generation every 10 days. But first place for short life cycles (although asexual) goes to bacteria and bacteriophages (or simply phages), i.e. viruses which infect bacteria. Under optimum conditions, a generation time of around 20 minutes is usual.

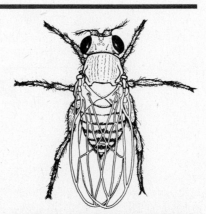

Figure 1-4. *The fruit fly,* **Drosophila melan-ogaster,** *an animal highly useful in genetic re-search.*

5. Large number of offspring. Genetic studies are greatly speeded if the organism chosen produces fairly sizable lots of progeny per mating. Cattle, with generally one calf per breeding, do not provide nearly as much information in a given number of generations as many lower forms of life, where offspring may number many thousands. Obviously, the larger the number of progeny, the greater the likelihood of detecting rare events.

6. Convenience of handling. For practical reasons, an experimental species should be of a type that can be raised and maintained conveniently and relatively inexpensively. Whales are obviously less useful in this context than are bacteria.

Methods of genetic study

Mendel's studies with garden peas illustrate an approach useful both in classical, descriptive genetics and in modern molecular studies. This approach is the *planned breeding experiment,* in which parents exhibiting contrasting expressions of the same trait or traits are mated or crossed and careful records of results are kept through several generations. A number of such experiments are examined in the next and succeeding chapters.

In addition to experimental breeding, pedigree analysis is used in cases where controlled breeding programs are impossible. Pedigrees of three different inherited conditions are shown in Figures 1-5 to 1-7. Figure 1-5 illustrates a pedigree of polydactyly, i.e, the occurrence of extra fingers. Shaded symbols represent individuals with this condition and unshaded ones represent persons with the usual five fingers. As is customary in pedigree diagrams, squares represent males, circles females. A polydactylous man and woman (generation I) have three children, a polydactylous daughter, a polydactylous son, and a "normal" son (generation II). The first and third individuals of generation II each marry "normal" persons; the resulting children are shown in generation III. Note in this case that affected individuals appear only when at least one parent is polydactylous; marriages between five-fingered persons appear to produce only "normal" offspring.

A different kind of situation is illustrated in Figure 1-6. Here the condition involved is the shape of the ear lobe, which may be either "free" or "attached"

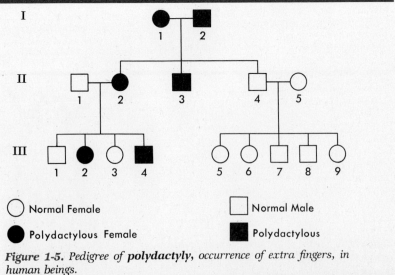

Figure 1-5. *Pedigree of **polydactyly**, occurrence of extra fingers, in human beings.*

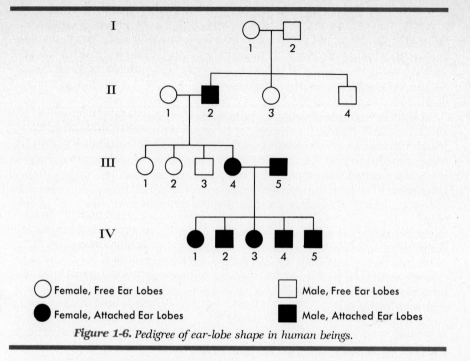

Figure 1-6. *Pedigree of ear-lobe shape in human beings.*

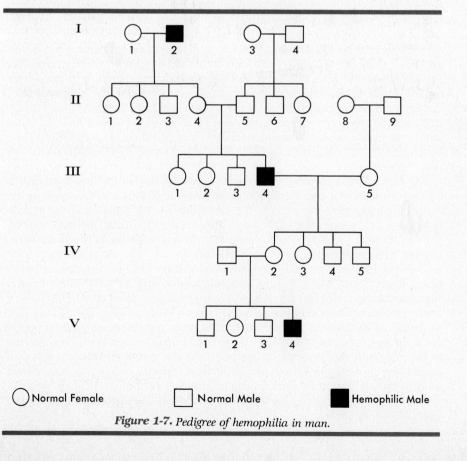

Figure 1-7. *Pedigree of hemophilia in man.*

(Fig. 1-1). In this pedigree, the shaded symbols represent persons with attached ear lobes. The pedigree indicates that an individual may display attached ear lobes without either parent showing this form of the trait; whereas, if both parents have attached ear lobes, all the children have the same condition. This does not mean that heredity is not operating in this case. Certainly, however, the genetic situation here is not the same as in the previous illustration of polydactyly.

As another example of pedigree, note the case shown in Figure 1-7. Here, the shaded symbols represent hemophilia, in which an impairment of the clotting mechanism causes severe bleeding from even minor skin breaks with life-threatening consequences. In this instance the trait appears to be confined to the males, but is transmitted from an affected male through his daughters to some of his grandsons. This pattern, repeated in many pedigrees of hemophilia, certainly suggests a genetic basis, but one that is quite different from either of the first two.

Finally, *statistical analyses* of several kinds are used (1) to predict the probability of certain results in untried crosses and (2) to provide degrees of confidence in a theory regarding the specific genetic mechanism operating in a given case. Geneticists primarily employ several statistics of probability, which may be applied either to the results of an experimental breeding program or to a particular pedigree.

Fields of study useful in genetics

After an initial and appropriate preoccupation with descriptive genetics, scientists turned naturally to problems of the mechanics of the processes they observed. The "what" of the early twentieth century rapidly gave way to a concern with "how." Parallels between inheritance patterns and the structure and behavior of cells were noted by a number of pioneer investigators. Thus *cytology* rapidly became an important adjunct to genetics. In fact, a pair of papers by Sutton as early as 1902 and 1903 (only two and three years, respectively, after the rediscovery of Mendel's landmark papers initiated the modern age of genetics) clearly pointed the way to a physical basis for the burgeoning science of heredity. Sutton concluded his 1902 paper with a bold prediction: "I may finally call attention to the probability that (the behavior of chromosomes) may constitute the physical basis of the Mendelian law of heredity." As a result, the door was opened to an objective examination of the physical mechanisms of the genetic processes.

The rapid development of the science of genetics during the first third of this century resulted in a considerable body of knowledge for such organisms as the fruit fly (*Drosophila*), corn (*Zea*), the laboratory mouse (*Mus*), and the tomato (*Lycopersicon*), concerning *what* traits are inherited and how different expressions of these are related to each other. Genetic *maps*, based on breeding experiments, were constructed for these and other species. The maps showed which genes were located on which chromosomes and the relative distances between *linked* genes. (Such maps and the methods used to acquire information needed to construct them will be examined in Chapter 6.) Geneticists next began to turn from concern with inheritance patterns of such traits as eye color in fruit flies to problems of *how* the observable trait is produced. Especially in the period since the beginning of World War II, a central question has been the *structure* of the gene and the mode of its operation. As the search for answers has proceeded ever more deeply into molecular levels, Chemistry and physics have played an increasingly important part in genetic study. Contributions of these sciences have enabled geneticists to gain a clear concept of the molecular nature and operation of the gene.

Figure 1-8. *Effects of breeding programs in cattle. (A) Texas longhorn.* (Photo courtesy of Dr. Charles E. Denman, Oklahoma State University.) *(B) Hereford, bred for mean production.* (Photo courtesy of National Hereford Association.)

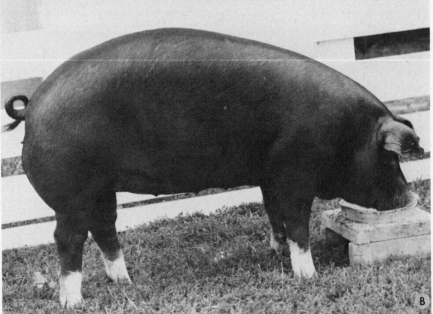

Figure 1-9. *Effects of breeding programs in swine. (A) Wild boar.* (Photo courtesy of Tennessee Game and Fish Commission. By permission.) *(B) Modern Poland China gilt.* (Photo courtesy Poland China Record Association. By permission.)

Practical applications of genetics

Genetics appeals to many people not only because they are part of an ongoing genetic stream, but also because it has had such an exciting history, in which theory has evolved out of observation and led, in turn, to experimental proof of fundamental operating mechanisms. Of course, any science may make the same claim, but the history of our knowledge and understanding of genetics is to other sciences as a time-lapse movie of a growth process is to a normal-speed film. A fraction of a century ago, the scientific community at large knew nothing of genetic mechanisms. Now, however, molecular models of genes can be constructed with considerable accuracy, atom by atom. In fact, some relatively simple genes have been synthesized in several laboratories. But besides being a fascinating intellectual discipline intimately related to ourselves, genetics has many important practical applications, and new ones are being developed at a rapid rate. Some of these are fairly familiar; others may be less so. The history of improvement of food crops and domestic animals by selective breeding is too well known to warrant detailed description here (Figs. 1-8, 1-9). Increases in yield of crops such as corn and rice, improvement in flavor and size, as well as the production of seedless varieties of fruits, and advances in meat production of cattle and swine have benefited humanity significantly. As the population of the world continues to increase, this practical utilization of genetics is likely to assume even greater significance. Appropriately, the 1970 Nobel Peace Prize was awarded to a scientist, Norman Borlaug, for more than a quarter century of successful work in breeding high-yield, stiff-stemmed varieties of Mexican wheat. These new varieties, incorporating genes from American, Japanese, Australian, and Colombian stocks, not only have much-improved yield, but also wide geographic, photoperiodic, and climatic adaptability, being successfully grown in such varied parts of the world as Mexico, Turkey, Afghanistan, Pakistan, and India. In a five-year period, introduction of the new varieties into India resulted in raising the wheat crop from 12 to 21 million tons, a rate of increase greater than that of the country's population. In fact, India had become self-sufficient in grain production by 1981. Borlaug's efforts have bought precious time in efforts to gain control of population growth. Similarly, the effort to breed disease-resistant plants must be never-ending because of the frequent occurrence of new mutant strains of disease-causing organisms to which widely used varieties of a given food crop may be susceptible. Not only are mutants of crop-disease organisms among the important factors, but the genetic homogeneity bred into crop plants, especially cereal grains, is important as well. The 1970 epidemic of southern corn blight in the United States is a case in point.

The ability to insert a gene or genes from one species into the genetic material of another is becoming easier with the passage of time. Experimental insertion of the human insulin gene into some of the genetic material of the bacterium *Escherichia coli*, an inhabitant of the gut of most mammals, including human beings, was achieved in late 1978. Prospects for abundant supplies of human insulin to replace the more costly cattle and pig hormone (to which some persons are allergic) are now excellent. We shall return to this kind of work in Chapter 19. In addition, scientists are exploring the feasibility of inserting genes for nitrogen fixation from certain bacteria into the genetic make-up of such crop plants as corn and wheat. When this endeavor is successful (and success appears to be much closer than believed earlier), tremendous savings in fertilizer and other agricultural costs can be realized.

Applications of genetics in the general field of medicine are numerous and growing. Many diseases and abnormalities are now known to have genetic bases. Hemophilia, some types of diabetes, an anemia known as hemolytic icterus, some forms of deafness and of blindness, several hemoglobin abnormalities, and Rh incompatibility are a few conditions that fall into this category. Recognition of the inherited nature of such disorders and of many malformations is important in anticipating their possible future occurrence in a given family, in order that appropriate preventive steps may be taken.

Closely related is the whole field of genetic counseling. An estimate of the likelihood of a particular desirable or undesirable trait appearing in the children of a given couple can be provided by one who has sound genetic training and some information on the ancestors of the prospective parents. Questions might range from the probability of a couple's having any red-haired children to the chance of muscular dystrophy appearing in the offspring.

The related matter of *screening* populations or large segments thereof for the occurrence of genes causing one or more of a variety of genetic disorders (e.g., **Tay-Sachs disorder,** principally in Ashkenazi Jews; **thalassemia,** a form of anemia occuring primarily in Greeks and Italians; **ankylosing spondylitis,** or "poker spine," a form of spinal bone deterioration most common in whites; **sickle-cell anemia,** a blood disorder commonest in blacks) involves ethical and racial questions. The purpose of screening has not always been made clear to members of the populations in which such disorders have higher rates of incidence.

So in this study of genetics, the same general route followed by other scientists in their search for an understanding of the mechanisms of inheritance will be followed. We will begin with simple observations that could be made by anyone, in many cases without a laboratory or special equipment; questions will be raised continually, and possibilities concerning the causes of observed phenomena will be examined. The specific principles that form the groundwork of modern genetics will be explored by inductive reasoning and tested deductively by application to still other cases. The quest for the "why" and "how" of genetics will become progressively more specific until answers on the deepest molecular levels yet penetrated have been reached. In the process, not only will some fundamental genetic knowledge about human beings be acquired, but also the powers of critical, analytical, and skeptical thinking will be sharpened.

References

Iltis, H., 1932. *Life of Mendel.* New York: W. W. Norton.

Martin, N. G., 1975. No Evidence for a Genetic Basis of Tongue Rolling or Hand Clasping. *Jour. Hered.,* **66:** 179–180.

Mendel, G., 1865. *Experiments in Plant Hybridization.* Reprinted in J. A. Peters, ed., 1959. *Classic Papers in Genetics.* Englewood Cliffs, N.J.: Prentice-Hall.

Sturtevant, A. H., 1965. *A History of Genetics.* New York: Harper & Row.

Problems

(Some of these problems may necessitate consultation of sources outside the textbook.)

1-1 Suggest a number of ways in which bacteria make good genetic subjects.

1-2 Why are organisms that have only a single set of chromosomes (monoploid) often more favorable for genetic experiments than those having two sets (diploid)?

1-3 There are several species of wheat, each with a different number of chromosomes in its somatic (body) cells: einkorn, with 14; durum, with 28; and the commonly cultivated species, with 42. (a) Can you see any pattern in these chromosome numbers? Explain. (b) The genetics of which of these three would be easiest to study?

1-4 Find out such facts about the garden pea as length of time from seed to harvest (i.e., length of life cycle), ease of cultivation, relative number of offspring, contrasting traits (e.g., flower color, seed pod color, smoothness of seed coat, height of plant, etc.), then decide how well Mendel's garden peas fit criteria for useful genetic organisms.

1-5 List as many reasons as possible why humans are not good experimental genetic organisms.

1-6 Why is the inheritance of acquired characters no longer accepted as fact?

1-7 Devise an experiment of your own to test the question of inheritance of acquired characters.

1-8 How could you, at this stage of your study of genetics, justify Sutton's prediction (page 8) that the chromosomes may serve as a physical basis of inheritance?

1-9 Some races of corn produce red kernels, some white, and others white kernels that turn red only if they are exposed to light during maturation. Does kernel color in corn appear to depend on heredity, environment, or both? Explain.

1-10 Poliomyelitis is a disease known to be caused by a specific virus. It has been observed to be more frequent in some families than in others, even when children appear to have had the same degree of exposure (in the absence of inoculation) to the disease. Explain.

1-11 Blue-green algae are primitive prokaryotic plants that reproduce only by cell division or fragmentation. They appear to have changed little over a very long period of geologic time. Moreover, all individuals of a given species appear to be morphologically and physiologically alike or to differ only slightly in a very few characters. Suggest a rational explanation for these latter two observations.

1-12 Huntington's chorea is a genetically based disorder of the central nervous system, leading to physical and mental deterioration and culminating in death. Its symptoms most often begin in middle age. Would this condition be easier or more difficult than ear lobes to study genetically? Why?

1-13 Considering variation in height in members of your family and in other families, would you say that height in human beings is determined by heredity, environment, or both? Why?

1-14 A child is born with six fingers on each hand (a form of the heritable condition known as polydactyly), but the extra fingers are removed by a simple surgical procedure shortly after birth. Polydactyly occurs only in persons who have at least one polydactylous parent. Will the removal of the extra fingers prevent this child from ever having any polydactylous children? Explain.

1-15 In diabetics insulin production is faulty, so that normal sugar metabolism does not occur. In some forms of diabetes symptoms can be prevented by daily injections of insulin. The various types of diabetes are inherited, although the disorder can appear in children both of whose parents are not diabetic. Will insulin administration to diabetics prevent their having any children with the disorder? Explain.

CHAPTER 2

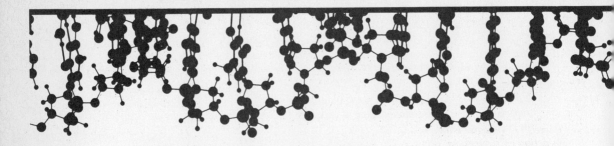

Monohybrid inheritance

As noted in the introduction, early attempts to determine fundamental genetic mechanisms frequently failed because investigators tried to examine simultaneously all discernible traits. Mendel's success in preparing the groundwork of modern understanding lay in (1) concentrating on one or a few characters at a time, (2) making controlled crosses and keeping careful quantified records of results, and (3) suggesting "factors" as the particulate causes of various genetic patterns. Similarly, in order to learn something of the inheritance of *vestigial* wings in the fruit fly, *Drosophila* (Fig. 2-1), an individual having normal, full-sized wings would be crossed with one having vestigial wings; all other traits would be ignored. The appearance of the offspring through several generations and the numbers of *normal* and *vestigial* individuals produced would be carefully recorded and evaluated.

Such a cross, involving contrasting expressions of one trait, is referred to as a **monohybrid cross.** Although this term is restricted by some geneticists to cases in which the parents *differ* detectably in one trait (as in the example of the fruit flies), it may also be extended to any cross in which only a single character is being considered, whether or not the parents differ discernibly. The latter, broader definition will be used in this text.

Let us begin our study of genetics along the general lines of Mendel's technique, but using a different organism for the sake of variety. Shortly, it will be seen that the same principles apply to human beings.

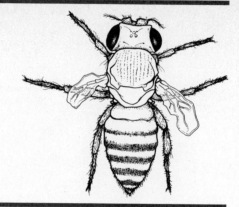

Figure 2-1. *Vestigial wing, resulting from the action of a recessive gene, in* **Drosophila melanogaster.**

The standard monohybrid cross

COMPLETE DOMINANCE

The common house plant, variegated coleus (*Coleus blumei*, of the mint family) is a frequently encountered flowering plant that lends itself readily to simple genetic experimentation. It has many of the desirable characteristics of an organism useful for genetic experimentation. Desirable characteristics of this plant include a relatively short reproductive cycle, easily made crosses between different individuals (yet self-pollination is also possible), and appearance of many of its genetically determined traits in the seedling stage.

One such character involves the shape of the leaf margins. Some plants have shallowly crenate edges; others have rather deeply incised leaves (Figs. 2-2 and 2-3). Plants displaying one or the other of these two traits may be referred to as "shallow" and "deep," respectively. Imagine two such plants, each the product of a long series of generations obtained by self-pollination, in which the same individual serves as both maternal and paternal parent. If, in such repeated breeding, deep always gave rise to deep offspring only, and shallow to shallow

Figure 2-2. *The common house plant* **Coleus,** *showing shallow-lobed leaves.*

*Figure 2-3. Deep-lobed leaves in **Coleus**. Compare with Figure 2-2.*

only, each such line of descent would be established as "pure-breeding" for this trait. The geneticist customarily uses the term **homozygous** in such cases. The precise genetic implication of this latter term will be seen shortly.

Suppose now a cross were to be made between two such homozygous individuals, one deep and the other shallow. In theory, the outcome might be any one of the following: (1) all the offspring may resemble one or the other parent only; (2) some of the progeny may resemble one parent, whereas the remainder may look like the other parent; (3) all the offspring, though like each other, may be intermediate in appearance between the two parents; (4) the offspring may look like neither parent and yet not be clearly intermediate between them; or (5) there may be a considerable range of types in the progeny, some being about as deeply lobed as one parent, some about as shallowly cut as the other, with the remainder forming a continuum of variation between the two parental types. When the offspring of this cross are examined, however, all of them are seen to be *deep*, like one parent. The *shallow* trait does not appear at all. Since the days of Mendel, a characteristic that expresses itself in all the offspring (as *deep* does in this case) has been termed **dominant,** and the trait that fails to be expressed (here, *shallow*) has been referred to as **recessive.**

If various individuals of this progeny are either intercrossed or self-pollinated, what will be the results? From similar work of Mendel (Table 2-1) it can be predicted that about three-fourths of the second generation should be deep and about one-fourth shallow. This predicted outcome is actually observed. (It should be noted, however, that only with quite large samples would a close approach to a 3:1 ratio be expected.)

According to standard terminology, results in the *Coleus* example may be summarized as follows:

(*Parental generation*) P   deep $\times$ shallow
(*First filial generation*) F_1 all deep
(*Second filial generation*) F_2 $\frac{3}{4}$ deep $+$ $\frac{1}{4}$ shallow

Table 2-1. Mendel's earliest experiments on peas

Parents	First generation		Second generation			Ratio
Round × wrinkled seed	all round	5,474	round	1,850	wrinkled	2.96:1
Yellow × green cotyledons	all yellow	6,022	yellow	2,001	green	3.01:1
Gray-brown × white seed						
coats	all gray-brown	705	gray-brown	224	white	3.15:1
Inflated × constricted pods	all inflated	882	inflated	299	constricted	2.95:1
Green × yellow pods	all green	428	green	152	yellow	2.82:1
Axial × terminal flowers	all axial	651	axial	207	terminal	3.14:1
Long × short stems	all long	787	long	277	short	2.84:1
TOTALS		14,949		5,010		Av. 2.98:1

F_2 progeny are obtained either by "selfing" or by interbreeding members of the preceding F_1.

MECHANISM OF THE MONOHYBRID CROSS

Genes and their location. The task now arises to determine what mechanism might account for these results. Obviously, what is transmitted from parent to offspring is not the trait itself, but rather something that *determines* the later development of that trait at an appropriate time and place in a particular environment. These determiners are called *genes*. Their actual structure and behavior will be examined later.

Because the sex cells or gametes constitute the only link in sexually reproducing organisms between parent and offspring, it is clear that genes must be transmitted from generation to generation via this gametic bridge. Where, then, are the genes located in the gametes? Cytological examination of the gametes of a great many sexually reproducing organisms, both plant and animal, discloses that, although the egg cell has a large amount of cytoplasm, the sperm is largely nucleus, having relatively little cytoplasm. This relationship breaks down, of course, in such isogamous forms as some of the algae, in which relative amounts of cytoplasm and nuclear material are quite similar. Though this observation does not furnish conclusive *proof* of the location of the genes within the nuclei of the sex cells, it does offer a *probability* that the genes in the *Coleus* example are more likely to be nuclear than cytoplasmic in location. The possibility of other genes being located in the cytoplasm is discussed in Chapter 21. For the present, until and unless contrary evidence becomes available, we will pursue the thesis that these genes are nuclear in location.

That sperm and egg in this kind of cross do make equal genetic contribution to the next generation is strongly indicated by the fact that, whether *deep* serves as the **pistillate** (egg-contributing) parent or as the **staminate** (sperm-contributing) parent, the result is the same: all the F_1 are deep-lobed (Fig. 2-4). Such **reciprocal crosses** do not, in every case, produce identical results. In fact, whether reciprocal crosses give the same or different results provides certain important additional information concerning the operation of the genes involved. (See Chapters 18 and 19.)

If genes are situated in the nucleus, it would be useful to locate them more precisely within the nucleus. Genes should be associated with some cell organelle that is (1) quantitatively distributed with complete exactness during nuclear division and (2) contributed equally by sperm and egg at gametic union, or syngamy. Although Chapter 4 deals more closely with the cellular processes involved, previous work that you as a student have had in the life sciences surely suggests the function of chromosomes as the vehicles of gene transmission in such cases. Pending a test of that hypothesis, this is the assumption on which

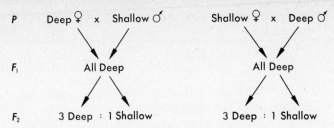

Figure 2-4. *Diagram of a reciprocal cross in* **Coleus.** *Note that, in this case, results in the* F$_1$ *and* F$_2$ *are the same regardless of which plant is used as the pistillate parent.*

attempts to explain these observations in *Coleus* (and, by implication, other organisms as well) will be based. Regardless of the attractiveness of a theory, however, acceptance must rest on testing, and revision must always be considered carefully in the light of new experimental evidence. *Coleus* has a discrete nucleus; that is, it is a **eukaryote.** Genetic hypotheses with regard to *Coleus* will have to be modified somewhat for prokaryotes (which do not have a discrete nucleus and in which chromosomes *as known in eukaryotes* are absent). Both bacteria and blue-green algae are prokaryotic organisms, the former being of great importance to geneticists and, indeed, to the human species. Viruses, which also have genetic material, will require additional refinements or widening of hypotheses that are based on eukaryotes.

For the present, then, it will be assumed that genes are situated on the chromosomes, transmitted via the gametes from one generation to the next, and contributed equally by both parents. If *D* represents the dominant gene for deep lobes and *d* the recessive, the gametic contribution of the P generation will be designated as

$$\textcircled{D} \text{ and } \textcircled{d}$$

where the two genes, one for deep *(D)* and one for shallow *(d)*, are said to be **alleles** of each other, or to form an allelic pair.

Based on (1) the observation that chromosomes of both fusing gametes become incorporated within the nucleus of the zygote, each maintaining a separate identity, and (2) the assumption that chromosomes serve as likely candidates for gene location, it then follows that the zygotes (which will become F$_1$ individuals) can be represented as *Dd.* So, also, the P individuals must be represented as *DD* and *dd*, respectively. Thus each of the parental individuals possesses two identical genes in each of their somatic cells; that is, each is **homozygous.** In the same way, the F$_1$ plants must have in each of their somatic cells one gene for deep *(D) and* one for shallow *(d)*; thus they are **heterozygous** with respect to this pair of alleles. Because in the heterozygous F$_1$ the effect of the dominant gene appears to mask completely the presence of the recessive allele, this case illustrates **complete dominance.**

The assumption is, therefore, that there are two genes for a given characteristic in the somatic cells and only one in the gametes. Some sort of nuclear division, prior to sex cell formation, which will reduce the gene number from two per trait to one—i.e., separate the alleles—is thus required. Chapter 4 explores this possibility further. The original *Coleus* cross may now be written as follows:

	P	deep	×	shallow
		DD		dd
P gametes		$\textcircled{D}$	×	$\textcircled{d}$
F$_1$			deep	
			Dd	
F$_1$ gametes		$\textcircled{D}$	+	$\textcircled{d}$

Phenotype and genotype. *Detectable* traits of the organism, with regard to the character or characters under consideration, constitute the **phenotype,** which is generally designated by a descriptive word or phrase. Phenotypic traits may be either morphological or physiological, though it is, in the final analysis, not always easy to separate the two. On the other hand, an individual's *genetic makeup* is termed the **genotype** and is customarily given by letters of the alphabet or other convenient symbols. In this case, *deep* and *shallow* represent *phenotypes*, whereas *DD, Dd, dd, D,* and *d* represent *genotypes* (somatic and gametic).

An individual's phenotype is not always a rigidly expressed "either-or" condition but is often modified by environmental influences. For example, although a quantitative character such as height is genetically determined in many plants (and in humans), even in those cases in plants in which only a single pair of genes can be shown to be operating, height variation in both the tall and short individuals often occurs. Such variations, usually clustering fairly closely around a mean, can be shown in suitable, controlled experiments, to be the results of environmental influences. In other words, genotype determines the phenotypic *range* within which an individual will fall; environment determines where in that range the individual will occur. In certain environments, too, it is possible that a particular genotype will not be expressed at all. In corn, for example, one gene ("sun-red") produces red grains if the ear is exposed to light (Fig. 2-5), but as long as the husks are intact the grains remain white, so that both sun-red and white genotypes remain phenotypically indistinguishable. Furthermore, as will soon be shown, phenotype may often be physiological and therefore "observable" only in the biochemical sense.

Probability method of calculating ratios. If a nuclear division, which segregates the alleles *D* and *d*, occurs in the F_1 prior to gamete formation, then *one half* of the F_1 gametes should carry the dominant *D* and *one half* the recessive *d*. If, furthermore, syngamy is random, so that a *D* egg has an equal chance to be fertilized by either a *D* or a *d* sperm, for example, mating of the F_1 to produce the F_2 can be represented as follows:

Figure 2-5. *"Sun-red" corn. Action of the sun-red gene is to produce red kernels when these are exposed to light; in the absence of light, kernels remain white. In this instance, the husks had been replaced by the black mask shown below.* (Courtesy Department of Plant Breeding, Cornell University.)

$$\begin{aligned}
&\textit{eggs from } \female \text{ F}_1 \ \tfrac{1}{2} \ \textcircled{D} + \tfrac{1}{2} \ \textcircled{d} \\
&\textit{sperms from } \male \text{ F}_1 \ \tfrac{1}{2} \ \textcircled{D} + \tfrac{1}{2} \ \textcircled{d} \\
\hline
&\text{F}_2 \ \tfrac{1}{4} DD + \tfrac{1}{4} Dd + \tfrac{1}{4} dD + \tfrac{1}{4} dd
\end{aligned}$$

or $\tfrac{1}{4} DD + \tfrac{2}{4} Dd + \tfrac{1}{4} dd$ for a 1:2:1 F_2 monohybrid genotypic ratio.

The basis for this kind of calculation is the **product law of probability.** Briefly stated, this law holds that the probability of the simultaneous occurrence of two independent events equals the product of the probabilities of their separate occurrences. Thus if one half the eggs are of genotype D and one half of the sperms have the same genotype, and if fertilization is a completely random event, then the probability of getting a DD zygote is $\tfrac{1}{2} \times \tfrac{1}{2} = \tfrac{1}{4}$.

As to phenotypes, we see that DD and Dd organisms appear indistinguishable on visual bases, so that the 1:2:1 genotypic ratio gives our observed 3:1 (F_2) monohybrid phenotypic ratio. A 3:1 ratio is shown in Fig. 2-6. (As will be seen, these ratios may occur in the F_1, given certain P individuals, and are, of course, not the only possible monohybrid ratios.) Thus assumptions developed so far have led to an explanation completely compatible with earlier observations. These assumptions remain to be tested further, both by cytological examination and by additional breeding experiments. Note that the so-called checkerboard or Punnett square method has not been used, but rather a *probability* method of calculating genotypic and phenotypic ratios. This latter system becomes extremely advantageous when calculating ratios involving several pairs of genes.

THE TESTCROSS

As indicated earlier, DD and Dd individuals in *Coleus* cannot be distinguished visually from each other. This raises two questions. First, can homozygous dominant and heterozygous individuals be distinguished in any manner? The answer is yes, and the testcross can supply the answer. In the testcross the individual having the dominant phenotype (which can be represented as having genotype $D-$) is crossed with one having a recessive phenotype that is, of course, homozygous. In the *Coleus* case the testcross of a homozygous dominant is as follows:

$$\begin{array}{ccc}
\text{P} & \text{deep} & \times & \text{shallow} \\
 & DD & & dd \\
\textit{P gametes} \quad \textit{eggs} & & (1) \ \textcircled{D} \\
\textit{sperms} & & (1) \ \textcircled{d} \\
\hline
\text{F}_1 \quad \text{all} & & Dd \ (\text{deep})
\end{array}$$

If the P dominant phenotype had been heterozygous, the result would be the classic **1:1 monohybrid testcross ratio:**

Figure 2-6. *A 3:1 purple: white ratio in corn. This is the* F_2 *of homozygous purple* $\times$ *white.*

$$\text{P} \quad \text{deep} \quad \times \quad \text{shallow}$$
$$Dd \qquad\qquad dd$$

$$\text{P gametes} \qquad eggs \;\; \tfrac{1}{2}\, \textcircled{D} \, + \, \tfrac{1}{2}\, \textcircled{d}$$

$$sperms \qquad 1\, \textcircled{d}$$

$$\text{F}_1 \qquad\quad \overline{\tfrac{1}{2}\, Dd \; + \; \tfrac{1}{2}\, dd}$$
$$\text{deep} \qquad \text{shallow}$$

Note that $DD \times dd$ produces an F_1 all of the dominant phenotype, whereas $Dd \times dd$ gives rise to a 1:1 ratio in the progeny. A 1:1 testcross ratio in corn is shown in Figure 2-7.

At this point use of the expression F_1 must be explained. It will be used to designate the first generation resulting from *any* given mating regardless of the parental genotypes. Therefore the term F_1 does not necessarily imply heterozygosity. In the two testcrosses just outlined, note that the F_1 genotype depends on the P genotypes and may, of course, be either heterozygous (as in the first instance) or homozygous (as in the *dd* individuals of the second case).

The second question suggested by the visual similarity of homozygous dominants and heterozygotes is the matter of how genes operate to exert their phenotypic effects. Although a complete answer must be deferred to later considerations of the molecular nature of the gene and the chemistry of its action, a genetic trait in Mendel's peas offers a tempting suggestion. It also furnishes an introduction to a variation in the classic 3:1 phenotypic ratio seen in *Coleus.* This case will be examined in the following section.

Modifications of the 3:1 phenotypic ratio

INCOMPLETE DOMINANCE

Peas. The data of Table 2-1 were first reported by Mendel in a pair of originally widely ignored papers read before the Natural History Society at Brünn on February 8 and March 8, 1865. Mendel's results led him to designate round seeds as (completely) dominant to wrinkled seeds; indeed, homozygous and heterozygous round seeds cannot be differentiated macroscopically. However, when cells of the cotyledons of the three genotypes (*WW*, *Ww* round, and *ww* wrinkled) are examined *microscopically*, an abundance of well-formed starch grains is seen in the *WW* plants, and very few in *ww* individuals. Heterozygotes show an intermediate number of grains, many appearing imperfect or eroded. Moreover, *ww* embryos test higher in reducing sugars (the raw material from which starch molecules are constructed) than do those of *Ww* genotype. *WW* embryos test lowest of all for sugar. In plant cells starch is synthesized from glucose-1-phos-

Figure 2-7. *A 1:1 testcross ratio in corn. This ear resulted from the cross white $\times$ heterozygous purple.*

phate under the influence of an enzyme system. Although a discussion of the chemistry of the sugar $\rightleftharpoons$ starch interconversion is outside the scope of a beginning course in genetics, a plausible explanation is that gene W is responsible for the production of one of the enzymes required for the reaction glucose-1-phosphate $\rightarrow$ starch. Likewise, gene w may produce either a smaller quantity of the enzyme or, perhaps, a molecule that functions only imperfectly and produces a less efficient enzyme molecule, or is unable to function at all because of sufficient structural differences. The larger amount of starch in WW plants is associated with higher water retention (starch is a hydrophilic colloid) and therefore with plump, distended, spherical seeds. Homozygous recessive embryos, on the other hand, retain considerably less water upon reaching maturity and hence appear shriveled. The starch content of heterozygotes, however, is apparently sufficient to produce embryos visually indistinguishable from those of WW genotype. Consistent with these differences in starch content, heterozygotes have an intermediate amount of functional enzyme. The relationship between genes and enzymes suggested here is a fundamental concern of modern genetics and offers much help in elucidating the nature and action of the gene. It will receive considerable attention in later chapters of this book.

For present purposes, however, it is clear that on the gross, macroscopic level, a case of complete dominance is operating, yielding the familiar 3:1 phenotypic and 1:2:1 genotypic ratios. But on the microscopic, chemical, or molecular level gene W must be considered only **incompletely dominant** to the w allele. Therefore, at the *physiological* level the cross $Ww \times Ww$ yields identical phenotypic and genotypic ratios of 1:2:1. Whether round versus wrinkled seeds is considered as complete dominance or as incomplete dominance depends entirely on the level of analysis. Because the physiological level more accurately reflects the actual operation of the genes involved, most geneticists would prefer to regard this as a case of incomplete dominance. It is important to note that in incomplete dominance the heterozygotes are phenotypically *intermediate* between the two homozygous types.

Other organisms. Examples of incomplete dominance at the macroscopic level occur in both plants and animals. For example, radishes may be long, oval, or round. Crosses of long $\times$ round produce an F_1 of wholly oval phenotype:

$$
\begin{array}{ccc}
\text{P} & \text{long} \quad \times & \text{round} \\
 & l^1l^1 & l^2l^2 \\
\end{array}
$$

$$
\begin{array}{cc}
F_1 & \text{all oval} \\
 & l^1l^2 \\
\end{array}
$$

$$
\begin{array}{ccc}
F_1 \; gametes & eggs & \tfrac{1}{2}\,\boxed{l^1} + \tfrac{1}{2}\,\boxed{l^2} \\
 & sperms & \tfrac{1}{2}\,\boxed{l^1} + \tfrac{1}{2}\,\boxed{l^2} \\
\end{array}
$$

$$
F_2 \quad \tfrac{1}{4}\,l^1l^1 + \tfrac{2}{4}\,l^1l^2 + \tfrac{1}{4}\,l^2l^2
$$

Standard practice is to use lower-case letters with numerical superscripts to designate incompletely dominant genes.

Many morphological and physiological traits in a wide variety of animals and plants give evidence of being the result of a single pair of genes, and new ones are continually being reported in many different organisms. These include resistance to blister rust in the sugar pine (dominant), coloboma in fowl "with profound effects on all body parts through effects on cartilage formation," (recessive lethal, Abbott et al., 1970) early seed stalk development, or "bolting," in celery (dominant), height in cultivated geranium, *Pelargonium*, (incomplete dominance), lack of anthocyanin pigment in cotton (recessive), resistance to downy mildew in grape (dominant), resistance to a mottling virus in pepper plants (recessive), resistance to stem rust in certain varieties of wheat (domi-

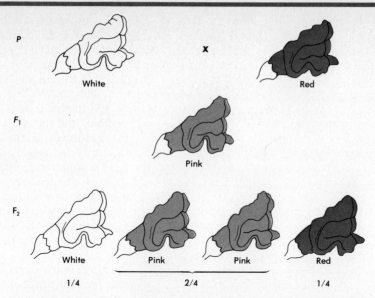

Figure 2-8. *A case of incomplete dominance in snapdragon. The pink F₁ hybrid is truly intermediate between the red and white parents, and the F₂ segregates in a characteristic 1:2:1 phenotypic ratio.*

nant), lactose tolerance (lactase production) in humans (dominant), and familial hypercholesterolemia in humans (incompletely dominant[1]). (See references at the end of this chapter.)

CODOMINANCE

Cattle. In shorthorn cattle genes for red and white coat occur. Crosses between red (r^1r^1) and white (r^2r^2) produce offspring (r^1r^2) whose coats appear at a distance to be reddish gray, or *roan*. Superficially this would seem to be a case of incomplete dominance, but close examination of roan animals reveals that the coat is composed of a *mixture of red hairs and white hairs*, rather than hairs all of a color intermediate between red and white. Instances such as this, where the heterozygote exhibits a mixture of the phenotypic characters of both homozygotes, instead of a single intermediate expression, illustrate **codominance.** Genotypic and phenotypic ratios are identical in incomplete dominance and codominance, but the distinction reflects something of the way in which genes operate, a topic that will be discussed in detail in later chapters.

Humans. Several illustrations of codominance are found in human genetics. One of these involves a pair of codominant alleles, *M* and *N*, responsible for production of antigenic substances M and N, respectively, on the surfaces of the red blood cells. Although the genetics of blood antigens is explored more fully in Chapter 7, it is interesting to note that, whereas persons of genotype *MM* produce antigen M and *NN* individuals produce the somewhat different antigen N, heterozygotes (*MN*) produce *both* M and N antigens, not a single intermediate

---

[1]This disorder, characterized by excessive blood levels of cholesterol and due to a decrease in the number of low density lipoprotein (LDL) receptors, is sometimes attributed to a dominant gene. A typical normal human cell has about 20,000 LDL receptors. Heterozygotes have a reduced number of these receptors, but cells of homozygotes for the lethal gene generally have no LDL receptors. Therefore, this disorder meets specifications used in this text for *incomplete* dominance.

substance. All persons belong to one of three phenotypic classes, i.e., M, MN, or N. However, because antibodies to these antigens rarely occur, they are not considered in transfusion, and most people do not know their M–N blood group. Accumulated data from families in which both parents have been identified as MN show a close approximation to the expected 1:2:1 ratio in the children.

Another and more important case is found in the sickle-cell phenotype, which is particularly prevalent in blacks in this country and in certain African peoples. The hemoglobin of most persons is of a particular chemical structure and is known as hemoglobin A (for adult hemoglobin). Many chemical variants of hemoglobin A are found in relatively small numbers of people; one of these, hemoglobin S, is involved in the sickle-cell disorder. Genes responsible for hemoglobin types here are Hb^A and Hb^S. Most persons belong to genotype Hb^AHb^A. Their erythrocytes contain only hemoglobin A and are biconcave disk-shaped (Fig. 2-9A). Persons with *sickle-cell anemia* are of the genotype Hb^SHb^S and are characterized by a collection of symptoms, chiefly a chronic hemolytic anemia. In the blood of such persons the erythrocytes become distorted, many being essentially sickle-shaped (Fig. 2-9B). Sickle-shaped cells not only impede circulation by blocking capillaries, but also cannot properly perform the function of carrying oxygen and carbon dioxide to and from the tissues.

In heterozygotes, Hb^AHb^S, some red cells contain hemoglobin A, others hemoglobin S. Because both types of hemoglobin, rather than a single intermediate form, are produced, this is another case of codominance. Microscopic examination of heterozygotes' blood under low oxygen tension discloses both normal and sickled erythrocytes (Fig. 2-10). Under normal conditions heterozygotes manifest none of the severe symptoms of Hb^SHb^S persons, though they may suffer some periodic discomfort and even develop anemia after a time at high altitudes. Genotypes and phenotypes in this disorder are as follows:

Hb^AHb^A normal (hemoglobin A only; no sickling of red cells)
Hb^AHb^S sickle-cell trait (hemoglobins A and S; sickling under reduced oxygen tensions)
Hb^SHb^S sickle-cell anemia (hemoglobin S only; sickling under normal oxygen tension)

The chemistry of sickle-cell and other hemoglobin disorders, as well as the operation of the genes involved will be examined in more detail in later chapters.

LETHAL GENES

A second major modification of the classic monohybrid ratios is produced by genes whose effect is sufficiently drastic to kill the bearers of certain genotypes. Here both the effect itself and the developmental stage at which the lethal effect is exerted are important.

Humans. In persons with sickle-cell anemia, blocking of the capillaries may produce infarction in almost any organ, leading to painful episodes, tissue destruction, and death, usually before attainment of reproductive age. Because homozygosity for Hb^S causes premature death, it is a lethal gene, and its lethality is recessive, inasmuch as heterozygotes are of generally good health. Marriages between heterozygotes may be expected to produce children in the ratio of 1 sickle-cell anemia:2 sickle-cell trait:1 normal at birth, but this phenotypic ratio becomes 2:1 after the sickle-cell anemia persons have died. A 2:1 phenotypic ratio, either at birth or later in life, is indicative of a pair of codominant or incompletely dominant genes in which one allele shows recessive lethality.

Corn. Most plants with which a nonbotanist is familiar are characteristically autotrophic; hence they are able to manufacture all required food from carbon

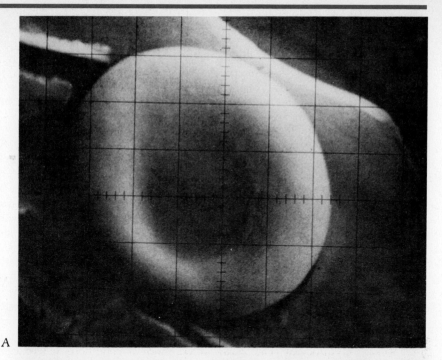

A

B

Figure 2-9. (A) Norman red blood cell and (B) sickled red blood cells. [Samples courtesy Dr. Patricia Farnsworth, Barnard College.] *The photomicrographs were taken at magnifications of (A) 10,000 × and (B) 5,000 × by Irene Piscopo of Philips Electronic Instruments on a Philips EM 300 Electron Microscope with Scanning Attachment.* (Photos courtesy Philips Electronic Instruments.)

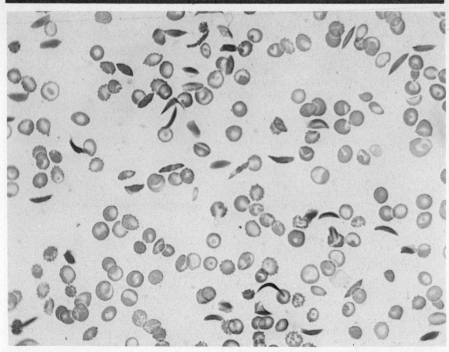

Figure 2-10. *Blood smear showing normal and sickle erythrocytes.* (Courtesy Carolina Biological Supply Co.)

dioxide and water in the process of photosynthesis. For this process, in all but the few autotrophic bacteria, the presence of a green, light-absorbing pigment, chlorophyll, is required. In corn (*Zea mays*) several pairs of genes affecting chlorophyll production have been described in the literature. One such gene, designated *G*, for normal chlorophyll production, is completely dominant to its allele *g*, so, as a result *G*− plants contain chlorophyll and are photosynthetic. On the other hand, *gg* plants produce no chlorophyll and are yellowish white (because they still produce the yellow carotenoid pigments). (See Fig. 2-11.)

On the average, about one-fourth of the progeny of two heterozygous parents are thus without chlorophyll, and seedlings show the classic 3:1 ratio. In corn, germination and early seedling development take place at the expense of a food-storage tissue, the endosperm, in the grain. In normal green plants, by the time this food reserve has been exhausted (in about 10 to 14 days' time), the seedling has developed a sufficient root system and amount of green tissue to be physiologically independent. Therefore, in the cross of two heterozygotes, the initial 3:1 phenotypic ratio becomes what might be termed a 3:0 or a "1:0" ratio after some two weeks:

$$\text{P} \quad \text{green} \quad \times \quad \text{green}$$
$$Gg \qquad\qquad Gg$$
$$\text{F}_1 \ \tfrac{1}{4} \text{ green} \ + \ \tfrac{2}{4} \text{ green} \ + \ \tfrac{1}{4} \text{ nongreen (die)}$$
$$GG \qquad\quad Gg \qquad\qquad gg$$

Note that after gene *g* has exerted its lethal effect, the genotypic ratio is converted from 1:2:1 to 2:1; as in the case of sickle-cell in humans. So whereas homozygous green plants, for example, initially make up one-fourth of the F₁, they later comprise *one-third* of the surviving progeny because of the death of *gg* individuals. This 2:1 genotypic ratio, or the occurrence of only one phenotypic

Figure 2-11. A lethal gene in corn. The chlorophyll-less plants are unable to manufacture their own food and will die as soon as food stored in the grain has been consumed. Photo shows progeny plants of the cross heterozygous green × heterozygous green; there are 36 green and 12 "albino" seedlings in the flat, a perfect 3:1 ratio.

class where two would be expected in a 3:1 ratio, is a clear indication of a (recessive) lethal gene.

In the instance just described, the occurrence and action of the lethal gene is easily discerned because the death of the homozygous recessives takes place only after nearly two weeks of growth following germination. Yet the effect of other lethals might conceivably be produced at almost any time between syngamy (i.e., in the zygote state) on through embryogeny to a very late point in life. Obviously, the effect of lethals killing very late in life might well be hard to separate from other causes of death.

Mouse. A classic case of a recessive lethal that kills early in embryo development, and one of the first to be reported in the literature, was "yellow" in mice. Early in this century Cuénot (1904, 1905) noted that black × black always produced black offspring, but that yellow × black produced yellow and black in a 1:1 ratio. He concluded correctly that yellow is heterozygous. Yet crosses of yellow × yellow always produced yellow and black in a ratio of 2:1, with litters of such matings being about one-fourth smaller than those from other crosses. Letting A^Y represent a gene for yellow and a one for black (as Cuénot did) Cuénot's crosses may be represented as follows:

Testcross:

$$\text{P} \quad \underset{A^Y a}{\text{yellow}} \quad \times \quad \underset{aa}{\text{black}}$$

$$\text{F}_1 \quad \underset{A^Y a}{\tfrac{1}{2} \text{ yellow}} \quad + \quad \underset{aa}{\tfrac{1}{2} \text{ black}}$$

or, **simple monohybrid:**

$$\text{P} \quad \underset{A^Y a}{\text{yellow}} \quad \times \quad \underset{A^Y a}{\text{yellow}}$$

$$\text{F}_1 \quad \underset{\text{die}}{\tfrac{1}{4} A^Y A^Y} \quad + \quad \underset{\text{yellow}}{\tfrac{2}{4} A^Y a} \quad + \quad \underset{\text{black}}{\tfrac{1}{4} aa}$$

Thus gene A^Y appears to be dominant with respect to coat color but recessive as to lethality. For some time the nature of the action of $A^Y A^Y$ was unknown. Although selective fertilization was considered a possible factor, it was suggested that $A^Y A^Y$ animals are conceived but die soon after. Robertson (1942) and later Eaton and Green (1962) were able to demonstrate that about one-fourth of the embryos of pregnant yellow ($A^Y a$) females that had been mated to yellow males did die soon after conception. In this case death usually occurs at gastrulation.

Fowl. The well-known "creeper" condition in fowl falls into the same category. Creeper birds have much shortened and deformed legs and wings, giving them a squatty appearance and creeping gait. Creeper × creeper always produces two creeper to one normal with the homozygous creepers having such gross deformities (greater than in heterozygotes) that death occurs during incubation, generally about the fourth day:

$$\text{P} \quad \underset{c^1 c^2}{\text{creeper}} \quad \times \quad \underset{c^1 c^2}{\text{creeper}}$$

$$\text{F}_1 \quad \underset{c^1 c^1}{\tfrac{1}{4} \text{ normal}} \quad + \quad \underset{c^1 c^2}{\tfrac{2}{4} \text{ creeper}} \quad + \quad \underset{c^2 c^2}{\tfrac{1}{4} \text{ (die)}}$$

It has been shown that the creeper gene produces general retardation of embryo growth, with the effect being greatest at the stage of limb bud formation.

A dominant lethal in man. Thus far, only cases in which lethality itself is recessive have been examined. Reasoning a priori, there is no cause not to expect lethal dominant genes, provided the death of the affected individual occurs somewhat after reproduction has taken place. **Huntington's chorea,** a hereditary disorder in man characterized by involuntary jerking of the body and a progressive degeneration of the nervous system, accompanied by gradual mental and physical deterioration, illustrates just such a situation. The mean age of onset of these symptoms is between 35 and 40 (though it is reported to occur as early as the first decade of life and as late as 60 or 70), by which time, of course, many afflicted persons have produced children. Affected offspring always have at least one parent who, sooner or later, is also choreic, though the variability of the age of onset (which may be due to the action of still other genes) makes this difficult to demonstrate in some cases and impossible when parents and grandparents die at early ages from other causes. Clearly, then, on available evidence, this disease is the result of a dominant gene, both as to lethality and as to its abnormal phenotype, even though many Huntington individuals reproduce before dying.

Table 2-2. Comparison of certain lethal genes

Organism	Phenotype	Dominance Phenotype	Dominance Lethality	F₁ phenotypic ratio of heterozygote × heterozygote Before lethality occurs	F₁ phenotypic ratio of heterozygote × heterozygote After lethality occurs	Age at death
Human	Sickle-cell	Codom.	Recessive	1 normal:2 sickle-cell trait:1 sickle-cell anemia*	1 normal:2 sickle-cell trait	Adolescence
Corn	Albinism	Recessive	Recessive	3 green:1 albino	All green ("1:0")	10–14 days
Mouse	Yellow	Inc. dom.?	Recessive	Unknown	2 yellow:1 black	Postzygote
Fowl	Creeper	Inc. dom.	Recessive	Unknown	2 creeper:1 normal	Early embryo
Human	Huntington's chorea	Dominant	Dominant	3 choreic:1 normal	All normal ("1:0")	Middle age, but somewhat variable

*The 1:2:1 ratio listed under F₁ phenotype for sickle-cell anemia is based on microscopic examination of blood samples.

The lethals discussed in this chapter are summarized in Table 2-2.

References

Abbott, U. K., R. M. Craig, and E. B. Bennett, 1970. Sex-linked Coloboma in the Chicken. *Jour. Hered.*, **61:** 95–102.

Bouwkamp, J. C., and S. Honma, 1970. Vernalization Response, Pinnae Number, and Leaf Shape in Celery. *Jour. Hered.*, **61:** 115–118.

Cuénot, L., 1904. L'Hérédité de la Pigmentation chez les Souris 3ᵐᵉ Note). *Arch. Zool. Exp. et Gén.*, 3ᵐᵉ Série, **10,** Notes et Revues, 27–30.

Cuénot, L., 1905. Les Races Pures et Leur Combinaisons chez les Souris. *Arch. Zool. Exp. et Gén.*, **3:** 123–132.

Eaton, G. J., and M. M. Green, 1962. Implantation and Lethality of the Yellow mouse. *Genetica*, **33:** 106–112.

Filippenko, I. M., and L. T. Shtin, 1973. Inheritance of Downy Mildew Hardiness (*Plasmopara viticola*) in European-American Grape Hybrids. *Genetika*, **9:** 53–60.

Green, E. L., 1967. Shambling, a Neurological Mutant of the Mouse. *Jour. Hered.*, **58:** 67–67.

Henault, R. E., and R. Craig, 1970. Inheritance of Plant Height in the Geranium. *Jour. Hered.*, **61:** 75–78.

Kerber, E. R., and P. L. Dyck, 1973. Inheritance of Stem Rust Resistance Transferred from Diploid Wheat (*Triticum monococcum*) to Tetraploid and Hexaploid Wheat and Chromosome Location of the Gene Involved. *Canad. Jour. Genet. Cytol.*, **15:** 397–409.

Konoplia, S. P., V. N. Fursov, A. A. Druzhkov, and G. N. Nuryeva, 1973. Independent Inheritance of the New Character "Lack of Anthocyanin" and of the Type of Generative Branches in the Cotton Species *Gossypium peruvianum. Genetika*, **9:** 154–156.

Kretchmer, N., 1972. Lactose and Lactase. *Sci. Amer.*, **227:** 71–78.

Parks, G. K., and C. W. Fowler, 1970. White Pine Blister Rust: Inherited Resistance in Sugar Pine. *Science*, **167:** 193–195.

Robertson, G. G., 1942. An Analysis of the Development of Homozygous Yellow Mouse Embryos. *Jour. Exp. Zool.*, **89:** 197–231.

Rosenfeld, A., 1981. The heartbreak gene. *Science 81*, **2:** 46–50.

Zitter, T. A., and A. A. Cook, 1973. Inheritance of Tolerance to a Pepper Virus in Florida. *Phytopathology*, **63:** 1211–1212.

Problems

2-1 Make a list of several phenotypic characters in your family for as many generations and individuals as possible. Consider traits singly, and try to determine the kind of inheritance involved. Save any that seem not to fit patterns developed in this chapter until somewhat later on.

2-2 In human beings a downward pointed frontal hairline ("widow's peak") is a heritable trait. A person with widow's peak always has at least one parent who also has this trait, whereas persons with a straight frontal hairline may occur in families in which one or even both parents have widow's peaks. When both parents have a straight frontal hairline, all children also have a straight hairline. Using W and w to symbolize genes for this trait, what is the genotype of an individual *without* widow's peak?

2-3 Some individuals have one whorl of hair on the back of the head, whereas others have two. In the following pedigrees sold symbols represent one whorl, open symbols two:

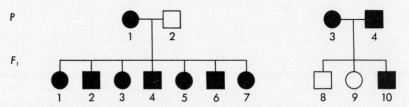

(a) Using the first letter of the alphabet, give the probable genotype of (1) P2; (2) P 3; (3) F_1 7; (4) F_1 8; (5) F_1 10.

(b) What should be the phenotypic ratio of the progeny produced by the marriage of F_1 7 × F_1 8, assuming a large family?

2-4 Radishes may have any of three shapes: long, spherical, or ovoid. Crosses of long × spherical always produce an F_1 consisting only of ovoid radishes. In terms of the genes discussed in this chapter, how should the pair of genes involved here be designated?

2-5 Albinism, the total lack of pigment, is due to a recessive gene. A man and a woman plan to marry and wish to know the probability that they will have any albino children. What could you tell them if (a) both are normally pigmented, but each has one albino parent; (b) the man is an albino, the woman is normal, but her father is an albino; (c) the man is an albino and the woman's family includes no albinos for at least three generations?

2-6 Cystic fibrosis of the pancreas is an inherited condition characterized by faulty metabolism of fats. Affected individuals are homozygous for the gene responsible and ordinarily die in childhood. Such individuals also produce a much higher concentration of chlorides in their sweat than homozygous normals, whereas heterozygotes, who live a normal lifespan, have an intermediate chloride concentration. In terms of the types of genes discussed in this chapter, how would you designate (a) this pair of alleles as to dominance; (b) the gene for cystic fibrosis?

2-7 One study has estimated the number of persons in the United States who are heterozygous for the cystic gene at about 8 million. A couple planning marriage decide to have a sweat test because a brother of the man died in infancy from the disorder. The tests show the man to be heterozygous and the woman homozygous normal. (a) What is the chance that any of their children might have cystic fibrosis? (b) Could any of the couple's grandchildren have it?

2-8 In a certain plant the cross purple × blue yields purple- and blue-flowered progeny in equal proportions, but blue × blue always gives rise only to blue. (a) What does this tell you about the genotypes of blue- and purple-flowered plants? (b) Which phenotype is dominant?

2-9 In cattle, the cross horned × hornless sometimes produces only hornless offspring, and in other crosses horned and hornless appear in equal numbers. A cattle owner has a large herd of hornless cattle in which horned progeny occasionally appear.

He has red, roan, and white animals and wishes to establish a pure-breeding line of red hornless animals. How should he proceed?

2-10 In corn, resistance to a certain fungus is conferred by gene h, which is completely recessive to its allele H for susceptibility. If a resistant plant ($♀$) is pollinated by a homozygous susceptible plant ($♂$), give the genotypes for (a) the pistillate parent, (b) the staminate parent, (c) sperm, (d) egg, (e) polar nucleus, (f) F_1 embryo, (g) endosperm surrounding the F_1 embryo, (h) epidermis of kernels that contain the F_1 embryos.

2-11 Two curly-winged fruit flies (*Drosophila*) are mated; the F_1 consists of 341 curly and 162 normal. Explain.

2-12 Using the sixth letter of the alphabet, give the genotype of each of the following persons in Figure 1-5: I-1, I-2, II-1, II-2, II-3, II-4, III-1, and III-2.

2-13 The marriage between II-1 and II-2 in Figure 1-5 represents what sort of genetic cross?

2-14 Using the first letter of the alphabet, give the genotype of each of the following persons from Figure 1-6: I-1, II-1, III-4.

2-15 No ancestry information is given for II-1 in Figure 1-6; how do you justify your designation of her genotype?

2-16 Rh negative children (those not producing rhesus antigen D) may be born to either Rh positive or Rh negative parents, but Rh positive children always have at least one Rh positive parent. Which phenotype is due to a dominant gene?

2-17 A normal couple has five children, two of whom suffer from a somewhat uncommon genetic disorder which has, however, appeared sporadically in this familial line. (a) What kind of gene is responsible in this case: completely dominant, completely recessive, codominant, or incompletely dominant? (b) What does the occurrence of affected children in this family tell you about the parental genotypes?

2-18 Among the many antigenic substances which may occur on human red blood cells are two, known as M and N. Individuals' phenotypes are referred to as M, N, or MN depending on which antigen or antigens they produce (one, the other, or both). In the case of an M father and an N mother, children are always MN. What kind of genes are responsible for these antigens? (Choose from among the choices given in problem 2-17a.)

2-19 **Thalassemia** is a hemoglobin defect in humans; it occurs in two forms, (a) **thalassemia minor,** in which erythrocytes are small (**microcytic**) and increased in number (**polycythemic**), but health is essentially normal, and (b) **thalassemia major,** characterized by early, severe anemia, enlargement of the spleen, microcytes, and polycythemia, among other symptoms. The latter form usually culminates in death before attainment of reproductive age. From the following hypothetical pedigree, determine the mode of inheritance:

Clear symbols represent normal persons, tinted symbols thalassemia minor, and the blackened thalassemia major.

2-20 In families in which both parents have sickle-cell trait, what is the probability of their having (a) a child with sickle-cell trait, (b) a normal child?

2-21 A normal individual, $Hb^A Hb^A$, receives a transfusion of blood from a person who has sickle-cell trait. Would this transfusion transmit the sickle-cell trait to the recipient? Explain.

2-21 Phenylketonuria (PKU) is a heritable condition in humans involving inability to metabolize the amino acid phenylalanine because of failure to produce the enzyme phenylalanine hydroxylase. If not diagnosed and treated very soon after birth, PKUs develop such severe mental retardation (among other symptoms) that they almost never reproduce. PKU children, therefore, are born to parents who are not PKUs.

(a) Is the gene responsible for phenylketonuria completely dominant, incompletely dominant, codominant, or recessive? (b) Inasmuch as PKUs so rarely reproduce, why does such a disadvantageous gene persist in the population?

2-22   The normal brother of a PKU seeks the advice of a genetic counselor before a contemplated marriage. (a) What is the probability that he is heterozygous? (b) If PKU occurs once in 25,000 live births in the United States, and he contemplates marrying a normal woman in whose family no cases of PKU have occurred since her ancestors came over on the *Mayflower,* what is the probability of their having a PKU child? (c) What else might the genetic counselor consider telling this couple?

2-23   In **juvenile amaurotic idiocy,** children are normal until about age six. Subsequently there is a progressive decline in mental development, an impairment of vision leading to blindness, and muscular degeneration, culminating in death, usually before age 20. The trait may appear in families in which both parents are completely normal. A couple, age 25, planning marriage, are first cousins; siblings of both parties have died of the disorder. (a) Knowing no more than you do at this point about the genotypes of these two persons, what is the probability of *both* of them being heterozygous? (b) On the basis of your answer to part a, what could you tell them about the chance of their having an affected child? (c) Heterozygotes can be detected by an increase in vacuolization of **lymphocytes** (a type of white blood cell). If such a test should disclose that both persons are actually heterozygotes, what then could you say about the probability of their having an affected child?

2-24   **Dentinogenesis imperfecta** (opalescent dentine) is an hereditary tooth disorder characterized by defective dentine that splits from stress due to normal biting and chewing, and occurs in about one in 8,000 persons. Teeth of affected persons range in color from amber to opalescent blue. Affected children occur only in families where one or both parents have opalescent dentine; children with normal dentine may have parents who are both normal, or one normal and one affected, or both affected. (a) Is the gene for this condition completely dominant, codominant, incompletely dominant, or recessive? (b) A man with opalescent dentine marries a woman who has normal teeth; one of his parents was affected, one was not, but neither parent of the woman had opalescent dentine. What ratio of affected to unaffected would be found in the children of a large number of such families? (c) What kind of cross does the marriage in part (b) represent?

2-25   In addition to the M and N antigens of human red blood cells referred to in this chapter, two more human blood antigens, A and B (among many others) may occur on the surfaces of the red blood cells. Persons belong to any one of four blood groups: A, having only antigen A; B, having only antigen B; AB, having both antigens; and O, having neither of these antigens. Examine the following blood group pedigrees (a + sign in the progeny indicates a child of that group can occur):

P	Progeny			
	A	B	AB	O
A × A	+			+
A × O	+			+
A × B	+	+	+	+
B × B		+		+
B × O		+		+
AB × A	+	+	+	
AB × B	+	+	+	
AB × O	+	+		
AB × AB	+	+	+	
O × O				+

Assuming three alleles, one for production of antigen A, another for production of antigen B, and a third that results in failure to produce either antigen, what is the dominance relationship among these three alleles? (This question deals with a genetic situation that will be taken up in a later chapter, but you might wish to try your powers of inductive reasoning at this point.)

CHAPTER 3

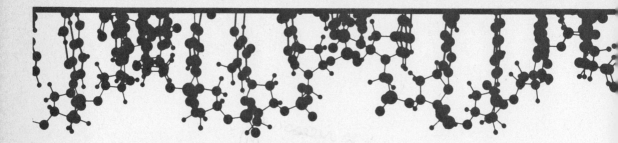

Dihybrid inheritance

The behavior of a single pair of genes through several generations has been followed in the preceding chapter. It will be interesting now to examine the course of two or more independently segregating pairs of genes through a number of generations. Behavior of nonindependently segregating genes will be considered in Chapter 6.

Classic two-pair ratios

COMPLETE DOMINANCE IN TWO PAIRS

In addition to the leaf margin trait of *Coleus* considered in Chapter 2, another vegetative character involves the venation pattern, more easily observed on lower surfaces of leaves. A typically encountered vein arrangement is the one shown in Fig. 3-1. Here a single midvein branches in standard pinnate fashion. An alternative expression is the highly irregular arrangement seen in Fig. 3-2. For convenience let the terms *regular* and *irregular*, respectively, refer to these two phenotypes. A simple monohybrid cross involving these two characters can easily demonstrate that *irregular* is completely dominant to *regular*.

Therefore a cross of a doubly homozygous *deep irregular* with a *shallow regular* individual (using D and d again to represent the deep and shallow genotypes, and I and i to denote irregular and regular) may be represented as follows:

P   deep irregular   ×   shallow regular

DDII   *ddii*

P gametes   (*DI*)   (*di*)

F₁   deep irregular

DdIi

Figure 3-1. *Regular venation pattern, a trait caused by a recessive gene, in* **Coleus.**

The F$_1$ individuals in this instance are referred to as dihybrid individuals because they are heterozygous for each of the two pairs of genes.

What will be the result of crossing two members of this F$_1$ to produce an F$_2$? The outcome of this cross can be predicted on the basis of the theses developed thus far. Recalling assumptions already made and the experimental evidence for them, when *considering one pair of genes at a time*, the phenotypic result will be the usual 3:1 ratio:

$$
\begin{array}{ll}
F_1 & Dd \times Dd \\[4pt]
F_1\ gametes & \left\{
\begin{array}{l}
eggs \quad \tfrac{1}{2}\,\textcircled{D} + \tfrac{1}{2}\,\textcircled{d} \\
sperms \quad \tfrac{1}{2}\,\textcircled{D} + \tfrac{1}{2}\,\textcircled{d}
\end{array}
\right. \\[12pt]
F_2\ genotypes & \tfrac{1}{4}\,DD + \tfrac{2}{4}\,Dd + \tfrac{1}{4}\,dd \\[8pt]
F_2\ phenotypes & \tfrac{3}{4}\ \text{deep} \quad + \tfrac{1}{4}\ \text{shallow}
\end{array}
$$

Likewise, *Ii* × *Ii* will produce the same $\tfrac{3}{4}$ irregular (*I* −):$\tfrac{1}{4}$ regular (*ii*) progeny. To determine the *phenotypic ratio* of the dihybrid cross *DdIi* × *DdIi*, the simplest and most logical (though not necessarily correct) assumption is that either the deep or shallow phenotype may be associated *at random* with either the irregular or regular phenotype if the two pairs of genes involved segregate independently.

Figure 3-2. *Irregular venation pattern, produced by the dominant allele of the gene for regular venation in* **Coleus.**

That is, in terms of the product law of probability noted in Chapter 2, the phenotypic ratio expected here is the *product* of its two component monohybrid ratios, thus

$$Dd \times Dd \text{ yields: } \tfrac{3}{4} \text{ deep } + \tfrac{1}{4} \text{ shallow}$$
$$Ii \times Ii \text{ yields: } \tfrac{3}{4} \text{ irregular } + \tfrac{1}{4} \text{ regular}$$

The F_2 result, based on random combination of phenotypic classes is as follows: ($\tfrac{3}{4}$ deep $\times$ $\tfrac{3}{4}$ irregular) $= \tfrac{9}{16}$ deep irregular $+$ ($\tfrac{1}{4}$ shallow $\times$ $\tfrac{3}{4}$ irregular) $= \tfrac{3}{16}$ shallow irregular $+$ ($\tfrac{3}{4}$ deep $\times$ $\tfrac{1}{4}$ regular) $= \tfrac{3}{16}$ deep regular $+$ ($\tfrac{1}{4}$ shallow $\times$ $\tfrac{1}{4}$ regular) $= \tfrac{1}{16}$ shallow regular.

This 9:3:3:1 ratio is the same as Mendel observed in his work with peas and is one of the classical phenotypic ratios. On this basis, the *genotypic* ratio of this cross is readily obtainable:

$$Dd \times Dd \text{ yields: } \quad \tfrac{1}{4} DD + \tfrac{2}{4} Dd + \tfrac{1}{4} dd$$
$$Ii \times Ii \text{ yields: } \quad \underline{\tfrac{1}{4} II \;\; + \tfrac{2}{4} Ii \;\; + \tfrac{1}{4} ii}$$

$$F_2 \text{ genotypes: } \quad \tfrac{1}{16} DDII + \tfrac{2}{16} DdII + \tfrac{1}{16} ddII$$
$$+ \tfrac{2}{16} DDIi + \tfrac{4}{16} DdIi + \tfrac{2}{16} ddIi$$
$$+ \tfrac{1}{16} DDii + \tfrac{2}{16} Ddii + \tfrac{1}{16} ddii$$

Closer examination of this 1:2:1:2:4:2:1:2:1 genotypic ratio reveals the 9:3:3:1 phenotypic ratio:

$$\frac{1}{16} DDII \qquad \frac{1}{16} ddII \qquad \frac{1}{16} DDii \qquad \frac{1}{16} ddii$$
$$\frac{2}{16} DdII \qquad \frac{2}{16} ddIi \qquad \frac{2}{16} Ddii$$
$$\frac{2}{16} DDIi$$
$$\frac{4}{16} DdIi$$

$\frac{9}{16} D - I -$	+	$\frac{3}{16} ddI -$	+	$\frac{3}{16} D - ii$	+	$\frac{1}{16} ddii$
deep		shallow		deep		shallow
irregular		irregular		regular		regular

Alternatively, this 9:3:3:1 ratio could have been calculated more quickly in this fashion:

$$Dd \times Dd \text{ yields } \tfrac{3}{4} D - + \tfrac{1}{4} dd$$
$$Ii \times Ii \text{ yields } \tfrac{3}{4} I - + \tfrac{1}{4} ii$$

to get the same result as obtained by collecting and summing the nine genotypes immediately above. *This ratio will be useful in later cases:*

$$9\ D - I -$$
$$3\ ddI -$$
$$3\ D - ii$$
$$1\ ddii$$

A 9:3:3:1 phenotypic ratio in corn is shown in Figure 3-3.

Note that in calculating our expected two-pair results, we once more followed the probability method and thus avoided the more cumbersome "checkerboard" or Punnett square. The probability approach reflects the random association of different pairs of genes and is more direct, especially in cases where one wishes to know what fraction of the progeny in a particular cross will be of a given genotype or phenotype. For instance, in this example with *Coleus*, the probability method permits an almost instant answer to such questions as, "What fraction of the offspring of the cross *DdIi* × *DdIi* will be (1) shallow irregular or (2) of the genotype *ddII?*" In the first case, both parents are doubly heterozygous and just those offspring are sought that combine the recessive phenotype of one trait with the dominant of the other character. Clearly the cross *DdIi* × *DdIi* will produce $\frac{1}{4}$ shallow × $\frac{3}{4}$ irregular, or $\frac{3}{16}$. The second question is calculated in the same manner, the P individuals given producing $\frac{1}{4}$ of their progeny having *each* of the genotypes *dd* and *II*, or $\frac{1}{4} \times \frac{1}{4} = \frac{1}{16}$. Compare these calculations, which, with a little practice and remembering the classic 3:1 and 1:2:1 ratios, can be determined mentally, with the results previously arrived at on pages 35–37. The same approach can be used to great advantage with trihybrid and other poly-hybrid crosses in which, as in the sample just given, only the specific information

Figure 3-3. *A 9:3:3:1 purple-starchy : purple-sweet : white-starchy : white-sweet* F_2 *phenotypic ratio in corn. Sweet kernels are shriveled; starchy are plump.*

sought is obtained with no need to cull out those facts from a mass of unneeded information about the entire range of progeny types.

It is most important to recognize the importance of one tacit assumption involved in this method of calculating dihybrid ratios. From the behavior of chromosomes at nuclear division, to be discussed in the next chapter, it will be clear (if genes are indeed located on chromosomes) that the calculation on page 36 is based on the expectation that, in gametes, association of D with I or i, and of d with I or i, is random. This then implies *equal* numbers of four possible gamete genotypes: *DI, dI, Di, and di* from a *DdIi* individual. The results of the cross *DdIi* × *DdIi* could be calculated either as on page 36 or by expected gamete genotypes:

$$DdIi \text{ eggs:} \quad \tfrac{1}{4}\widehat{DI} + \tfrac{1}{4}\widehat{dI} + \tfrac{1}{4}\widehat{Di} + \tfrac{1}{4}\widehat{di}$$
$$DdIi \text{ sperms:} \quad \tfrac{1}{4}\widehat{DI} + \tfrac{1}{4}\widehat{dI} + \tfrac{1}{4}\widehat{Di} + \tfrac{1}{4}\widehat{di}$$

A useful exercise would be to verify that, if gamete union is also random, the phenotypic and genotypic ratios arrived at in this way will be identical with those calculated on page 36. Production of four kinds of gametes in equal number by a doubly heterozygous individual will occur, of course, if each pair of genes is on a different pair of chromosomes. That is, each pair of genes here behaves exactly as in a one-pair cross. Cytological justification for this conclusion will be detailed in the next chapter.

GAMETE AND ZYGOTE COMBINATIONS

As seen earlier, a monohybrid such as *Dd* produces two kinds of gametes (D and d) in equal numbers, which can combine by syngamy to form three different zygote genotypes (*DD, Dd, dd*), from which two phenotypes (deep and shallow) will be discernible in the progeny. Likewise, a doubly heterozygous individual (*DdIi*) produces four kinds of gametes (*DI, Di, dI, di*), again in equal numbers if the two pairs of genes are located on different chromosome pairs. With random syngamy, nine different zygote genotypes are produced:

DDII   Ddii
DDIi   ddII
DdII   ddIi
DdIi   ddii
DDii

from which four phenotypes (deep irregular, deep regular, shallow irregular, shallow regular) will be apparent in the young seedlings.

A third pair of genes in *Coleus* can be symbolized as follows:

W (no white area at the base of the leaf blade)
w (white area at the base of the leaf blade)

What gamete combinations can be produced by the trihybrid *DdIiWw*? If each of the three pairs of genes is on a different chromosome pair, then either allele of any pair can combine with either allele of any other pair. A simple application of the probability method indicates eight possible gamete genotypes. As just seen, the dihybrid *DdIi* produces four gamete genotypes, $\tfrac{1}{4} DI + \tfrac{1}{4} Di + \tfrac{1}{4} dI + \tfrac{1}{4} di$. With the addition of the W, w pair, two additional gamete genotypes, W and w, become possible, and may combine with any of the dihybrid gamete genotypes:

$$\tfrac{1}{4}\widehat{DI} + \tfrac{1}{4}\widehat{Di} + \tfrac{1}{4}\widehat{dI} + \tfrac{1}{4}\widehat{di}$$
$$\tfrac{1}{2}\widehat{W} + \tfrac{1}{2}\widehat{w}$$

$\tfrac{1}{8}$ each of   *DIW, DiW, dIW, diW, DIw, Diw, dIw, diw*

Again applying the probability method used for monohybrid and dihybrid cases, these eight gamete genotypes may be expected to combine randomly into 27 zygote genotypes, producing eight phenotypic classes in the progeny.

Thus, considering the number of possible gamete genotypes per pair of heterozygous alleles in which dominance is complete and which are located on different chromosome pairs, the emergence of these mathematical relationships can be detected (where n = number of pairs of chromosomes with single gene differences):

number of gamete genotypes produced by parents = 2^n
number of progeny phenotypic classes = 2^n
number of progeny genotypic classes = 3^n

These relationships are summarized and extended in Table 3-1.

TESTCROSS

The testcross is a useful way of determining homozygosity or heterozygosity of a dominant phenotype in one-pair cases and can be equally valuable in determining genotypes in situations with two or more pairs of genes. In a two-pair cross, in which each pair of genes is on a different chromosome pair (i.e., the genes are not linked), the resulting progeny ratio is the product of two one-pair ratios. Thus, as we have seen, a monohybrid testcross ratio is 1:1, and a dihybrid testcross produces a 1:1:1:1 ratio when one parent is heterozygous for both gene pairs:

P   *DdIi* × *ddii*

P gametes $\frac{1}{4}$(DI) + $\frac{1}{4}$(dI) + $\frac{1}{4}$(Di) + $\frac{1}{4}$(di)

                1(di)

F$_1$ $\frac{1}{4}$ *DdIi* + $\frac{1}{4}$ *ddIi* + $\frac{1}{4}$ *Ddii* + $\frac{1}{4}$ *ddii*

        deep   shallow   deep   shallow

      irregular   irregular   regular   regular

The testcross is a very useful technique in mapping gene locations on chromosomes as will be discussed in Chapter 6.

With the information arrived at thus far it will be useful to determine testcross ratios in (1) cases in which only one of two pairs is heterozygous and (2) trihybrids or other polyhybrids. Problems at the end of this chapter explore cases such as these.

Table 3-1. Parental gamete genotypes, gamete combinations, and progeny phenotypes and genotypes (dominance complete in all pairs)

Number of pairs of heterozygous genes	Number of gamete genotypes	Number of progeny phenotypes	Number of progeny genotypes	Number of possible combinations of gametes
1	2	2	3	4
2	4	4	9	16
3	8	8	27	64
4	16	16	81	256
n	2^n	2^n	3^n	4^n

Modifications of the 9:3:3:1 ratio

INCOMPLETE DOMINANCE

Having examined dihybrid crosses in which dominance is complete in both pairs, it should now be determined whether incomplete dominance in one or both pairs has any effect on ratios and on numbers of phenotypic classes.

In tomato, two pairs of genes, located on different pairs of chromosome pairs are

$D-$	tall plant	h^1h^1	hairless stems
dd	dwarf plant	h^1h^2	scattered short hairs
		h^2h^2	very hairy stems

Crossing two individuals of genotype Ddh^1h^2 produces progeny as follows:

$\frac{3}{16}$ tall, hairless

$\frac{6}{16}$ tall, scattered hairs

$\frac{3}{16}$ tall, very hairy

$\frac{1}{16}$ dwarf, hairless

$\frac{2}{16}$ dwarf, scattered hairs

$\frac{1}{16}$ dwarf, very hairy

Note that this is precisely what would be expected. Tall and dwarf segregate in a 3:1 ratio, and hairless, scattered hairs, and very hairy segregate in a 1:2:1 ratio, producing a dihybrid 3:6:3:1:2:1 phenotypic ratio here. In such a case as this, because $Dd \times Dd$ produces two phenotypic classes in the offspring, and $h^1h^2 \times h^1h^2$ is responsible for three phenotypic classes, the cross $Ddh^1h^2 \times Ddh^1h^2$ produces offspring of $2 \times 3 = 6$ phenotypic classes. Adding to the mathematical expressions developed on page 39, it is clear that, if dominance is *in*complete, the number of phenotypic classes is 3^n (where again $n = $ the number of chromosome pairs with a single gene difference). In a dihybrid situation in which one pair of genes exhibits complete dominance and the other incomplete dominance, as in this example from tomato, the number of F_1 phenotypic classes from two doubly heterozygous parents is $2^n \times 3^n$.

Thus, again, incomplete dominance increases the number of phenotypic classes. Further possibilities of this sort are suggested in some of the problems at the end of this chapter.

EPISTASIS

Mouse. The laboratory mouse occurs in a number of colors and patterns. The wild type,[1] or "agouti," is characterized by color-banded hairs in which the part nearest the skin is gray, then a yellow band, and finally the distal part is either black or brown. The wild type has rather obvious selection value in natural surroundings, enhancing concealment of the individual. Two other colors are albino and solid black. In albinos there is a total lack of pigment, producing white hair and pink eyes (the latter results when blood vessel color shows through unpigmented irises).

The cross *black* × *albino* produces a uniform F_1 of agouti which, in certain instances, when inbred, results in an F_2 of 9 *agouti*:3 *black*:4 *albino*. The segregation of the F_2 into sixteenths immediately suggests two pairs of genes, and this particular ratio imples a 9:3:3:1 ratio in which the $\frac{1}{16}$ class and one of the $\frac{3}{16}$ classes are indistinguishable.[2] These results would then indicate further that the

---

[1]That is, the customary or most frequently encountered phenotype in natural populations, often used as a standard of comparison.

[2]A way to demonstrate the likelihood that this is *not* an approximation of a 1:2:1 ratio is explored in Chapter 5.

F_2 individuals in this case are heterozygous for both pairs of genes. Assume one of the two pairs of genes to include one allele for color production and another allele for color inhibition (the latter perhaps responsible for either a defective enzyme or the absence of a particular enzyme required for a specific intermediate biochemical step in pigment production). Further assume the other pair of genes to include one allele for agouti and one for black. Represent the genes as follows:

A	agouti	C	color
a	black	c	color inhibition

Note that dominance of agouti over black is suggested by the $\frac{9}{16}$ agouti class versus the $\frac{3}{16}$ black in the F_2 (this is tantamount to a 3:1 segregation). On these assumptions, the cross in mouse may be diagramed in this way:

$$P \quad \text{black} \quad \times \quad \text{albino}$$
$$aaCC \qquad AAcc$$

$$F_1 \quad AaCc \text{ agouti}$$
$$F_2 \quad \frac{9}{16} A - C - \quad \text{agouti}$$
$$\frac{3}{16} aaC - \quad \text{black}$$

$$\left. \begin{array}{l} \frac{3}{16} A - cc \\ \frac{1}{16} aacc \end{array} \right\} \quad \text{albino}$$

In this particular example, gene c (which is recessive to its own allele C) actually masked the effect of either $A-$ or aa so that any $-cc$ individual is albino. Such a gene, which masks the effect of one or both members of a *different* pair of genes, is said to be *epistatic*; the masked gene or genes may be termed *hypostatic*. Here c is epistatic to A and a, and this case illustrates recessive epistasis because the recessive of one pair is epistatic to another pair. Note that epistasis is quite different from dominance, in that the masking operates between different pairs of alleles rather than between members of one pair.

Clover. An interesting example in white clover (*Trifolium repens*, the clover so frequently seen in lawns), reported in 1943 by Atwood and Sullivan, furnishes presumptive evidence that epistasis depends on a gene-enzyme relationship in a series of sequential biochemical steps.

Some strains of white clover test high in hydrocyanic acid (HCN), whereas others test negatively for this substance. HCN content is associated with more vigorous growth; it does not harm cattle eating such varieties. Often the cross *positive* × *negative* results in an F_1 testing uniformly positive for HCN and an F_2 segregating 3 positive: 1 negative, suggesting a single pair of genes with *positive* the dominant trait.

One series of crosses reported by Atwood and Sullivan, however, produced unexpected totals:

$$P \quad \text{positive} \times \text{negative}$$
$$F_1 \qquad \text{positive}$$
$$F_2 \quad 351 \text{ positive} + 256 \text{ negative}$$

A 3:1 expectancy in the F_2 would be approximately 455 positive: 152 negative; the actual results differ sufficiently from a 3:1 ratio to cast doubt on its relevance here. (Statistical tests, described in Chapter 5, give objective support to the hypothesis that, although a ratio of 351:256 *could* occur by chance alone in a 3:1 expectancy, this result is unlikely enough to cause one to look for a better explanation.) Note that these results are very close to a 9:7 ratio, which would be 342 positive : 265 negative. A 9:7 ratio immediately suggests an epistatic expression of the 9:3:3:1, which, in turn, indicates two pairs of genes. One such 9:7 ratio (purple:yellow) in corn is illustrated in Figure 3-4.

Figure 3-4. *A 9:7 purple:yellow epistatic ratio in corn.*

HCN formation follows a path that may be represented thus:

$$\longrightarrow (\text{precursor}) \xrightarrow[\text{"}\alpha\text{"}]{\text{enzyme}} \text{cyanogenic glucoside} \xrightarrow[\text{"}\beta\text{"}]{\text{enzyme}} \text{HCN}$$

Each of the conversions indicated by the arrows is enzymatically controlled. Tests of F_2 individuals of the cross just described for (1) HCN, (2) enzyme "β," and (3) cyanogenic glucoside revealed four classes of individuals:

Class	HCN	Enzyme "β"	Glucoside
1	+	+	+
2	0	+	0
3	0	0	+
4	0	0	0

Each of these four classes was then tested for HCN after adding either enzyme "β" or glucoside to their leaf extracts, with the following results:

Class	Control test for HCN	HCN test after adding enzyme "β"	HCN test after adding glucoside
1	+	+	+
2	0	0	+
3	0	+	0
4	0	0	0

Results show that class 1 plants produce both glucoside and enzyme "β" (and, by inference, also enzyme "α"); class 2 plants produce enzyme "β" but no glucoside (hence, by inference, no enzyme "α"); class 3 plants produce glucoside (and, there, enzyme "α") but no enzyme "β"; class 4 plants produce neither enzyme "β" nor glucoside (therefore, by inference, no enzyme "α"). Conclusions are summarized in tabular form:

Class	Production			Substance Accumulating
	Enzyme "α"	Glucoside	Enzyme "β"	
1	+	+	+	HCN
2	0	0	+	precursor
3	+	+	0	glucoside
4	0	0	0	precursor

It therefore seems highly likely that production of enzymes "α" and "β" is determined by two different pairs of genes:

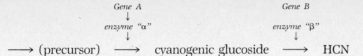

So an individual must have at least one dominant of each of two pairs of genes, *A* and *B*, in order to carry the process from precursor to HCN. Addition of genotypes to the cross developed on page 41 produces the following:

P   positive   ×   negative
       AABB   *aabb*

F_1 positive
           AaBb

F_2 $\frac{9}{16}$ HCN positive $A-B-$ ("class 1")   (9)
     $\frac{3}{16}$ HCN negative $aaB-$ ("class 2")
     $\frac{3}{16}$ HCN negative $A-bb$ ("class 3") $\Big\} \ (7)$
     $\frac{1}{16}$ HCN negative *aabb*   ("class 4")

A little reflection will serve to indicate that this situation may be considered in any of the following ways:

9:7   HCN : no HCN
12:4   glucoside : no glucoside
9:3:4   HCN : glucoside : precursor
9:3:3:1   enzymes "α" and "β" : enzyme "β" only :
           enzyme "α" only : neither enzyme

The ratio chosen depends, of course, on the level of chemical analysis to which consideration is carried. Thus we are talking about phenotype in terms of chemical reaction and content or of enzyme production. Because enzymes are proteins in whole or in part, this suggests again a relationship between gene and enzyme and, therefore, between protein synthesis and phenotype. This will be a promising avenue to explore later in this book.

Incidentally, an explanation for the occurrence of all positive cyanide content progeny from crosses of negative × negative (which is also reported) works out on these bases quite readily:

P   negative   ×   negative
       aaBB   *AAbb*

F_1 positive
           AaBb

The cross (P) positive × negative that produces an F_2 of 3 positive:1 negative, referred to on page 41 genotypically would look like this:

P   positive   ×   negative
       AABB   *aaBB*

F_1 positive
           AaBB

F_2 3 positive   :   1 negative
     $A-BB$ *aaBB*

Epistasis (which can operate in any cross that involves two or more pairs of genes) can be recognized *by the reduction in number of expected phenotypic classes* in which two or more of the classes become indistinguishable from each other.

Many other ratios are, of course, possible and have been reported in the literature. Additional examples are included in the problems at the end of this chapter, as well as in Table 3-2. One of these, however, requires additional explanation. In certain breeds of domestic fowl (e.g., White Leghorn), individuals are white because of a dominant color-inhibiting gene *I*. Even though birds may carry genes for color, such genes cannot be expressed in the presence of *I* −. On the other hand, such breeds as the White Silkie are white because of homo-

Table 3-2. Summary of dihybrid ratios in F² of the cross AABB × aabb

		AABB	AABb	AaBB	AaBb	AAbb	Aabb	aaBB	aaBb	aabb
More than four phenotypic classes	A and B both incompletely dominant	1	2	2	4	1	2	1	2	1
	A incompletely dominant; B completely dominant	3		6		1	2	3		1
Four phenotypic classes	A and B both completely dominant (classic ratio)	9				3		3		1
Fewer than four phenotypic classes	aa epistatic to B and b Recessive epistasis	9				3		4		
	A epistatic to B and b Dominant epistasis	12						3		1
	A epistatic to B and b; bb epistatic to A and a Dominant and recessive epistasis	13*						3		
	aa epistatic to B and b; bb epistatic to A and a Duplicate recessive epistasis	9				7				
	A epistatic to B and b; B epistatic to A and a Duplicate dominant epistasis	15								1
	Duplicate interaction	9				6				1

*The 13 is composed of the 12 classes immediately above, plus the one *aabb* from the last column.

zygosity for the recessive gene c, which blocks synthesis of a necessary pigment precursor.

Crosses of White Leghorn ($IICC$) × White Silkie ($iicc$) produce an F_2 ratio of 13:3, as follows:

9 $I - C -$ white (because of color-inhibitor I)
3 $iiC -$ colored
3 $I - cc$ white (because of both I and cc)
1 $iicc$ white (because of cc)

The actual color of the $iiC -$ individuals depends on the presence of additional genes for particular colors. These are not shown here.

Lethal genes also reduce the number of expected phenotypic classes in a two-pair cross, but the change is somewhat different from that caused by epistasis. For example, consider the case of corn, in which tall ($D -$) and (dwarf) (dd) phenotypes are known in addition to the green and "albino" condition described in Chapter 2. Note that the cross $DdGg$ × $DdGg$ will produce the classic 9:3:3:1 phenotypic ratio of seedlings, but because of the lethal effect of gg, this becomes 9 tall green:3 dwarf green (that is, 3 tall:1 dwarf) after gg has exerted its lethal effect. In humans, for example, consider the dominant lethal for Huntington's chorea (Chapter 2) together with another pair of genes, free ear lobes ($A -$) versus attached ear lobes (aa) (Fig. 1-1). A marriage involving two double heterozygotes, $HhAa$ (where H represents the dominant lethal for chorea and h its recessive allele for "normal"), would, statistically (or actually in collections of family data), produce a 9:3:3:1 ratio initially, which would ultimately become 3 free (normal):1 attached (normal) after the choreic individuals had died.

References

Atwood, S. S., and J. T. Sullivan, 1943. Inheritance of a Cyanogenic Glucoside and Its Hydrolyzing Enzyme in *Trifolium repens. Jour. Hered.*, **34:** 311–320.

Steward, R. N., and T. Arisumi, 1966. Genetic and Histogenic Determination of Pink Bract Color in Poinsettia. *Jour. Hered.*, **57:** 217–220.

Problems

3-1 How many different matings can be made in a population in which only one pair of genes is considered?

3-2 If two $DdIi$ *Coleus* plants are crossed, what fraction of the offspring will be (a) shallow irregular, (b) deep regular, (c) $DDIi$, (d) $ddII$?

3-3 How many progeny phenotypic classes result if two *Coleus* plants (a) of genotype $DdIiWw$ are crossed; (b) heterozygous for one pair of genes (showing complete dominance) on each of its pairs of chromosomes are crossed?

3-4 What is the phenotypic ratio of the testcross (a) $DdII$ × $ddii$ in Coleus, (b) $DdIiWw$ × $ddiiww$ in *Coleus?*

3-5 How many different gamete genotypic classes are produced by the tetrahybrid $AaBbCcDd$?

3-6 If two tetrahybrids like that of the preceding problem are crossed, how many of each of the following can be expected in the progeny: (a) phenotypic classes, (b) genotypic classes?

3-7 In how many ways can gametes of two tetrahybrids ($AaBbCcDd$) be combined to form zygotes?

3-8 How many phenotypic classes are produced by a testcross in which one parent is heterozygous for (a) two pairs of genes, (b) three pairs of genes, (c) four pairs of genes, (d) n pairs of genes?

3-9 Suggest a mathematical formula for determining the probability of a completely homozygous recessive progeny individual resulting from selfing an individual heterozygous for n pairs of genes, all of which exhibit complete dominance.

3-10 Assume the following sets of genes in human beings:

$A-$	free ear lobes	h^1h^1	straight hair
aa	attached ear lobes	h^1h^2	wavy hair
		h^2h^2	curly hair
$R-$	nonred hair	$P-$	polydactylous
rr	red hair	pp	nonpolydactylous

(a) A husband and wife, $Aah^1h^2PpRr \times Aah^1h^2ppRr$, want to know the probability of their having a child with attached earlobes, wavy red hair, and nonpolydactylous. What would you tell them? (b) What is the chance that they might have a child with curly red hair (without regard to the other traits)?

3-11 A few of the many known genes (each on a different chromosome pair) in tomato are

P	smooth-skinned fruit	p	"peach" (pubescent fruit)
W	yellow flowers	w	white flowers
h^1	hairless stems and leaves	h^2	hairy stems and leaves

The h^1, h^2 pair shows incomplete dominance. A tetrahybrid smooth, yellow, cut, scattered-hair plant is self-pollinated. (a) How many phenotypic classes can occur in the progeny? (b) What fraction of the offspring can be expected to be peach, white, potato, hairless? (c) What fraction of the progeny can be expected to be peach, white, potato, with scattered hairs?

3-12 You raise 100 tomato plants from seed received from a friend and find 37 red-fruited plants with scattered short hairs on stems and leaves, 19 red hairless, 18 red very hairy, 13 yellow-fruited with scattered short hairs, 7 yellow very hairy, and 6 yellow hairless. Suggest genotypes and phenotypes for the unknown parent plants from which the 100 seeds were obtained.

3-13 Coat in guinea pigs may be either long or short; matings of short $\times$ short may produce long-haired progeny, but long $\times$ long gives rise only to long. Additionally, coat color may be yellow, cream, or white. The mating cream $\times$ cream produces progeny of each of the three colors. Given the following incomplete pedigree:

P   long yellow $\times$ short white
F_1 all short $-$

(a) What is the coat color in the F_1? (b) If members of the F_1 were interbred, what fraction of their progeny would be long cream?

3-14 What progeny phenotypic ratio results from the cross $AaBbc^1c^2 \times AaBbc^1c^2$ if bb individuals die during an early embryo stage?

3-15 In addition to the genes for flower color in snapdragon described in Chapter 2, leaves in this plant may be broad, narrow, or intermediate. From the cross red broad $\times$ white narrow this F_2 was obtained: 10 red broad, 20 red intermediate, 10 red narrow, 20 pink broad, 40 pink intermediate, 20 pink narrow, 10 white broad, 20 white intermediate, and 10 white narrow. (a) How many pairs of genes are involved and what kind of dominance is demonstrated by this case? (b) Which of these F_2 phenotypic classes can you recognize as homozygous?

3-16 The Christmas poinsettia (*Euphorbia pulcherrima*) produces colored modified leaves (bracts) below the clusters of small flowers. These bracts may be red, pink, or white. Work of Steward and Arisumi (1966) shows that red pigmentation results from a two-step, enzymatically controlled biochemical process from a colorless precursor via an intermediate pink pigment that is converted to a red pigment in the second step. Let W represent the completely dominant gene that is responsible for production of the enzyme catalyzing the first step, from colorless to pink, and P the completely dominant gene for producing the enzyme bringing about conversion of the pink pigment to a red one. If a $wwpp$ white-bracted plant is crossed to a doubly homozygous red-bracted one, what phenotypic ratio can be expected in the F_2?

3-17 Normal hearing depends on the presence of at least one dominant of each of two pairs of genes, D and E. If you examined the collective progeny of a large number of $DdEe \times DdEe$ marriages, what phenotypic ratio would you expect to find?

3-18 In sweet pea the cross white flowers × white flowers produced an F_1 of all purple flowers. An F_2 of 350 white and 450 purple was then obtained. (a) What is the phenotypic ratio in the F_2? Using the first letter of the alphabet and as many more in sequence as needed, give (b) the genotype of the purple F_2, (c) the genotype of the F_1, (d) the genotypes of the two P individuals.

3-19 The fruit of the weed shepherd's purse (*Capsella bursa-pastoris*) is ordinarily heart-shaped in outline and somewhat flattened, but occasionally individuals with ovoid fruits occur. Crosses between pure-breeding heart and ovoid yield all heart in the F_1. Selfing this F_1 produces and F_2 in which 6 percent of the individuals are ovoid. Starting with the first letter of the alphabet, and using as many more as necessary, give the genotypes of (a) ovoid, (b) F_1 heart, (c) F_2 heart.

3-20 In certain breeds of dog the genotype $C-$ produces a pigmented coat, whereas cc gives rise to a white coat (*not* albino). Another pair of genes (B and b) determines the color of the coat in $C-$ dogs such that $C-B-$ animals are black and $C-bb$ animals are brown. Assume two animals of genotype $CcBb$ are crossed. What phenotypic ratio results in the pups of a large number of such matings?

3-21 Instead of the gene action described in the preceding problem, assume $C-$ animals are white, whereas cc animals have pigmented coats. Assume also that gene pair B, b produces coat color (in the presence of cc) as in problem 3-20. If two $CcBb$ dogs are crossed, what phenotypic ratio is expected in the F_1?

3-22 Give the F_1 phenotypic ratio resulting from the cross of two $CcBb$ dogs if $C-$ is a color inhibiting genotype, cc a genotype producing a pigmented coat, $B-$ animals have brown coats, and bb black.

3-23 If, in another breed of dogs, $C-$ is responsible for a pigmented coat and cc for an unpigmented coat, $B-$ produces brown coat and bb black, what is the phenotypic ratio in the F_1 of the cross $CcBb × CcBb$?

3-24 In some plants cyanidin, a red pigment, is synthesized enzymatically from a colorless precursor; delphinidin, a purple pigment, may be made from cyanidin by the enzymatic addition of one $-OH$ group to the cyanidin molecule. In one cross in which these pigments were involved, purple × purple produced F_1 progeny as follows: 81 purple, 27 red, and 36 white. (a) How many pairs of genes are involved? (b) What is the genotype of the purple parents? (c) What is the genotype of each of the three F_1 phenotypic classes? (Use as many letters of the alphabet as needed, starting with A.)

3-25 In terms of *enzyme production*, instead of flower color, as a phenotypic character in the data of problem 3-24, what is the F_1 phenotypic ratio? The enzyme catalyzing conversion of precursor to cyanidin may be designated enzyme 1, and that controlling the production of delphinidin from cyanidin may be designated enzyme 2.

3-26 In the plants of problem 3-24, one cross of white × red produced all purple progeny, whereas another white × red cross gave rise to 1 purple:2 white:1 red. What were the parental genotypes in each of these two crosses?

3-27 In addition to the round, oval, and long radishes mentioned earlier, radishes may be red, purple, or white in color. Red × white produces progeny all of which are purple. If purple oval were crossed with purple oval, how many pure-breeding types would occur in the progeny?

3-28 In cattle, "short spine" is lethal shortly after birth; it is caused by the homozygous recessive genotype ss. Heterozygotes are normal. A series of matings between roan animals heterozygous for the short spine gene produces what phenotypic ratio (a) at birth and (b) after several days?

3-29 In the summer squash, fruits may be white, yellow, or green. In one case, the cross of yellow × white produced an F_1 of all white-fruited plants that, when selfed, gave an F_2 segregating 12 white:3 yellow:1 green. (a) Suggest genotypes for the white, yellow, and green phenotypes. (b) Give genotypes of the P, F_1, and F_2 of this cross. Use genotype symbols starting with the first letter of the alphabet.

3-30 Summer squash fruit shape may be disk, sphere, or elongate. The cross of sphere × sphere produced an F_1 with all disk-shaped fruits. Selfing the F_1 gave 9 disk:6

sphere:1 elongate. (a) Suggest genotypes for disk, sphere, and elongate. (b) Give genotypes of the P, F_1, and F_2 of this cross. Use genotype symbols beginning with the first letter of the alphabet *after* those used for problem 3-29.

3-31 Considering the facts suggested by the two preceding problems, how many different genotypes are responsible for the (a) yellow sphere, (b) elongate green, (c) disk white phenotypes?

3-32 A tetrahybrid white disk plant is selfed. (a) How many phenotypic classes could occur in its F_1? (b) What fraction of the F_1 will be white disk?

3-33 In Duroc Jersey pigs, two pairs of interacting genes, R and S, are known. (a) The cross of red $\times$ red sometimes produces an F_1 phenotypic ratio of 9 red:6 sandy:1 white. What is the genotype of each of these F_1 phenotypes? (b) For each of the crosses that follow, give the parental genotypes:

	P	F_1	F_2
Case 1.	Red $\times$ red	All red	All red
Case 2.	Red $\times$ red	3 red:1 sandy	Not reported
Case 3.	Red $\times$ white	All red	9 red:6 sandy:1 white
Case 4.	Sandy $\times$ sandy	All red	9 red:6 sandy:1 white
Case 5.	Sandy $\times$ sandy	1 red:2 sandy:1 white	Not reported

3-34 Determine the genotypic and phenotypic ratios resulting from each of the following dihybrid crosses (assume lethals to exert their effect during early embryo development):

Parental genotypes	Gene characteristics		Progeny ratios	
	First pair	Second pair	Genotypic	Phenotypic
(a) $AaBb \times AaBb$	Complete Dominance	Complete Dominance		
(b) Aab^1b^2 $\times Aab^1b^2$	Complete Dominance	Incomplete Dominance		
(c) $a^1a^2b^1b^2$ $\times a^1a^2b^1b^2$	Incomplete Dominance	Incomplete Dominance		
(d) $AaBb \times AaBb$	Complete Dominance	Recessive Lethal		
(e) $a^1a^2Bb \times$ a^1a^2Bb	Incomplete Dominance	Recessive Lethal		
(f) $AaBb \times AaBb$	Recessive Lethal	Recessive Lethal		

3-35 In a hypothetical flowering plant assume petal color to be due to a pair of codominant genes, a^1 and a^2, heterozygotes being purple, and homozygotes either red (a^1a^1) or blue (a^2a^2). Assume also another pair of genes, completely dominant B for color, and recessive b for color inhibition. This pair segregates independently from the a^1, a^2 pair. Two doubly heterozygous purple plants, a^1a^2Bb, are crossed. What phenotypic ratio results in the progeny?

CHAPTER 4

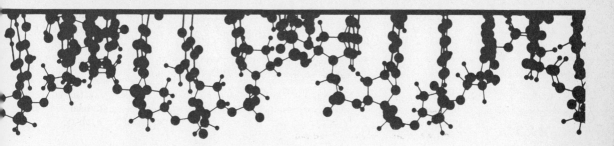

Cytological bases of inheritance

In the two preceding chapters genes were assumed to be located in the nucleus. If this assumption is true, some confirming, objective evidence should be found. Such evidence does exist and can be conveniently designated as of two kinds, cytological and chemical. In the history of the science of genetics the former preceded the latter by many years. The cytological evidence will be examined here, with the chemical deferred to a later and more logical point in the development of our concepts of genetic mechanisms.

The interphase nucleus

As your previous experience in the life sciences has shown you, living cells of most organisms (*eukaryotes*) are characterized by the presence of a discrete, often spherical body, the **nucleus** (Fig. 4-1). Notable exceptions are such *prokaryotes* as the blue-green algae and the bacteria in which, although "nuclear material" can be shown to occur, the visible, structural organization so typical of the cells of higher organisms is lacking. Prokaryotes, of course, lack the conventional mitotic and meiotic nuclear divisions that characterize eukaryotes. In addition to these two types of genetic systems, we can recognize still a third in the viruses. The structure of an important type of virus, the bacteriophages, is described more fully in Chapter 6.

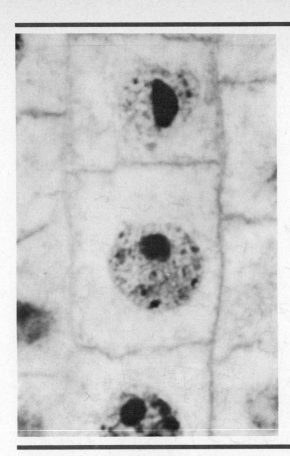

Figure 4-1. *Interphase nucleus of onion root tip, as seen with the light microscope. Compare with Figure 4-2. Note the prominent nucleolus.* (Courtesy Carolina Biological Supply Co.)

The interphase (nondividing) nucleus is bounded by an interface, the **nuclear membrane,** which is not clearly visible with the light microscope. Electron micrographs, however, reveal this membrane to be a double layer, provided with numerous "pores" of about 20 to 80 nm in inside diameter.[1] Each pore is surrounded by a thickened, electron-dense ring in the nuclear membrane, giving an outside diameter of up to 200 nm. These pores function in active control of the passage of macromolecules between nucleus and cytoplasm. The nuclear membrane is continuous with a cytoplasmic double membrane system, the **endoplasmic reticulum,** to which dense granular structures, the **ribosomes,** some 17 by 22 nm in size, can usually be seen attached in electron micrographs (Fig. 4-2). Ribosomes are rich in **ribonucleic acid** (RNA) and play an important part in protein synthesis (see Chapters 17 and 18).

Within the interphase nucleus three major components can be distinguished. The first of these is nuclear sap, or **karyolymph,** a clear, usually nonstaining, largely proteinaceous, colloidal material. The second intranuclear component is a generally spherical, densely staining body, the **nucleolus.** Many nuclei contain two or more nucleoli. The general size range is 2 to 5 μm, but size varies with the tissue, the degree of protein synthesizing activity of the cell (the nucleolus is larger when this activity is high), and nutritional factors. Nucleoli contain RNA, as well as some DNA and protein. Typically each nucleolus is produced by and is physically associated with a *nucleolus organizing region* (which is the site of synthesis of ribosomal RNA, as described in Chapter 17) of a particular **chromosome,** the third major nuclear component.

---

[1]Twenty to 80 nanometers (nm); a nanometer is 1×10^{-9} meter, or 10 Angstrom units. In older usage a nanometer was called a millimicron (mμ).

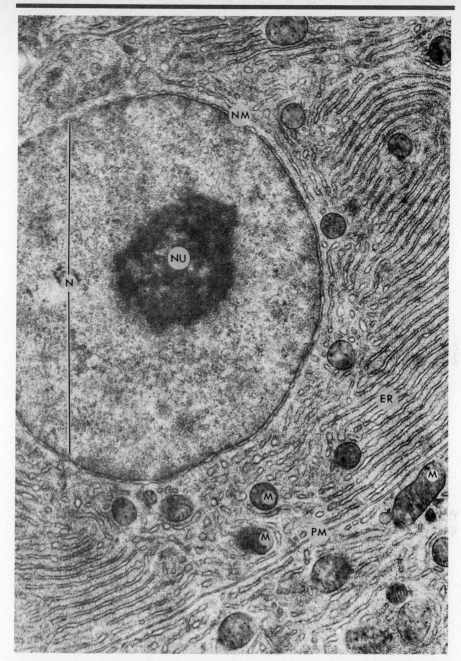

Figure 4-2. *Electron micrograph of interphase cell from bat pancreas. N, nucleus; NU, nucleolus; NM, nuclear membrane; ER, endoplasmic reticulum with ribosomes attached; M, mitochondria; PM, plasma membrane delimiting the cell. Note doubleness of the membrane systems,* (From *Cell Ultrastructure* by William Jensen and Roderic Park. © 1967 by Wadsworth Publishing Co., Inc., Belmont California. Reproduced by permission.)

The remainder of the nucleus consists of **chromatin,** fine, threadlike strands of deoxyribonucleic acid (DNA) intimately complexed with various proteins to form *nucleoproteins.* These include (1) five kinds of basic (i.e., positively charged) low molecular weight proteins called **histones,** and (2) a variety of **nonhistone**

proteins, numbering perhaps 100 kinds (Watson, 1976) if all eukaryotic species are considered.

At somewhat unequal intervals of ~100 Å (Angstrom units; 1 Å = 100 nm) there occur nucleoprotein units, the **nucleosomes,** as shown in Fig. 4-3A. The nucleosome core is roughly cylindrical, but has a wedge shape when viewed from certain angles in electron microscopy ("side view"), 100 Å in external diameter and 55 Å in height. In the words of Chambon (1978), ". . . the organization of eukaryotic DNA into nucleosomes, the repeating nucleoprotein units of chromatin, constitute the first level in a stepwise compaction mechanism which leads to the packaging of DNA at high concentration in the interphase nucleus or metaphase chromosome." The problem of packaging DNA in eukaryotic chromosomes and in bacterial cells will be discussed in greater detail in later chapters.

Nucleosome cores consist of an octamer of two molecules each of four histones, designated as H2A, H2B, H3, and H4 (Fig. 4-3B). The surface of the core particle is surrounded by a flat, superhelical strand of DNA that makes about 1.75 turns totaling 496 Å (= 146 deoxyriboncleotide pairs; see Chapter 16).

Internucleosomal, or **linker DNA,** has a linear measurement of about 50 to 340 Å, depending on the cell type, tissue, and species. A fifth kind of histone, H1, is bound to a part of the linker DNA (Fig. 4-3C). Recent evidence suggests that H1 may be located at or near the points at which linker DNA enters and leaves the core particle. H1 is larger than any of the core histones, and much more variable in component amino acids. It appears to have been less rigorously conserved in evolution. In nucleated erythrocytes of birds, fish, and amphibians the role of H1 is assumed by a sixth histone, H5, which may be merely an extreme variant (in terms of constituent amino acids) of H1. In sperms histones are replaced by another basic protein, protamine, as spermatids mature into sperms. The diameter of the H1-DNA complex is approximately 30 to 35 Å, of which 20 Å are accounted for by the diameter of the DNA molecule. The structure and function of nucleoprotein, as well as the role of the nonhistone proteins, is explored in detail in Chapter 18.

During nuclear division the chromatin strands contract and appear thicker, to become the **chromosomes,** which are visible with an ordinary light microscope. The number and morphology of the chromosomes are specific, distinct, and ordinarily constant for each species, although, especially in plants, subspecific taxonomic categories with multiple sets (polyploids) are not infrequent. In some cases, however, morphological differences among different chromosomal races is sufficient to give them species or subspecific rank (Table 4-1 and Chapter 13). Some details of the gross morphology of chromosomes are pointed out later in this chapter, and their molecular structure is described in Chapter 16.

THE INTERPHASE CYCLE

The interphase cycle may be defined as the entire sequence of events transpiring from the close of one nuclear division to the beginning of the next one. The total time varies with the species, maturation, tissue, and temperature, among other factors. Duration of as little as about three hours as to as much as 174 hours has been reported for various organisms. In a number of kinds of human tissue in vitro the interphase cycle typically occupies some 18 to 24 hours. For convenience, the following stages are recognized:

G_1 The first *growth* (or gap[2]) stage of interphase in which nucleus and cytoplasm are enlarging toward mature size. Chromatin is fully extended and not distinguishable as discrete chromosomes with the light microscope (Fig. 4-4). This is a time of active synthesis of ribonucleic acid

[2]*Gap* signifies a gap in DNA synthesis during interphase.

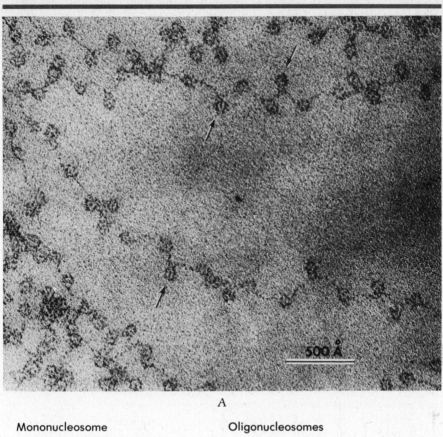

A

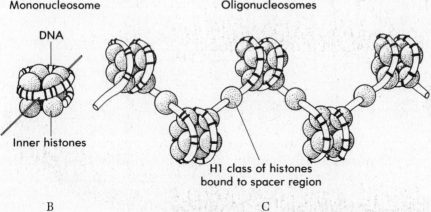

Mononucleosome   Oligonucleosomes

DNA

Inner histones

H1 class of histones
bound to spacer region

B   C

Figure 4-3. (A) Electron micrograph of spread chicken erythrocyte nucleus showing nucleosomes (arrows) and spacer DNA connecting them. (B) Model of a single nucleosome showing a core particle of eight inner histone molecules, surrounded by 1.75 turns of DNA. (C) Model of a small portion of chromatin showing five nucleosomes with histone 1 bound to the spacer DNA. [(A) from A. L. Olins, J. P. Breilatt, R. D. Carlson, M. B. Senior, E. B. Wright, and D. E. Olins, 1977. On Nu Models for Chromatin Structure. In P. O. P. T.'So, ed., *The Molecular Biology of the Mammalian Genetic Apparatus*, vol. 1, published by Elsevier/North-Holland, New York. Used by permission of Dr. A. L. Olins (who kindly supplied the electron micrograph) and the publisher. (B) and (C) redrawn from D. E. Olins and A. L. Olins, 1978. "Nucleosomes: The Structural Quantum in Chromosomes." *Amer. Scientist, 66:*704–711. Used by permission of authors and publisher.]

Table 4-1. Chromosome numbers in some plants and animals*

	Common name	Scientific name	Chromosome number Monoploid†	Diploid†
		Plants		
I Chlorophyta	Chlamydomonas	*Chlamydomonas moewussi*	8	
	Chlamydomonas	*C. reinhardi*	16	
II Phycomycota	Bread mold	*Mucor spp.*	2	
III Ascomycota	Bread mold	*Neurospora crassa*	7	
	Penicillium	*Penicillium spp.*	2,4,5	
IV Bryophyta	Liverwort	*Sphaerocarpos donnelii*	♂ 7 + Y ♀ 7 + X	
V Pterophyta	Adder's tongue fern	*O. reticulatum*		1,262
VI Anthophyta				
Dicotyledonae	Watermelon	*Citrullus vulgaris*		22§
	Coleus	*Coleus blumei*		24
	Jimson weed	*Datura stramonium*		24
	Tomato	*Lycopersicon esculentum*		24
	Evening primrose	*Oenothera spp.*		14‡§
	Pea	*Pisum sativum*		14
	Swamp saxifrage	*Saxifraga pensylvanica*		56‡§
	Red clover	*T. pratense*		14
	White clover	*T. repense*		32§
Monocotyledonae	Slender oat	*Avena barbata*		28
	Cultivated oat	*A. sativa*		42
	Rye	*Secale cereale*		14
	Smooth cordgrass	*Spartina alterniflora*		62
	Cordgrass	*S. maritima*		60
	Cordgrass	*S. anglica*		124
	Emmer wheat	*Triticum dicoccum*		28
	Wild emmer	*T. dicoccoides*		28
	Durum wheat	*T. durum*		28
	Einkorn	*T. monococcum*		14
	Spelt	*T. spelta*		42
	Common wheat	*T. vulgare*		42
	Corn	*Zea mays*		20

(RNA) and protein, especially (1) enzymes necessary for the DNA replication of the next stage, (2) possibly a protein that acts to trigger nuclear division, and (3) tubulin and mitotic apparatus proteins (see page 59). This is the most variable stage as to duration; it may occupy 30 to 50 percent of the total time of the interphase cycle or be entirely lacking in rapidly dividing cells (e.g., those of the early mammalian embryo and those in such lower forms as slime molds (*Physarum*) and a yeast, (*Schizosaccharomyces*). On the other hand, G_1 may last as long as 151 hours in mature cells of corn (*Zea*) roots. Differentiated somatic cells that no longer divide (e.g., neurons) can usually be found in the G_1 stage. Some authorities prefer to designate this last situation as the G_0 stage. Even though changes in synthesis patterns may occur in such differentiated cells, the condition of their DNA is the same as that of typical G_1 cells. So the question has been raised as to whether G_0 represents any more than a modification of G_1. Watson (1976) aptly refers to G_1 as a sequence of "preprogrammed operations" in preparation for DNA replication. This may include cytoplasmic synthesis of DNA polymerases (an enzyme system catalyzing DNA replication).

Table 4-1. (Continued)

	Common name	Scientific name	Chromosome number Monoploid†	Diploid†
		Animals		
I Arthropoda	Honeybee	*Apis mellifica*	♂ 16	♀ 32
	Fruit fly	*Drosophila affinis*		10
		D. hydei		12
		D. melanogaster		8
		D. prosaltans		6
		D. pseudoobscura		10
		D. virilis		12
		D. willistoni		6
	Grasshopper	*Melanoplus differentialis*		♀ 24, ♂ 23
II Chordata	Cattle	*Bos taurus*		60
	Ass	*Equus asinus*		62
	Horse	*Equus caballus*		64
	Mule	*E. asinus* × *E. caballus*		63
	Human	*Homo sapiens*		46
	Rhesus monkey	*Macaca mullatta*		42
	Mouse	*Mus musculus*		40
	Chimpanzee	*Pan troglodytes*		48

*An extensive list of chromosome numbers is given in P.L. Altman and D.S. Ditmer, eds., 1962, and 1972. In each plant example here the chromosome number of the *dominant* generation (gametophyte or sporophyte) is cited. See also Appendix B.
†*Diploid*, from the Greek *di* (as a prefix, "two") and *ploid* ("unit") refers to any nucleus, cell, or organism that has two "units" or sets of chromosomes that, in sexually reproducing organisms, are normally composed of one paternal and one maternal set. The often used but anomalous *haploid* (literally "half-unit" or "set") for structures having but one set of chromosomes is here replaced by the more logical *monoploid* (Greek *monos*, "only" or "alone," hence "one," and *ploid*).
‡Tetraploid (4*n*) forms also known.
§Triploid (3*n*) forms also known.

S During the *synthesis* stage, replication of DNA and synthesis of histones occur, the former doubling in amount during S. The chromosomes are each composed of two *sister chromatids* (sharing a common centromere, as described later in this chapter) by the end of this stage. This is the most important activity of the S stage. Early in S there occurs a rapid rise in DNA polymerases (this requires transport from cytoplasm to nucleus) and in an RNA needed for later degeneration of the nuclear membrane, along with other RNAs. This stage may occupy roughly 35 to 45 percent of the interphase cycle; in cultured human cells it extends over some six to eight hours.

G₂ A second *growth* (gap) period, not as well understood as G_1, but in which new DNA is rapidly complexed with chromosomal proteins, and in which synthesis of RNA (in lesser amounts than in G_1) and proteins continues. It may occupy about 10 to 20 percent of the interphase cycle.

Duration of the several stages of the interphase cycle in cultured HeLa[3] cells is approximately as follows: G_1, 8.2 hours; S, 6.2 hours; and G_2, 4.6 hours. An idealized diagram of the relative amounts of time occupied by each stage of the interphase cycle and nuclear division is shown in Figure 4-5. As the G_2 stage draws to a close, the cell gradually enters division, though governing factors for this transition are unclear.

[3]HeLa designates a cell line obtained in 1952 from a carcinoma of the cervix in a patient whose pseudonym is Helen Lane, but whose real name was Henrietta Lacks.

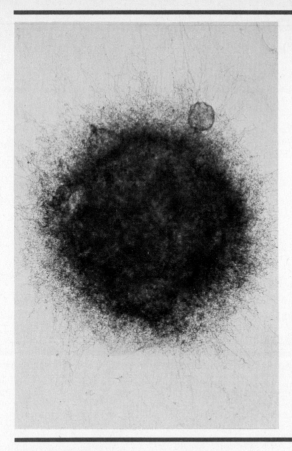

Figure 4-4. Lymphocyte at G_1 with many 200 Å nucleoprotein fibers (chromatin) extending outward from the periphery. (Electron micrograph courtesy Dr. G. F. Bahr. From G. F. Bahr, 1977. Chromosomes and Chromatin Structure. In J. J. Yunis, ed., *Molecular Structure of Human Chromosomes,* published by Academic Press, New York. Used by permission of author and publisher.)

Cell division

Growth and development of every organism depend in large part on multiplication, enlargement and differentiation of its cells, beginning with the zygote. In multicellular individuals attainment of adult form depends on a coordinated sequence of increase in cell number, size, and differentiation from zygote to maturity. In unicellular organisms cell division serves also as a form of reproduction, often the only one. Sexually reproducing forms also depend in most instances directly on cell division for the formation of sex cells, or gametes.

Division of nucleate cells consists of two distinct but integrated activities, nuclear division (**karyokinesis**) and cytoplasmic division (**cytokinesis**). In general, cytokinesis begins after nuclear division is well underway, but in many instances may be deferred or, indeed, entirely lacking. For example, in the development of the female gametophyte of pine (see Appendix B), about 11 nuclear divisions occur before cytokinesis begins, and in some algae and fungi the plant body is a coenocyte, without any walls separating nuclei except for the reproductive cells.

Two types of nuclear division, **mitosis** and **meiosis,** are characteristic of most plant and animal cells. Mitosis is regularly associated with nuclear division of vegetative or *somatic* cells; meiosis occurs in conjunction with formation of reproductive cells (either gametes or meiospores) in sexually reproducing species.

MITOSIS

As a process, mitosis is remarkably similar in all but relatively minor details in both plants and animals, from the least specialized to the most highly evolved

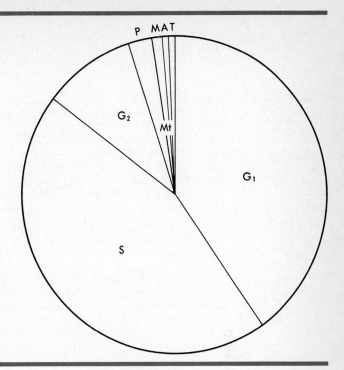

Figure 4-5. Idealized diagram of the relative amounts of time occupied by each stage of interphase and the mitotic cycle. **G_1**, first growth stage; **S**, synthesis of DNA and histones; **G_2**, second growth stage; **P**, prophase of mitosis; **M**, metaphase of mitosis; **A**, anaphase of mitosis; **T**, telophase of mitosis; **Mt**, mitotic period.

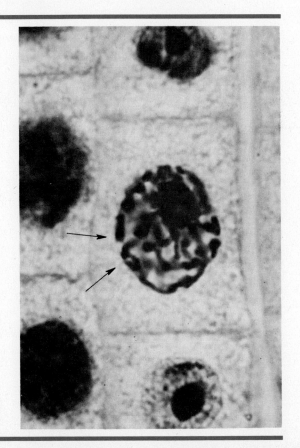

Figure 4-6. Mitosis in onion root tip cell; early prophase. Longitudinal doubleness is evident in a few areas (e.g., lower left portion of the nucleus); the nuclear membrane is becoming indistinct. (Courtesy Carolina Biological Supply Co.)

forms. Mitosis is a smoothly continuous process, but is divided arbitrarily into several stages or phases for convenient reference. The following description of the mitotic process in plant cells will adequately serve the purpose of determining whether the mechanism provides a reasonable basis for the genetic assumptions developed in Chapters 2 and 3.

Prophase. As stage G_2 of interphase gives way to prophase of mitosis, chromosomes progressively shorten and thicken to form individually recognizable, elongate, longitudinally double structures (Fig. 4-6) arranged randomly in the nucleus. This contraction of the chromosomes is one of the most conspicuous features of prophase. The two **sister chromatids** of each chromosome are closely aligned and somewhat coiled on themselves. The tightening of these coils contributes in large measure to the shortening and thickening of the chromosomes.

The two sister chromatids of each chromosome are held together by strands in a specialized region, the **centromere** (Fig. 4-7). In well-prepared stained material under the light microscope the centromere appears as a small, relatively

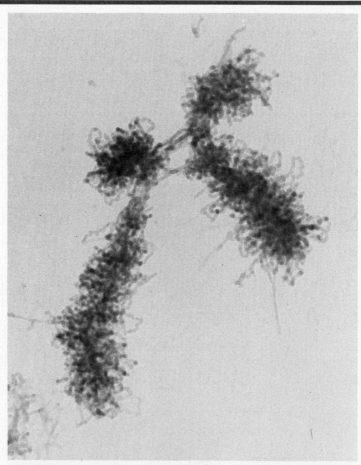

Figure 4-7. *Electron micrograph showing two sister chromatids held together by chromatin fibers in the centromere region. This is one of the human chromosomes.* (Photo courtesy Dr. E. J. Du Praw. From E. J. Du Praw, 1970. *DNA and Chromosomes.* New York, Holt, Rinehart and Winston, Inc. Used by permission.)

clear, spherical zone, but appears dark in electron micrographs. The centromere has been shown by electron microscopy to be a complex region of the chromosome. It includes two **kinetochores,** one for each sister chromatid, to which the microtubules of the spindle (see later) are attached. Kinetochores appear to become organized in prophase only after chromosome contraction has taken place. In the absence of the centromere, chromosome fragments (acentric fragments) such as those resulting from radiation-induced breakage fail to move normally during nuclear division and are generally, therefore, lost from one or both of the reorganizing daughter nuclei. This illustrates the essential role of the centromere.

During prophase the nucleolus gradually disappears in most organisms (Fig. 4-8 and 4-15). In some grasses, the green alga *Spirogyra,* and in *Euglena,* the nucleolus apparently ceases to function and is discarded into the cytoplasm.

Another common feature of prophase is the degeneration and disappearance of the nuclear membrane (Figs. 4-8 and 4-15). The mechanism is incompletely understood, but an aggregating of mitochondria around the membrane (especially in animal cells) suggests enzymatic processes. Physical stresses exerted by microtubules (see next paragraph), which are attached to the nuclear membrane, may also play a role in membrane degeneration.

As prophase progresses an important and often conspicuous component, a football-shaped **mitotic apparatus** (*spindle* and associated structures) begins to form (see Figs. 4-11 and 4-15). By the end of prophase this structure occupies a large portion of the cell volume and often extends very nearly from one "end"

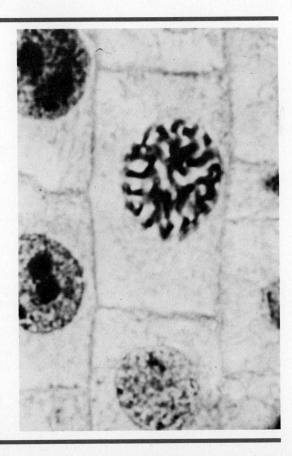

Figure 4-8. Mitosis in onion root tip cell; late prophase. Chromosomes are shorter and thicker than in early prophase; note the chromatids. (Courtesy Carolian Biological Supply Co.)

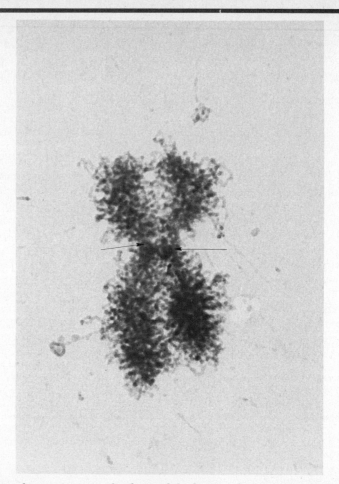

Figure 4-9. *Electron micrograph of one of the human chromosomes showing the two kinetochores of the centromere.* (Photo courtesy Dr. E. J. Du Praw. From E. J. Du Praw, 1970, *DNA and Chromosomes.* New York, Holt, Rinehart and Winston, Inc. Used by permission.)

of the cell to the other. The mitotic apparatus consists of slender microtubules arranged along the long axis of the spindle. Some of the microtubules are continuous from pole to pole of the mitotic apparatus; these are the *continuous fibers.* Groups of others, the *chromosomal fibers,* extend from one pole or the other to each chromosomal kinetochore to which they are attached (Figs. 4-9 and 4-15). A third type of microtubule, the *interzonal fibers,* is also recognized. These run between the centromeres of the separating daughter chromosomes later on in anaphase.

As prophase draws to a close, the longitudinally double chromosomes move or are moved in the direction of the midplane, or equator, of the developing spindle, a period of time often designated as prometaphase or metakinesis (Fig. 4-10). In most organisms prophase comprises the bulk of the time consumed by mitosis (Fig. 4-5 and Table 4-2).

Metaphase.   Metaphase (Fig. 4-11) is that period of time in which the centromeres of the longitudinally double chromosomes occupy the plane of the

Figure 4-10. Chromosomes nearing equatorial plane of developing spindle; prometaphase. (Courtesy Carolina Biological Supply Co.)

equator of the mitotic apparatus, although the chromosomal arms may extend in any direction. At this stage the sister chromatids are still held together by connecting chromatin fibers at the centromere regions (Figs. 4-7 and 4-9). Electron microscopy shows that the kinetochores of the two sister chromatids face opposite poles; this will permit proper separation in the next phase (anaphase).

During metaphase the chromosomes are shortest and thickest. Polar views furnish good material for chromosome counts; both lateral and polar views are useful for studying chromosome morphology.

Anaphase. Anaphase (Fig. 4-12) is characterized by the separation of the metaphase sister chromatids and their passage as **daughter chromosomes** to the spindle poles. It begins at the moment when the centromeres of each of the sister chromatids become *functionally* double and ends with the arrival of the daughter chromosomes at the poles (Fig. 4-12). For clarity, it is important to distinguish between sister chromatids and daughter chromosomes. A metaphase cell contains a number of sister chromatids, associated in pairs through centromeric connection, equal to twice the number of chromosomes. On the other hand, the anaphase cell contains no sister chromatids but does, by definition, contain a number of daughter chromosomes twice as great as the metaphase or prophase chromosome number. Anaphase therefore accomplishes the *quantitatively* equal distribution of chromosomal material to two developing daughter nuclei. That distribution is also *qualitatively* equal if the replication process of the preceding S stage has been exact. The replication process will be more closely examined at the molecular level in Chapter 16.

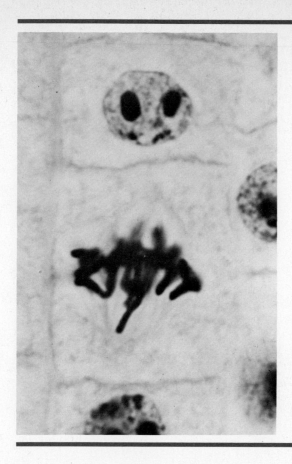

Figure 4-11. *Mitosis in onion root tip cell; metaphase. Note the spindle.* (Courtesy Carolina Biological Supply Co.)

The mechanism of the anaphasic movement of chromosomes is not understood in spite of intensive experimentation and study.

Telophase. The arrival of the (longitudinally single) daughter chromosomes at the spindle poles marks the beginning of telophase; it is, in turn, terminated by the reorganization of two new nuclei and their entry into the G_1 stage of interphase (Figs. 4-13 and 4-15). In general terms, the events of prophase occur in reverse sequence during this phase. New nuclear membranes are constructed from materials that may be remnants of the original membrane, or derived from the endoplasmic reticulum, or newly synthesized from appropriate cellular components. The mitotic apparatus gradually disappears, the nucleoli are reformed at the nucleolar organizing sites of specific chromosomes, and the chromosomes resume their long, slender, extended form as their coils relax. Replication of chromosomal material, by which each chromosome comes again to consist of two sister chromatids, then occurs in the S stage of the succeeding interphase.

CYTOKINESIS

If cytokinesis occurs, the process takes place during telophase, though it may be initiated during anaphase. In cells of higher plants cytokinesis is typically accomplished by formation of a **cell plate** (Figs. 4-13B and 4-15). Early steps in cytokinesis include formation of vesicles in the midplane of the mitotic apparatus and the coalescence, of these vesicles starting at the center of the spindle, to

Figure 4-12. *Mitosis in onion root tip cell; anaphase. The daughter chromosomes are nearing the spindle poles, which are clearly evident.* (Courtesy Carolina Biological Supply Co.)

form a **phragmoplast.** The cell plate forms within the phragmoplast, gradually extending centrifugally, dividing the phragmoplast into two parts before the latter disappears. The cell plate is then converted into a **middle lamella,** with new cross walls between the daughter cells deposited on each side of the middle lamella (Fig. 4-14). In animal cells cytokinesis is accomplished by **furrowing.** This is a process of progressive constriction of a contractile ring, perhaps of contractile fibers. Figure 4-15 summarizes the entire process of mitosis and cytokinesis.

SIGNIFICANCE OF MITOSIS

The process of mitosis and the subsequent replication of chromosomal material in the succeeding interphase has the inevitable result, if the cell "makes no mistakes," of creating from one cell two new ones that are chromosomally identical. (Actually, "mistakes" do sometimes occur; these, however, shed considerable light on the nature of the gene, as detailed in Chapter 18). Certainly in mitosis the chromosomal material is distributed in equal quantity to two daughter cells in a strikingly precise manner.

If the replication process during the S stage of interphase is qualitatively equal, then the chromosomes will serve quite adequately as physical bearers of the genes considered in Chapters 2 and 3. The pattern of inheritance followed for *Coleus*, for example, requires that genes be transmitted in cell division from the zygote to every somatic cell of the mature individual. The same is also true, of

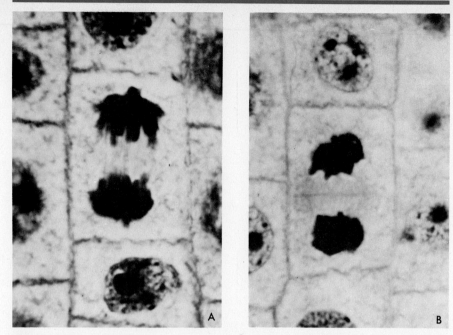

Figure 4-13. *Mitosis in onion root tip; telophase. (A)* **Early telophase.** *Chromsomes have just arrived at the poles. The cell plate has begun to form. (B)* **Later telophase.** *The cell plate is prominent, and will later extend from side wall to side wall.* (Courtesy Carolina Biological Supply Co.)

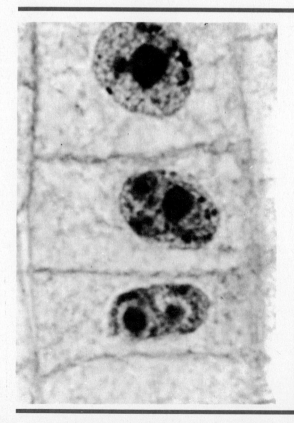

Figure 4-14. After completion of mitosis in an onion root tip cell, two new interphase nuclei have been formed. In plant cells these are typically separated by a wall as a result of **cytokinesis** *begun during telophase.* (Courtesy Carolina Biological Supply Co.)

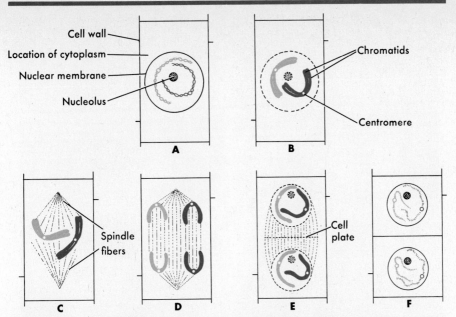

Figure 4-15. *Diagrammatic summary of mitosis in onion root tip cell. Only two of the total complement of 16 chromosomes are shown; the dark and light shading represent a "paternal" and a "maternal" chromosome, respectively, of one of the eight pairs. (A)* **Prophase;** *nuclear membrane and nucleolus still evident. (B)* **Late prophase;** *chromosomes becoming shorter and thicker, nucleolus and nuclear membrane disappearing. Note that each chromosome consists of two chromatids. (C)* **Metaphase;** *centromeres aligned on the midplane of the spindle, chromatids still present. (D)* **Anaphase;** *daughter chromosomes moving poleward. (E)* **Telophase;** *chromosomes have reached poles, nucleoli and nuclear membranes reappearing, cytokinesis by cell plate underway. (F)* **Interphase.** *See text for details.*

Table 4-2. Duration of mitosis in living cells

Common	Scientific	Tissue	Temp. °C	P	M	A	T	Total
\multicolumn{9}{c}{Plants (angiosperms)}									
Onion	*Allium cepa*	Root tip	20	71	6.5	2.4	3.8	83.7
Oatgrass	*Arrhenatherum sp.*	Stigma	19	36–45	7–10	15–20	20–35	78–110
Pea	*Pisum sativum*	Endosperm	—	40	20	12	110	182
Pea	*Pisum sativum*	Root tip	20	78	14.4	4.2	13.2	110
Spiderwort	*Tradescantia sp.*	Stamen hair	20	181	14	15	130	340
Broad bean	*Vicia faba*	Root tip	19	90	31	34	34	155
\multicolumn{9}{c}{Animals}									
Fowl	*Gallus sp.*	Fibroblast culture	—	19–25	4–7	3.5–6	7.5–14	34–52
Grasshopper	*Melanoplus differentialis*	Neuroblast	—	102	13	9	57	181
Mouse	*Mus musculus*	Spleen mesenchyme	38	21	13	5	20	59
Salamander	*Salamandra maculosa*	Embryo kidney	20	59	55	6	75	195

Name of organism spans Common and Scientific columns; Duration, minutes spans P, M, A, T, Total.

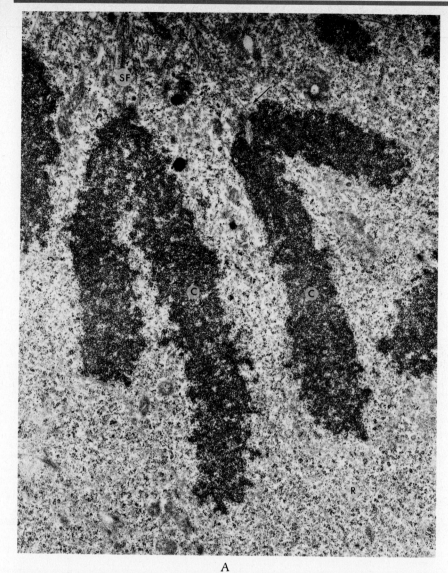

A

Figure 4-16. (A) Electron micrograph of chromosomes of Tasmanian wallaby. C, chromosome; SF, spindle fibers; R, ribosomes. Note spindle fibers at centromere (arrow). (B) One of the human chromsomes showing centromere (arrow) and DNA fibrils. [(A), from W. Jensen and R. Park, 1967, Cell Ultrastructure. © 1967 by Wads-

course, for human beings. In mitosis there exists a process by which the precise, equal distribution of structures called chromosomes can be carried through cell generation after cell generation. The assumptions of the two preceding chapters regarding genes are borne out if the genes are indeed located on the chromosomes; certainly the behavior of these bodies makes them ideal vehicles for genes in terms of the speculations previously raised. Before accepting the chromosomal locations of genes, however, additional parallels between the behavior of genes as deduced from breeding experiments and the behavior of chromosomes as seen under the microscope must be noted. The mechanism of mitosis provides good

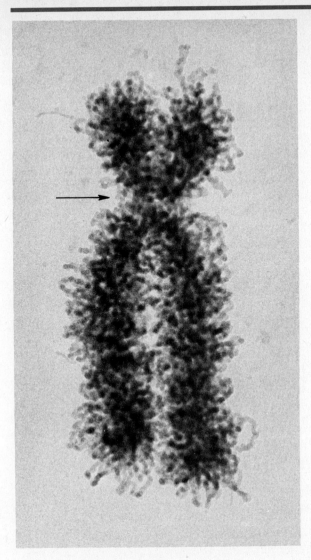

B

worth Publishing Co., Inc., Belmont, California. Reproduced by permission. (B) Courtesy Dr. E. J. DuPraw, who took the electron micrograph; from E. J. DuPraw, 1970. *DNA and Chromosomes*. New York, Holt, Rinehart and Winston, Inc. Used by permission.]

presumptive evidence that many of the theories raised in the two preceding chapters regarding gene location are sound. However, physical and chemical *proof* are still required, but the theory looks sufficiently promising to retain for the present.

Duration of mitosis. Although not directly germane to present purposes, it is interesting to note that the rather complicated physical and chemical changes in mitotic nuclei often occur in a surprisingly short time. Table 4-2 summarizes a few studies from the literature.

General structure. In stained preparations, as seen in the light microscope, chromosomes appear to possess few definitive morphological characteristics by which individual members of a set may be distinguished from each other. The only useful features in this context are (1) length, (2) centromere location and relative arm lengths, and (3) presence of satellites (if any), set off by secondary constrictions. During mitotic metaphase chromosomes are ordinarily at their shortest and thickest, as has been pointed out, varying from as short as a fraction of a micrometer[4] to as long as around 400 μm in exceptional cases, and between about 0.2 and 2 μm in diameter (Fig. 4-16). Each chromosome of a complement has its own characteristic length (within relatively narrow limits) and centromere location. However, in some organisms, especially those with large numbers of chromosomes, there is often considerable size similarity among some of the members in a set. This is true for some of the human chromosomes, and until recently, members of different homologous pairs could not easily be distinguished. This difficulty has been especially troublesome in assigning groups of genes to the proper chromosome, as well as in distinguishing various chromosomal aberrations. Before the development of new techniques, results of which will be described in the next section, it was possible only to photograph smears of metaphase cells (Fig. 4-17), cut the chromosomes apart, and arrange them in as nearly matching pairs as could be determined from length, centromere location, and presence or absence of satellites. Photographs so prepared (or drawings made from them) are referred to as **karyotypes** (Fig. 4-18).

Although no such thing as a "typical" chromosome exists, a composite one is diagramed in Figure 4-19. Many chromosomes, however, show a distressing lack of morphological landmarks such as satellites. Every normal one does possess a centromere; its position and therefore the lengths of the arms are relatively constant. Such properties as these do, of course, permit to some degree the identification of some of the chromosomal aberrations that are described in Chapters 13 and 14. On the basis of centromere location cytologists recognize four major types of chromosomes (Fig. 4-20):

Metacentric: centromere central; arms of equal or essentially equal length.
Submetacentric: centromere submedian, giving one longer and one shorter arm.
Acrocentric: centromere very near one end; arms very unequal in length.
Telocentric: centromere terminal; only one arm.

Heterochromatin and euchromatin. In interphase and early prophase chromatin can be seen to take two forms: **heterochromatin,** tightly coiled (condensed) and either more darkly or less darkly staining than **euchromatin** (which is noncondensed). Heterochromatin is often in close contact with the nucleolar organizer of a specific chromosome. During nuclear division heterochromatin may occur in the vicinity of the centromeres, at the ends of the arms (*telomeric*), or in other, intercalary positions.

Constitutive heterochromatin may be found in the centromeric region or almost anywhere else along a chromosome, as just noted. However, special locations of intercalary constitutive heterochromatin, though consistent for cytologically "normal" individuals of a given species, vary among different species. Heterochromatin, once thought to be wholly inert genetically, is now known to include some important genetically active regions. These include repetitive DNA (or satellite DNA), nucleolar organizers, and genes for some of the RNA molecules.

---

[4]A micrometer (μm) is 1×10^{-6} meter; in older usage, a micron (μ).

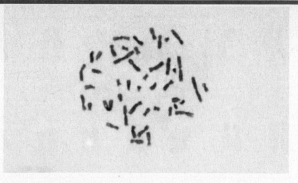

Figure 4-17. *Mitotic metaphase of human male chromosomes.* (Photomicrograph by Mr. John Derr.)

Centromeric constitutive heterochromatin probably plays an unknown role in anaphasic separation of sister chromatids.

With *facultative heterochromatin* one chromosome of a homologous pair becomes partially or wholly heterochromatic and inactive genetically. This is the case with one of the X chromosomes of the human female.

Chromosome banding. Progress in positive identification of chromosomes came in a fast-breaking series of developments starting in 1968 and 1969 with the work of Caspersson and his colleagues in Sweden. Briefly stated, the basis for this breakthrough lies in the fact that many dyes that have an affinity for DNA fluoresce under ultraviolet light. After suitable treatment with such dyes, each chromosome shows bright and dark zones, or *bands*, which are specific in location and extent for that chromosome. This feature is best seen in metaphase chromosomes. Considerable success has been achieved with quinacrine mustard, which produces fluorescent bands of various degrees of brightness. Bands produced by quinacrine mustard are referred to as **Q-bands** (Fig. 4-21A). Banding patterns are unique for each member of the human chromosomal complement, therefore, positive identification and construction of a karyotype are easy (Fig. 4-21B). There are, however, some disadvantages to the Q-banding technique, particularly the impermanence of the fluorescence. That difficulty can be overcome by staining with the Giemsa dye mixture, which produces a unique set of bands, the **G-bands.** Figure 4-22 presents both a metaphase spread and a karyotype for the same cell from a human male comparing G-bands (Figs. 4-22A and B) with Q-bands (Figs. 4-22C and D). In general, darkly stained Giemsa bands correspond in large degree to the brightly fluorescing Q-bands, though agreement is not complete. However, virtually every chromosome in a complement can be positively identified, and structural alterations of parts of chromosomes clearly determined, as in the case of Q-banding. One variation of the Giemsa method produces band patterns that are the reverse of the G-bands, that is, darkly stained G-bands are lightly stained **R-bands** (Fig. 4-23) and vice versa. Figure 4-24 shows diagrammatically Q-, G-, and R-band patterns of the human chromosome complement.

In a modification of the technique using Giemsa stain, treatment of cells (fixed to slides) with alkali leads to dense, bright staining in the centromeric region. This produces the **C-banding,** which is specific for constitutive heterochromatin. The centromeric region, which includes repetitive DNA, responds only to the C-banding technique, as does also any intercalary region that also contains repetitive DNA.

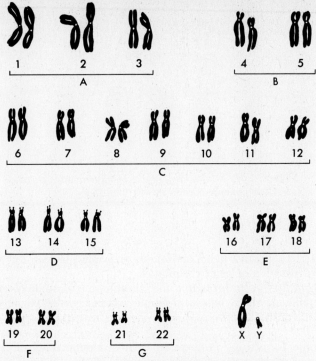

Figure 4-18. *The chromsomes of a normal human male ar-ranged as a karyotype. Note the knobs or satellites on several pairs. Because of morphological similarities among several of the human chromosomes, groups are assigned letter disigna-tions as shown here.*

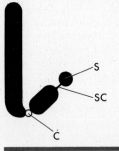

Figure 4-19. *Diagram of a composite chromosome as it might appear at high magnification with a light microscope. For sim-plicity, chromatids are not represented. C, centromere; S, sat-ellite or knob; SC, secondary constriction.*

Figure 4-20. *Four major morphological chro-mosome types are recognized on the basis of centromere position and consequent relative arm length, (A)* **metacentric;** *(B)* **submetacentric;** *(C)* **acrocentric;** *(D)* **telocentric.**

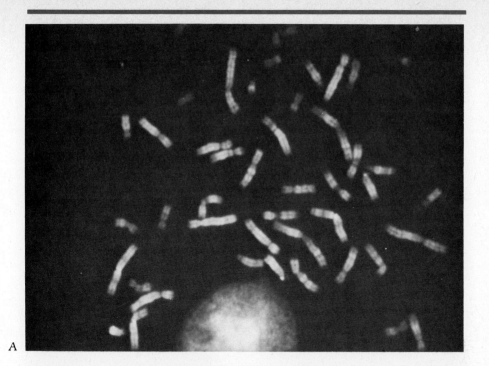

A

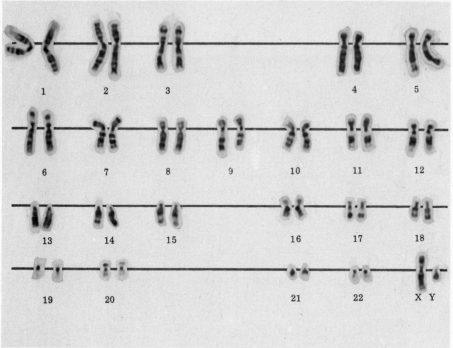

B

Figure 4-21. (A) *Metaphase spread of normal human male chromsomes showing Q-banding.* (B) *Karyotype of normal human male, Q-banding.* (Photos courtesy Dr. C. C. Lin.)

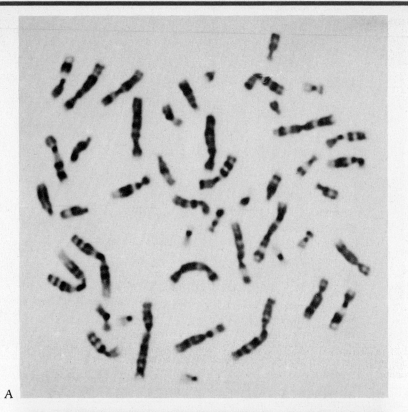

A

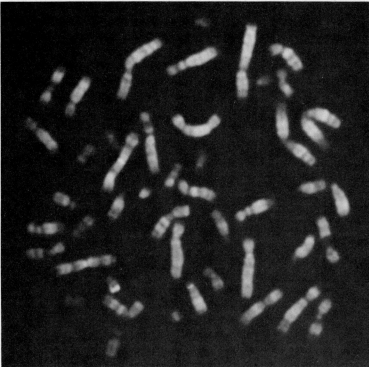

B

Figure 4-22. (A) Metaphase spread of normal human male chromosomes, G-banding. (B) Metaphase spread of normal human male, Q-banding. (Photos courtesy Dr. Herbert A. Lubs.)

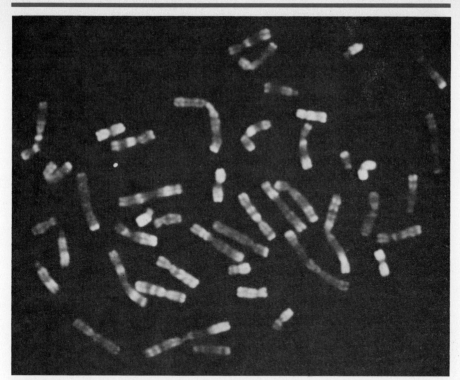

Figure 4-23. *Reverse (R-) banding with* acridine orange. *Metaphase spread of chromosomes of a normal human female.* (Courtesy Dr. C. C. Lin.)

Chromosome banding and taxonomic relationship.   Banding techniques have now been applied to a wide range of eukaryotes. As might be suspected, where species have been assigned close taxonomic relationship on the grounds of morphology and physiology, banding similarities have been found. One of the most interesting sets of findings has come from a comparison of banded karyotypes of man (*Homo sapiens*) where 2n = 46, with those of the chimpanzee (*Pan troglodytes*), the gorilla (*Gorilla gorilla*), and the orangutan (*Pongo pygmaeus*). In these latter three species the diploid chromosome number is 48. Some chromosomes of the four karyotypes have been evolutionarily stable; for example, human autosome 1 is closely similar to autosome 1 of the three other species. Others, like human autosome 3, are closely similar in all but the orangutan. Human autosome 2 has been found to have resulted from the fusion of two acrocentric ape chromosomes, numbers 12 and 13, which accounts for the difference in somatic chromosome number between humans and the three ape species. Relationships are summarized in Table 4-3 and in the karyotypes shown in Figure 4-25.

MEIOSIS

One of two fundamental cytological and genetic events in the life cycle of sexually reproducing plants and animals is the union of gametes, or sex cells, to form a zygote, a process to which the term **syngamy** is applied. Studies from as long ago as the last century clearly indicate that in gametic union the chromosomes contributed by each gamete retain their individual identities in the zygote nucleus. The zygote thus contains twice as many chromosomes as does a gamete or, more accurately, all the chromosomes of both gametes.

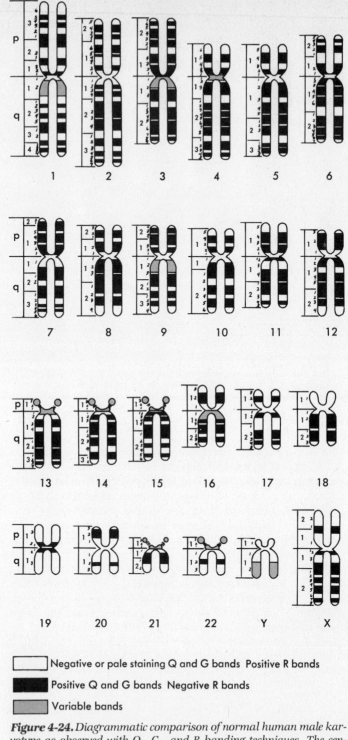

Negative or pale staining Q and G bands Positive R bands

Positive Q and G bands Negative R bands

Variable bands

Figure 4-24. *Diagrammatic comparison of normal human male kar-yotype as observed with Q-, G-, and R-banding techniques. The cen-tromere is represented as observed in Q-banding only.* [*Redrawn from Paris Conference (1971): Standardization in Human Cytogenetics.* In *Birth Defects: Orig. Art. Ser.,* D. Bergsma, ed. Published by The Na-tional Foundation-March of Dimes, White Plains, New York, Vol. VIII (7), 1972. Used by permission.]

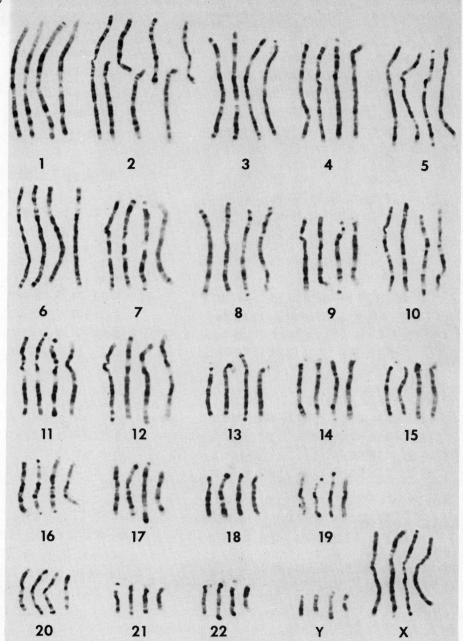

Figure 4-25. (A) G-banded late prophase chromosomes in the 1000-band stage of (left to right) human, chimpanzee, gorilla, and orangutan, showing clearly the extensive chromosomal homology among the four species. **On following pp. 76–77:** (B) Schematic representation of G-banded, late prophase chromosomes from the same four species (left to right) human, chimpanzee, gorilla, and orangutan. The banding homology is even more evident here. (Courtesy Dr. Jorge J. Yunis. Reprinted from J. J. Yunis and O. Prakash, "The Origin of Man: a Chromosomal Pictorial Legacy." *Science* **215:**1525–1530. © 1982 by the American Association for the Advancement of Science. Used by permission of Dr. Yunis and the American Association for the Advancement of Science.)

Figure 4-25. (Continued)

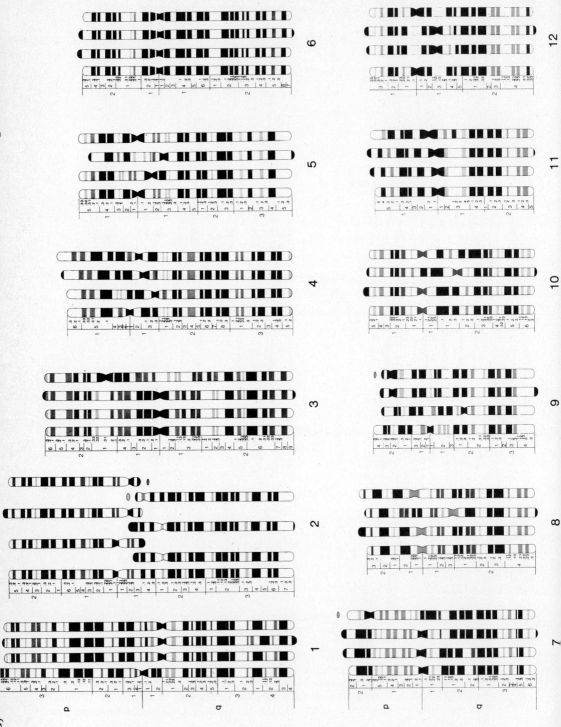

(B)

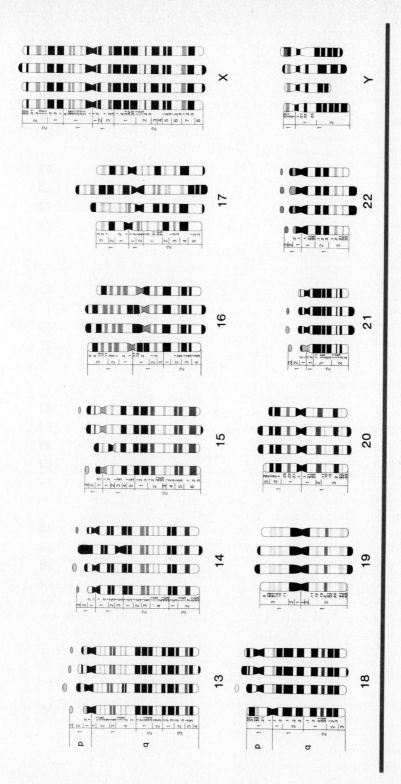

This fact, of course, is responsible for the occurrence in diploid or $2n$ cells of matching pairs of chromosomes; each member of a given pair is the homolog of the other. In each diploid nucleus, then, there occurs the monoploid number of homologous *pairs* of chromosomes, one member of each pair having been contributed by the paternal parent, the other by the maternal parent. Thus, if among the chromosomes of the sperm, there is a metacentric chromosome (Fig. 4-20) of, for example, an average metaphase length of 5 μm, there will be an identical chromosome in the chromosomal complement of the egg. This statement applies to all the **autosomes** (those chromosomes not associated with the sex of the bearer). In the zygote and all subsequent cells derived by mitosis, *two* metacentric autosomes having an average metaphase length of 5 μm will be found. As will be shown later, these statements must be modified for the sex chromosomes where these occur.

The result of syngamy is the incorporation into a zygote nucleus of all the chromosomes of each gamete. Such facts would seem to require a counterbalancing event whereby the doubling of chromosome quantity at syngamy is offset by a nuclear division that halves the amount of chromatin per nucleus at some point prior to gamete formation. Such a "reduction division" does occur in all sexually reproducing organisms that have discrete nuclei. This is **meiosis,** the second of the two fundamental cytological and genetic events in the sexual cycle.

Meiosis as a form of nuclear division differs greatly from mitosis. It consists of two successive divisions, each with its own prophase, metaphase, anaphase, and telophase; therefore, it results in *four* daughter nuclei instead of two as in mitosis (although in many species not all the cellular products are functional). Furthermore, because of fundamental "procedural" differences between mitosis and meiosis, not only do the nuclear products of a meiotic division have one set of chromosomes each (monoploid or "haploid") as opposed to the two sets (diploid) of the parent nucleus, but the nuclear products are, if genes and chromosomes have any relationship, also genetically unlike the original diploid nucleus and often genetically unlike each other.

The term **meiocyte** (Fig. 4-26) is a convenient general one to designate any diploid cell whose next act will be to undergo meiosis. In seed plants and other heterosporous forms, for example, meiocytes are represented by the megasporocytes and microsporocytes ("spore mother cells"), in homosporous plants (those producing only one kind of spore by meiosis) by sporocytes or "spore mother cells," and in higher animals by primary spermatocytes and primary oocytes. The position of the meiocyte in the life cycle varies greatly from one major group to another. For instance, in the vascular plants the meiocytes give rise directly to megaspores and microspores (i.e., the **meiospores**), which, by mitosis, produce gamete-bearing plants. In the seed plants these gamete-bearing plants (the gametophytes) are much reduced in size, number of cells, and complexity; refer to Appendix B. In humans, on the other hand, meiocytes give rise directly to gametes. In many algae and fungi the zygote nucleus itself functions as a meiocyte, giving rise to monoploid cells that ultimately, by mitosis, produce monoploid, gamete-bearing plants. In summary, meiosis may be sporic (vascular plants and others), or gametic (human and many other animals), or zygotic, and may be performed by (1) sporocytes ("spore mother cells," Fig. 4-26), (2) primary spermatocytes and oocytes, or even (3) the zygote itself.

First division: prophase-I. Although most cytologists recognize and name at least five stages of prophase-I, because of its complexities of chromosome behavior, it will suffice for our purposes merely to emphasize events in sequence. The first meiotic prophase often persists for a very lengthy time, and may be measured in weeks or months or even longer. In the human female it persists in each primary oocyte from a fetal age of 12 to 16 weeks until ovulation, generally once each 28 days, after sexual maturity. Some human oocytes, therefore, persist in arrested prophase-I until the end of the reproductive cycle, at

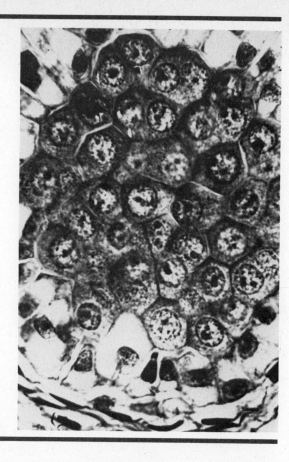

Figure 4-26. Meiocytes (microsporocytes, microspore mother cells, or pollen mother cells) in anther of lily stamen. These cells are diploid (2n) and are about to undergo meiosis. (Courtesy Carolina Biological Supply Co.)

menopause. This may be some 45 years, more or less. In very early prophase-I, the diploid number of chromosomes gradually becomes recognizable under the light microscope as long, slender, threadlike structures (Fig. 4-27). In the light microscope the chromosomes *appear* to be longitudinally single in earliest prophase-I (*leptotene*), although chemical and autoradiographic studies show that replication has occurred in the S stage of the preceding interphase. By definition, then, no *visible* chromatids exist at this time in a meiocyte. Chromosomes shorten and thicken progressively, presumably by the same mechanism as for mitosis.

While this contraction is underway, a second event that characterizes meiosis (as opposed to mitosis) takes place. Recall that each diploid nucleus (including meiocytes) contains *pairs* of homologous chromosomes. In early prophase-I (*zygotene*) the homologs begin to pair, or **synapse.** This **synapsis** is remarkably exact and specific, taking place point for point, with the two homologs usually somewhat twined about each other (Fig. 4-28). However, nothing definitive is known about the attractive forces which bring together homologs, as opposed to random chromosomes. Synapsis begins by an unknown mechanism when the ends of the homologs are within ~300 nm of each other.

Electron microscopy shows at synapsis a feature of dimensions too small to be discernible with the light microscope, the **synaptonemal complex.** This intricate and complex structure is the physical mechanism by which (1) homologs are held together at synapsis and (2) crossing-over (to be described) is made possible. The synaptonemal complex is composed of three parallel bands, interconnected by fine strands arranged at right angles to them. The two outer bands are axial components of the homologous chromosomes and are some 30 to 60 nm in diameter; the central one may be preformed in the nucleolus. The central

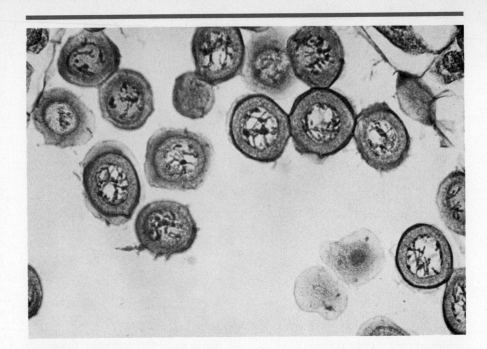

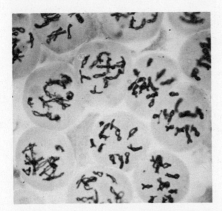 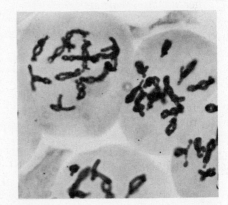

Figure 4-27. Meiosis in microsporocytes of lily (Lilium); *prophase I, low power. Synapsis is under way.* (Courtesy Carolina Biological Supply Co.)

component gives rise to the lateral elements and is about 100 nm in width. These elements may be amorphous (both central and lateral in most higher plants and animals, including humans), banded (lateral elements in ascomycete fungi and some insects), or latticelike with fine rods spaced about 10 nm apart (central element in insects). The whole structure is some 140 to 170 nm in width, just below the limit of resolution in light microscopy. Formation of the synaptonemal complex is initiated at several points along the pair of homologs. There is good chemical evidence that breakage and repair of DNA molecules of the chromosomes take place within the synaptonemal complex; this can and often does result in exchange of material between nonsister chromatids (*crossing-over*). Furthermore, synaptonemal complexes are not found in organisms in which crossing-over does not occur (e.g., the male fruit fly, *Drosophila*).

Synapsed chromosomes continue to shorten and thicken, and in the process, their longitudinal doubleness becomes apparent in light microscopy.

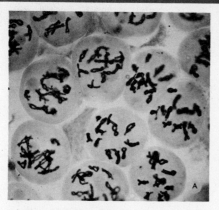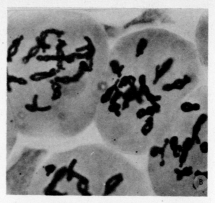

Figure 4-28. *Prophase-I in microsporocytes of the* fritillary *plant* (Fritillaria) *showing synapsis and chiasmata.* (Courtesy Dr. L. F. LaCour, John Innes Institute.)

The synaptonemal complex is completed at the next stage of prophase-I (*pachytene*), during which the synapsed homologs now are clearly seen to be composed of two chromatids each. The points at which exchange of material occurs between nonsister chromatids is evidenced by more or less X-shaped configurations, the **chiasmata** (singular, **chiasma**), as seen in Figure 4-28. The longer the chromosome pair, the greater likelihood of more than one chiasma, although one chiasma appears to interfere with the formation of another in a closely adjacent region of the chromosomes on the same side of the centromere. The basis for this **interference** is not altogether clear.

Next, in the *diplotene* stage of prophase-I, separation of homologs (except at points where chiasmata occur) is initiated. Chromosomes continue to contract and the nucleolus begins to disappear.

Finally, in the last stage of prophase-I (*diakinesis*), chromosomes reach maximum contraction. At the outset of diakinesis the synapsed homologs become well spaced out in the nucleus, often near the nuclear membrane. Chiasmata gradually *terminalize*, that is, they appear to move toward the ends of the arms and finally "slip-off," owing to the continued shortening of the chromosomes. Lastly, the nucleolus disappears, the nuclear membrane degenerates, and a spindle is formed.

Metaphase-I. The arrival of synapsed homologous chromosome pairs at the equator of the spindle begins metaphase-I. This phase differs from mitotic metaphase in (1) the arrangement of the monoploid number of synapsed chromosome *pairs* on the equatorial plane and (2) the tendency for the centromere of each homolog to be directed somewhat toward one of the poles (Fig. 4-29). An important point to be noted here is the *randomness* of arrangement of the paired homologs; that is, **for a given pair, it is just as likely that the paternal member be directed toward the "north" pole as for the maternal member to be so oriented.** Recall the gamete genotypes produced by a polyhybrid individual (Chapter 3).

Anaphase-I. In anaphase-I actual **disjunction** of synapsed homologs occurs, one longitudinally double chromosome of each pair moving to each pole, thereby completing the process of terminalization. Here is an additional difference from mitosis, mitotic anaphase being marked by separation of sister chromatids, which then move poleward as longitudinally single daughter chromosomes. Thus in mitosis one of each of the chromosomes present (i.e., of the entire chromosomal complement) travels to each pole, with the result that each new nucleus has just

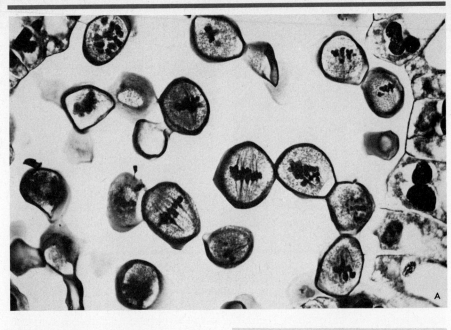

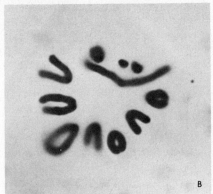

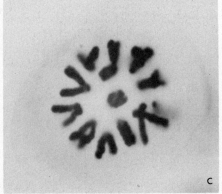

Figure 4-29. *Meiosis, metaphase-I. (A) In lily microsporocytes. Cell in the upper left of the anther cavity is seen in polar view; most of the remainder are seen in side view with spindles clearly evident. (Courtesy Carolina Biological Co.) (B) Polar view in spermatocytes of the orthopeteran* Mecostethus grossus. *(C) Microsporocytes of* Fritillaria, *a plant of the lily family.* [(B) and (C) courtesy Dr. L. F. LaCour, John Innes Institute.]

as many chromosomes as had the parent nucleus, whether monoploid or diploid. In meiosis, however, whole chromosomes of each homologous pair (as modified by any crossingover that has occurred in prophase-I) separate, so that each pole receives either a paternal or maternal, longitudinally double chromosome of each pair. This ensures a change in chromosome number from diploid to monoploid in the resultant reorganized daughter nuclei. In short, whereas mitotic anaphase was marked by the separation of sister chromatids, anaphase-I of meiosis is characterized by separation of homologous entire chromosomes (Fig. 4-30).

Telophase-I The arrival of chromosomes at the poles of the spindle signals the end of anaphase-I and the beginning of telophase-I. During this phase the chromosomes may persist for a time in the condensed state, the nucleolus and

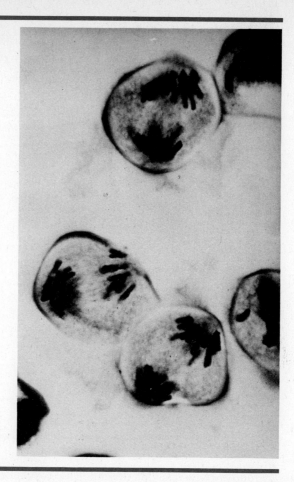

Figure 4-30. Anaphase-I in lily microsporocytes. (Courtesy General Biological Supply House, Inc., Chicago.)

nuclear membranes may be reconstituted, and cytokinesis may also occur (Fig. 4-31). In some cases, as in the flowering plant genus *Trillium*, meiocytes are reported to progress virtually directly from anaphase-I to prophase-II or even metaphase-II; in other organisms there may be either a short or fairly long interphase between the first and second meiotic divisions. In any event, the first division has thus accomplished the separation of the chromosomal complement into two monoploid nuclei.

Second division: prophase-II. Prophase-II is generally short and superficially resembles mitotic prophase, except that sister chromatids of each chromosome are usually widely divergent, exhibiting no relational coiling.

Metaphase-II. On two spindles, generally oriented at right angles to the first division spindle and often separated by a membrane or wall, the monoploid numbers of chromosomes, each consisting of two chromatids joined by a common centromere, are arranged in the equatorial plane. This stage is generally brief (Fig. 4-32).

Anaphase-II. Centromeres now separate and the sister chromatids of metaphase-II move poleward as daughter chromosomes, much as in mitosis. Their arrival at the poles marks the close of this phase.

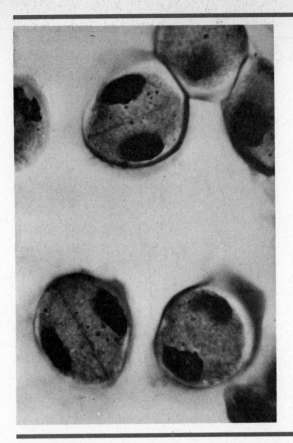

Figure 4-31. Telophase-I in lily microsporocytes. Note the prominent cell plate. (Courtesy General Biological Supply House, Inc., Chicago.)

Telophase-II. Following arrival of the monoploid number of daughter chromosomes at the poles, the chromosomes return to a long, attenuate, reticulate conformation, nuclear membranes are reconstituted, nucleoli reappear, and cytokinesis generally separates each nucleus from the others (Fig. 4-33).

The process of meiosis is summarized diagrammatically in Figure 4-34.

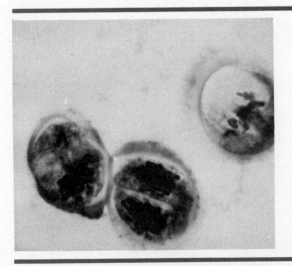

Figure 4-32. Metaphase-II in lily microsporocytes. (Courtesy General Biological Supply House, Inc., Chicago.)

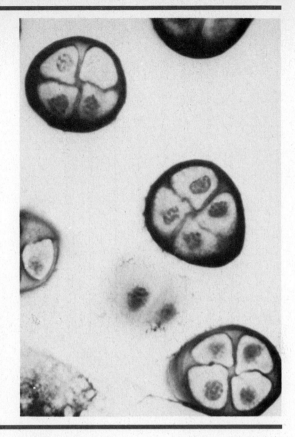

Figure 4-33. *Tetrads of lily microspores within the old microsporocyte wall. In some of the nuclei the chromosomes retain the telophase-II morphology, not yet having assumed the long, attenuate form of interphase.* (Courtesy General Biological Supply House, Inc., Chicago.)

Concluding view Electron microscopy has supplied answers to some questions about meiosis, but others remain unanswered. For instance, such tools permit the observation that in anaphase-I *homologs* separate, whereas in mitotic anaphase *sister chromatids* move apart to the poles. In mitotic metaphase the kinetochores of *sister chromatids* face in opposite directions, one toward each pole, whereas in metaphase-I, the two centromeres of each pair of *synapsed homologs* face different poles. Because of the attachment of microtubules to the kinetochores of the centromeres, then, homologous whole chromosomes are separated in anaphase-I, but sister chromatids separate in mitotic anaphase.

Note that the second meiotic division is quite different from mitosis, so that the two meiotic divisions are integral parts of one overall process of nuclear division. On the spindles in meiosis-II the chromosomes are always present in the monoploid number; in mitosis either the monoploid or the diploid chromosome number may occur, depending on the tissue in which division is taking place. Chromatids in the second division are widely separated, exhibit no relational coiling, and, in addition, have often been modified by crossing-over.

Although electron microscope observations of the synaptonemal complex have given some information on the mechanism and time of crossing-over, a full explanation of the precise physical and chemical details of synapsis remains to be elucidated. Finally, the triggering stimulus for meiosis is still unknown, although it appears likely to be hormonal, or at least chemical. Meiocytes excised in interphase often undergo mitosis instead of meiosis in vitro; excision of cells successively later in prophase-I results in progressively larger number of cells that do undergo meiosis. In all probability commitment to meiosis, as opposed to mitosis, may occur as early as the S or G_2 stage of interphase.

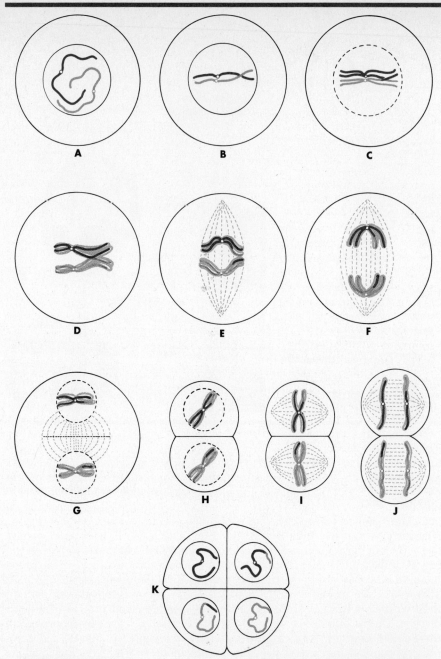

Figure 4-34. *Diagrammatic representation of meiosis in a meiocyte showing one pair of homologous chromosomes* (2n = 2) **based on light microscopy.** *(A)* **Prophase-I,** *chromosomes long and slender,* appearing *longitudinally single, although sister chromatids had been produced in the preceding interphase; (B)* **Prophase-I,** *homologous chromosomes synapsing, and still* appearing *longitudinally single; (C)* **Prophase-I,** *chromatids now visually evident, with one chiasma; (D)* **Prophase-I,** *disjunction underway, chaisma still evident; (E)* **Metaphase-I** *chromsomes on midplane of spindle with centromeres divergent; (F)* **Anaphase-I,** *poleward separation of previously synapsed homologs; note that each chromosome is still composed of two chromatids but some of these have been modified by crossing-over; (G)* **Telophase-I,** *each reorganizing nucleus now contains the monoploid number of chromosomes that are still composed of two chromatids each; (H)* **Prophase-II;** *(I)* **Metaphase-II;** *(J)* **Anaphase-II;** *(K), a postmeiotic tetrad of monoploid cells. Note that each in this case contains a genetically unique chromosome because of crossing-over.*

Cytologically, the basic significance of meiosis is the formation of four monoploid nuclei from a single diploid nucleus in two successive divisions, thus balancing off, as it were, the doubling of chromosome number that results from syngamy. Note that the first meiotic division accomplishes the reduction in chromosome number from diploid to monoploid, whereas the second is equational in distributing equal numbers of daughter chromosomes to developing new nuclei. In higher animals the cellular products of meiosis directly become gametes and/or polar bodies (Appendix B-8), but in vascular plants the meiotic products are meiospores that give rise to reduced gamete-bearing plants (Appendix B-6).

But how does meiosis relate to the hypotheses raised concerning probable gene distribution in somatic and gametic cells in the preceding chapter? If genes are located on chromosomes, the process of meiosis generates genetic variability in two important ways: (1) random assortment of paternal and maternal chromosomes and (2) crossing-over. Assume, for example, an organism heterozygous for three pairs of genes (a trihybrid), $AaBbCc$ in which ABC was derived from its paternal parent and abc from its maternal parent. If these three pairs of genes are located on three *different* chromosome pairs (which might be designated as pair 1, pair 2, and pair 3), then in prophase-I, the paternal and maternal number 1 chromosomes (bearing genes A and a, respectively) will synapse, as will the number 2s (bearing genes B and b) and the number 3s (with genes C and c). Arrangement of each pair of synapsed homologs on the metaphase-I spindle is random; i.e., the paternal member of each pair has an equal chance of being oriented toward either pole, as does the maternal member. In anaphase-I each (longitudinally double) chromosome moves toward the nearer pole as it separates from its homolog. Therefore each telophase-I nucleus has an equal chance of receiving a paternal or a maternal chromosome. The same chance exists for each remaining chromosome pair. Hence, for three pairs of genes on three different pairs of chromosomes, eight possible telophase-I genotypic combinations are possible:

Daughter Nucleus No. 1	Daughter Nucleus No. 2
ABC	abc
ABc	abC
AbC	aBc
Abc	aBC
aBC	Abc
aBc	AbC
abC	ABc
abc	ABC

The second division will simply increase the number of each genotype from one to two. The number of possible gamete genotypes occurring after the second meiotic division is here 2^3, or 8. That is, a given gamete in this example has $(\frac{1}{2})^3$ chance of receiving any particular arrangement of paternal and/or maternal chromosomes and genes. Furthermore, in a sample of several hundred gametes from such an individual, each of the eight genotypes would be expected to occur in approximately equal number. For organisms with many chromosomes the number of possible combinations becomes very large. In humans, for instance, with 23 pairs of chromosomes, the probability that any particular gamete will have a specific combination of chromosomes (excluding crossing-over) is $(\frac{1}{2})^{23}$, or about one in 8 million. Possible zygote genotypes resulting from random fusions of these gametes rise to about 64 trillion. A good deal of the variation in natural populations is due to this kind of **independent segregation** of genes normally occurring in the breeding population.

What would be the situation if these three gene pairs are, instead, located on a single pair of homologous chromosomes (i.e., **linked**)? Assume, for the purpose of illustration, that one member of the pair of homologs bears genes *A*, *B*, *C* on a given arm and that the other homolog bears genes *a*, *b*, and *c*. If no chiasmata are formed, then only two kinds of gametes (*ABC* and *abc*) will be produced, these occurring in equal number. If chiasmata do form between *A* and *B* and between *B* and *C* in at least some meiocytes, as is more often the case, then eight gamete genotypes will again be produced. However, in this case *the fraction of each type produced will depend on the frequency with which crossing-over occurs* between *A* and *B* and between *B* and *C*. This has considerable importance in the mapping of genes, which will be considered in detail in Chapter 6.

This examination of two types of nuclear division, which are universal in organisms having discrete nuclei and sexual reproduction, reveals some impressive similarities between the visually observable chromosome behavior and postulated gene behavior discussed in preceding chapters. Of course, this could conceivably be an unusual case of multiple coincidences, but the very multiplicity of these parallels furnishes a strong presumptive basis for the **chromosomal theory** of genetics. Further parallels and, more importantly, some positive experimental *proofs* are necessary. These proofs will come not only from the science of genetics itself, but also from cytology, physics, and chemistry. Much of the remainder of this book is devoted, then, to the quest for a sound understanding of what a gene is (even at the molecular level), how it operates, and where in a cell it is located.

References

Altman, P. L., and D. S. Dittmer, eds., 1962. *Growth Including Reproduction and Morphological Development*. Washington, Federation of American Societies for Experimental Biology.

Altman, P. L., and D. S. Dittmer, eds., 1972. *Biology Data Book*, 2nd ed., volume I. Washington, Federation of American Societies for Experimental Biology.

Bahr, G. F., 1977. Chromosomes and Chromatin Structure. In J. J. Yunis, ed., *Molecular Structure of Human Chromosomes*. New York: Academic Press.

Bergsma, D., ed., 1972. *Paris Conference (1971): Standardization in Human Cytogenetics*. In *Birth Defects: Orig. Art. Ser.*, volume VIII. White Plains, N.Y., The National Foundation–March of Dimes.

Caspersson, T., L. Zech, and C. Johansson, 1970a. Differential Binding of Alkylating Fluorochromes in Human Chromosomes. *Exp. Cell Res.*, **60:** 315–319.

Caspersson, T., L. Zech, and C. Johansson, 1970b. Analysis of the Human Metaphase Chromosome Set by Aid of DNA-Binding Fluorescent Agents. *Exp. Cell Res.*, **62:** 490–492.

Caspersson, T., L. Zech, C. Johansson, and E. J. Modest, 1970c. Identification of Human Chromosomes by DNA-Binding Fluorescing Agents. *Chromosoma*, **30:** 215–227.

Chambon, P., 1978. Summary: The Molecular Biology of the Eukaryotic Genome Is Coming of Age. *Cold Spring Harbor Symposia Quant. Biol.*, **42:** 1209–1234.

Felsenfeld, G., 1978. Chromatin. *Nature*, **271:** 115–122.

Kedes, L. H., 1979. Histone Genes and Histone Messengers. In E. E. Snell, P. D. Boyer, A. Meister, and C. C. Richardson, ed., *Annual Review of Biochemistry*, vol. 49. Palo Alto, Calif.: Annual Reviews, Inc.

King, R. D., 1974. *A Dictionary of Genetics*, 2nd ed. revised. New York: Oxford University Press.

Kornberg, R. D., 1977. Structure of Chromatin. In P. O. P. T'so, ed. *The Molecular Biology of the Mammalian Genetic Apparatus*, vol. 1. New York: Elsevier/North-Holland.

Kornberg, R. D., 1977. Structure of Chromatin. In E. E. Snell, P. D. Boyer, A. Meister, and C. C. Richardson, eds. *Annual Review of Biochemistry*, vol. 46. Palo Alto, Calif.: Annual Reviews, Inc.

Laemmli, U. K., S. M. Cheng, K. W. Adolph, J. R. Paulson, J. A. Brown, and W. R. Baumbach, 1978. Metaphase Chromosome Structure: The Role of Nonhistone Proteins. *Cold Spring Harbor Symposia Quant. Biol.*, **42:** 351–360.

Lewin, B., 1980. *Gene Expression: 2 Eucaryotic Chromosomes*, p. 1160. New York: John Wiley & Sons.

McGhee, J. D., and G. Felsenfeld, 1980. Nucleosome Structure. In E. E. Snell, P. D. Boyer, A. Meister, and C. C. Richardson, eds., *Annual Review of Biochemistry*, vol. 49. Palo Alto, Calif.: Annual Reviews, Inc.

Olins, A. L., J. P. Breilatt, R. D. Carlson, M. B. Senior, E. B. Wright, and D. E. Olins, 1977. On Nu Models for Chromatin Structure. In P. O. P. T'so, ed. *The Molecular Biology of the Mammalian Genetic Apparatus*, vol. 1. New York: Elsevier/North-Holland.

Olins, D. E., and A. L. Olins, 1978. Nucleosomes: The Structural Quantum in Chromosomes. *Amer. Scientist*, **66:** 704–711.

Pardee, A. B., R. Dubrow, J. L. Hamlin, and R. F. Kletzien, 1978. Animal Cell Cycle. In E. E. Snell, P. D. Boyer, A. Meister, and C. C. Richardson, eds., *Annual Review of Biochemistry*, vol. **47.** Palo Alto, Calif.: Annual Reviews, Inc.

Ris, H., and D. F. Kubai, 1970. Chromosome Structure. In H. L. Roman, ed., *Annual Review of Genetics*, vol. **4.** Palo Alto, Calif.: Annual Reviews, Inc.

T'so, P. O. P., ed., 1977. *The Molecular Biology of the Mammalian Genetic Apparatus*, vol. 1. New York: Elsevier/North-Holland.

Watson, J. D., 1976. *Molecular Biology of the Gene*, 3rd ed. Menlo Park, Calif.: W. A. Benjamin, Inc.

Wray, W., M. Mace, Jr., Y. Daskal, and E. Stubblefield, 1978. Metaphase Chromosome Architecture. *Cold Spring Harbor Symposia Quant. Biol.*, **42:** 361–365.

Yunis, J. J., ed., 1977. *Molecular Structure of Human Chromosomes.* New York: Academic Press.

Problems

4-1   In *Coleus* the somatic cells are diploid, having 24 chromosomes. How many of each of the following are present in each cell at the stage of mitosis or meiosis indicated? (Assume cytokinesis to occur in mid-telophase.) (a) Centromeres at anaphase, (b) centromeres at anaphase-I, (c) chromatids at metaphase-I, (d) chromatids at anaphase, (e) chromosomes at anaphase, (f) chromosomes at metaphase-I, (g) chromosomes at the close of telophase-I, (h) chromosomes at telophase-II.

4-2   Corn is a flowering plant whose somatic chromosome number is 20. How many of each of the following will be present in *one* somatic cell at the stage listed: (a) centromeres at prophase, (b) chromatids at prophase, (c) kinetochores at prophase, (d) chromatids in G_1, (e) chromatids in G_2?

4-3   Either from information you already have or based on facts in Appendix B, for a corn plant how many chromosomes are present in each of the following: (a) leaf epidermal cell, (b) antipodal nucleus, (c) endosperm cell, (d) generative nucleus, (e) egg, (f) megaspore, (g) microspore mother cell?

4-4   Based on your present knowledge, or after consulting Appendix B, how many human eggs will be formed from (a) 40 primary oocytes, (b) 40 secondary oocytes, (c) 40 ootids?

4-5   From your present knowledge, or after consulting Appendix B, how many human sperms will be formed from 40 primary spermatocytes?

4-6   From your present knowledge, or from Appendix B, 20 microsporocytes of a flowering plant would be expected to produce how many (a) microspores, (b) sperms?

4-7   Consult Table 4-1 in answering the following questions. (a) What is the probability in cattle that a particular egg cell will contain only chromosomes derived from the maternal parent of the cow producing the egg? (b) If this cow is mated to its brother, what is the probability that the calf will receive only chromosomes originally contributed by the calf's grandmother? (*Suggestion:* Can the product law of probability be of help here?)

4-8   Triploid watermelons have the advantage of being seedless. (a) What is the somatic chromosome number of such plants? (b) What explanation can you offer for their lack of seeds?

4-9 A cell of genotype *Aa* undergoes mitosis. What will be the genotype(s) of the daughter cells?

4-10 A cell of genotype *Aa* undergoes meiosis. What will be the genotype(s) of the daughter cells if all of them are functional?

4-11 A student examining a number of onion root tips counted 1,000 cells in some phase of mitosis. He noted 692 cells in prophase, 105 in metaphase, 35 in anaphase, and 168 in telophase. From these data what can be concluded about relative duration of the different stages of the process?

4-12 What is the probability that any ascospore of the orange-pink bread mold (*Neurospora crassa*) will have all its chromosomes derived from the + parent? (You may wish to consult Table 4-1 and Appendix B before trying to answer.)

4-13 How many different gamete genotypes will be produced by individuals of the following genotypes if all genes shown are on different chromosome pairs: (a) *AA*, (b) *Aa*, (c) *AaBB*, (d) *AaBb*, (e) *AAbbCc*, (f) *AaBbCcDdEe*?

4-14 A given individual is heterozygous for two pairs of genes that are located on the *same* chromosome pair, *A* and *B* one chromosome, with *a* and *b* located on the homologous chromosome. (a) How many gamete genotypes can such an individual produce if there is no crossing-over? (b) What will be those genotypes?

4-15 Use the same individual as in the preceding problem, but assume that crossing-over between the two pairs of genes occurs with a total frequency of 0.2 (i.e., 2 in each 10 gametes will have crossover genotypes). (a) How many gamete genotypes can this individual produce? (b) What will be those genotypes? (c) Using the product law of probability with what frequency will *each* of the *crossover* gamete genotypes occur? (d) With what frequency will *each* of the *noncrossover* gamete genotypes occur?

4-16 In corn, recessive gene *hm* determines susceptibility to the fungus *Helminthosporium*, which produces lesions on leaves and in kernels, reducing yield, vigor, and market value of the crop. Its dominant allele determines resistance to the fungus. Another recessive gene, br_1, produces shortened internodes (segments of stem between successive leaves) and stiff, erect leaves (brachytic plants). Its dominant allele is responsible for normal plant form. These two pairs of genes are linked on chromosome 1. If a resistant, normal plant, heterozygous for both pairs of genes (with dominants of each pair on one homologous chromosome) is pollinated by a susceptible, brachytic plant, (a) would you expect any resistant, brachytic and susceptible, normal individuals in the progeny? (b) Explain the cytological basis of your answer.

4-17 Assume a particular metaphase human chromosome has an average length of 5 μm. Give the equivalent in (a) meters, (b) millimeters, (c) nanometers, (d) Angstrom units.

4-18 What possible advantages to a species can you see in crossing-over?

4-19 Can you see possible disadvantages that might result from crossing-over?

4-20 Notice (Table 4-1) that diploid chromosome numbers are, with rare exceptions such as the male grasshopper, even numbers. Explain.

CHAPTER 5

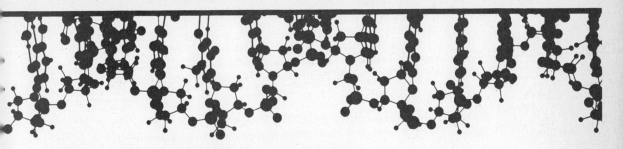

Probability and goodness of fit

An understanding of the laws of probability is of fundamental importance in (1) appreciating the operation of genetic mechanisms, (2) predicting the likelihood of certain results from a given cross, and (3) assessing how well a progeny phenotypic ratio fits a particular postulated inheritance pattern. One of the fundamental laws of probability (i.e., the law of the probability of coincident independent events) has already been applied in studies of dihybrid and polyhybrid ratios based on transmission of two pairs of genes that are assumed to be on different chromosome pairs, as well as gene pairs that are on a single chromosome pair. This "product law of probability" states in essence, that **the chance (or probability) of the simultaneous occurrence of two or more independent events is equal to the product of the probabilities that each will occur separately.**

Two independent, nongenetic events

SINGLE-COIN TOSSES

The laws of probability can be applied to *any* chance or random event. For example, a coin tossed into the air and allowed to come to rest is likely to land either "heads" or "tails," if the highly improbable chance of its landing on edge is neglected. On the basis of certain implicit assumptions, therefore, one could predict that, in the total array of possibilities (heads plus tails), the probability of either a head or a tail is 1 in 2, or $\frac{1}{2}$. But if one coin is tossed four successive times, a ratio other than 2 heads : 2 tails would not be surprising. Occasionally, such combinations as four heads, or three heads and one tail, or one head and three tails, or even four tails would be expected. If, however, a very large number of tosses were to be made, a result quite close to a 1:1 ratio would be expected. Note that successive tosses are assumed to be independent of each other; that is, the result of one toss has no effect on any succeeding toss.

TWO-COIN TOSSES

What can be expected if two coins are tossed simultaneously for, say, 50 tosses? Assume two students were each asked to toss two coins simultaneously 50 times by shaking the coins in closed, cupped hands and letting them fall lightly on a table. Ignoring the possibility of a coin landing on edge, only two results can occur: each coin will come to rest flat on the table, with either one side (which will be called a head) up or the other side (which will be designated a tail) up. With the two coins, then, the outcome of any single toss will be HH, HT, or TT. The result of the series by student *A* was

HH	12
HT	27
TT	11

Now one has to ask, "Are these the results I could have expected, based on the theory that each coin has an equal chance of landing either heads or tails?" Or, in a genetic experiment, after having made a given cross and obtained certain results, one may ask "What possible mechanism could be operating in order to produce these results?" In either a coin toss or a genetic breeding experiment, one of the first necessary steps is to formulate certain hypotheses to predict or explain results, and then ask whether, *in terms of those hypotheses, deviation from the predicted result is within limits set by chance alone.* If so, the hypotheses may be used in predicting outcomes of untried cases; if not, the hypotheses will have to be altered. (The chi-square test, to be discussed later in this chapter, will permit a judgment on the validity of hypotheses.)

In order to make a judgment on the question detailed in the preceding paragraph, several **simplifying assumptions** about the coins, as well as the general conditions of the experiment itself, must be made. First, one must ask *whether the coins themselves were unbiased;* that is, were they so constructed physically that each coin had an equal chance of landing either head or tail? Because there is no real evidence to the contrary, it may be assumed that the coins are indeed unbiased. If this assumption should turn out to be untrue, then the ultimate evaluation of the results will have to be changed.

A second question must also be examined; that is, whether the "headness" or "tailness" of the fall of one of the two coins will have any effect on the fall of the second coin. That is to say, does *each* coin have an *equal* chance of coming to rest heads or tails? Certainly, in this experiment there should be no more than a remote chance that the ultimate fall of one coin affects the other. Therefore, a *second assumption* will be that *the coins themselves are independent of each other.*

A *third assumption* will also have to be made, at least at the outset, namely that *successive tosses (events) are also independent of each other.* This would imply, for example, that a toss resulting in two heads will in no way affect the outcome of the next or any other toss.

If it is assumed, therefore, that the **coins are (1) unbiased and (2) independent of each other and also that (3) the several trials or tosses are likewise independent of each other,** what kind of results would be *expected?*

Obviously, each coin has one chance in two, or a probability of $\frac{1}{2}$, of coming to rest heads and a probability of $\frac{1}{2}$ of landing tails. By the product law of probability, the chance of *both* coins showing heads in the same toss is $\frac{1}{2} \times \frac{1}{2}$, or $\frac{1}{4}$. Student *A* did, indeed, get two heads in almost exactly one fourth of his tosses. Likewise, the probability of having two tails simultaneously is also $\frac{1}{2} \times \frac{1}{2} = \frac{1}{4}$. This result was not quite obtained in the example under consideration.

What should be expected with regard to one head and one tail? The chance of coin 1 landing heads is $\frac{1}{2}$; the chance of coin 2 coming to rest tails is also $\frac{1}{2}$. Now $\frac{1}{2} \times \frac{1}{2} = \frac{1}{4}$, but the total array of possibilities thus arrived at ($\frac{1}{4}$ HH + $\frac{1}{4}$ TT + $\frac{1}{4}$ HT) equals only $\frac{3}{4}$, leaving $\frac{1}{4}$ of the possibilities unaccounted for. Note that the chance of one head plus one tail is really the sum of the probability of coin 1 being heads and coin 2 being tails, *plus* the probability of coin 1 being tails and coin 2 being heads; i.e., the HT category is really HT + TH, or 2 HT. Therefore the probability of one head and one tail is $2(\frac{1}{2} \times \frac{1}{2})$, or $\frac{1}{2}$.

The general statement for two independent events of known probability may be written as

$$a^2 + 2ab + b^2$$

where a represents the probability of a head and b the probability of a tail. If, as in this instance, a and b each equal $\frac{1}{2}$, then the value of this expression upon substitution becomes $(\frac{1}{2})^2 + 2(\frac{1}{2} \times \frac{1}{2}) + (\frac{1}{2})^2$ or $\frac{1}{4} + \frac{2}{4} + \frac{1}{4} = 1$. Notice that the total array of probabilities here is 1, and also that $a^2 + 2ab + b^2$ is the expansion of the binomial $(a + b)^2$.

The observations and expectations for this two-coin toss can therefore be summarized as follows:

Class	Observed	Expected
HH	12	12.5
HT	27	25.0
TT	11	12.5
	50	50.0

Student *B*'s results, however, were a little different:

Class	Observed	Expected
HH	10	12.5
HT	33	25.0
TT	7	12.5
	50	50.0

Obviously the first set of data is closer to the results expected under the assumptions previously made, yet the second set of results was actually obtained. Each of these sets of data will be examined to see if the departure from the expected results is too great for *chance alone* to have operated within the framework of the three assumptions stated on page 92.

FOUR-COIN TOSSES

Consider an experiment involving four coins tossed together 100 times. The possible combinations of heads and tails, and the results actually obtained in a particular trial, are as follows:

Class	Observed
HHHH	9
HHHT	32
HHTT	29
HTTT	25
TTTT	5

After a set of ideal results in a two-coin toss has been determined from the expansion of $(a + b)^2$, the binomial $(a + b)^4$ must be expanded here to determine the expected results. The *power* of the binomial used is determined by the number of "things" (coins, children, and so forth) involved. Again, let

$$a = \text{probability of a head for any coin} = \tfrac{1}{2}$$

and

$$b = \text{probability of a tail for any coin} = \tfrac{1}{2}$$

Expand $(a + b)^4$:

$$a^4 + 4a^3b + 6a^2b^2 + 4ab^3 + b^4$$

Substitute the numerical values of a and b:

$$(\tfrac{1}{2})^4 + 4[(\tfrac{1}{2})^3 \cdot \tfrac{1}{2}] + 6[(\tfrac{1}{2})^2 \cdot (\tfrac{1}{2})^2] + 4[(\tfrac{1}{2}) \cdot (\tfrac{1}{2})^3] + (\tfrac{1}{2})^4 = 1$$

or

$$\tfrac{1}{16} + \tfrac{4}{16} + \tfrac{6}{16} + \tfrac{4}{16} + \tfrac{1}{16} = 1$$

The first term of the expression, $(\tfrac{1}{2})^4$, gives us the probability of all four coins coming up heads simultaneously; the second term, the probability of three heads (a^3) plus one tail (b), and so on. The observed results can now be compared with those calculated for expectancy in a four-coin toss:

Class	Observed	Calculated	
HHHH	9	6.25	$(= \tfrac{1}{16}$ of 100)
HHHT	32	25.00	$(= \tfrac{4}{16}$ of 100)
HHTT	29	37.50	$(= \tfrac{6}{16}$ of 100)
HTTT	25	25.00	$(= \tfrac{4}{16}$ of 100)
TTTT	5	6.25	$(= \tfrac{1}{16}$ of 100)
	100	100.00	

Again, the problem of determining *how well* these observations compare with the expected results must be considered on the basis of the three simplifying assumptions previously discussed—that is, whether these results should be accepted as *chance deviations* from the results of two- and four-coin tosses where the coins *are* unbiased and independent of each other, as are successive tosses.

The binomial expression

Answers to many genetic problems involving "either-or" situations, including those with which members of the medical or legal professions must deal from time to time, are easily provided by the binomial approach. Therefore, it is necessary to understand the method of binomial expansion before going on. Instead of multiplying out algebraically, one can apply the following simple rules for expansion of $(a + b)^n$.

1. *The power of the binomial chosen, that is, the value of* n, *is determined by the number of coins, individuals, and so on.* Thus, for two coins the expression $(a + b)^2$ is used, even if the coins are tossed 50 times; for a four-coin toss, $(a + b)^4$ is used, and so on. The power of the binomial remains constant for a given number of events even though the number of tosses is changed. The same principle applies to families of n children displaying one or the other of two phenotypic expressions.

2. *The number of terms in the expansion is* n + 1. Thus $(a + b)^2$ expands to $a^2 + 2ab + b^2$, having three terms, $(a + b)^4$ expanded contains five terms, and so on.

3. *Every term of the expansion contains both a and b.* The power of a in the first term equals n, the power of the binomial, and descends in units of 1 to 0 in the last term (and a^0, which equals 1, is not written in); likewise, b increases from b^0 (not written) in the first term to b^n in the last. Furthermore, the sum of the powers *in each term* equals n. Check this against the expansion of $(a + b)^4$.

4. *The coefficient of the first term* is 1 (not written); the coefficient of the second term is found by multiplying the coefficient of the preceding term by the exponent of a and dividing by the number indicating the position of that preceding term in the series:

coefficient of next term =

$$\frac{\text{coefficient of preceding term} \times \text{exponent of preceding term}}{\text{ordinal number of preceding term}}$$

The same principle may be represented graphically in the form of Pascal's triangle, where each coefficient is shown as the sum of two numbers immediately above. Note that the second coefficient in each horizontal line also represents the power of the binomial. Thus the second number of the last line is 6, so that this line represents the series of coefficients in the expansion of $(a + b)^6$.

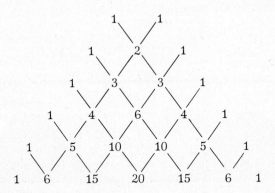

Thus, in expanding $(a + b)^4$, the first term is a^4 (for $1a^4b^0$). The second is $4a^3b$; i.e., from the preceding term, a^4,

$$\frac{4 \times 1}{1} = 4$$

which is the coefficient of the second term. The third term, which from rule 3, we know contains a^2b^2, becomes $6a^2b^2$ in the same way:

$$\frac{4 \times 3}{2} = 6$$

and so on.

In this way the binomial expansions listed in Table 5-1 are obtained.

Table 5-1. Expansions of binomials

$(a + b)^1\ a + b$
$(a + b)^2\ a^2 + 2ab + b^2$
$(a + b)^3\ a^3 + 3a^2b + 3ab^2 + b^3$
$(a + b)^4\ a^4 + 4a^3b + 6a^2b^2 + 4ab^3 + b^4$
$(a + b)^5\ a^5 + 5a^4b + 10a^3b^2 + 10a^2b^3 + 5ab^4 + b^5$
$(a + b)^6\ a^6 + 6a^5b + 15a^4b^2 + 20a^3b^3 + 15a^2b^4 + 6ab^5 + b^6$

Another useful way of thinking of binomial expansions is to remember that the expression $(a + b)^2$ means, of course, $(a + b) \times (a + b)$:

$$\begin{array}{r} a + b \\ \times\ a + b \\ \hline a^2 + ab + ab + b^2 \end{array}$$

Collecting like terms from this last expression gives us $a^2 + 2ab + b^2$. Similarly, $(a + b)^3$ can be thought of as $(a + b) \times (a + b) \times (a + b)$, or

$$\begin{array}{r} a^2 + 2ab + b^2 \\ \times\ a + b \\ \hline a^3 + 2a^2b + ab^2 + a^2b + 2ab^2 + b^3 \end{array}$$

which may be rewritten as

$$a^3 + 3a^2b + 3ab^2 + b^3$$

If a and b are each equal to $\frac{1}{2}$, symmetrical *probability curves*, as shown in Figure 5-1, are produced. On the other hand, if $a = \frac{3}{4}$ and $b = \frac{1}{4}$, a skewed curve is obtained, as indicated in Figure 5-2.

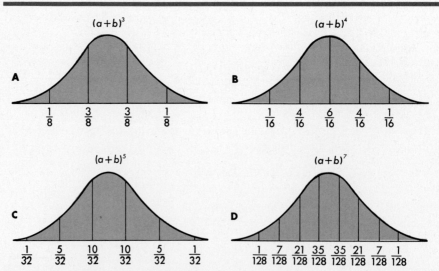

Figure 5-1. *Probability curves for binomial expansions where* $a = b = \frac{1}{2}$. *(A)* $(a + b)^3$; *(B)* $(a + b)^4$; *(C)* $(a + b)^5$; *(D)* $(a + b)^7$.

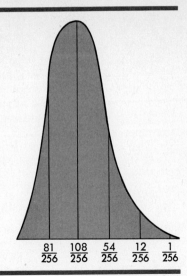

Figure 5-2. *Skewed curve for expansion of the binomial $(a + b)^4$ where $a = \frac{3}{4}$ and $b = \frac{1}{4}$.*

$$\frac{81}{256} \quad \frac{108}{256} \quad \frac{54}{256} \quad \frac{12}{256} \quad \frac{1}{256}$$

GENETIC APPLICATIONS OF THE BINOMIAL

Just as the binomial is used to determine expected results in coin tosses, it may also be used to determine the probability of children showing a given heritable trait in a particular family.

For instance, *ptosis* (drooping eyelids) is an inherited trait in which affected persons are unable to raise the eyelids, so that only a relatively small space between upper and lower lids is available for vision, giving them a "sleepy" appearance. Most pedigrees indicate that ptosis is due to an autosomal dominant. Suppose a man with ptosis, whose father also displayed the trait but whose mother did not, wishes to marry a woman with normal eyelids. They consult a physician to determine the likelihood of the occurrence of the defect in their children. If, for example, they plan to have four children, what is the probability of three of those being normal and one having ptosis?

From the known facts, it is evident that the young man is heterozygous because he has ptosis (letting P represent the trait, his phenotype alone tells us he is $P-$), although his mother was normal and, therefore, pp. So having the trait and having necessarily received a recessive gene from his mother, he must be Pp. The woman, on the other hand, is pp, as determined by her normal phenotype. So this is the cross $Pp \times pp$. This is recognized as a testcross, so that the probability of normal children is $\frac{1}{2}$. But what is the chance for a family of *three normal children and one affected child* if they have four children? Given the binomial $(a + b)^4$ where

$$a = \text{probability of a normal child} = \tfrac{1}{2}$$

$$b = \text{probability of a child with ptosis} = \tfrac{1}{2}$$

the expansion of the second term, $4a^3b$ will, upon substitution, yield the information:

$$4a^3b = 4[(\tfrac{1}{2})^3 \times \tfrac{1}{2}] = \tfrac{4}{16} = \tfrac{1}{4}$$

The second term ($4a^3b$) in this expansion is chosen because it contains a^3 (representing three children of the phenotype denoted by a) and b (for one child of the phenotype represented by b). Therefore there is a probability in this case of 1 in 4, or 0.25, that *if they have four children*, three will be normal and one will have ptosis. Similarly, there is one chance in sixteen (a^4), or 0.0625, that none of the four will exhibit ptosis or (b^4) that all four will have ptosis. Or it can be said that, of all families of four children born of parents with these genotypes, it is expected that one out of four will consist of three children without ptosis and one with the condition.

On the other hand, consider a young man and woman, each of whom has ptosis and is heterozygous. If such a couple should marry and have four children, what is the probability of three being normal and one having ptosis? The potential parents are represented as $Pp \times Pp$. From previous experience with monohybrid crosses of two heterozygotes, it is known that, in cases of complete dominance, the expected progeny phenotypic ratio is 3:1. That is, the probability here of an affected child should be expected to be $\frac{3}{4}$, and of a normal one to be $\frac{1}{4}$. As in the preceding example, let

$$a = \text{probability of a normal child} = \tfrac{1}{4}$$

$$b = \text{probability of a child with ptosis} = \tfrac{3}{4}$$

Because this case again concerns a family of four children, the binomial $(a + b)^4$ is used once more. From the expansion of this binomial one must again substitute in the second term and solve:

$$4a^3b = 4[(\tfrac{1}{4})^3 \times \tfrac{3}{4}] = \tfrac{12}{256}$$

So with two heterozygous parents the probability of three normal children and one affected child is 12/256 or 0.047. This is quite different from the 0.25 probability for the preceding case.

Had the question been, say, two normal and two affected children, the third term, $6a^2b^2$, in the expansion of $(a + b)^4$ would have been utilized. In any such problem it is imperative to select both a binomial to the proper power and the correct term within the expansion. The former is determined by the number of events (children, heads, and so on), the latter by the number of individuals of each of the two alternative types (e.g., affected versus normal, heads versus tails). Of course, ptosis, in itself, is not a serious departure from "normal," but this method of calculating probability might be of considerable importance to parents, both of whom are heterozygous for some recessive disabling or lethal gene.

PROBABILITY OF SEPARATE OCCURRENCE OF INDEPENDENT EVENTS

Just as the probability of the simultaneous occurrence of two independent events is the product of their separate probabilities, **the probability of the separate occurrence of either of two independent events equals the square root of their simultaneous occurrence,** provided, of course, the two events are of equal probability. In the preceding section we saw that the probability of tossing two heads simultaneously is $\frac{1}{4}$. The probability of tossing a head in one toss is $\sqrt{\frac{1}{4}}$, or $\frac{1}{2}$.

The same approach can be applied in genetics. Suppose one wished to determine the frequency of a gene for albinism in a particular population. Studies show, for example, that about 1 in 10,000 babies in Norway and Ireland is an albino, a condition that is the result of the action of a recessive autosomal gene. Each albino child represents the simultaneous occurrence of independent events of equal frequency, namely the union of two gametes, each of which carries the recessive gene. Inasmuch as the frequency of albinos in this case is 0.0001, the probability that any given gamete in the population carries the gene for albinism is equal to $\sqrt{0.0001}$, or 0.01. Because this gene is one member of a *pair* of alleles whose total frequency, therefore, must be 1, the frequency of the dominant gene for normal pigmentation is $1 - 0.01 = 0.99$.

Having these two values, the frequencies of the three possible genotypes can be calculated readily. Let

$$a = \text{frequency of the gene for normal pigmentation } (A) = 0.99$$

$$b = \text{frequency of the gene for albinism } (a) = 0.01$$

Substituting the expansion of $(a + b)^2 = 1$ provides genotypic frequencies:

binomial
expansion: a^2 + $2ab$ + b^2
genotype: AA Aa aa
frequency: $(0.99)^2 = 0.9801 + 2(0.99 \times 0.01) = 0.0198 + (0.01)^2 = 0.0001$

This operation has applications in population genetics (Chapter 15).

Determining "goodness of fit"

NONGENETIC EVENTS

In the earlier discussion of calculating expected results of coin tosses, one as yet unanswered question was raised: *In terms of the assumptions made concerning the coins and the tosses, are the results obtained valid by chance alone?* This question implies that one does not expect always to achieve results *exactly* equal to one's calculations and raises the problem of how much deviation from calculated probabilities can be accepted as likely being due purely to chance. Reference to Table 2-1 shows that even Mendel's results, involving hundreds or thousands of individuals, did not reflect *exactly* the expected ratios (although they were surprisingly close!). In other words, a mathematical tool is needed to determine "goodness of fit." Such a tool is the chi-square (χ^2) test.

To understand how to use this important statistical test, reexamine the earlier coin tosses. Recall the two sets of two-coin tosses made by students *A* and *B*:

	A's tosses			*B's tosses*	
Class	Observed	Calculated	Class	Observed	Calculated
HH	12	12.5	HH	10	12.5
HT	27	25.0	HT	33	25.0
TT	11	12.5	TT	7	12.5

Obviously, *A*'s tosses are closer to the expected results, but they do not correspond exactly. Therefore, are the deviations of magnitude in the two sets of results acceptable as chance deviations when the assumptions previously made are correct and complete, and only chance is operating? How large a departure from results expected under a given set of assumptions can be tolerated as representing merely chance, nonsignificant deviations? The chi-square test will make possible a judgment in this question.

The formula for calculating chi-square is

$$\chi^2 = \Sigma \left[\frac{(o - c)^2}{c} \right]$$

where o = observed frequencies, c = calculated frequencies, and Σ indicates that the bracketed quantity is to be summed for all classes. Both o and c must *be calculated in actual numbers and not in percentages*. Moreover, *each class must contain more than 5 individuals*.

To calculate chi-square for student *A*'s coin tosses (1:2:1 expectation), his data may conveniently be set up as follows:

Class	Observed o	Calculated c	Deviation $o - c$	Squared deviation $(o - c)^2$	$\dfrac{(o - c)^2}{c}$
HH	12	12.5	−0.5	0.25	0.02
HT	27	25.0	+2.0	4.00	0.16
TT	11	12.5	−1.5	2.25	0.18
TOTALS	50	50.0	0		$\chi^2 = 0.36$

With the value $\chi^2 = 0.36$, our question can now be stated as, "How often, *by chance alone*, will a deviation *this large or larger* be found when a 1:2:1 ratio is expected?" or, "How often, by chance, can one expect a value of $\chi^2 \geq 0.36$?" If this probability is quite high, then one can accept the validity of the assumptions under which the expectancy was determined as well as the thesis that the observed deviation was produced by chance.

The answer to our question may be obtained by consulting a table of chi-square (Table 5-2).

Table 5-2. Table of chi-square

Degrees of freedom	$P = 0.99$	0.95	0.80	0.70	0.50	0.30	0.20	0.05	0.01
1	0.00016	0.004	0.064	0.148	0.455	1.074	1.642	3.841	6.635
2	0.0201	0.103	0.446	0.713	1.386	2.408	3.219	5.991	9.210
3	0.115	0.352	1.005	1.424	2.366	3.665	4.642	7.815	11.341
4	0.297	0.711	1.649	2.195	3.357	4.878	5.989	9.488	13.277
5	0.554	1.145	2.343	3.000	4.351	6.064	7.289	11.070	15.086
6	0.872	1.635	3.070	3.828	5.348	7.231	8.558	12.592	16.812
7	1.239	2.167	3.822	4.671	6.346	8.383	9.803	14.067	18.475
8	1.646	2.733	4.594	5.527	7.344	9.524	11.030	15.507	20.090
9	2.088	3.325	5.380	6.393	8.343	10.656	12.242	16.919	21.666
10	2.558	3.940	6.179	7.267	9.342	11.781	13.442	18.307	23.209

Taken from Table 3 of Fisher, *Statistical Methods for Research Workers*, published by Oliver and Boyd, Ltd., Edinburgh, by permission.

To use the table it is necessary only to know the "degrees of freedom" operating in any particular case. This is one less than the number of classes involved and represents the number of *independent* classes that contribute to the calculated value of χ^2. In A's coin tosses, two of the classes may have any value (between 0 and 50), but once values for these two are set, the third is automatically determined as the difference between the total for *all* classes and the sum of all *other* classes. The values of P across the top in the table indicate the probability of obtaining a value of chi-square (and, therefore, a deviation) as large or larger, *purely by chance*.

In Table 5-2, then, with two degrees of freedom, the table is read across until either the calculated value of chi-square or two values between which it lies are found. In the case at hand, the value of $\chi^2 = 0.36$ does not appear in the table, but for two degrees of freedom there are two values, 0.103 and 0.446, between which it lies. Reading up to values of P, it is seen that a chi-square value of 0.36 corresponds to a probability value of between 0.95 and 0.80. This means that, for an expectancy of 1:2:1, one can expect a deviation as large or larger than that experienced in between 80 and 95 percent of repeated trials. The observed deviation could, therefore, easily be a result of chance, and both the expectancy and the assumptions on which it was based appear good, that is, there is a *good fit* between observed results and the calculated expectancy. Note that the *larger the value of P* (and, therefore, *the smaller the value of* χ^2), the more closely the data approximate the expected ratio, and the greater the evidence that the original hypotheses are correct. In the same way, the *smaller the value of P*, the "less" the evidence that the hypotheses are correct. Finally, at a level of $P \le 0.05$, the likelihood of a correct set of hypotheses is so small that the hypotheses are rejected. Two courses of action are open at that point: (1) secure a larger sample, or (2) revise the hypotheses and test the data against the revised hypotheses.

In the same way, χ^2 for *B*'s coin tosses (again a 1:2:1 expectation) is calculated in this way:

Class	Observed o	Calculated c	Deviation $o - c$	Squared deviation $(o - c)^2$	$\dfrac{(o - c)^2}{c}$
HH	10	12.5	-2.5	6.25	0.50
HT	33	25.0	$+8.0$	64.00	2.56
TT	7	12.5	-5.5	30.25	2.42
TOTALS	50	50.0	0		$\chi^2 = 5.48$

In Table 5-2 it is seen that one may expect, by chance alone, a value of $\chi^2 = 5.48$ in between 5 and 20 percent of such trials. This is obviously not as good a fit between observed and calculated results as was obtained by student *A*, but is it close enough to accept? The answer is "yes," because *P* is greater than 0.05. At the level of $P = 0.05$ only 1 in 20 such trials will show this large a deviation by chance alone. A chi-square value equal to or greater than that for a probability of 0.05 does not *tell* one that the large deviation necessary to produce such a high value of chi-square could *not* occur purely by chance, only that it is highly unlikely to do so. In the two-coin tosses, then, values of χ^2 up to (but not including) 5.991 indicate a sufficient probability of chance alone to be accepted as the cause of the deviation. Larger values of chi-square for two degrees of freedom, however, would require a careful review of the assumptions under which (in this case) a 1:2:1 expectancy was calculated.

In the same way, χ^2 for the four-coin toss, also reported in this chapter, is calculated to be 5.35. Table 5-2 shows that, for this value of chi-square and four degrees of freedom (remember, five different combinations are possible), $P = 0.30$ to 0.20, which shows this value of chi-square to be well below the level of significance.

GENETIC APPLICATIONS OF CHI-SQUARE

Recall the case of hydrocyanic acid in clover discussed in Chapter 3. A series of crosses (pages 41–43) of two parental strains, only one of which produced this substance in its leaves, gave rise to an F_2 of 351 HCN:256 no HCN. Although the 607 F_2 individuals reported by Atwood and Sullivan resulted from some 23 different crosses, it will suffice to treat them as though they were all sister progeny of the same cross. Chi-square calculation for a 3:1 expectancy gives the value $\chi^2 = 95.49$:

Class	o	c	$o - c$	$(o - c)^2$	$\dfrac{(o - c)^2}{c}$
HCN positive	351	455.25	-104.25	10,868	23.87
HCN negative	256	151.75	$+104.25$	10,868	71.62
TOTALS	607	607.00	0.0		$\chi^2 = 95.49$

Reference to Table 5-2 shows that, for one degree of freedom, such a value is highly significant, since it is likely to occur by chance in far fewer than one in a hundred trials. While the great deviation that produces this extremely high value of chi-square is not impossible when one expects a 3:1 ratio, it is so improbable that the assumptions on which that expected ratio was based must be reexamined. For a 3:1 ratio, these would, of course, include the concept of a single pair of genes with one allele completely dominant, or (for a 12:4 ratio) two pairs in which, say, A and a are both epistatic to B and b, though A is dominant to a:

$$\left. \begin{array}{l} 9\ A - B - \\ 3\ A - bb \end{array} \right\} 12 \text{ if } A \text{ is epistatic to } B \text{ and } b$$

$$\left. \begin{array}{l} 3\ aaB - \\ 1\ aabb \end{array} \right\} 4 \text{ if } a \text{ is epistatic to } B \text{ and } b \text{ but recessive to } A$$

Since 351:256 is considerably closer to a 9:7 expectancy, it is worth determining χ^2 for this ratio:

Class	o	c	$o - c$	$(o - c)^2$	$\dfrac{(o - c)^2}{c}$
HCN positive	351	341.4	+9.6	92.16	0.27
HCN negative	256	265.6	−9.6	92.16	0.35
TOTALS	607	607.0	0.0		$\chi^2 = 0.62$

With one degree of freedom, Table 5-2 indicates, for this value of χ^2, a probability of between 0.30 and 0.50. This value of χ^2 is well below the level of significance, and interpolation in Table 5-2 indicates that such a value of χ^2 will occur in slightly more than 44 percent of similar trials when the assumptions underlying a 9:7 ratio are operating. Therefore, until conflicting data are turned up, these assumptions are acceptable. As described in Chapter 3, these include the concept of two pairs of genes with "duplicate recessive epistasis" where only the $A - B -$ individuals are phenotypically distinguishable from other possible genotypes resulting from the crosses studied.

The chi-square test is a very useful one for obtaining an objective approximation of goodness of fit, but, as pointed out earlier, it is reliable only when the observed or calculated frequency in any class is more than 5 and numbers of individuals (not percentages) are used. Its proper use in genetic situations can, as seen in this chapter, clarify the operation of mechanisms in particular crosses.

References

Danks, D. M., J. Allan, and Anderson, C. M., 1965. A Genetic Study of Fibrocystic Disease of the Pancreas. *Ann. Hum. Genet.*, **28:** 323–356.

Froggatt, P., 1960. Albinism in Northern Ireland. *Ann. Hum. Genet.*, **24:** 213–238.

Problems

5-1 In tossing three coins simultaneously, what is the probability, in one toss, of (a) three heads, (b) two heads and one tail?

5-2 A couple has two girls and is expecting a third child. They hope it will be a boy. What is the probability that their wish will be realized?

5-3 Another couple has eight children, all boys. What is the chance that their ninth child would be another boy?

5-4 What is the probability of getting (a) a 5 with a single die, (b) a 5 on each of two dice thrown simultaneously, (c) any combination totaling 7 on two dice thrown simultaneously?

5-5 In crossing two heterozygous deep *Coleus* plants, what is the probability of the occurrence in the F_1 of (a) deep, (b) shallow?

5-6 In crossing two *Coleus* plants of genotype *DdIi*, what is the probability in the F_1 of (a) *DdIi*, (b) *ddII*, (c) deep irregular, (d) shallow irregular?

5-7 (a) Give the third term in the expansion of $(a + b)^6$. (b) If $a = b = \frac{1}{2}$, what is the numerical value of this term? (c) If $a = \frac{1}{4}$ and $b = \frac{3}{4}$, what then is the numerical value of this term?

5-8 For this problem refer back to problems 3-29 through 3-32. You plant 256 squash seeds from a cross between two tetrahybrid white disk plants (*AaBbCcDd*) and hope for some F_1 plants that bear yellow sphere fruits. If all 256 seeds give rise to fruiting plants, how many of them would be expected to produce yellow sphere fruits?

5-9 Astigmatism is a vision defect produced by unequal curvature of the cornea, causing objects in one plane to be in sharper focus. It results from a dominant gene. Wavy hair appears to be the heterozygous expression of a pair of alleles for straight (h^1) or curly hair (h^2). A wavy-haired woman who has astigmatism, but whose mother did not, marries a wavy-haired man who does not have astigmatism. What is the probability that their first child will be (a) curly-haired and nonastigmatic, (b) wavy-haired and astigmatic? (c) How many different phenotypes could appear in their children with respect to these hair and eye conditions?

5-10 Free ear lobes (*A—*) and clockwise whorl of hair on the back of the head (*C—*) are dominant to attached lobes and counterclockwise whorl, respectively. A husband and wife know their genotypes to be *AaCc* and *aaCc*. They expect to have five children. What is the probability that three will be "free-clockwise" and two "attached-counterclockwise"?

5-11 Red hair (*rr*) and possession of two whorls of hair (*ww*) on the back of the head both appear to be inherited as recessive traits in most pedigrees. (a) How many times in families of three children, in which both parents are *RrWw* (nonredhaired, one whorl), will these consist of one red-haired boy with two whorls and two nonred-haired girls with one whorl? (b) Does $a + b = 1$ in this case? Why?

5-12 **Multiple telangiectasia** in humans is the heterozygous expression of a gene that is lethal when homozygous. Heterozygotes have enlarged blood vessels of face, tongue, lips, nose, and/or fingers and are subject to unusually frequent, serious nose bleeding. Homozygotes for the trait have many fragile and abnormally dilated capillaries; because of severe multiple hemorrhaging these individuals die within a few months after birth. Two heterozygotes married forty years ago and now have four grown children. What is the probability that two of these are normal and two have multiple telangiectasia?

5-13 A study by Danks et al. (1965) shows that in Australia four in 10,000 live births are individuals who have **cystic fibrosis** of the pancreas, an inherited recessive metabolic defect in digestion of fats, which is fatal in children homozygous for the gene. This figure is similar to that of other studies in the United States. (a) What is the probability that any given gamete in these populations carries this recessive gene? (b) What is the expected frequency of heterozygotes in these populations?

5-14 A value of $\chi^2 = 0$ indicates what degree of correspondence between calculated results and those actually observed?

5-15 What is the value of P in the case described in the preceding problem?

5-16 A certain cross yields a progeny ratio of 210:90. Chi-square for a 2:1 expectancy is 1.5, and that for a 3:1 expectancy is 4.0. (a) How many degrees of freedom are there in this case? (b) Is deviation significant in either case? (c) What genetic explanation would you therefore prefer?

5-17 A certain cross produces an F_1 ratio of 157:43. By means of the chi-square test, determine the probability of a chance deviation this large or larger on the basis of a 13:3 expectancy.

5-18   Another cross involving different genes gives rise to an F_1 of 110:90. By means of the chi-square test determine the probability of a chance deviation this large or larger on the basis of (a) a 1:1 expectancy and (b) a 9:7 expectancy. (c) Is the deviation to be considered significant in either case? (d) What do you do with these results?

5-19   Suppose, with the genes involved in problem 5-18, the F_1 ratio had been 1,100:900. Try a chi-square test with this sample to determine whether there is a significant deviation from (a) 1:1 and (b) 9:7 expectancy. (c) Is the deviation now significant in either case? (d) What is the effect of sample size on the usefulness of the chi-square test?

5-20   Following are listed some of Mendel's reported results with the garden pea. Test each for goodness of fit to the given hypothesis:

Cross	Progeny	Hypothesis
(a) Yellow × green cotyledons	(F_2) 6,022:2,001	3:1
(b) Green × yellow pods	(F_2) 428:152	3:1
(c) Violet red × white flowers	(F_1) 47:40	1:1
(d) Round yellow × wrinkled green seeds	(F_1) 31:26:27:26	1:1:1:1

CHAPTER 6

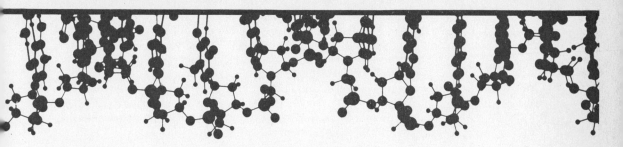

Linkage, crossing-over, and genetic mapping of chromosomes

From the discussion of dihybrid inheritance (Chapter 3) recall that the *Coleus* testcross *DdIi* (deep irregular) × *ddii* (shallow regular) produced four progeny phenotypic classes in a 1:1:1:1 ratio. Also note that this result is to be expected on the basis of the behavior of chromosomes in meiosis, if each of the two pairs of genes is on a different pair of chromosomes. With this *unlinked* arrangement of genes, either member of one pair of genes can combine at random with either member of the other pair, resulting in the production of four kinds of gametes, *DI, Di, dI,* and *di, in equal number,* by the doubly heterozygous individual. Random fusion of these four gamete genotypes with the *di* gametes of the completely recessive parent results in the 1:1:1:1 progeny phenotypic ratio. As knowledge of the genetics of various organisms increases, clearly the number of genes per species exceeds the number of chromosome pairs by a considerable margin. For example, it has been estimated on both genetic and molecular grounds that the number of genes in the fruit fly (*Drosophila melanogaster*) is about 5,000. Other estimates have been even higher, so this figure may well be conservative. Because this species has only four pairs of chromosomes, each chromosome must bear many genes. The same expectation holds for all eukaryotic organisms.

All the genes carried on a given pair of autosomes constitute a **linkage group,** and would be expected to be inherited as a block were it not for crossing-over

106

CHAPTER 6
LINKAGE,
CROSSING-OVER,
AND GENETIC
MAPPING OF
CHROMOSOMES

(Chapter 4). Therefore, it should be expected that the number of linkage groups for any organism is equal to the monoploid chromosome number. This expectation will have to be amended when the sex chromosomes are considered (Chapter 11). Interestingly enough, linkage was anticipated before it was actually demonstrated. Just three years after the rediscovery of Mendel's pioneer papers, Sutton (1903) suggested that each chromosome must bear more than a single gene and that genes "represented by any one chromosome must be inherited together." However, Sutton was unable to support his hypothesis experimentally. Only a few years later, Bateson and Punnett (1905–1908) did have the data with which to do so, but failed to recognize that they were dealing with gene pairs located on the same chromosome. Bateson and Punnett's method of interpretation is examined in the next section.

Linkage and crossing-over

BATESON AND PUNNETT ON SWEET PEA

In the sweet pea (*Lathyrus odoratus*) two pairs of genes affecting flower color and pollen grain shape occur, each pair exhibiting complete dominance:

R	purple flowers	Ro	long pollen grains
r	red flowers	ro	round pollen grains

Bateson and Punnett crossed a completely homozygous purple long with a red round. The F_1 was, as expected, all purple long, and a 9:3:3:1 phenotypic ratio was expected in the F_2. Results, however, were quite different, as seen in Table 6-1.

Table 6-1. Bateson and Punnett sweat pea cross

	Observed		Expected (9:3:3:1)	
Phenotype	Number	Frequency	Number	Frequency
Purple long	296	0.6932	240	0.5625 $(= \frac{9}{16})$
Purple round	19	0.0445	80	0.1875 $(= \frac{3}{16})$
Red long	27	0.0632	80	0.1875 $(= \frac{3}{16})$
Red round	85	0.1991	27	0.0625 $(= \frac{1}{16})$
	427	1.0000	427	1.0000

Satisfy yourself that chi-square for a 9:3:3:1 expectancy is 219.28. Inspection of Table 5-2 shows that, for three degrees of freedom, the basis for this expected ratio (independent segregation of the two pairs of alleles) is, therefore, not acceptable. From the arithmetic standpoint Bateson and Punnett recognized that the observed ratio was "explicable on the assumption that . . . the gametes were produced in a series of 16, viz., 7 purple long, 1 purple round, 1 red long, and 7 red round." For example, if each parent produced *r ro* gametes with a frequency of $\frac{7}{16}$ ($= 0.4375$), then $(\frac{7}{16})^2$, or $\frac{49}{256}$ ($= 0.1914$), of the progeny should be red round. The observed frequency of such plants, 0.1991, is quite close to the calculated frequency of 0.1914. Or, to look at it another way, the red round plants represent the fusion of two *r ro* gametes, and, by the product law of probability, the frequency of such gametes must equal the square root of the frequency of the red round plants. The square root of 0.1991, i.e., the observed frequency of the red round individuals, is 0.4462, which is acceptably close to 0.4375, the Bateson and Punnett figure of $\frac{7}{16}$. Expressed as decimals, the 7:1:1:7

array of gamete genotypes then would be 0.4375 *R Ro*, 0.0625 *R ro*, 0.0625 *r Ro*, and 0.4375 *r ro*. However, Bateson and Punnett were unable to explain why or how this gamete ratio came about because they did not relate it to the behavior of chromosomes in meiosis. Later genetic studies provided evidence that genes for flower color and shape of pollen grains are linked.

ARRANGEMENT OF LINKED GENES

When two pairs of genes are linked, the linkage may be of either of two types in an individual heterozygous for both pairs: (1) the two dominants, *R* and *Ro*, may be located on one member of the chromosome pair, with the two recessives, *r* and *ro* on the other, or (2) the dominant of one pair and the recessive of the other may be located on one chromosome of the pair, with the recessive of the first gene pair and the dominant of the second gene pair on the other chromosome. The first arrangement, with two dominants on the same chromosome, is referred to as the *cis* arrangement; the second, having one dominant and one recessive on the same chromosome is called the *trans* arrangement. Figure 6-1 illustrates these possibilities.

Thus the genotypes of the parental and first filial generations of the Bateson and Punnett cross may be written in standard fashion to reflect linkage:

$$P \qquad R \; Ro/R \; Ro \; \times \; r \; ro/r \; ro$$
$$F_1 \qquad R \; Ro/r \; ro$$

Without crossing-over, the F_1 would produce but two types of gametes, *R Ro* and *r ro*. Crossing-over, however, produces two additional gamete genotypes, *R ro* and *r Ro*. That these four genotypes are *not* produced in equal frequency could easily be seen from a testcross in which the double heterozygote carries these genes in the *cis* linkage:

$$P \; ♀ \; \text{purple long} \; \times \; \text{red round} \; ♂$$
$$R \; Ro/r \; ro \qquad r \; ro/r \; ro$$

If the two pairs of genes were not linked, a 1:1:1:1 dihybrid testcross ratio would, of course, result. Based, however, on the gamete frequencies that Bateson and Punnett recognized, testcross progeny should occur in these frequencies:

Phenotype	Genotype	Frequency		Type	
1. Purple long	*R Ro/r ro*	0.4375	} 0.875	{	Noncrossovers or
2. Red round	*r ro/r ro*	0.4375			nonrecombinants
3. Purple round	*R ro/r ro*	0.0625	} 0.125	{	Crossovers or
4. Red long	*r Ro/r ro*	0.0625			recombinants

The first two phenotypic classes have received from the pistillate parent an unaltered chromosome carrying either *R Ro* or *r ro*. The latter two classes, however, have received a *crossover* chromosome, either *R ro* or *r Ro*. Gametes

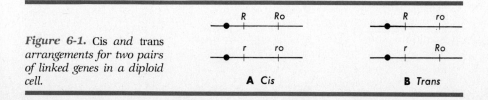

Figure 6-1. Cis *and* trans *arrangements for two pairs of linked genes in a diploid cell.*

108

CHAPTER 6
LINKAGE,
CROSSING-OVER,
AND GENETIC
MAPPING OF
CHROMOSOMES

carrying each of these altered chromosomes must occur with a frequency of only 0.0625 (= $\frac{1}{16}$), rather than 0.25, as would be the case with unlinked genes. With genes R and Ro, the *crossover frequency* is 12.5 percent. This frequency is remarkably constant, regardless of the type of cross or the kind of linkage (*cis* or *trans*).

To understand how this happens, recall the occurrence of chiasmata in prophase-I, as described in Chapter 4. If a chiasma occurs between these two pairs of genes, *crossing-over* has taken place so that the original linkage, R Ro/r ro, becomes R ro/r Ro, as indicated in Figure 6-2. For each meiocyte in which this happens, the result is four monoploid nuclei of genotypes R Ro, R ro, r Ro, and r ro. For each meiocyte in which a crossover fails to occur between these two pairs of genes, the result is four monoploid nuclei, two each of genotypes R Ro and r ro. If the meiocyte carries these genes in the *cis* configuration, as in this case, R ro and r Ro gametes are referred to as *crossover gametes*, and R Ro and r ro sex cells as *noncrossover* types. Progeny such as classes 1 and 2, which receive one or the other *intact chromosome* from the heterozygous parent, are referred to as **parental types**. Progeny making up classes 3 and 4 in this testcross incorporate new combinations of linked genes and are referred to as **recombinants**.

Linkage, linkage groups, and mapping

The chromosomal basis of heredity became clearly established in the second decade of this century, thus verifying Sutton's earlier hypothesis and supplying experimental data to explain and extend cases that had puzzled Bateson and Punnett. Discovery followed discovery in rapid succession. In 1910 Thomas Hunt Morgan was able to provide evidence for the location of a particular gene of *Drosophila* on a specific chromosome (Morgan 1910a). Within a short time he demonstrated clearly that linkage does exist and that linked genes are often inherited together, but may be separated by crossing-over (Morgan 1910b, 1911a). Even more exciting than these proofs of earlier hypotheses was Morgan's conclusion that a definite relation exists between recombination frequency and the linear distance separating genes within a chromosome. He wrote (Morgan 1911b), "in consequence, we find coupling in certain characters, and little or no evidence at all of coupling in other characters, the difference *depending on the linear distance apart of the chromosomal materials that represent the factors*" (italics added). Morgan used the term *coupling* in the sense developed by Bateson, Saunders, and Punnett (1905), that is, referring to a greater frequency of gametes carrying two dominants (or two recessives) than would occur by random segregation. They called similar association of one dominant and one recessive,

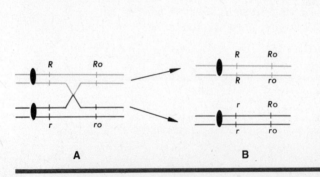

A **B**

Figure 6-2. *Result of crossing-over. A chiasma occurring between linked genes R and Ro, and involving two nonsister chromatids as shown in (A) may result in "repair" such that the original cis linkage is converted to a trans linkage in two of the four chromatids, as depicted in (B).*

repulsion. Largely through cytological work of Morgan, the concept of coupling and repulsion was replaced by that of linkage and crossing-over. The terms *cis* and *trans* came to be employed later (Haldane, 1942).

Such techniques made it possible to assign genes to particular chromosomes, as well as to begin construction of *chromosome maps* that show linkage groups and relative distances between successive genes. Linkage groups were developed rapidly for *Drosophila melanogaster* and also for a variety of animals and plants. Perhaps the most remarkable aspect of this burgeoning knowledge was that geneticists could now assign relative positions on chromosomes to genes that could not be seen in the light microscope and whose nature and precise function were not to be clarified for another 40 or 50 years.

As this kind of information was accumulated, it became clear that the number of linkage groups in any species is ultimately found to equal its monoploid chromosome number, or, in the sex having dissimilar sex chromosomes, one more than the monoploid number. Thus, four linkage groups are known in females of *Drosophila melanogaster*, where $n = 4$, five in males of the same species, whose sex chromosomes are nonhomologous (see Chapter 11), and 10 in corn (*Zea mays*), which has 10 pairs of chromosomes. In species whose genetics is less completely worked out, the number of *known* linkage groups is temporarily smaller than its monoploid chromosome number. These statements apply only to nuclear genes of eukaryotes; the matter of extranuclear genes will be examined later (Chapter 21).

Determination of linkage groups and chromosomal assignments in human beings is proceeding at a rapid pace (see Table 6-2). As early as 1976 geneticists were able for the first time to assign at least one gene to each of the human chromosomes. By 1978 the number of genes assigned to the autosomes (nonsex chromosomes) had risen to 1,300 (McKusick, 1978). At the same time 107 were known to be on the X chromosome (plus many "probables"), and one had been assigned to the Y chromosomes. (The X and Y chromosomes are the sex chromosomes; they are treated in Chapters 11 and 12.) When one realizes that the estimate for the total (diploid) gene number in *Homo sapiens* may be as high as 100,000 (McKusick, 1978), it is clear that much mapping work remains to be done.

Table 6-2. Some linkage groups and chromosome assignments in human beings (*indicates provisional assignment)

Chromosome	Genes	Chromosome	Genes
1	Amylase (pancreatic)	2	Acid phosphatase-1
	Amylase (salivary)		*Interferon-1
	*Antithrombin III deficiency		Isocitrate dehydrogenase-1
	*Cataract (zonular pulverulent)		Malate dehydrogenase-1
	Duffy blood antigen		*MNSs blood antigens
	Elliptocytosis-1	3	*Galactose-1-phosphate
	Fumarate hydratase		uridyltransferase
	Peptidase-C		*Mitochondrial aconitase
	6-phosphogluconate hydrogenase	4	*Hemoglobin α or β chain
	Phosphoglucomutase-1		*Phosphoglucomutase-2
	Rhesus blood antigens	5	*Diphtheria toxin sensitivity
	5S RNA		Hexosaminidase B
	Uncoiler-1 (uncoiled long arm of		*Interferon-2
	chromosome 1)	6	Chido blood group
	*Uridyl diphosphate glucose		*Hemochromatosis (abnormalities in
	pyrophosphorylase		iron metabolism)

Table 6-2. Some linkage groups and chromosome assignments in human beings (*indicates provisional assignment) (Continued)

Chromosome	Genes	Chromosome	Genes
	Malic enzyme-1		*Interferon
	MHC (Major histocompatibility complex or HLA system)		Lecithin-cholesterol acyltransferase
	*P blood group		*Mitochondrial thymidine kinase
	Phosphoglucomutase-3	17	Galactokinase
	Rodgers blood group		Cytoplasmic thymidine kinase
	Urinary pepsinogen-5	18	Peptidase-A
7	*Hageman factor (blood coagulation factor XII)	19	*Peptidase-D
	Hexosaminidase A		Phosphohexose isomerase
	*Kidd blood group		Polio sensitivity
	Mitochondrial malate dehydrogenase-2	20	Adenosine deaminase
			Inosine triosephosphatase
	*SV-40 T antigen		Desmosterol → cholesterol enzyme
8	*Glutathione reductase	21	Antiviral protein
9	ABO blood groups		*Glutathione peroxidase
	*Aconitase (cytoplasmic)		Ribosomal RNA
	*Adenylate kinase-1	22	*β galactosidase
	*Adenylate kinase-3		Ribosomal RNA
	Nail-patella syndrome	X	Adrenal hypoplasia
10	*Adenosine kinase		Agammaglobulinemia
	Glutamate oxaloacetic transaminase-1		Cleft palate (X-linked form)
	Hexokinase-1		Color blindness, partial (two types: 1. red and 2. green; two loci involved)
	Inorganic pyrophosphate		Diabetes insipidus
11	*Acid phosphatase-2 (lysosomal)		Ectodermal dysplasia (anhidrotic) (absence of sweat glands)
	Esterase-A4		Glucose-6-phosphate dehydrogenase (G6PD)
	Lactate dehydrogenase-A		
	Lethal antigen (3 loci)		Hemophilia A (classical type; due to defective antihemophilic globulin, or Factor VIII)
12	*Enolase-2		
	Glyceraldehyde-3-phosphate dehydrogenase		Hemophilia B (Christmas disease; due to defective thromboplastic component, or Factor IX)
	Lactate dehydrogenase-B		
	*Mitochondrial citrate synthetase		Lesch-Nyhan syndrome
	*Peptidase-B		Muscular dystrophy (Duchenne and two other types)
	Triosephosphate isomerase		
13	Esterase-D		Night blindness (probably more than one form)
	*Retinoblastoma-1		
	Ribosomal RNA		Nystagmus (uncontrollable oscillatory eyeball movement)
14	*Nucleoside phosphorylase		
	Ribosomal RNA		Tooth enamel, defective (enamel hypoplasia)
	Tryptophanyl transfer-RNA synthetase		
15	Hexosaminidase-A		Xg blood group
	Mannosephosphate dehydrogenase		Xm blood group
	Pyruvate kinase-3	Y	H - Y histocompatibility antigen
	Ribosomal RNA		Testis determining factor
16	Haptoglobin α		

Based in part on data in de Grouchy and Turleau (1977), and Bergsma (1974).

CYTOLOGICAL EVIDENCE FOR CROSSING-OVER

Whenever a particular chromosome pair (bearing certain genes) can be clearly identified because of some structural characteristic, experimental evidence shows that when an interchange of material between two homologs occurs, there is likewise an interchange of genes (i.e., genetic crossing-over). In an analysis of such a situation in corn, Creighton and McClintock (1931) furnished such a convincing correlation between cytological evidence and genetic results that their work has rightly been called a landmark in experimental genetics.

In corn, chromosome 9 (the second shortest in the complement of 10 pairs) ordinarily lacks a small knob (satellite). But in one particular strain investigated, a single plant was found to have dissimilar ninth chromosomes. One possessed a satellite at the end of the short arm and also an added segment that had been translocated from chromosome 8 to the long arm. The other member of the pair was normal, lacking both the satellite and the added segment of number 8. Chromosome 9 bears, among other genes, the following:

C	colored aleurone	Wx	starchy endosperm
c	colorless aleurone	wx	waxy endosperm

Aleurone and endosperm are parts of the triploid food storage tissue in the grain. The plant having these dissimilar ninth chromosomes was heterozygous for both the aleurone and endosperm genes. From earlier work, Creighton and McClintock knew which chromosome of the pair carried which gene. Thus the two ninth chromosomes of this plant, which they used as the pistillate parent in their crosses, could be identified visually and may be diagramed thus:

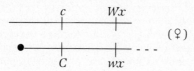

($♀$)

The dashed portion indicates the translocated segment of chromosome 8. An individual possessing such a dissimilar ninth pair of chromosomes was crossed with a *c Wx/c wx* plant possessing two knobless ninth chromosomes that also did not have the added segment of chromosome 8 (i.e., "normal" ninth chromosomes):

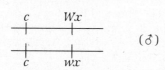

($♂$)

Although only 28 grains resulted from this cross, cytological examination of microsporocytes from adults grown from these grains confirmed the predicted relation between cytology and genetics (except for one class that was not found).

This cross, with the ninth chromosome in each parent and progeny class, is diagramed in Figure 6-3.

The fact that endosperm and aleurone are triploid tissues need not complicate our understanding of this masterful research because Creighton and McClintock used chromosomes of the *diploid* microsporocytes to demonstrate the correlation between cytological and genetic crossing-over.

A similar verification that genetic crossing over is accompanied by a physical exchange between homologous chromosomes was also established in 1931 by Stern for *Drosophila*. He utilized a strain in which the females had a portion of the Y chromosome attached to one of their X chromosomes. Work by Stern and by Creighton and McClintock indicated a clear relationship between the interchange of material between homologs and genetic crossing-over.

The genetic aspects of localization of genes to the chromosomes covered thus far may be summarized as follows:

1. Certain genes assort *at random*; these are unlinked genes (Chapter 3).
2. *Other genes do not assort randomly, but are linked* (Chapter 6). These *linkage groups* tend to be transmitted in unitary groups.

112

CHAPTER 6
LINKAGE,
CROSSING-OVER,
AND GENETIC
MAPPING OF
CHROMOSOMES

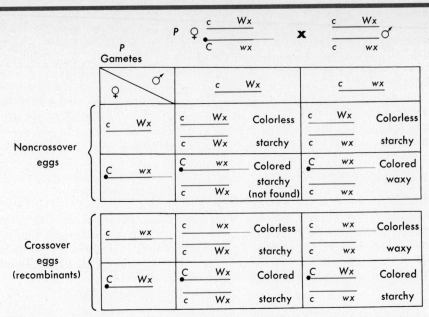

Figure 6-3. *A diagrammatic representation of parallelism between cytological and genetic crossing-over, showing the phenotypes of aleurone and endosperm of F_1 grains and the chromosome morphology and genotypes for microsporocytes produced by F_1 plants.* (Based on the work of Creighton and McClintock. See text for full explanation.)

3. In diploid cells *chromosomes also occur in pairs* that are normally transmitted as units to daughter nuclei (Chapter 4).
4. Linked genes do not always "stay together" but are often exchanged reciprocally (genetic *crossing-over*) (Chapter 6).
5. Chromosomes may be observed forming *chiasmata* and exchanging parts reciprocally (Chapters 4 and 6). Such exchange is reflected in genetic crossing-over. Furthermore, chiasma formation and genetic crossing-over occur with similar frequencies.

Thus linkage is an exception to the pattern of random segregation of genes, and crossing-over results in an exception to the usual consequences of linkage.

Mapping nonhuman chromosomes

As Morgan (1911b) predicted, frequency of crossing-over is governed largely by distances between genes. That is, the probability of crossing-over occurring between two *particular* genes increases as the distance between the genes becomes larger, so that crossover frequency appears to be directly proportional to distances between genes. As will be described later, this relationship is quite valid, though not equally so in all parts of a chromosome, because the proximity of one crossover to another decreases the probability of another very close by. The centromere has a similar interference effect; frequency of crossing-over is also reduced near the ends of the chromosome arms. Because of this general relationship between intergene distance and crossover frequency, and because such distances cannot be measured in the customary units employed in light microscopy, geneticists use an arbitrary unit of measure, the **map unit,** to describe distances between linked genes. A *map unit is equal to 1 percent of crossovers* (recombinants); that is, it represents the linear distance along the chromosome for which a recombination frequency of 1 percent is observed. These distances can also be expressed in **morgan units;** one morgan unit represents 100 percent crossing-over. Thus, 1 percent crossing-over can also be expressed as 1 *centimorgan* (1 cM), 10 percent crossing-over as 1 *decimorgan*, and so on. The morgan unit is obviously named in honor of T. H. Morgan (page 108). So for sweet peas, the distance from *R* to *Ro* would be described as 12.5 map units, or 12.5 centimorgans.

Interestingly enough, it is now possible to calculate the sizes of many genes, as well as the distances separating them, and to photograph genes in the electron microscope. Such approaches require use of information about the molecular structure of the genetic material, which will be explored in later chapters.

THE THREE-POINT CROSS

The most commonly used method in genetic mapping of *Drosophila* chromosomes and those of many other nonhuman eukaryotes is the trihybrid (or "three-point") testcross. Such a cross in *Drosophila melanogaster*, the little fruit fly whose genetics is so well known, will demonstrate this approach. Using a plus sign to denote the so-called wild type, as is customary in mapping problems, a three-point testcross will be examined, using these genes:

Gene symbol	Phenotype
+	Normal wing (*dominant*)
cu	Curled wing (*recessive*)
+	Normal thorax (*dominant*)
sr	Striped thorax (*recessive*)
+	Normal bristles (*dominant*)
ss	Spineless bristles (*recessive*)

The number of individuals in each progeny class in the following illustration are hypothetical, but the map distances and the genes are real. For the moment, an arbitrary gene sequence will be chosen; the results may either confirm that order or dictate a different one. How to determine the correct gene sequence will be demonstrated after examining the cross:

P   ♀ normal normal normal   ×   curled spineless striped ♂
     + + + /cu ss sr   cu ss sr/cu ss sr

F_1 phenotype	Maternal chromosome	#	%	Type
Normal normal normal	+ + +	430		
Curled spineless striped	cu ss sr	452	88.2	noncrossovers
Normal spineless striped	+ ss sr	45		
Curled normal normal	cu + +	38	8.3	cu-ss single crossovers
Normal normal striped	+ + sr	16		
Curled spineless normal	cu ss +	17	3.3	ss-sr single crossovers
Normal spineless normal	+ ss +	1		
Curled normal striped	cu + sr	1	0.2	double crossovers
		1,000	100.00	

114

CHAPTER 6
LINKAGE,
CROSSING-OVER,
AND GENETIC
MAPPING OF
CHROMOSOMES

The first two classes, normal-normal-normal and curled-spineless-striped, *in this example* are each phenotypically like one parent or the other. For this reason, they are sometimes referred to as "parental types." However, in some of the problems at the end of this chapter, it will be noted that the progeny classes that resemble the parents *phenotypically* differ from them *genotypically*. For this reason it is best to refer to these two classes merely as **noncrossover** types.

By grouping F_1 data as in the table, several important considerations are apparent:

1. **The maternal chromosome received by members of each of the two numerically largest classes** (noncrossover flies), determinable from the phenotypes, **discloses whether** *cis* **or** *trans* **linkage obtained in the maternal parent.** Here, for example, normal normal normal individuals must have received $+ + +$ from the maternal parent and *cu ss sr* from the paternal. Because noncrossover gametes will be more frequent than crossover sex cells, $+ + +$/*cu ss sr* and *cu ss sr*/*cu ss sr* flies will occur in the majority if the linkage is in the *cis* configuration.

2. **Double-crossover individuals** (those that would not be noted in a two-pair cross involving only *cu* and *sr*) **show up.** These are individuals that, as the term implies, result from the occurrence of two crossovers between the first and third genes in order on the maternal chromosome. *Even numbers of crossovers between two successive genes will not be picked up if the same two chromatids are involved in both crossovers:*

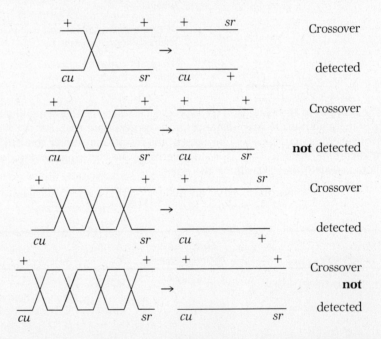

3. **The double crossovers are recognizable as the numerically smallest progeny groups and may be used to determine gene sequence.** Double crossovers are numerically the smallest groups because the product law of probability applies; that is, if the single crossover frequency between genes 1 and 2 is 0.1, and the crossover frequency between genes 2 and 3 is 0.2, then the probability of two crossovers in the *same chromatids*, one between genes 1 and 2 and another between genes 2 and 3 (thus, a double crossover) should be 0.1 × 0.2, or 0.02—a much smaller frequency than for either single crossover class. Because $+ ss +$ and *cu + sr* are, therefore, here identifiable as double crossovers, there is only one

sequence of genes in the *cis* configuration in the maternal parent that would yield these two gene combinations following a double crossover. This becomes clear if we think of the original *cis* arrangement and how the + *ss* + and *cu* + *sr* chromosomes can be derived therefrom. This will be possible only by a crossover between the first and second genes in the sequence, *plus* another between the second and third:

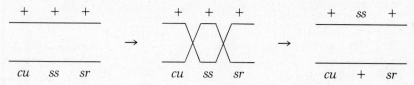

If, on the other hand, these genes were arranged in either of the other two possible sequences (*sr cu ss* or *ss sr cu*), the outcome of double crossing-over would be incompatible with the results observed in this example:

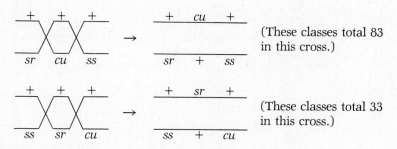

(These classes total 83 in this cross.)

(These classes total 33 in this cross.)

Thus the true sequence here can only be *cu ss sr*.

4. **The true distance** between *cu* and *ss* is, therefore, 8.3 + 0.2 = 8.5 (single crossovers + double crossovers).
5. **The true distance** between *ss* and *sr* is, therefore, 3.3 + 0.2 = 3.5 (single crossovers + double crossovers).
6. **The true distance** between *cu* and *sr* is 8.3 + 0.2 + 3.3 + 0.2 = 12.0 (*cu ss* single crossovers + *ss sr* single crossovers + *twice* the double crossovers). This is true because the double crossovers represent just what their name implies: *two* crossovers, one between *cu* and *ss*, plus a second one between *ss* and *sr*.

The genes here described constitute three members of a linkage group in *Drosophila*. With the cross outlined above, a beginning of genetically mapping one chromosome of *Drosophila* can be made. If these are thought of as the first three to be known in a new linkage group, they can be placed arbitrarily at particular *loci:*

```
        cu           ss           sr
    ────┼────────────┼────────────┼────
       0.0          8.5          12.0
```

Of course, *sr* may be placed at the "left" end and the sequence reversed. If later work shows another gene, *W* (dominant for wrinkled wing), to be located 4.0 map units to the "left" of *cu*, then the map is redrawn and each locus renumbered accordingly:

```
        W           cu           ss             sr
    ────┼───────────┼────────────┼──────────────┼────
       0.0         4.0          12.5           16.0
```

Figure 6-4 shows the locations of these and some of the other genes as presently assigned. Knowledge of which chromosome bears which genes in *Drosophila*

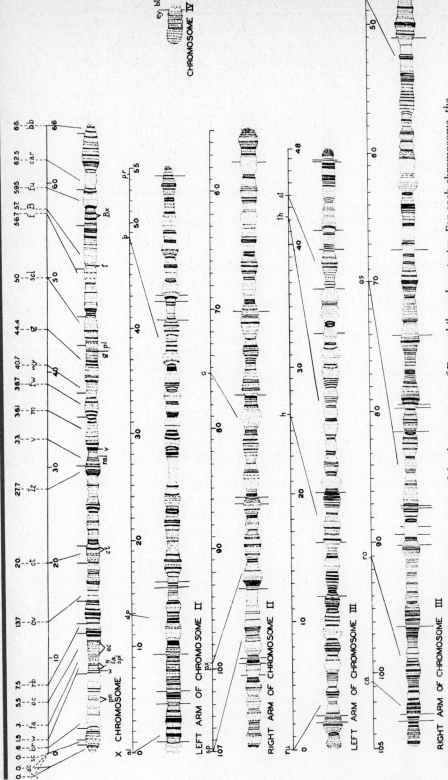

Figure 6-4. *Comparison of cytologic and genetic maps of the chromosomes of Drosophila melanogaster. For each chromosome, the genetic map is above the cytologic.* (From T. S. Painter, 1934, *Journal of Heredity,* **25:** 465–476. By permission.)

(and other dipterans) is aided by a study of giant chromosomes (Chapter 13). A similar linkage map for corn (*Zea mays*) is shown in Figure 6-5. Some of the problems at the end of this chapter further explore the techniques of genetic mapping.

If the reciprocal of the trihybrid testcross given on page 113, namely,

$$\female \; cu \; ss \; sr/cu \; ss \; sr \; \times \; + \; + \; +/cu \; ss \; sr \; \male$$

had been made only the two noncrossover phenotypes (normal normal normal and curled spineless striped) would have been recovered in the progeny. This is so because crossing-over does not occur in the male flies. Dipterans are unusual in this respect.

INTERFERENCE AND COINCIDENCE

The discussion of mapping techniques thus far would seem to imply that crossing-over in one part of a chromosome is independent of crossing-over elsewhere in that chromosome. The fact that this is not true was demonstrated as long ago as 1916 by Nobel-Prize-winning geneticist H. J. Muller.

Consider for a moment a chromosome bearing three genes, *a*, *b*, and *c*:

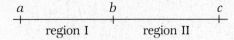

Let the *a-b* portion be designated region I and the *b-c* segment as region II. Then, recalling the now familiar product law of probability, the frequency of double crossovers between genes *a* and *c* should equal the crossover frequency of region I times the crossover frequency of region II. Actually this is seldom true unless the distances involved are large.

For example, in the three-point mapping experiment in *Drosophila*, the following crossover frequencies were obtained:

"Region"	Genes	Frequency of crossovers	Percentage crossovers	Map distance (in map units)
I	*cu ss*	0.083	8.3	8.3 + 0.2 = 8.5
II	*ss sr*	0.033	3.3	3.3 + 0.2 = 3.5
Double Crossovers	*cu ss sr*	0.002	0.2	

If crossing-over in regions I and II were independent, 0.085 × 0.035, or almost 0.3 percent double crossovers would be predicted, whereas only 0.2 percent was observed. A disparity of this kind, in which the number of actual double crossovers is less than the number calculated on the basis of independence, is very common, suggesting that, once a crossover occurs, the probability of another in an adjacent region is reduced. This phenomenon is called **interference.**

Interference appears to be unequal in different parts of a chromosome, as well as among the several chromosomes of a given complement. In general, interference appears to be greatest near the centromere and at the ends of a chromosome. Degrees of interference are commonly expressed as **coefficients of coincidence** or simply as coincidence:

$$\text{coincidence} = \frac{\text{actual frequency of double crossovers}}{\text{calculated frequency of double crossovers}}$$

In the *Drosophila* example coincidence is

$$\frac{0.002}{0.003} = 0.67$$

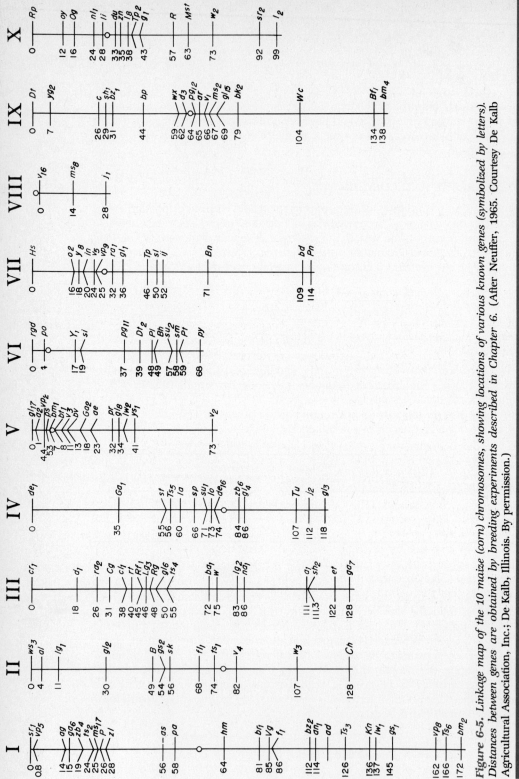

Figure 6-5. Linkage map of the 10 maize (corn) chromosomes, showing locations of various known genes (symbolized by letters). Distances between genes are obtained by breeding experiments described in Chapter 6. (After Neuffer, 1965. Courtesy De Kalb Agricultural Association, Inc.; De Kalb, Illinois. By permission.)

As interference decreases, coincidence increases. Coincidence values ordinarily vary between 0 and 1. Absence of interference gives a coincidence value of 1, whereas complete interference results in a coincidence of 0. Coincidence is generally quite small for short map distances. In *Drosophila*, coincidence is zero for distances of less than 10 to 15 map units, but gradually increases to 1 as distances exceed 15 map units. Furthermore, there seems to be no interference across the centromere from one arm of the chromosome to the other.

Similarly, interference is reported from a wide variety of organisms. For example, Hutchison (1922), who discovered the *c-sh* linkage in corn, reported map distances for three genes, *c* (colorless aleurone), *sh* (shrunken grains), and *wx* (waxy endosperm). His data indicated the following crossover frequencies:

"Region"	Genes	Frequency of crossovers	Percentage crossovers	Map distance (in map units)
I	c sh	0.034	3.4	3.4 + 0.1 = 3.5
II	sh wx	0.183	18.3	18.3 + 0.1 = 18.4
Double Crossovers	c sh wx	0.001	0.1	

Again, if crossing-over in regions I and II were independent, $0.035 \times 0.184 = 0.6$ percent double crossovers would be anticipated, whereas only 0.1 percent was observed, giving a coefficient of coincidence of 0.167. On the other hand, especially in bacterial and bacteriophage genetics, double crossovers may be encountered in excess of random expectation, giving coincidence values > 1. This is referred to as *negative interference.*

Mapping human chromosomes

Because human beings are not bred in laboratories, it is necessary to resort to other procedures in order to map the human genome. A highly successful method, somatic cell hybridization, began around 1960 as an unexpected dividend from a French experiment on cancer cell lines. It was discovered that cells from widely disparate species can and do fuse in vitro, even though the evolutionary characteristics of these species are so divergent that mating is not possible. The particular work of concern here involves the production of human-mouse hybrid cells. Mouse cells are particularly valuable in this kind of experiment because (1) a large number of murine mutant lines is readily available from commercial laboratories; (2) mouse and human chromosomes can be easily distinguished from each other through their morphology and their different banding patterns; and (3) because both human and mouse genes are expressed in the hybrid cells.

When mixed together, a very few (about one in a million) mouse and human cells fuse. Fusion extends even to the nuclei, so that the hybrid cells initially contain all 40 mouse chromosomes and all 46 human chromosomes. In mapping experiments the rate of fusion can be increased by adding Sendai virus (related to the influenza viruses) that has been inactivated, but not destroyed, by chemical treatment or by ultraviolet irradiation. This virus alters cell surfaces so that the rate of fusion is increased by a factor of as much as 1,000. Mouse and human chromosomes can also be differentiated by the fact that all those of the former species are acrocentric, whereas all human chromosomes except 13, 14, 15, 21, and 22 are metacentric or submetacentric.

As human-mouse hybrid cells are cultured further, human chromosomes are progressively and preferentially lost with each division cycle. (In mouse-rat and mouse-Chinese hamster hybrid cells, however, the *mouse* chromosomes are preferentially lost.) Approximately one to fifteen human chromosomes are preferentially lost. Only one to fifteen human chromosomes are retained beyond the

120

CHAPTER 6
LINKAGE,
CROSSING-OVER,
AND GENETIC
MAPPING OF
CHROMOSOMES

first several cell generations, and after 100 generations many cells have lost all their human chromosomes. Use of a variety of selective media determines, in many cases, which human chromosomes will more readily be lost or retained, and acts to select for or against growth of strains unable to synthesize certain substances, including enzymes. For example, 5-bromodeoxyuridine (5-BDU) kills cells that produce the enzyme thymidine kinase. Growth of hybrid cells derived from thymidine kinase deficient (TK^-) murine cells and TK^+ human cells occurs in the presence of 5-BDU only if human chromosome 17 is lacking.

Thus, it is concluded that gene *TK* is located somewhere in human chromosome 17. Of course, each human chromosome can be easily identified by its unique banding pattern. Although it has not yet been possible to assign gene *TK* to a specific locus in chromosome 17, this can be achieved in general terms through the study of chromosomal aberrations, whereby part of one chromosome has been transferred to a member of a different (nonhomologous) pair. This is called translocation and is treated, along with other kinds of chromosomal structural aberrations, in Chapter 14. Most of the human chromosome assignments in Table 6-2 have been made through the somatic cell hybridization technique.

If two or more specific human gene products and a given human chromosome are both present in the same hybrid cells, then those genes are located in the same chromosome; that is, they are **syntenic.** The term **synteny** refers to genes that are located in the same chromosome; *linkage* refers only to chromosomes that have been shown *by recombination studies* to be in the same chromosome. Syntenic genes may be so far apart in their chromosome that they seem to segregate independently; that is, they may show as much as 50 percent recombination—as would be exhibited by nonsyntenic genes.

Further differentiation of mouse versus human enzymes is often possible through electrophoretic patterns of the respective enzymes. *Gel electrophoresis* studies of enzymes and other cell substances involves placing the material in a homogeneous gel, such as starch. An electrical field is applied to the gel, and the cell substance moves toward either the negative or positive pole, depending upon its own net charge. Enzymes are protein in whole or in part, and proteins are composed of amino acids. Some of these amino acids carry a positive charge, others a negative charge, but most carry none, and thus determine direction of electrophoretic mobility.

Single metaphase chromosomes can be isolated and added to cultured cells, resulting in a product change that can be associated with the particular chromosome transferred. These cells, however, are somewhat unstable, thus limiting the general usefulness of the method.

Still another method of determining gene distances on the X chromosome, one of the sex chromosomes, will be described briefly in the treatment of sex determination (Chapter 11).

Linkage studies in bacteria

Present evidence indicates that such bacteria as the common colon bacillus, *Escherichia coli*, contain one to several *nucleoids*. These are rich in deoxyribonucleic acid (Chapter 16) and appear in electron micrographs as areas of lesser density. Internal organization is not clearly discernible, but studies show that each nucleoid contains long, continuous DNA fibrils with no free ends. Such a closed DNA structure of the nucleoid is *functionally* comparable to the chromosome of higher organisms, but is *not* a chromosome in the *structural* sense of eukaryotes. As described in Appendix B1, during conjugation between donor and recipient, one such ring "chromosome" opens at a particular point and passes as a filament in part or in whole into the body of the recipient cell, the

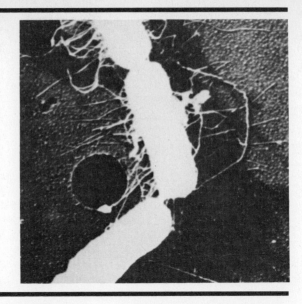

Figure 6-6. Electron micrograph of conjugating E. coli cells. The bridge connecting the two conjugants may be seen at the angle between upper and lower partners. It happens that the conjugating cells are both in late stages of fission as well, and one of them is coincidentally being attacked by bacteriophages. (Photo courtesy Dr. Thomas F. Anderson, Institute for Cancer Research, Philadelphia.)

length of the transferred segment depending on the duration of the conjugation process.

EVIDENCE FROM CONJUGATION

In such monoploid organisms, the usual technique of determining linkage distances by means of recombinational frequencies cannot be employed. Instead, donor cells of a known genotype (for several loci) are mixed with a large number of recipients of a different genotype, allowed to **conjugate** (Figure 6-6), then separated at predetermined times by agitating with a Waring blender. After separation, progeny of the recipient cells are tested by inoculation on different *deficiency media* to detect various physiologically deficient strains (*auxotrophs*). Genes are thus found to be located sequentially along the "chromosome" and are transferred in order to the recipient cell. For example, if donor cells of *Escherichia coli* K-12 (see Figure 6-7) are used, with these four known genes:

leu A   production of the enzyme α-isopropylmalate (affects synthesis of the amino acid leucine).

lac Y   production of the enzyme galactoside permease (affects metabolism of lactose).

tsx   resistance or susceptibility to phage (virus; see next section) T6.

gal K   production of the enzyme glactokinase (affects metabolism of galactose).

By interrupting conjugation after different time intervals, genes are observed to be transferred in this time sequence:

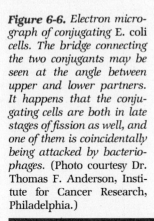

Figure 6-7. Linkage map of Escherichia coli K-12. *The numbers represent map positions in minutes determined from time-of-entry interruptions during conjugation at 37°C. Because of the large number of genes mapped, the circular chromosome has been represented as linear. More than 1,000 genes are now known in this organism.* (Redrawn from B. J. Bachmann and K. B. Low, 1980. Linkage Map of *Escherichia coli K-12*, Edition 6. Microbiological Reviews, 44 (1):1–56. Used by permission of authors and publisher.)

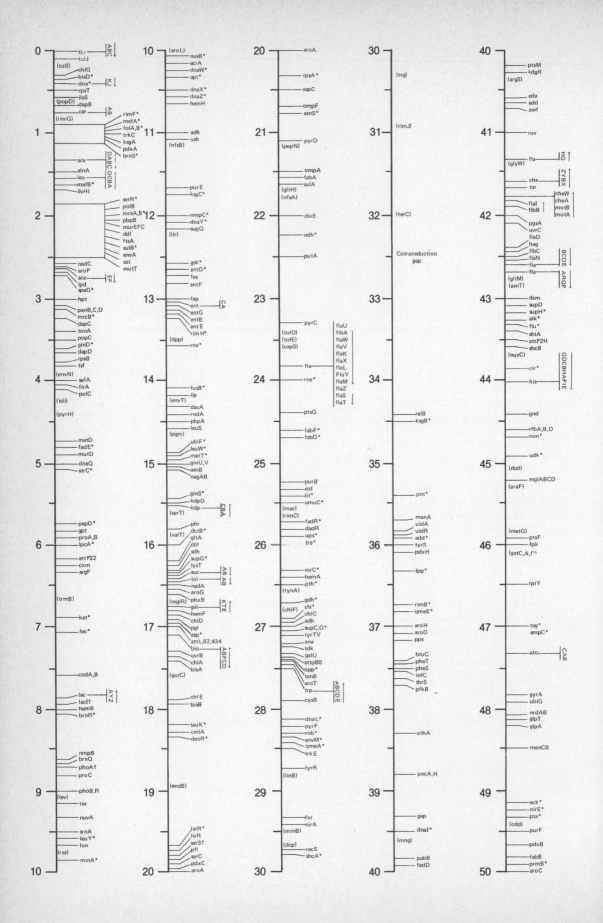

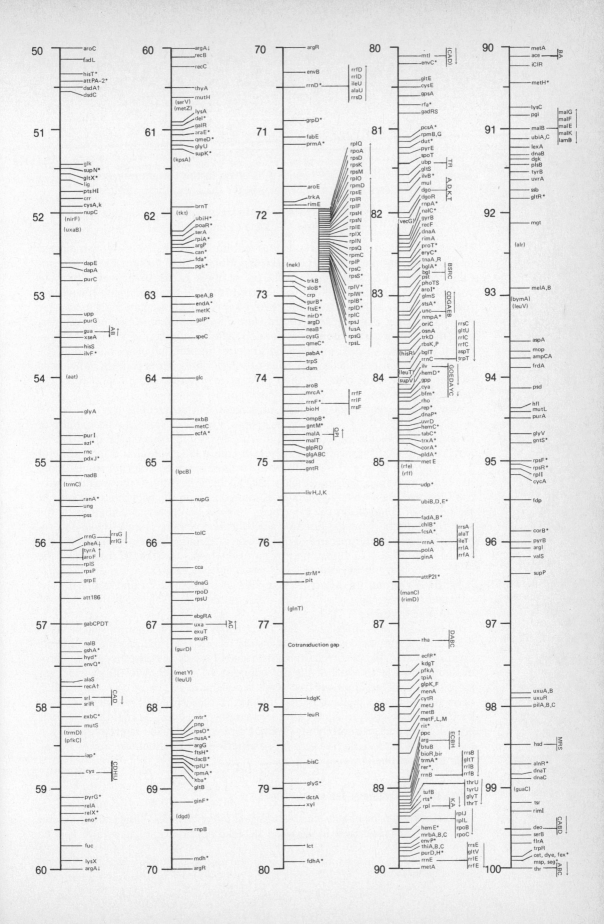

124

CHAPTER 6
LINKAGE,
CROSSING-OVER,
AND GENETIC
MAPPING OF
CHROMOSOMES

Duration of conjugation (min.)	Genes transferred
1.6	leu A
7.9	leu A, lac Y
9.2	leu A, lac Y, tsx
16.7	leu A, lac Y, tsx, gal K

Therefore, the order of these genes is as given in the 16.7-minute sequence.

A particular experiment might involve, for instance, these strains:

(donor) $leu\ A^-$, $lac\ Y^-$, tsx^+, $gal\ K^+$ ...
(recipient) $leu\ A^+$, $lac\ Y^+$, tsx^-, $gal\ K^-$ ...

Here a minus sign superscript indicates inability to synthesize the enzyme listed in the preceding paragraph (i.e., **auxotrophic** for that substance) or susceptibility to phage T6 (tsx^-). Such genotypic symbols as $leu\ A^+$ indicate ability to synthesize the enzyme involved (**prototrophic**). The two strains are mixed in a tube of liquid (complete) medium, then, after conjugation, plated out in an agar plate containing a complete medium. Here a large number of progeny colonies, both auxotrophs and prototrophs, develop. After incubation, colonies are transferred to a series of deficiency media to detect recombinations. This is done by pressing the master plate onto a sheet of velvet whose fibers pick up individuals of each colony. The velvet is then pressed to a series of replica plates of a deficiency medium.

Many such experiments, usually utilizing triple auxotrophs (e.g., $-\ -\ -$ $+\ +\ +\ \times\ +\ +\ +\ -\ -\ -$) to reduce to a very low value the probability of mutation as a factor, have resulted in a fairly complete genetic map of *Escherichia coli* K-12, for example, as shown in Figure 6-7. Note that a total time of 100 minutes is indicated for transfer of the entire nucleoid during conjugation. Reference to Appendix B1 will show that in the *Hfr* strain (high frequency recombination) the *F* (fertility) factor[1] is integrated into the bacterial nucleoid. At the outset of conjugation the *F* factor is "nicked," i.e., part of it then leads the way through the conjugation tube into the recipient (F^-) cell, while the remainder trails at the far end of the donor nucleoid. Ordinarily, however, the conjugation process is interrupted by natural stresses far short of transfer of the complete donor chromosome.

In the F^+ strain, the *F* factor exists free in the cytoplasm and is usually the only genetic material transferred in conjugation—thus converting the F^- recipient to an F^+ strain.

If the *F* factor is integrated into the nucleoid its replication is synchronous with that of the nucleoid so that it also is passed from the parental cell to its fission progeny. In some cells the integrated *F* factor is excised from its chromosome, often with some of the nucleoid genes attached. A fertility factor carrying some attached chromosomal genes is designated F' ("F prime"). Conjugation between F' and F^- results in partial diploidy (for the chromosomal genes carried by F'). Occasionally, in such cases, crossing-over between homologous regions occurs, producing recombinants. Partial diploids are called **merozygotes** and the process just described is termed **sexduction.** Sexduction also makes it possible to determine dominance of alleles in these otherwise monoploid cells.

Linkage studies in viruses

Bacteriophages, or phages, are viruses that infect bacteria. Phage structure and "life cycle" are described in Chapter 16. Like all viruses, phages have genetic

---

[1] The *F* factor is also composed of deoxyribonucleic acid which, however, bears genes differing from those of the nucleoid.

Moreover, changes or mutations of phages can occur in such genetically controlled properties as virulence toward a particular host or in the kind of coat proteins produced. During the 1940s and 1950s a number of studies showed convincingly that bacteria and viruses indeed do possess genetic material and that it is DNA in some viruses and RNA in other viruses.

Virulent phages destroy, or lyse, their hosts. One mutant in T2 and T4 phages (the T series infects the common colon bacterium, *Escherichia coli*) accomplishes this host destruction more rapidly than does the "wild-type" phage. Rapid lysing strains are designated as T2r, T4r, and so on. If a culture of *E. coli* is infected by a mixture of T2 and T4r phages, *four* progeny types are recovered: the "parental" T2 and T4r and also the recombinants T2r and T4.

Such recombination does not result from a sexual process, but rather by recombination and exchange of genetic material between the phages in conjunction with involvement of the bacterial host's genetic material. The details of this process will be discussed when the molecular nature of the genetic material itself is described, but for present purposes, it is important to note (1) that viruses can be "crossed," and (2) as a result, a map of the viral genome can be constructed. Rather early in phage studies three different linkage groups were proposed for T2, but later work has shown that the phage map, like that of bacteria, is "circular," in that each marker is linked to another on either side. One whole complex of T4r mutants, known as rII (because they were originally assigned to linkage group II of early investigators) has provided important information on the fine structure of the gene.

MECHANISM OF RECOMBINATION

The detailed mechanism whereby the donated nucleoid segment becomes integrated into the recipient's nucleoid in bacterial conjugation is not completely understood, as are the fine points of crossing-over in eukaryotes. A segment of donor chromosome is not necessarily incorporated as a whole although a mechanism analogous to crossing-over is assumed. A full statement of present understanding must be deferred until the nature and operation of the genetic material has been considered at the molecular level. It does, however, appear that some sort of break-and-exchange mechanism is operative. For immediate purposes it will suffice to point out that the segment of donor nucleoid synapses with the homologous segment of the recipient's nucleoid, following which breaks in both donor and recipient may occur. A segment of the donor nucleoid then may replace a corresponding section of the recipient's nucleoid. If genes in the two segments differ (e.g., *leu A*$^+$ versus *leu A*$^-$), recombination will have occurred in the recipient cell.

In eukaryotes, the most suggestive evidence concerning the events of recombination is derived from certain plants that have a dominant monoploid phase— e.g., many algae, most true fungi, and all bryophytes (liverworts and mosses). In bryophytes, for example, meiosis results in a spherical tetrad of four *unordered* meiospores. But in such ascomycete fungi as *Neurospora*, the cells resulting from meiosis are *ordered*; that is, situated in line in the ascus (see life cycle, Appendix B2). Such ordering reflects the pattern of chromosomal arrangement at each meiotic stage. Although meiosis in *Neurospora* produces the usual four meiospores, each divides once by mitosis to produce a total of eight ascospores, sequential pairs of which are genotypically identical. Each ascospore may be removed in order from the ascus and germinated to determine physiological phenotype, or examined visually for such morphological traits as color. Such an analysis is termed **tetrad analysis;** it clearly indicates that recombination via crossing-over must occur at the four-chromatid stage, and thus supports the concept of a break-and-exchange mechanism.

From a cross between an auxotroph (e.g., *prolineless*) and the wild-type pro-

126

CHAPTER 6
LINKAGE,
CROSSING-OVER,
AND GENETIC
MAPPING OF
CHROMOSOMES

totroph, both + and *pro* ascospores are found in equal numbers in each ascus. However, the sequential arrangement of these spores may be either + + + + *pro pro pro pro*, or + + *pro pro* + + *pro pro*. The latter arrangement is possible only if recombination by crossing-over occurs during the four-strand stage (Fig. 6-8).

If a double auxotroph for the linked traits *prolineless/serineless* (*pro/ser*) and

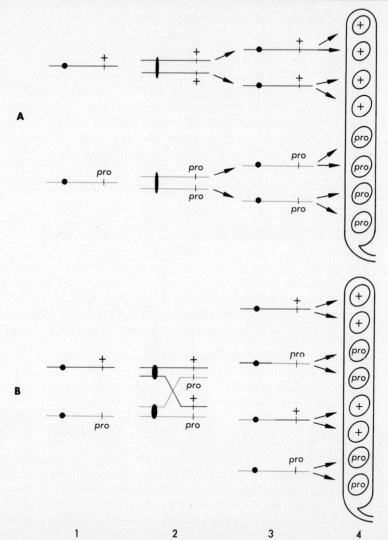

Figure 6-8. *Alternative arrangements of ascospores in ascus of* Neurospora. *In (A) crossing-over does not occur or does not involve genes + and* pro; *in (B) crossing-over occurs between the centromere and the + pro alleles (B 2). Only if crossing-over occurs in the four-strand stage, involving nonsister chromatids, can the sequence of ascospores shown at (B 4) be attained. In 1 of both (A) and (B), the chromosomes contributed to the zygote by each parental strain are shown; in 2, replication has taken place and it is at this stage that synapsis and crossing-over (if any) occurs; in 3, the chromosomes of each of the meiospores resulting from meiosis are depicted; in 4, a mature ascus and the genotypes of each of its ascospores are shown. Meiosis is taking place between 1 and 3; mitosis occurs between 3 and 4.*

the wild type $(+/+)$ are crossed, the reciprocal recombinants $(+/ser$ and $pro/+)$ are produced with equal frequency, along with the nonrecombinant types.

Even more can be learned from studies of segregation of blocks of three linked markers. Tetrad analysis clearly shows that each crossover can involve either of the two chromatids of each homologous chromosome so that three different basic types of double crossover tetrads are possible:

1. *Two-strand doubles*, in which the same two chromatids are involved in both crossovers (Fig. 6-9A).
2. *Three-strand doubles*, in which three chromatids are involved, one of them participating twice (Fig. 6-9B).
3. *Four-strand doubles*, in which each of the crossovers involves a different pair of chromatids (Fig. 6-9C).

At any given moment a visible chiasma does not necessarily indicate the point at which crossing-over occurs, because after formation, chiasmata move toward the ends of the chromatids involved. There is, nevertheless, a close correspondence between number of chiasmata and the frequency of crossovers in organisms whose cytology and genetics have been extensively investigated. Three- and four-strand doubles (Figs. 6-8B and C) can be explained only if breaks occur before chiasmata are formed and after replication of chromatids has occurred.

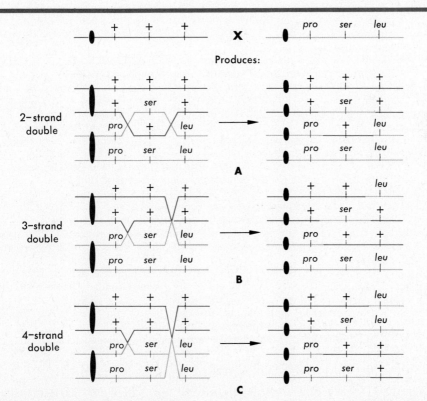

Figure 6-9. Diagram showing possible types of double crossing-over involving (A) two chromatids only, (B) three chromatids. (C) all four chromatids, as inferred from tetrad analysis.

128

CHAPTER 6
LINKAGE,
CROSSING-OVER,
AND GENETIC
MAPPING OF
CHROMOSOMES

Summary

Thus, current evidence demonstrates that

1. The number of genes exceeds the number of pairs of chromosomes as our knowledge of the organism's genetics develops.
2. Therefore certain blocks of genes are linked, and
3. The number of linkage groups is equal to the number of pairs of chromosomes (with appropriate exception for nonhomologous sex chromosomes), but
4. Linkage is not inviolable.
5. A reciprocal exchange of material between homologous chromosomes in heterozygotes is reflected in crossing-over.
6. The frequency of crossing-over appears to be closely related to physical distance between genes on a chromosome and serves as a tool in constructing genetic maps of chromosomes.
7. Crossing-over results basically from an exchange of genetic material between nonsister chromatids by break-and-exchange following replication.

References

Bachman, B. J., and K. B. Low, 1980. Linkage Map of *Escherichia coli* K–12, Edition 6. *Microbio. Rev.*, **44** (1): 1–56.

Bateson, W., and R. C. Punnett, 1905–1908. Experimental studies in the Physiology of Heredity, Reports to the Evolution Committee of the Royal Society, 2, 3, and 4. Reprinted in J. A. Peters, ed., 1959. *Classic Papers in Genetics*. Englewood Cliffs, N.J.: Prentice-Hall.

Bateson, W., E. R. Saunders, and R. G. Punnett, 1905. Experimental Studies in the Physiology of Heredity. *Rep. Evol. Comm. Roy. Soc.*, II, 1–55 and 80–99.

Bergsma, D., ed. 1974. *Human Gene Mapping*. New York: Intercontinental Medical Book Corp.

Blixt, S., 1975. Why Didn't Gregor Mendel Find Linkage? *Nature*, **256:** 206.

Conneally, P. M. and M. L. Rivas, 1980. Linkage Analysis in Man. In H. Harris and K. Hirschhorn, eds., *Advances in Human Genetics*, vol. 10. New York: Plenum Press.

Creighton, H. S., and B. McClintock, 1931. A Correlation of Cytological and Genetical Crossing-Over in *Zea mays*. *Proc. Nat. Acad. Sci.* (U.S.), **17:** 492–497. Reprinted in J. A. Peters, ed., 1959. *Classic Papers in Genetics*. Englewood Cliffs, N.J.: Prentice-Hall.

Hutchison, C. B., 1922. The Linkage of Certain Aleurone and Endosperm Factors in Maize, and Their Relation to Other Linkage Groups. *Cornell Agr. Exp. Sta. Mem.*, 60.

McKusick, V. A., 1971. The Mapping of Human Chromosomes. *Sci. Amer.*, **224:** 104–113.

McKusick, V. A., 1978. *Mendelian Inheritance in Man*, 5th ed. Baltimore: The Johns Hopkins Press.

Morgan, T. H., 1910a. Sex-Limited Inheritance in *Drosophila*. *Science*, **32:** 120–122.

Morgan, T. H., 1910b. The Method of Inheritance of Two Sex-Limited Characters in the Same Animal. *Proc. Soc. Exp. Biol. Med.*, **8:** 17.

Morgan, T. H., 1911a. The Application of the Conception of Pure Lines to Sex-Limited Inheritance and to Sexual Dimorphism. *Amer. Nat.* **45:** 65.

Morgan, T. H., 1911b. Random Segregation Versus Coupling in Mendelian Inheritance. *Science*, **34:** 384.

Renwick, J. H., 1971. The Mapping of Human Chromosomes. In H. L. Roman, ed., *Annual Review of Genetics*, vol. 5. Palo Alto, Calif.: Annual Reviews, Inc.

Sutton, W. S., 1903. The Chromosomes in Heredity. *Biol. Bull.*, **4:** 231–251. Reprinted in J. A. Peters, ed., 1959, *Classic Papers in Genetics*. Englewood Cliffs, N.J.: Prentice-Hall.

Problems

6-1 If, in sweet pea, the cross $R\ Ro/r\ ro \times R\ Ro/r\ ro$ is made, what would be the expected frequencies of (a) parental gametes of each of the possible genotypes, (b) $R\ Ro/R\ Ro$ progeny, (c) purple long progeny?

6-2 **Elliptocytosis,** a rare but harmless condition in which the erythrocytes are ellipsoidal instead of the more common disk shape, is due to the presence of either of two completely dominant genes, El_1 or El_2. Production of rhesus antigen D is also due to a dominant gene, D; inability to produce this antigen is associated with genotype dd. Persons of genotypes DD and Dd are referred to as being Rh positive (Rh+). Loci El_1 and D are linked. They have been assigned to chromosome 1. In the following case consider only gene pairs El_1, el_1 and D, d. An Rh + man exhibits elliptocytosis as did his Rh − mother. His Rh + father had normal disk-shaped erythrocytes. (a) Neglecting the small probability of crossing-over in his parents, give this man's genotype. (b) What kind of linkage configuration is this?

6-3 As noted in the preceding problem, gene pairs El_1, el_1, and D, d, are linked on human chromosome 1. Some studies suggest a distance of three map units between these loci. If a doubly heterozygous man, known to have these two genes linked in the *cis* configuration, marries an Rh negative woman who has normal disk-shaped red blood cells, and if coincidence is assumed to be 1, what is the probability of each of the following phenotypes among their children: (a) Rh + with elliptocytosis, (b) Rh + without elliptocytosis?

6-4 The following four pairs of genes are linked on chromosome 2 of tomato:

Aw, aw purple, green stems
Dil, dil normal green, light green leaves
O, o oval, spherical fruit
Wo, wo wooly, smooth leaves

Crossover frequencies in a series of two-pair testcrosses were found to be: *wo-o*, 14 percent; *wo-dil*, 9 percent; *wo-aw*, 20 percent; *dil-o*, 6 percent; *dil-aw*, 12 percent; *o-aw*, 7 percent. (a) What is the sequence of these genes on chromosome 2? (b) Why is the *wo-aw* two-pair crossover frequency not greater?

In the next four questions, the following facts will have to be used. In *Drosophila*, the following genes occur on chromosome III: *e*, ebony body; *fl*, fluted or creased wings; *jvl* ("javelin"), bristles cylindrical and crooked; *obt* ("obtuse"), wings short and blunt.

6-5 A series of dihybrid testcrosses shows the following crossover frequencies: *jvl-fl*, 3 percent; *jvl-e*, 13 percent; *fl-e*, 11 percent. (a) What is the gene sequence? (b) How do you account for the fact that the sum of the *fl-e* and *fl-jvl* frequencies exceeds the *jvl-e* frequency?

6-6 Another cross discloses a crossover frequency of 19 percent between *jvl* and *obt*. How well can you locate *obt* in the sequence established in problem 6-4?

6-7 If the *e-obt* crossover fequency is next found to be 7 percent, where should gene *obt* be located in the sequence?

6-8 The *fl-obt* crossover frequency is determined by the cross $+\ +\ /fl\ obt\ (♀) \times fl\ obt/fl\ obt\ (♂)$ to be 17.5 percent. (a) Does this confirm your answer to problem 6-6? (b) What should be the frequency of double crossovers in a trihybrid testcross involving genes *fl*, *e*, and *obt* if there is no interference?

6-9 As pointed out in the explanatory note preceding problem 6-5, genes *e*, *fl*, *jvl*, and *obt* are located on chromosome III, along with many other known genes, thus constituting part of one linkage group. How many linkage groups are there altogether in *Drosophila melanogaster* females?

6-10 Each individual of the Jimson weed (*Datura stramonium*) produces both sperms and eggs. After consulting Table 3-1, give the number of linkage groups in this plant.

130

CHAPTER 6
LINKAGE,
CROSSING-OVER,
AND GENETIC
MAPPING OF
CHROMOSOMES

6-11 How many linkage groups are there in the (a) female grasshopper, (b) male grasshopper, (c) human female, (d) human male?

6-12 Mendel studies seven pairs of contrasting characters in the garden pea. Why, do you suppose, did he not discover the principle of linkage? (It would be worthwhile to consult Blixt's paper before you reach a final conclusion.)

6-13 (a) Does crossing-over take place in the ♀ *Drosophila* in oogenesis? (b) Does crossing-over take place in the ♂ *Drosophila* during spermatogenesis?

6-14 How many different gamete genotypes are produced by (a) the female in the cross + + /jvl fl (♀) × jvl fl/jvl fl (♂), (b) the male in the cross + + /jvl fl (♀) × + + /jvl fl (♂)? (Refer back to problems 6-5 through 6-8 and your answers to problem 6-13.)

6-15 How many different gamete genotypes are produced by (a) the female in the cross + + + /jvl fl e (♀) × jvl fl e/jvl fl e (♂), (b) the male in the cross + + + /jvl fl e (♀) × + + + /jvl fl e (♂)?

6-16 Two of the pairs of alleles known in tomato are

Cu, "curl" (leaves curled)
cu, normal leaves
Bk, "beakless" fruits
bk, "beaked" fruits, having sharp-pointed protuberance on blossom end of mature fruit.

The cross of two doubly heterozygous "curl beakless" plants yields four phenotypic classes in the offspring, of which 23.04 percent are "normal beaked." Are these two pairs of genes linked? How do you know?

6-17 From the data of the preceding problem can you determine (a) whether, in the parents, *cu* and *bk* were in the *cis* or *trans* configuration; (b) the distance between them in map units (assuming no interference)?

In the next four problems, use this information. Two of the many known pairs of genes in corn are

Pl purple plant
pl green plant
Py tall plant (normal height)
pl pigmy (very dwarf)

These genes are 20 map units apart on chromosome 6. The cross *Pl Py/pl py* × *Pl Py/pl py* is made. Now answer the following four questions.

6-18 What is the gamete genotype ratio produced by each parent?

6-19 What percentage of the offspring has the genotype *pl py/pl py*?

6-20 What percentage of the progeny is purple pigmy?

6-21 What percentage of the offspring will be "true-breeding"?

6-22 Look again at the *Drosophila* trihybrid testcross on page 113. If the cross + ss + /cu + sr × cu ss sr/cu ss sr had been made instead, what would be the percentage of (a) + ss + and cu + sr, (b) + + + and cu ss sr flies in the progeny, assuming the same crossover frequencies and the same interference?

6-23 In *Drosophila*, these genes occur on chromosome III:

+ wild *h* hairy (extra hairs on scutellars and head)
+ wild *fz* frizzled (thoracic hairs turn inward)
+ wild *eg* eagle (wings spread and raised)

The cross + + + /h fz eg ♀ × h fz eg/h fz eg ♂ yielded this F₁:

wild wild wild	393	wild wild eagle	28
hairy frizzled eagle	409	hairy frizzled wild	30
wild frizzled eagle	58	wild frizzled wild	1
hairy wild wild	80	hairy wild eagle	1

(a) Give the sequence of genes and the distances between them. (b) What is the coincidence?

6-24 The cross $+ + + /a\ b\ c\ ♀ \times a\ b\ c/a\ b\ c\ ♂$ in the fruit fly gives the following crossover results:

a-b single crossovers	5.75 percent
b-c single crossovers	8.08 percent
a-c double crossovers	0.25 percent

What is the coincidence?

6-25 In tomato the following genes are located on chromosome 2:

$+$ tall plant	d dwarf plant
$+$ uniformly green leaves	m mottled green leaves
$+$ smooth fruit	p pubescent (hairy) fruit

Results of the cross $+ + + /d\ m\ p \times d\ m\ p/d\ m\ p$ were

$+ + +$	470		$+\ m\ p$	1
$+ + p$	14		$d\ +\ p$	25
$d\ +\ +$	0		$d\ m\ p$	441
$+\ m\ +$	19		$d\ m\ +$	30

(a) Which groups in the progeny represent double crossovers? (b) What is the correct gene sequence? (c) What are the distances in map units between the first and second, and between the second and third genes? (d) Is there interference?

6-26 The following genes are linked on chromosome III of *Drosophila:*

$+$ wild	b	black body
$+$ wild	cn	cinnabar eyes
$+$ wild	vg	vestigial wings

A trihybrid cross between a heterozygous ♀ and a homozygous recessive ♂ produced the following 1,000 progeny:

$+$	$+$	$+$	39	b	cn	$+$	1
b	$+$	$+$	416	b	cn	vg	48
$+$	cn	$+$	42	b	$+$	vg	50
$+$	cn	vg	402	$+$	$+$	vg	2

(a) Which are the noncrossover progeny classes? (b) Which are the double crossover classes? (c) What kind of linkage occurs in the ♀? (d) Which is the middle gene in the sequence? (e) What is the distance in cM between cn and b; between vg and cn? (f) Is there interference? (g) What is the coefficient of coincidence to the nearest second decimal place (carry the *expected* frequency to 5 decimal places)?

6-27 From the chromosome map for maize (corn), Figure 6-5, note that genes pg_{12}, gl_{15}, and bk_2 are all on chromosome 9. The testcross $+ + + /pg_{12}\ gl_{15}\ bk_2 \times pg_{12}\ gl_{15}\ bk_2 / pg_{12}\ gl_{15}\ bk_2$ is made. If there is complete interference, what is the frequency of (a) noncrossovers, (b) $pg_{12}\ gl_{15}$ single crossovers, (c) $gl_{15}\ bk_2$ single crossovers, (d) double crossovers in the progeny? (Assume all genotypes to be equally viable.)

6-28 With the cross of the preceding problem, what would be the frequencies of each of the eight progeny classes if (a) there is no interference, (b) coincidence is 0.5? (Assume all genotypes to be equally viable and crossover probabilities to be equal in all parts of the chromosome.)

CHAPTER 7

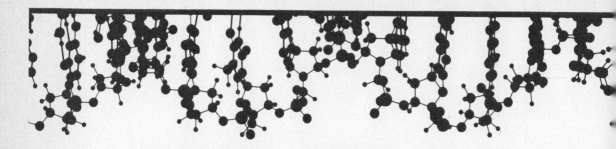

Multiple alleles and blood group inheritance

In the discussions so far, a basic assumption has been that a particular chromosomal position, or **locus,** is occupied by either of two alleles. Many instances are known, however, in which a given locus may bear any one of a series of several alleles, so that a diploid individual possesses any two genes of the series. *When any one of three or more genes may occupy a given locus in a particular pair of homologous chromosomes, such genes are said to constitute a series of* **multiple alleles.**

The concept of multiple alleles

COAT COLOR IN RABBITS

The coat of the ordinary (wild-type) rabbit is referred to as "agouti" or full color, in which individuals have banded hairs, the portion nearest the skin being gray, succeeded by a yellow band, and finally a black or brown tip (Fig. 7-1). Albino rabbits, which are totally lacking in pigmentation, have also long been known (Fig. 7-2). Crosses of homozygous agouti and albino individuals produce a uniform agouti F_1; interbreeding of the F_1 produces an F_2 ratio of 3 agouti:1 albino. Two-thirds of the F_2 agouti individuals can be shown by testcrosses to be het-

Figure 7-1. *Wild-type Agouti rabbit.* (Photo courtesy American Genetic Association.)

erozygous. Clearly then, this is a case of monohybrid inheritance, with agouti dominant to albino.

Other individuals, lacking yellow pigment in the coat, have a silvery-gray appearance because of the optical effect of black and gray hairs. This phenotype is referred to as chinchilla (Fig. 7-3). Crosses between chinchilla and agouti produce all agouti individuals in the F_1 and a 3 agouti:1 chinchilla ratio in the F_2. Thus genes determining chinchilla and agouti appear to be alleles, with agouti

Figure 7-2. *Albino rabbit. Albino animals are totally lacking in pigment.* (Photo courtesy American Genetic Association.)

Figure 7-3. Chinchilla rabbit. (Photo courtesy American Genetic Association.)

again dominant. If, however, the cross chinchilla × albino is made, the F_1 are all chinchilla, and the F_2 shows 3 chinchilla:1 albino. Therefore, genes for chinchilla and albino are also alleles, and agouti, chinchilla, albino are said to form a multiple allele series.

Still another phenotype is often encountered in pet shops. This is Himalayan (Fig. 7-4), in which the coat is white except for black extremities (nose, ears,

Figure 7-4. Himalayan rabbit. (Photo courtesy American Genetic Association.)

feet, and tail). Eyes are pigmented, unlike albino. By appropriate crosses it can be shown that the gene for Himalayan is dominant to that for albino, but recessive to genes determining agouti and chinchilla. Some of the possible crosses, with F_1 and F_2 progeny, are shown in Figure 7-5. Gene symbols often assigned are c^+ (agouti), c^{ch} chinchilla, c^h (Himalayan), and c (albino). From the crosses shown in Figure 7-5, the following dominance interrelationships can be established:

$$c^+ > c^{ch} > c^h > c$$

It is easy, of course, to predict F_1 and F_2 progeny for two crosses not shown in Figure 7-5, such as agouti × Himalayan or chinchilla × albino.

Phenotypes and their associated genotypes, therefore, for this series in rabbit are as follows:

Phenotype	Genotype
Agouti	c^+c^+, c^+c^{ch}, c^+c^h, c^+c
Chinchilla	$c^{ch}c^{ch}$, $c^{ch}c^h$, $c^{ch}c$
Himalayan	c^hc^h, c^hc
Albino	cc

Ten different genotypes occur in this series. Earlier (Chapter 2), it was seen that a single pair of alleles at a given locus produces three genotypes. By the same token, a series of three multiple alleles produce six genotypes. Note that as the number of genes in a series of multiple alleles increases, the variety of genotypes rises still more rapidly:

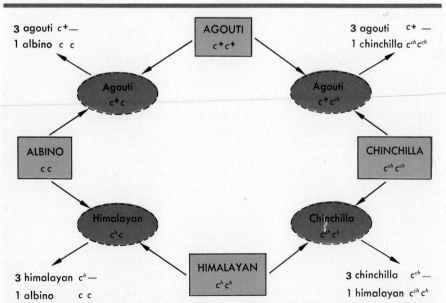

Figure 7-5. *Diagram representing several crosses in the multiple-allele coat series in the rabbit. Parental generations in solid rectangles. F_1 generations within dashed ovals. F_2 generations not enclosed.*

Number of alleles in series	Number of genotypes
2	3
3	6
4	10
5	15
n	$\dfrac{n}{2}(n + 1)$

With the number of possible genotypes increasing more rapidly than the number of alleles, a considerable increase in genetic variability ensues. Consider, for example, a hypothetical organism having only 100 loci, with a series of exactly four multiple alleles at each locus. Ten genotypes are possible at the first locus; these 10 can be combined with any of the 10 at the second locus, and so on. The total number of possible genotypes becomes 10^{100}, or 1 followed by 100 zeros!

Available evidence indicates that a given locus may mutate in several directions many times in the history of a species. The various members of the series in rabbit undoubtedly arose at different times and places as mutations of an ancestral gene, quite possibly c^+. Small wonder, then, that many apparent cases of one-pair differences ultimately turn out to involve series of multiple alleles! Other instances in animals and plants are referred to in the problems at the close of this chapter, though some interesting series in humans are described in the following sections.

The blood groups in humans

THE ANTIGEN-ANTIBODY REACTION

Blood consists of two principal components: cells (red, white, and platelets) and liquid (plasma). Plasma, minus the clotting protein fibrinogen, is referred to as serum. In early attempts at transfusion, in fact as long ago as the eighteenth century, death of the recipient sometimes ensued for no determinable reason. But in 1901, Dr. Karl Landsteiner, working in a laboratory in Vienna, observed that red blood cells (erythrocytes) of certain individuals would clump together into macroscopically visible groups when mixed with the serum of some, but not all, other persons.

The basis for this clumping is the **antigen-antibody reaction.** Injection of a foreign substance (**antigen**) into the bloodstream of an animal brings about the production by some component of the blood of a characteristic **antibody** that reacts with the antigen. The antigen is usually a protein, at least in part, and may be some plant or animal protein, a bacterial toxin, or may be derived from pollen; however in the A and B blood antigens, the antigenicity is specified by two different sugar components. An antibody is highly specific for a particular antigen (though cross reactions of varying degree may occur between one antibody and other closely similar antigen molecules). Such antibodies are termed *acquired*, because their production depends on the entry of the foreign antigen; they are not otherwise produced. Such a system forms the basis of immunization practices as well as of allergic reactions. On the other hand, in a few cases antibodies are produced naturally and normally by the blood, even in the absence of the appropriate antigen. These *natural antibodies* include several involved in human blood groups, particularly the important A-B-AB-O groups, which will be discussed shortly.

Depending on the nature of the antigen and of its antibody, numerous sorts of each can be differentiated, and each of which produces a typical antigen-

Table 7-1. Antigens and antibodies of the human blood groups

Blood group	Antigen on erythrocytes	Antibody in serum
A	A	anti-B
B	B	anti-A
AB	A and B	neither
O	neither	anti-A and anti-B

antibody reaction. If, for example, the antigen is a *toxin* (such as produced by typhoid, cholera, staphylococcus, whooping cough, and many other bacteria, or such substances as snake venom), neutralizing antibodies are called **antitoxins.** If the antigen is cellular in nature, the antibody may be a **lysin,** which lyses or disintegrates the invading cells, or an **agglutinin,** which causes clumping or agglutination of the cells. These are but a few of the recognized antibody types.

Following Landsteiner's discovery of agglutination of red blood cells and an understanding of the antigen-antibody reaction, further studies revealed the occurrence of two natural antibodies in blood serum and two antigens on the surfaces of the erythrocytes. With regard to antigens, an individual may produce either, both, or neither; he may produce either, neither, or both antibodies. After some early and confusing multiplicity of nomenclature for these substances, the system in most general use today designates the antigens as A and B, and the corresponding antibodies as anti-A (or α) and anti-B (or β). Chemically, the A and B antigens are mucopolysaccharides, consisting of a protein and a sugar. The protein portion is identical in both antigens; the sugar is the basis for the antigen-antibody specificity. An individual's blood group is designated by the type of antigen he or she produces, as indicated in Table 7-1; blood tests for each of the four major groups are shown in Figure 7-6.

A rather uncommon subgroup of A was discovered in 1911 so that group A was subdivided into A_1 and A_2. In 1936 a still rarer subgroup, A_3, was found; an even less common subgroup, A_4, is now known. These subgroups of A react weakly with anti-A_1 antiserum so that the presence of an A antigen can be recognized. Several slightly different variants of group B are also reported. Groups O and A are the most common in the United States population (Table 7-2). Group A_1 is by far the most frequently encountered in A and AB persons; for example, of 1,698 genetics students at Ohio Wesleyan University tested over a twelve-year period, only one possessed antigen A_2 and that was in an A_2B person. No A_3 or A_4 antigens have been encountered in this sample.

Although the A, B, AB, and O groups are important in transfusions and other medical situations, subgroups are relevant to certain kinds of legal problems. Therefore, attention will be given to the genetics of these four principal groups and some of their subgroups.

Table 7-2. Frequencies of the four major blood groups in United States samples

Sample description	A	B	AB	O
Rochester, N.Y. (mixed black and white) ($n = 23,787$)*	0.418	0.100	0.038	0.444
Blacks (Iowa)†	0.265	0.201	0.043	0.491
Whites (Massachusetts)†	0.397	0.106	0.034	0.463

*From Altman and Dittmer, eds., 1964.
†From Mourant, 1954 (sample size not available).

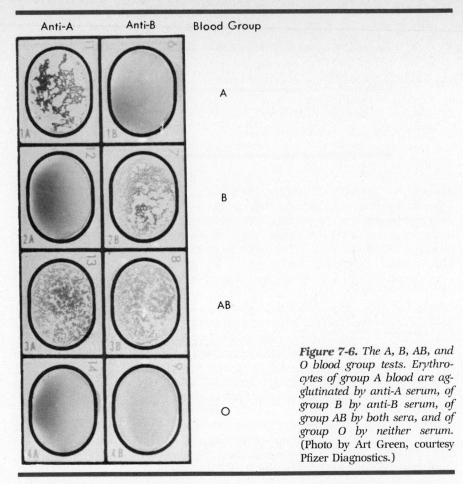

Figure 7-6. *The A, B, AB, and O blood group tests. Erythrocytes of group A blood are agglutinated by anti-A serum, of group B by anti-B serum, of group AB by both sera, and of group O by neither serum.* (Photo by Art Green, courtesy Pfizer Diagnostics.)

INHERITANCE OF A, B, AB, AND O BLOOD GROUPS

Studies of large numbers of human pedigrees have shown that children produce the A antigen only if at least one parent also produced it. Similarly, the B antigen is found only in individuals when at least one parent has it. However, group O individuals may occur in the progeny of A and/or B parents, but O parents have only O children, suggesting recessiveness of the gene for group O. In some cases, marriages of A and B parents produce children having both A and B antigens, indicating that these latter antigens are codominant. Table 7-3 presents a summary of progeny phenotypes.

Pedigree analysis clearly shows that an individual possesses, in either the homozygous or heterozygous state, any two of a series of *multiple alleles*. Because the antigens involved are of the type known as isoagglutinogens (or isohemagglutinogens), these genes are often designated as I^A, I^B, and i. Neglecting for the moment the subgroups, dominance relationships of these three alleles may be represented thus: $(I^A = I^B) > i$. Additional studies that take into account the subgroups of the A antigen indicate that I^A may occur in at least four allelic forms. These are symbolized I^{A_1}, I^{A_2}, I^{A_3}, and I^{A_4}. I^{A_1} is dominant to all other I^A alleles, I^{A_2} is recessive to I^{A_1} but dominant to the other three, and so on. Considering four forms of I^A, one of I^B, and one of i, dominance within the series may be shown in this way:

$$[(I^{A_1} > I^{A_2} > I^{A_3} > I^{A_4}) = I^B] > i$$

Table 7-3. Inheritance of blood group phenotypes in humans*

Parental groups →	A_1	A_2	A_3	B	O	A_1B	A_2B	A_3B
A_1	A_1 A_2 A_3 O	A_1 A_2 A_3 O	A_1 A_2 A_3 O	A_1B A_2B A_1 A_3B B A_2 O A_3	A_1 A_2 A_3 O	A_1 A_2B B A_1B A_3B	A_1 A_1B A_3B A_2 A_2B B	A_1 A_2 A_3 A_1B A_2B B A_3B
A_2		A_2 A_3 O	A_2 A_3 O	A_2 A_2B A_3 A_3B B O	A_2 A_3 O	A_1 A_2B A_3B B	A_2 A_2B A_3B B	A_2 A_3 A_2B A_3B B
A_3			A_3 O	A_3 A_3B B O	A_3 O	A_1 A_3B B	A_2 A_3B B	A_3 A_3B B
B				B O	B O	A_1 A_1B B	A_2 A_2B B	A_3 A_3B B
O					O	A_1 B	A_2 B	A_3 B
A_1B						A_1 A_1B B	A_1 A_1B A_2B B	A_1 A_1B A_3B B
A_2B							A_2 A_2B B	A_2 A_2B A_3B B
A_3B								A_3 A_3B B

*(Group A is arbitrarily shown as consisting of three subtypes. Phenotypes of parents are listed across the top and at the left; phenotypes of possible children are shown in the body of the table.)

Omitting the very rare I^{A4}, this series of multiple alleles produces 15 genotypes and 8 phenotypes:

Genotype	Phenotype	Genotype	Phenotype
$I^{A1} I^{A1}$		$I^{A1} I^B$	A_1B
$I^{A1} I^{A2}$			
$I^{A1} I^{A3}$	A_1	$I^{A2} I^B$	A_2B
$I^{A1} i$		$I^{A3} I^B$	A_3B
$I^{A2} I^{A2}$		$I^B I^B$	B
$I^{A2} I^{A3}$	A_2	$I^B i$	
$I^{A2} i$			
		ii	O
$I^{A3} I^{A3}$	A_3		
$I^{A3} i$			

Curiously enough, apparent changes in one's A-B-O phenotype are associated with certain pathologic conditions. Best known is a weakening of A or B antigenic reaction by a variable proportion of erythrocytes in persons suffering acute myelocytic leukemia. This is especially well documented in the case of group A persons; in a few instances the diminished response to anti-A antiserum is reported to be accompanied by a weak positive reaction to anti-B antiserum in leukemic persons of group A. During remissions normal antigenic response of the red blood cells returns, only to fall again during relapses. The basis of this change is not known, but it may be related to characteristic chromosome changes[1] occurring in myelocytic leukemia (Salmon et al., 1967). In addition to changes associated with leukemia, certain bacterial infections sometimes cause group A cells to acquire a weak response to anti-B serum. This has been explained on the basis of adsorption of a bacterial polysaccharide, chemically similar to the B antigen, on the A red cells (Race and Sanger, 1968). Except for such unusual situations, one's blood antigen-antibody traits are constant and accurate reflections of genotype.

THE H ANTIGEN

Antigens A and B are synthesized from a precursor mucopolysaccharide in the presence of the dominant allele of another pair designated as *H* and *h*. With genotypes *HH* or *Hh*, the precursor is converted to an H antigen, which, in turn, in the presence of I^A and/or I^B, is *partly* converted to antigens A and/or B (Fig. 7-7). Gene *h* is termed an **amorph** because it is responsible for no demonstrable product. So long as individuals are of genotype $H-$, group A persons produce antigens A and H, group B persons test positively for antigens B and H, and group AB persons produce antigens A, B, and H. Individuals of group O produce only antigen H if of the genotype $iiH-$. On the other hand, blood of persons of genotype $--hh$ (Fig. 7-7) does not react with anti-A, anti-B, or anti-H; this is the very rare *Bombay phenotype*, so named because it was first described in a family from that city. A similar case was described in 1955 in an American family of Italian descent, part of whose pedigree is shown in Figure 7-8. Gene *h* has an extremely low frequency; the Bombay phenotype has been described in the literature for fewer than 30 cases. The frequency of the Bombay phenotype has been estimated from large samples as only 1 in 13,000 in the Bombay area, and no cases have been found in England in more than 1 million persons tested.

Because the H antigen is only partly converted to antigens A and/or B in the presence of suitable genotypes, as shown in Figure 7-7, it should be expected that not all $--H-$ genotypes would exhibit an equally strong reaction to anti-

[1] These are described in Chapters 13 and 14.

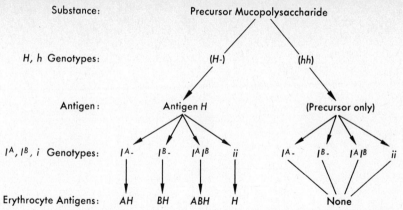

Figure 7-7. *Pathways leading to production of antigens on the red blood cells. The precursor mucopolysaccharide is converted into H substance in the presence of genotype H− and is, itself, partly converted into A and/or B antigens in the presence of genes I^A and/or I^B, with I^{A_1} more effective than I^{A_2}, and I^{A_3} least effective. The very rare gene h (for lack of H substance) is epistatic to the multiple alleles at the A-B-O locus. Cells of persons of − −hh genotype give no reaction with anti-A or anti-B sera (even though they possess genes I^A or I^B; this is the rare Bombay phenotype).*

H serum, and this is true. Blood of persons of genotype *iiH* − (group O) shows the strongest anti-H reaction, those of AB the weakest. Within the subtypes of A, A_3 reacts more strongly than A_2, which, in turn, reacts more strongly than A_1. A_1 blood gives a stronger reaction to anti-H than does group B.

MEDICOLEGAL ASPECTS OF THE A-B-O SERIES

From Table 7-3 and a knowledge of the dominance relationships of the multiple allele series involved, from which the data of the table are derived, applications to cases of disputed parentage can readily be seen. Although mixups are rare today in hospitals, situations have occurred in the past in which one or more sets of parents believed they were given someone else's child upon discharge of mother and new baby from the hospital. For example, a court case of some years ago involved just such a situation. Two sets of parents had taken babies home from a particular hospital at about the same time; in the process, the identification bracelet of the infant taken by family 1 had become detached. Family 2 soon discovered family 1's name on the child they had received, but the latter family would not agree to an exchange. Fortunately, blood tests quickly demonstrated that neither child could have belonged to the family that had taken it home, but each *could* belong to the other parents:

	Parental blood groups	Blood group of child taken home
Family 1	A × AB	O
Family 2	O × O	B

An exchange thereupon satisfied both families. Obviously, the tests in this case did not *prove* that the child received by family 1 belonged to family 2, but only that it *could*. Suppose the two families had been A × B and O × B, and that the children given to each had been found to be O and B, respectively. With no additional information from other tests (some of which are described later in

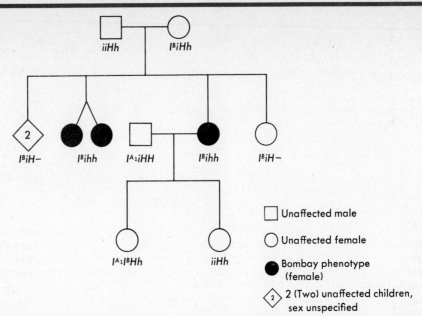

Figure 7-8. *Partial pedigree of a family tree with three children having the Bombay phenotype. The fact that these three are all girls is merely coincidental.* (Based on work of Levine, Robinson, Celano, Briggs, and Falkinburg, 1955.)

this chapter and in the next) it would be manifestly impossible to make a valid decision, because either family could have produced either child.

Quite clearly, too, blood tests are of considerable value in cases of illegitimacy. Again, tests cannot *prove* a man to be the father, but can in some instances show that he could *not* be. Based only on the A-B-O system, probability of exclusion of a wrongly accused man in a paternity case is only about 0.18. Adding tests for MNSs antigens and the Rh system[2] raises this probability to about 0.53. Adding the Kell, Lutheran, Duffy, and Kidd systems, plus the secretor trait (all described briefly later in this chapter), raises the probability to about 0.71. Making use of human leukocyte antigen tests (HLA; these antigens are discussed later in this chapter) raises the accuracy of exclusion to at least 98 percent. In California, where a battery of genetic tests may be required in paternity suits, the accuracy level is reported as 99.98 percent. In that state, only 1 percent of paternity legal actions go to trial.

In 1982 Ohio joined the few states with a sound scientific basis for hearing paternity actions. Under that state's "Uniform Parentage Act"[3] the court "may, upon its own motion or upon the motion of any party to the action," require a series of sophisticated genetic tests, the results of which may be used *either to establish or to exclude* parentage. The wording of the law regarding these tests is unequivocal:

> (Section 3111.09) (E) As used in this chapter, "genetic tests" means a series of serological tests that are either immunological or biochemical or both immunological and biochemical in nature, and that are specifically selected because of their known genetic transmittance. "Genetic tests" include, but are not limited to, tests for the presence or absence of the common blood group antigens, the red blood cell antigens, human leukocyte antigens, serum enzymes, and serum proteins.

[2]The MNSs system is described later in this chapter; the Rh, in Chapter 8.

[3]Amended Substitute to House Bill 245.

The law further stipulates (Section 3111.10) that:

In an action brought under this chapter, evidence relating to paternity may include:

(A) Evidence of sexual intercourse between the mother and alleged father at any possible time of conception;

(B) An expert's opinion concerning the statistical probability of the alleged father's paternity, which opinion is based upon the duration of the mother's pregnancy;

(C) Genetic test results, weighted in accordance with evidence, if available, of the statistical probability of the alleged father's paternity;

(D) Medical evidence relating to the alleged father's paternity of the child based on tests performed by experts. If a man has been identified as a possible father of the child, the Court may, and upon the request of a party shall, require the child, the mother, and the man to submit to appropriate tests. . . .

Before legislatures and courts had "caught up" with scientific fact, many injustices had occurred. One of the most celebrated cases of this kind occurred in 1944 in California. A widely known movie star was accused by a former starlet-protégé of being the father of her young daughter. The plaintiff's case rested on a remarkable memory for dates and details, a memory so precise that all other potential fathers were eliminated. Three physicians made blood tests of the alleged father, the mother, and the baby with these results:

	Blood group
Alleged father	O
Mother	A
Daughter	B

A moment's reflection on the genetics of these phenotypes, or reference to Table 7-3, shows clearly that the defendant could not possibly have been the father (barring a highly improbable mutation). Rather, the real father must have belonged either to group B or AB. In spite of such scientific evidence, the jury in a second trial (the first ended in a "hung jury") found the defendant *guilty!* As far as can be determined, the defendant was required to contribute for twenty-one years to the support of a child not his own.

The quantitative degree to which paternity cases occupy the time and attention of the courts is decreasing. The advent and increased use of oral contraceptives and other safe devices, as well as vasectomies for males and tying off the fallopian tubes in females, together with a decline in stigmatizing the "uniparental child" or the single parent increase the likelihood that, with the exception of rape, only ignorant or negligent couples will have unwanted children. Nevertheless, rights of a wrongfully accused man, or a child whose future security is doubtful, should be safeguarded.

Blood tests involving the A-B-O series, as well as others to be described, may, of course, also be used to good advantage in cases of claimants to estates or in certain kinds of criminal proceedings, particularly since blood group may usually be determined from corpses. In fact, blood type can often be determined from mummies, and this has become an important anthropological tool in some investigations. Problems at the end of this chapter explore some hypothetical possibilities.

OTHER BLOOD PHENOTYPES

The secretor trait. As study of the A, B, AB, and O blood groups continued, it was noted that in *some* persons whose erythrocytes bear A, B, and/or H antigens, these antigens could also be detected in aqueous secretions such as those from eyes, nose, and salivary glands. Persons with this trait are referred to as *secretors*; they produce water-soluble antigens. Several reports in the literature indicate that about 77 to 78 percent of all persons tested are secretors. Individuals lacking this trait are termed *nonsecretors,* and their antigens are only alcohol soluble. Note that the secretor-nonsecretor phenotype can be determined only for *HH* or *Hh* genotypes. The blood group of *H*− secretors can be determined even from dried saliva.

Pedigree studies indicate a single pair of genes, *Se* and *se*, to be responsible. The secretor trait is completely dominant. This pair of genes markedly increases the number of blood phenotypes.

The MNSs system.   In the course of their investigations of human blood antigens, Landsteiner and Levine in 1927 discovered two, M and N, which, when injected into rabbits or guinea pigs, stimulated antibody production in the serum of the experimental animal. Because human beings do not produce antibodies for M and N, these antigens are of no importance in transfusion. They are, however, of some interest in genetics, since inheritance of the trait depends on a pair of codominant genes sometimes referred to as L^M and L^N (for Landsteiner), but more frequently now simply as *M* and *N*, producing phenotypes as follows:

Phenotype	Antigen produced
M	antigen M only
MN	antigens M *and* N
N	antigen N only

No allele for the absence of either antigen is known. However, a large variety of alleles of the M-N-S-s loci, varying from uncommon to exceedingly rare, do exist and their antigens can be detected by appropriate antisera. Statistical analyses often reflect the 1:2:1 phenotypic ratio which occurs in cases of codominance (or incomplete dominance) when the frequencies of the two alleles are equal or nearly equal. For example, one study of 6,129 individuals in this country included 29.2 percent M, 49.6 percent MN, and 21.2 percent N type persons. Obviously, the frequencies of genes *M* and *N* must have been close to 0.5 each in the parents of these 6,129 individuals in order for this progeny ratio to be produced. Frequencies of the members of a pair of alleles, of course, need not necessarily be close to equality in a randomly mating group, and progeny ratios depend in large part on the frequencies with which the alleles occur. This topic is explored and extended in Chapter 15.

Another pair of antigens, S and s, with intimate genetic relation to the *M-N* series, was discovered in 1947. Unfortunately, the designation of antigens has developed rather randomly, so that we have here two sets of antigens, one denoted by two different capital letters (M and N), the other by the same letter in upper and lower case (S and s). Studies show that all human beings produce either antigen S, antigen s, or both; therefore another pair of codominant genes (now generally designated as *S* and *s*) is involved.

As indicated, there is a close genetic relationship between the *M-N* and *S-s* pairs of genes. For example, in families where the parents are phenotypically MNSs and NS, the children, with very rare exceptions, fall into either of two categories: (1) MNS and NSs, or (2) MNSs and NS. Clearly in the first of these two cases, children received genes for either MS or Ns from the heterozygous parent, whereas in the second they received either genes for Ms or NS. Earlier explanations favored a series of four multiple alleles at a single locus: M^S, M^s, N^S, and N^s and, indeed, such results as those just cited could rest on a multiple allele mechanism. However, Race and Sanger (1968), in their extensive studies on human blood groups, prefer an alternate explanation, i.e., two pairs of very closely linked codominant genes, *M-N*, and *S-s*. This conclusion is based on rare instances of recombination between the two suggested loci. On the basis of the latter assumption, the two cases described in this paragraph may be diagramed as follows:

$$Case\ 1:\quad P\quad MS/Ns \times NS/NS$$
$$F_1\quad \tfrac{1}{2}\,MS/NS + \tfrac{1}{2}\,Ns/NS$$

$$Case\ 2:\quad P\quad Ms/NS \times NS/NS$$
$$F_1\quad \tfrac{1}{2}\,Ms/NS + \tfrac{1}{2}\,NS/NS$$

Upward of a dozen and a half other antigens, most of them quite uncommon, have been shown to be related to the M-N and S-s substances. The explanation probably lies in a sequential series of gene-mediated reactions on one or a few precursor substances.

Other antigens.   Many other blood antigens have been described in the literature, some quite rare. These are usually designated by the family name of the individual in whom the antigen or antibody was first demonstrated. Examples include the Kidd factor (about 77 percent of the tested United States population is Kidd-positive) and the Cellano, Duffy, Kell, Lervis, and Lutheran factors (all named for the family in which each was first discovered), to list but a few.

These and a large number of additional antigens and/or antibodies produce a great diversity of human blood groups. In one test of 475 persons in London for A_1-A_2-B antigens, the M-N-S-s series, Rh, Kell, Lutheran, and Lervis groups, 269 types were reported, of which 211 included only a single person each. Considering the presently known groups and the fact that additional ones continue to be reported, the already large number of blood phenotypes may someday rise to the point where an individual's blood group may identify him as certainly as do his fingerprints. Only identical twins, triplets, etc., who have identical genotypes, would then be indistinguishable by appropriate tests.

The major histocompatibility complex in humans

Organ transplants.   Since the 1950s, it has been demonstrated that transplantation of organs (e.g., kidney, heart, liver, lungs, pancreas) is both safe and practical. However, although life span for recipients is, in many cases, extended beyond that expected with the malfunctioning organ, rarely is the life span similar to that of persons with essentially normally functioning organs. Disregarding for the moment tissue compatibility between donor and recipient, kidney transplants are the most successful, lung transplants the least. In a study of some 21,000 kidney transplant cases about half were alive after the passage of many years, but one-third died after an average of just under three years. As time passes, with more and more recipients still alive, this average may be expected to rise. In a much smaller sample of lung transplantation cases, the average survival time is less than a year.

Tissue rejection.   Although the same condition that made the transplantation necessary originally often appears later on in the transplanted organ, an early rejection occurs unless donor and recipient have a high degree of tissue compatibility. This histocompatibility is highest in monozygotic sibs (who have identical genotypes) and lowest in unrelated subjects. In general, tissue compatibility decreases as closeness of relationship decreases.

Genetic basis.   The genetics of histocompatibility is one of the most complex of all fields of human genetics and is by no means fully understood at present. Many million different HLA phenotypic classes are theoretically possible. Briefly, tissue compatibility/incompatibility rests on cellular immune reactions that involve lymphocyte (one type of leukocyte) antigens. These are designated **human leukocyte antigens,** or **HLA.** Such antigens occur on the surfaces of the lymphocytes and enable these cells to "recognize" tissues whose cells carry "foreign" antigens. The lymphocytes then attack and kill foreign tissues or organs, and thus bring about rejection. Histoincompatibility causes rejection about as rapidly as does failure to match donor and recipient for the A-B-O blood groups.

The genetic basis of the immune reaction derives from at least four loci on the sixth human autosome. After some early disagreement as to the precise location of the HLA genes on autosome 6, Pearson et al. (1979) conducted an interesting study of an unusual inversion duplication of chromosome 6, with trisomic codominant expression of the HLA antigens, and presented convincing evidence that the HLA loci are in the segment between p21 and the terminus of the short arm of the sixth autosome (6p21→pter). The four (or more) loci together comprise the **major histocompatibility complex (MHC)** and are designated HLA-A, HLA-B, HLA-C, and HLA-D. At each of these loci there may occur any of a series of codominant multiple alleles, each of which is responsible for producing a particular lymphocyte antigen. These four loci map within an interval of less than 2 cM, and the sequence (reading from the terminus of the short arm of the chromosome toward the centromere) is HLA-A, HLA-C, HLA-B, HLA-D.

Because of the large number of alleles, any of which may occur at each locus, the probability that any two persons (except for monozygotic sibs) have identical MHC genotypes is so low that it is effectively zero. The MHC loci are so closely linked that crossing-over occurs only rarely. Every individual, therefore, ordinarily inherits from each parent an unaltered number 6 autosome bearing four particular codominant HLA alleles. Each such chromosomal contribution to the next generation is termed a **haplotype.** Everyone thus has two haplotypes, one from each parent and written as, for example, *A1B8C2D4/A7B27C4D7*. Assume for the sake of illustration that any of 20 multiple alleles may occur at each MHC locus (at least 20 had been identified at the HLA-A locus, 40 at HLA-B, 8 at HLA-C, and 12 at HLA-D by 1980 according to Ryder et al.) With this hypothetical figure of 20, the number of possible combinations *within a single haplotype* would be 20^4, or 160,000. This figure, of course, makes some assumptions about the number of multiple alleles at each locus. Although these assumptions are not necessarily accurate, the figure of 160,000 does suggest the very high degree of polymorphism that exists at the MHC loci. Any of such a large number of haplotypes may be combined with the haplotype from the other parent to yield a tremendously large number of genotypes and a smaller, but still very sizable number of phenotypic classes. Adding tests for MHC antigens to other blood tests for determining paternity would raise the probability of excluding a wrongly accused man to 1 or very close to that value. Unfortunately, HLA typing is involved and expensive, and so it is not likely to be utilized in most cases of disputed parentage.

One of the most exciting and valuable spinoffs from HLA research is the discovery that certain of the MHC antigens are associated with particular disorders (Table 7-4).

A word of caution is needed in interpreting data such as those in Table 7-4. The occurrence of a given HLA type does not *guarantee* development of a particular disease, but does indicate a statistically significant probability of developing the associated condition. For example, persons with HLA B27 are 87.4 percent more at risk of developing ankylosing spondylitis (an arthritic spinal condition sometimes referred to as "poker spine") than are persons without HLA B27. In one study HLA B27 was present in 90 percent of persons afflicted with ankylosing spondylitis, whereas only 9.4 percent of nonsufferers possessed that antigen. In Table 7-4 relative risk of developing the particular disease listed and the associated HLA type ranges from 1.4x (Hodgkin's disease) to 87.4x (anky-

Table 7-4. *Association of some human diseases with HLA antigens**

Disease	HLA	Relative risk
Hodgkin's disease	A1	1.4
Ankylosing spondylitis	B27	87.4
Subacute thyroiditis	B35	13.7
Psoriasis vulgaris	**Cw6	13.3
Dermatitis herpetiformis	†D/DR3	15.4
Coeliac disease	D/DR3	10.8
Insulin-dependent diabetes	D/DR3, D/DR4, D/DR2	3.3, 6.4, 0.2
Myasthenia gravis	D/DR3	2.5
Multiple sclerosis	D/DR2	4.1
Rheumatoid arthritis	D/DR4	4.2

*Adapted from Ryder et al. (1980).
**The designation *w* (e.g., Cw6) stands for "workshop" and indicates a provisionally identified specificity.
†D and DR designations refer to the method of identification of the antigen; the R signifies "related."

losing spondylitis). An individual with given HLA markers will develop the associated disorder under appropriate environmental, dietary, or infection conditions. What one inherits, then, is not the disorder itself, but a genetic liability to develop it. A determination of the MHC genotype would permit identification of individuals who are genetically at risk for a number of serious disabling or lethal conditions. As noted earlier, the necessary tests are expensive and complex, keeping identification of human leukocyte antigens out of reach of most persons for the present.

References

Altman, P. L., and Dittmer, D. S., eds., 1962, 1972. *Biology Data Book*. Bethesda, Maryland, Federation of American Societies for Experimental Biology.

Amos, D. B., and D. D. Kostyu, 1980. HLA—A Central Immunological Agency of Man. In H. Harris and K. Hirschhorn, eds., *Advances In Human Genetics*, vol. 10. New York: Plenum Press.

Landsteiner, K., 1901. Über Agglutinationserscheinungen Normalen Menschlichen Blutes. *Wien Klin. Wochenschr.*, **14:** 1132–1134. Reprinted (in English) in S. H. Boyer, ed., 1963. *Papers on Human Genetics*. Englewood Cliffs, N.J.: Prentice-Hall.

Landsteiner, K., and P. Levine, 1927. A New Agglutinable Factor Differentiating Individual Human Bloods. *Proc. Soc. Exp. Biol. N.Y.*, **24:** 600–602.

Mourant, A. E., 1954. *The Distribution of Human Blood Groups*. Oxford: Blackwell Scientific Publications.

Pearson, G., J. D. Mann, J. Bensen, and R. W. Bull, 1979. Inversion Duplication of Chromosome 6 with Trisomic Codominant Expression of HLA Antigens. *Am. Jour. Human Genet.*, **31:** 29–34.

Race, R. R., and R. Sanger, 1968. 5th ed. *Blood Groups in Man*. Philadelphia: F. A. Davis Company.

Ryder, L. P., A. Svejgaard, and J. Dausset, 1980. Genetics of HLA Disease Association. In H. L. Roman, A. Campbell, and L. M. Sandler, eds., *Annual Review of Genetics*, vol. 15. Palo Alto, Calif.: Annual Reviews, Inc.

Salmon, C., B. Dreyfus, and R. André, 1958. Double Population de Globules, Différant Seulement par l'Antigène de Groupe ABO, Observée chez un Malade Leucémique. *Revue Hémat.*, **13:** 148–153.

Walsh, R. J., and C. Montgomery, 1947. A New Isoagglutinin Subdividing the MN Blood Group System. *Nature*, **160:** 504.

Watkins, W. M., 1980. Biochemistry and Genetics of the ABO, Lewis, and P Blood Group Systems. In H. Harris and K. Hirschhorn, eds., *Advances in Human Genetics*, vol. **10.** New York: Plenum Press.

Problems

7-1 Is it possible to cross two agouti rabbits and produce both chinchilla and Himalayan progeny?

7-2 A series of rabbit matings, chinchilla × Himalayan, produced a progeny ratio of 1 Himalayan:2 chinchilla:1 albino. What were the parental genotypes?

7-3 In the ornamental flowering plant nasturtium, flowers may be either single, double, or superdouble. These differ in number of petals, superdouble having the largest number. Crosses of superdouble × double sometimes yields 1 superdouble:1 double, and sometimes all superdouble. Superdouble × superdouble produces all superdouble, or 3 superdouble:1 double, or 3 superdouble:1 single. Single × single produces only single. (a) How many multiple alleles occur in this series? (b) Arrange the phenotypes in order of relative dominance. (c) Another cross of superdouble × double produces progeny in the ratio of 1 double:2 superdouble:1 single. What do you know about the parental genotypes?

Use the following information in answering the next four problems. In the Chinese primrose the flower has a center, or "eye," of a color different from the remainder of the petals. Normally this eye is of medium size and yellow in color. These variants also occur: very large yellow eye ("Primrose Queen" variety), white eye ("Alexandra"), and blue eye ("Blue Moon"). Results of certain crosses are the following:

P	F_1	F_2
Normal × Alexandra	Alexandra	3 Alexandra:1 Normal
Alexandra × Primrose Queen	Alexandra	(not reported)
Blue Moon × Normal	Normal	3 Normal:1 Blue Moon
Primrose Queen × Blue Moon	Blue Moon	(not reported)

7-4 Arrange these phenotypes in order of relative dominance.

7-5 How many genotypes can produce the "Alexandra" phenotype?

7-6 How many genotypes can produce the "Primrose Queen" phenotype?

7-7 How many different combinations of parental genotypes will produce a progeny ratio of 3 Alexandra:1 normal?

7-8 The frequency of gene I^{A1} was found in one study involving 3,459 persons to be approximately 0.21, of I^{A2} 0.07, and of I^B 0.06. What is the calculated frequency of each of these phenotypes: (a) A_1B; (b) A_2B? (Carry answers to four decimals.)

7-9 How many of the 3,459 persons in the sample of the preceding problem should be A_2B? (Carry answer to the nearest *whole* number.)

7-10 A man has genotype $I^A I^B Hh$, whereas that of his wife is *iihh* (the *ii* was determined by pedigree analysis). What is the probability of their having a child who exhibits the Bombay phenotype?

7-11 Assume that you do not know your M-N phenotype, or whether you are a secretor or a nonsecretor (most people do not know this information about themselves). What is your *most probable* phenotype?

7-12 A hypothetical series of 20 multiple alleles is known for a certain locus. How many phenotypic classes are possible?

7-13 How many different genotypic classes are possible for the locus referred to in problem 7-12?

7-14 Considering human blood group A to include three subtypes, and groups B and O to include one each, how many phenotypes are included in the A-B-O series?

7-15 Considering three subtypes of group A, which of the blood groups making up your answer to the preceding question would you expect to give the weakest anti-H response?

7-16 A paternity case involves these facts: the woman is A_1, her child is O, and the alleged father is B. Could he be the father? Explain.

7-17 In a paternity case it is determined that the woman is A_1, MS/NS, and a secretor. The man she charges is O, MS/NS, and a secretor. The child is A_2, MS/Ns, and a nonsecretor. Using only this information, can you eliminate the man as the father of this child? Explain.

7-18 In another paternity case it is determined that the woman is group A, MNS, a secretor, and A2B1C2D1/A3B4C3D4; the man is group B, Ms, a secretor, and A1B2C1D2/A4B3C4D3. Her child is group O, MSs, a nonsecretor, and A2B1C2D1/A4B5C6D7. (a) Could he be the father? (b) On which test(s) is the correct answer based?

7-19 Arrange the following relatives in order of decreasing suitability as kidney donors in a contemplated transplant operation: uncle, parent, dizygotic twin, sibling, monozygotic twin, first cousin.

7-20 In the following case consider that the HLA loci are sufficiently closely linked that crossing-over can be eliminated as a possibility. With regard to the HLA-A and HLA-B loci, a woman's *genotype* is A2B5/A9B12. She and her husband have three children whose *phenotypes* are (1) A1A2 B5B8; (2) A1A9 B8B12; (3) A2A3 B5B7. What is the father's *genotype*?

7-21 The HLA-A and HLA-C loci have been mapped 0.6cM apart. If an individual's genotype for these two loci is A6B12/A9B5, what would be the expected frequency of A6B5 gametes per thousand produced by that person?

CHAPTER 8

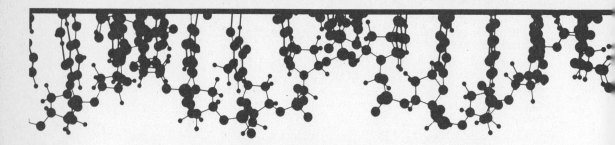

Pseudoalleles and the Rh blood factor

Mid-twentieth-century research has uncovered evidence in a number of different organisms that many traits, originally thought to be governed by multiple-allele series, are due to separate but extremely closely linked loci. Some of the earliest and best-known of these cases are found in *Drosophila*. An important trait in humans, the Rh blood factor, can be explained on the basis of at least three such closely linked loci. Both the Rh factor and some early illustrations from *Drosophila* will be examined in this chapter; from such examples a new concept of the nature of the gene emerges.

DROSOPHILA

Eye color. In *Drosophila* the wild-type eye is red; this phenotype is produced by a completely dominant gene at about locus 1.5 on chromosome I (the X chromosome; see also Chapter 11). Many other eye colors occur, including coral, cherry, apricot, eosin, ivory and white. Early work suggested that genes for these and some other eye colors formed a multiple-allele series, and all mapped at the same locus. Under this hypothesis, $+$ (or w^+) designated the gene for wild-type red; w^a, apricot; and w, white. As data were accumulated for crosses where progeny were expected to be of only two phenotypic classes, apricot and white, an occasional wild-type red appeared. The frequency of such red-eyed flies was low, i.e., less than one per thousand, but too high to be explained by mutation. By using the "marker genes" y (yellow body) and *spl* (split thoracic bristles) at loci 0.0 and 3.0, respectively, in the heterozygous condition, it was found that genes for apricot and white were at different, although very close, loci and thus could not be part of a multiple-allele series. The symbol for apricot was therefore changed to *apr*. From this work it was soon determined that (1) *apr* +/+ *w* flies had apricot eyes, (2) + +/*apr w* flies had red eyes, and (3) linkage relationships of *apr* and *w* did not change unless *y* and *spl* did also. Thus + + + +/*y apr w spl* flies were red-eyed, whereas + + *w spl*/*y apr* + + flies, for example, had apricot eyes, demonstrating separate loci for *apr* and *w*. Frequency of crossing-over between *apr* and *w* was found to be less than 0.001. Loci *apr* and *w* are thus *functionally allelic* in that they both affect the same general phenotypic character (eye color). They are, however, *structurally non-allelic*, that is, they can be shown by recombination to occupy different loci. In the *cis* configuration, the red-eye phenotype is produced, but the *trans* configuration results in a different phenotype, apricot. The two loci exhibit **complementation** in the *cis* position, but **noncomplementation** in the *trans* position. This is referred to as a *cis-trans* position effect. If the *cis* and *trans* configurations produce different phenotypes, the two forms are parts of the same gene; but should the *cis* and *trans* configurations produce identical phenotypes, the two are considered to be different genes (Fig. 8-1).

Genes that (1) govern different expressions of the same trait, (2) map at different but closely adjacent loci, (3) show a low frequency of recombination by crossing-over, and (4) exhibit the *cis-trans* position effect are called **pseu-**

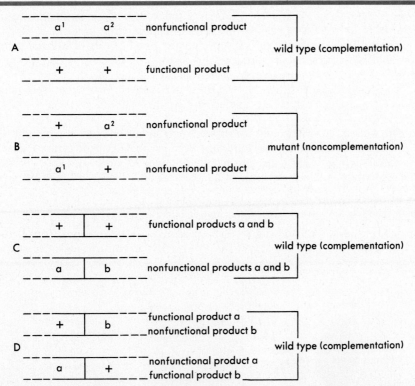

Figure 8-1. Cis-trans effect. A functional product is produced only when the two loci are in the cis position (A); in the trans position no functional product is produced (B). Loci a^1 and a^2 exhibit complementation only in the cis configuration and are functionally allelic but structurally nonallelic. In (C) and (D) loci a and b exhibit complementation in both the cis and trans positions and are, therefore, considered to be different genes. See text for further details.

doalleles. A cluster of pseudoalleles is referred to as a *complex locus*. Pseudoallelism suggests that the gene cannot be defined as a unit of *both* function and structure or, to put it another way, the structural gene may be smaller than the functional gene. Evidently some product of the functional gene must be produced *in toto* in the *cis* configuration, whereas the separate parts of this product, produced in the *trans* configuration, cannot determine a normally functioning substance. The classical gene may therefore be subdivisible, consisting of many intragenic units of recombination and mutation. This concept and the light it sheds on the nature and operation of the gene will be more fully explored after we have examined the molecular nature of the genetic material.

The lozenge loci. Study of complex loci began in 1940 with work on the so-called lozenge locus in *Drosophila*. In this organism the wild-type eye is broadly oval in shape (Fig. 8-2); in the recessive phenotype lozenge the eyes are narrowly ovoid with an irregular surface. The genes responsible are on chromosome I, mapping at about locus 27.7. Subsequent work showed four separate, closely linked loci (Fig. 8-3). These loci exhibit the *cis-trans* effect as in the preceding case. For example, the *cis* heterozygote $lz^8 \ lz^k \ + \ +/+ \ + \ + \ +$ has the wild-type phenotype, but the *trans* heterozygote $lz^8 \ + \ + \ +/+ \ lz^k \ + \ +$ is lozenge. Altogether more than a dozen alleles are known to occur at these four loci.

HUMAN BEINGS: THE Rh FACTOR

History. The now well-known Rh factor was discovered in 1940 by Landsteiner and Wiener, who reported that if a rabbit was injected with blood of the *Macaca rhesus* monkey, antibodies were formed by the rabbit. These antibodies would agglutinate the red blood cells of all rhesus monkeys. Thus surfaces of monkey erythrocytes bear a specific antigen, designated as Rh. Tests of human beings show that most persons also produce this antigen; in fact, some 85 percent of white Americans and more than 91 percent of black Americans do.[1] Such persons are designated as Rh-positive (Rh +); the much smaller percentage who do not produce the rhesus antigen are termed Rh-negative (Rh −).

Genetic bases. Evidence almost immediately indicated a genetic basis to the Rh phenotype, with Rh + the dominant trait. A single pair of genes, *R* and *r*, was postulated, with Rh + persons having genotype *RR* or *Rr* and Rh − persons having genotype *rr*. Since the work of Landsteiner and Wiener, more Rh antigens

---

[1]Reasons for this kind of disparity are explored in Chapter 15. Actually, the rabbit and the human anti-Rh antibodies are not the same, a fact not important for our purposes.

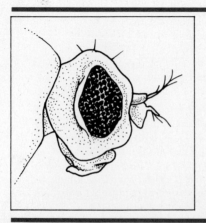

Figure 8-2. *Wild-type eye of* Drosophila melanogaster.

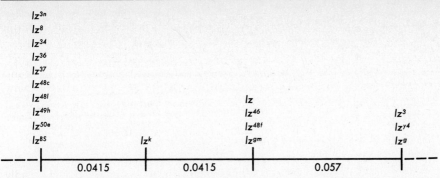

Figure 8-3. *The lozenge loci of* Drosophila melanogaster. *The horizontal line represents a portion of chromosome I, and the numbers below it show recombination frequencies. See text for details.*

have been discovered; the number is now over 30, and the genetics is much more complex than originally believed. Summaries of the complexity of the Rh blood groups are well represented in the literature. Two major explanatory theories have been developed, one by the American investigator Wiener, and a second, in England, by Fisher and others.

The **Wiener** system postulates a series of at least 10 multiple alleles (Table 8-1), some common, others very rare, at a single locus. As noted in Chapter 6, this locus has since been located on chromosome 1. Each, except the completely recessive gene r, is responsible for production of one or more of the Rh antigens.

The **Fisher** system, as elaborated by Race and Sanger (1968) and others, proposes a group of at least three[2] very closely linked pseudoalleles, D, d, C, c, E, and e. Evidence supports the sequence D-C-E. Instances of compound antigens (e.g., ce), a single reported case of possible crossing-over (Steinberg, 1965), and some apparent cases of position effect are more readily explained by the pseudoallele concept. For example, Race and Sanger point out that the *cis* heterozygote *Dce/DCE* produces compound antigen ce (among others), but in the *trans* configuration (e.g., *DCe/DcE*) this antigen is not produced.[3] Loci c and e thus illustrate the *cis-trans* position effect and are noncomplementary.

The case reported by Steinberg is interesting in that it can be explained either by recombination between closely linked pseudoalleles or by mutation within a single site. In this instance, the husband was determined to have genotype *DCe/dce* and the wife *dce/dce*. There were eight children, four *dce/dce*, three *DCe/dce*, and one *dCe/dce*. Steinberg was able to eliminate the possibility of illegitimacy of the *dCe/dce* child. Because evidence has become increasingly clear that the correct sequence of loci under the pseudoallele concept must be *D-C-E*, as first suggested by Fisher, birth of this child must have been preceded by a crossover in the father.

Although difficulties are inherent in both the Wiener and Fisher systems, each also has its advantages, and both are in current use. A final judgment between the two cannot yet be made. A comparison of the two systems is shown in Table 8-1.

In the Wiener system each gene listed is reponsible for formation of more than one antigen. Responsibility of a single gene for more than one phenotypic effect, often seemingly unrelated, is known as **pleiotropy.**

Antigen D (= Rh_0) is the one that most commonly causes problems in trans-

---

[2]As new antigens and antibodies are discovered, the Fisher system has had to be expanded to include at least two new gene pairs, F, f and V, v. These are not included here.

[3]Antigen ce is sometimes designated as antigen f.

Table 8-1. A comparison of the Wiener and Fisher Rh gene systems and their antigens

Genes			Antigens Rh_0	rh'	rh''	hr'	hr''	rh^w	hr	rh			(Wiener)
Wiener	Fisher	Frequency*	D	C	E	c	e	C^w	ce	Ce	CE	cE	(Fisher)
r	dce	0.385	−	−	−	+	+	−	+	−	−	−	
r'	dCe	0.007	−	+	−	−	+	−	−	+	−	−	
r''	dcE	0.007	−	−	+	+	−	−	−	−	−	+	
r^y	dCE	rare	−	+	+	−	−	−	−	−	+	−	
R^0	Dce	0.022	+	−	−	+	+	−	+	−	−	−	
R^1	DCe	0.405	+	+	−	−	+	−	−	+	−	−	
R^2	DcE	0.154	+	−	+	+	−	−	−	−	−	+	
R^z	DCE	0.002	+	+	+	−	−	−	−	−	+	−	
R^{1w}	DC^we	0.016	+	+	−	−	+	+	−	+	−	−	
$R^{0''}$	D^uce	rare	weak	−	−	+	+	−	+	−	−	−	

*Approximate, in white populations of western European affinities.
In this table a plus sign indicates production of a given antigen as well as agglutination of red cells by the corresponding antiserum. Some very rare genes are not included here.
Current practice designates as Rh+ any genotype that includes ability to produce antigen D (Rh_0); all others designated Rh−.
Antigen C^w was first described by Callender and Race (1946).

fusions and in certain pregnancies (see next section); therefore, under ordinary circumstances, persons are typed as Rh + if their erythrocytes are agglutinated by anti-D (anti-Rh_0) antiserum. Thus an Rh − person (e.g., of genotype dce/dce, dCE/dce, and so on) does not produce antigen D (Rh_0), but does produce other Rh antigens, as shown in Table 8-1. Some interesting results of Rh typing can usually be observed when, say, a genetics class is tested. Erythrocytes of some persons are strongly agglutinated by anti-D antiserum (the one most commonly used, and alone, for this purpose), whereas blood of some other persons reacts much more weakly against anti-D. For example, red blood cells of persons of genotype DCe/dce (R^1/r) are strongly agglutinated by anti-D, but cells of persons having genotype DCe/dCe (R^1/r') react only weakly. The reason for this sort of difference is unknown; however, under the Fisher system, it can be classed as a position effect. The strong reaction given by DCe/dce could, then, be due to the fact that d and c are adjacent; the weaker reaction of DCe/dCe would result from d being next to C. It is not clear how such a position effect operates, however, although such evidence would support the Fisher concept. No demonstrable antigen is produced by gene d.

Some of the commoner genotypes for samples from the American population are listed in Table 8-2. Heiken and Rasmuson (1966) have published an extensive list of genotypic frequencies for a large sample of Swedish children (Table 8-3).

Table 8-2. Rh phenotypes and genotypes with percent frequencies in the American population

		Genotypes		$n = 135$	$n = 105$	$n = 766$
Phenotype		Wiener	Fisher	Black	Oklahoma Indian	White
Rh+		R^0/r	Dce/dce	45.9	2.9	2.2
Rh+		R^1/R^1	DCe/DCe	0.9	34.3	20.9
Rh+		R^1/r	DCe/dce	22.8	5.7	33.8
Rh+		R^2/R^2	DcE/DcE	16.3	17.1	14.9
Rh+		R^1/R^2	DCe/DcE	4.4	36.2	13.9
Rh+		R^1/R^z	DCe/DCE	0.0	2.9	0.1
Rh−		r/r	dce/dce	9.6	0.9	13.9

Table 8-3. Rh phenotypes, genotypes, and percent frequencies in 8,297 Swedish children

| Phenotype | Representative probable genotype | | Percent of sample |
	Wiener	Fisher	
Rh +	R^0/r	Dce/dce	1.48
Rh +	R^2/R^2	DcE/DcE	3.08
Rh +	R^2/r	DcE/dce	12.50
Rh +	R^1/r	DCe/dce	32.72
Rh +	R^{1w}/r	DC^we/dce	1.45
Rh +	R^1/R^2	DCe/DcE	14.46
Rh +	R^z/R^2	DCE/DcE	0.05
Rh +	R^{1w}/R^2	DC^we/DcE	0.66
Rh +	R^1/R^1	DCe/DCe	16.16
Rh +	R^{1w}/R^z	DC^we/DCE	1.86
Rh +	R^z/R^1	DCE/DCe	0.05
		Total Rh +	84.47
Rh −	r/r	dce/dce	14.90
Rh −	r″/r	dcE/dce	0.22
Rh −	r'^w/r	dC^we/dce	0.02
Rh −	r′/r	dCe/dce	0.39
		Total Rh −	15.53

Based on work of Heiken and Rasmuson, 1966.

Erythroblastosis fetalis.   No cases are known of persons whose blood naturally contains anti-Rh antibodies, though Rh − individuals can and do develop them if exposed to the corresponding Rh antigen. Such exposure can occur by transfusion, and this, as noted earlier, is the reason the Rh type (ordinarily for production or nonproduction of the D, or Rh_0, antigen only) is now routinely determined for blood donors and recipients. But anti-Rh antibody development can also occur in certain pregnancies, often resulting in a fetal condition known as **erythroblastosis.** This is a hemolytic anemia, often accompanied by jaundice, as liver capillaries become clogged with red blood cell remains and bile is absorbed by the blood. The damaged erythrocytes are imperfect oxygen carriers and resemble the immature cells of the marrow, where they are formed. Death may occur before birth or soon after unless appropriate corrective measures are taken in time.

This disorder occurs only when a number of coincident conditions are met. The mother must be Rh −, the fetus Rh +; therefore only marriages of Rh − women and Rh + men are involved. There must also be a placental defect whereby fetal blood, the red cell surfaces of which carry the Rh antigen, passes from the embryo into the maternal circulation. This occurs largely just before or during birth. As a result, Rh antibody concentration is gradually built up in the mother and she becomes *sensitized.* In a second or subsequent pregnancy involving an Rh + child, these antibodies may return to the fetus, where they destroy the antigen-carrying red cells. The buildup of antibodies in the mother is gradual; moreover, she is sensitized only at or just before birth of her first R + child (unless, of course, she previously received an Rh + transfusion—not a likely event any longer).

Although fetal erythroblastosis is a severe and tragic condition, occurrence is, fortunately, relatively infrequent. One study in Chicago showed only 92 affected children in 22,742 births in a seven-year period. This is a frequency of 0.004, a fairly average figure for erythroblastosis resulting from Rh incompatibility. The expected frequency of erythroblastotic births, in the absence of any complicating factors, can be calculated from the frequencies of genes D and d, following the methods introduced in Chapter 5. Disregarding loci C and E, Rh$-$ persons are of genotype dd, and in the white American population occur with a frequency of about 0.15. This latter figure is the value of b^2 in the expansion of $(a + b)^2$. The frequency of the recessive gene d is then equal to $\sqrt{b^2}$ or $\sqrt{0.15} = 0.39$. Because the example here deals with only one pair of alleles, $(a + b)^2 = 1$, and $a = 1 - b$, or 0.61, which is, then, the frequency of gene D. Only two types of marriages, $DD\male \times dd\female$ and $Dd\male \times dd\female$, can result in Rh$+$ children by Rh$-$ women.[4] Frequencies of these marriages, based on the expectation of random mating, would be

$$DD\male \times dd\female: \quad a^2 \times b^2 = 0.61^2 \times 0.39^2 \qquad\qquad = 0.0566$$
$$Dd\male \times dd\female: 2ab \times b^2 = 2(0.61 \times 0.39) \times 0.39^2 = 0.0723$$

In the $DD\male \times dd\female$ marriages, all the children are Rh$+$, whereas in the second the probability of an Rh$+$ child is 0.5; therefore the expected frequency of all erythroblastotic births (if all pregnancies come to term) is $0.0566 + 0.0361 = 0.0927$, or nearly 10 percent of all pregnancies, neglecting the fact that in most cases the first child is unaffected. The 10 percent figure is far greater than the observed frequencies of most studies.

The relative infrequency of erythroblastosis is in part due to other probably genetically based traits, such as the defective placenta. But another important factor is *ABO incompatibility*. ABO-compatible marriages are those in which the husband is of suitable ABO group to donate blood to this wife; incompatible marriages are the reverse, namely, those in which the husband is not of an ABO group suitable for transfusion to his wife; that is,

	$\female$	$\male$
Compatible	A	A, O
	AB	A, B, AB, O
	B	B, O
	O	O
Incompatible	A	B, AB
	B	A, AB
	O	A, B, AB

Some studies show far fewer Rh-erythroblastotic children are born to ABO-incompatible marriages than to compatible ones. In ABO-incompatible marriages fetal erythrocytes, which bear an antigen for which the mother's blood contains the corresponding antibody and which cross the placenta, will be quickly destroyed before anti-Rh antibody formation can occur.

The discovery of two types of anti-D (anti-Rh$_0$) antibodies almost simultaneously in 1944 by Wiener and by Race has, fortunately, provided a simple preventive measure for Rh-erythroblastosis so that the condition need no longer be a cause of infant death or anemia. These two types of anti-D antibody are (1) *complete*, which agglutinates red cells carrying antigen D, and (2) *incomplete*, which does not. The incomplete antibody, however, attaches to receptor sites on Rh$+$ erythrocytes, "sensitizes" them and prevents them from acting antigenically. Intramuscular injection of human D (Rh$_0$) immune globulin, containing incomplete anti-D antibody, into Rh$-$ women within 72 hours after giving birth to a $D-$ or D^u- child effectively blocks the child's red blood cells already in the mother's circulatory system from inducing antibody production on her part. The incomplete antibody is often obtained from Rh$-$ males who have been injected with Rh$+$ blood.

---

[4]The rare gene D^u is also implicated, but is not included in these calculations because of its very low frequency.

Should a chromosome sustain two or more breaks with repair deferred until after replication, a repeated internal segment may result. Later these initially identical segments, perhaps consisting of a single gene each, might become functionally somewhat different through mutation and so constitute pseudoalleles. Other explanations are, of course, also possible. Mutation in one or both of two closely adjacent genes that produce different effects could result in two that perform similar functions. On the other hand, rearrangements of chromosomal segments following breaks might bring together in the same chromosome genes for similar products, i.e., genes originally located in different chromosomes.

References

Bach, F. H., 1976. Genetics of Transplantation: The Major Histocompatibility Complex. In H. L. Roman, A. Campbell, and L. M. Sandler, eds. *Annual Review of Genetics*, vol. 10. Palo Alto, Calif.: Annual Reviews, Inc.

Callender, S. T., and R. R. Race, 1946. A Serological and Genetic Study of Multiple Antibodies Formed in Response to Blood Transfusion by a Patient with Lupus Erythematosus Diffusus. *Ann. Eugen.*, **13**: 103–117.

Fisher, R. A., 1947. The Rhesus Factor: A Study in Scientific Method. *Am. Scien.*, 35: 95–103.

Green, M. M., 1961. Phenogenetics of the Lozenge Locus in *Drosophila melanogaster*. II. Genetics of Lozenge-Krivshenko (lz^k). *Genetics*, **46**: 1169–1176.

Green, M. M., 1963. Pseudoalleles and Recombination in *Drosophila*. In W. J. Burdette, ed. *Methodology in Basic Genetics*. San Francisco: Holden-Day.

Green, M. M., and K. C. Green, 1949. Crossing Over Between Alleles at the Lozenge Locus in *Drosophila melanogaster*. *Proc. Nat. Acad. Sci., (U.S.)*, **35**: 586–591.

Heiken, A., and M. Rasmuson, 1966. Genetical Studies on the Rh Blood Group System. *Hereditas Lund*, **55**: 192–212.

Landsteiner, K., and A. S. Wiener, 1940. An Agglutinable Factor in Human Blood Recognized by Immune Sera from Rhesus Blood. *Proc. Soc. Exp. Biol. Med. N.Y.*, **43**: 223.

Lewis, E. B., 1952. The Pseudoallelism of White and Apricot in *Drosophila melanogaster*. *Proc. Nat. Acad. Sci. (U.S.)*, **38**: 953–961.

Lewis, E. B., 1955. Some Aspects of Pseudoalleles. *Amer. Natur.*, 89: 73.

Steinberg, A. G., 1965. Evidence for a Mutation or Crossing-over at the Rh Locus. *Vox Sang.*, **10**: 721.

Wiener, A. S., 1944. A New Test (Blocking Test) for Rh Sensitization. *Proc. Soc. Exp. Biol. Med. N.Y.*, **56**: 173–176.

Problems

8-1 What is the phenotype of each of the following genotypes in *Drosophila melanogaster*: (a) lz^{BS} + + +/+ lz^k + +; (b) lz^{BS} lz^k + +/+ + + +?

8-2 Consider the following genotypes and phenotypes in a hypothetical diploid organism:

Genotype	Phenotype
+ +/ab	wild
+ b/a +	wild
+ +/cd	wild
+ d/c +	mutant

Do any of these pairs of alleles appear to represent pseudoalleles? Explain.

8-3 Recall that in erythroblastosis the important genes are D and d and that, therefore, Rh+ persons are of either genotype DD or Dd, and that Rh− persons are dd. In the following pedigree the marriage shown is the first for both the man and the woman, and the woman has never had a blood transfusion, nor has she ever received an injection of D immune globulin. Solid symbols represent cases of erythroblastosis fetalis. After examining the pedigree, give the Rh genotype of each member of the family.

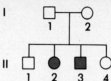

8-4 In the preceding problem why can you be sure of II-1's genotype and phenotype?

8-5 How should the pedigree of problem 8-3 be different had the mother received an injection of D-immune globulin after the births of II-1 and II-2?

8-6 In which of the following marriages is the risk of erythroblastosis fetalis greater: A $-$ ♀ × O + ♂, or O $-$ ♀ × A + ♂?

8-7 Considering human blood group A to include three subtypes, and groups B and O to include one each, how many phenotypes can be recognized if the A-B-O classes and the two Rh classes are all taken into account?

8-8 How many phenotypes are possible if A-B-O, Rh, and MN classes are all considered?

8-9 How many phenotypic classes are possible if the A-B-O, Rh, MN, Ss, and Hh phenotypes are considered together?

8-10 A woman of group A_2 charges that a man of group A_2 is the father of her group A_3Rh+ child. Could he be?

8-11 Both the man and the woman of the preceding problem are Rh $-$. Does this change your judgment?

8-12 A couple believes they have brought the wrong baby home from the hospital. The wife is O +, her husband B +, and the child O $-$. Could the child be theirs?

8-13 An additional test discloses the husband and wife of the preceding problem to be of blood type M, whereas the child is MN. Does this added fact change your judgment?

8-14 An elderly couple is killed in an accident; no survivors are known. Their estate has been willed to charity. Later a man claims the estate on the grounds that he is their son who left home at an early age. It is known by friends that this couple had had a son, but he was thought to have died while quite young. Unfortunately, birth records for the period have been destroyed in a courthouse fire. Their child was born at home, the attending physician has been dead for some time, and his records are no longer available. From various hospital and medical records it is determined that the dead man was of blood type A, MS/Ns, Rh+; his wife was B, MS/NS, Rh $-$. The claimant's blood tests as O, MS/Ns, Rh+. Does it appear that his claim is valid? Justify your conclusion.

8-15 In a court action a man claims that some of the six children purportedly belonging to him and his wife are not his children. Blood tests of the husband, wife, and six children gave the following information:

Husband:	O,	*Dce/DcE,*	*MS/Ms*
Wife:	A_1,	*DcE/dce,*	*MS/Ns*
Child 1:	A_1,	*DcE/DcE,*	*MS/MS*
Child 2:	O,	*Dce/dce,*	*MS/Ns*
Child 3:	O,	*DcE/dce,*	*Ms/Ns*
Child 4:	A_1,	*DcE/dce,*	*NS/Ns*
Child 5:	O,	*dce/dce,*	*MS/NS*
Child 6:	A_1B,	*DCe/DcE,*	*MS/NS*

Assuming all were, in fact, born to the wife, could all these children belong to the husband? Explain.

8-16 A student in a genetics class is blood typed as B+. She reports that her father is B$-$, her mother O+. Using standard gene symbols I^A, I^B, and i for the A-B-O system and restricting Rh type to the *D, d* gene pair, give this student's *genotype* for blood group and Rh type.

CHAPTER 9

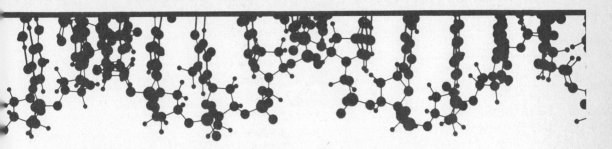

Polygenic inheritance

Of the heritable traits examined thus far in text and problems, recall that phenotypic classes have always been *discontinuous*. Such traits may be termed *qualitative* ones. Thus, *Coleus* leaves have either regular or irregular venation; cattle may or may not have horns, and may be red, roan, or white; rabbits may be distinguished by coat color, as may a host of other animals; people belong to one blood group or another, and so on. This has been the case whether the traits involved form and structure, pigments (i.e., an aspect of physiology), antigens and antibodies, and so on; and whether the genes involved show complete dominance, incomplete dominance, or codominance.

Not all inherited traits are expressed in this discontinuous fashion, however. In humans for example, height is a genetically determined trait. But if an attempt were made to classify a random sample of students on campus according to height, results would quickly disclose a trait showing essentially continuous phenotypic variation. It may be asked, then, where does one draw the line between phenotypic classes? Should these classes be one inch apart, one-half inch, one-tenth inch, a millimeter, etc.? Many other traits are expressed in a similar fashion, including intelligence, skin and eye color in people, color and food yield in various plants, size in many plants and animals, as well as degree of coat spotting in animals or of seed coat mottling in some plants. The essential difference between continuous and discontinuous inheritance is illustrated with generalized data from crosses of each type compared graphically in Figure 9-1. Clearly, concepts of simple Mendelian inheritance, with the reappearance of the parental phenotypes as distinct and separate classes in the F_1 and F_2 generations, must be modified in order to explain continuous variation in *quantitative* characters, such as height, weight, intelligence, or color.

Concern is not with just tall *versus* dwarf, but with *how* tall, that is, with *continuous characters of degree rather than discontinuous characters of kind.*

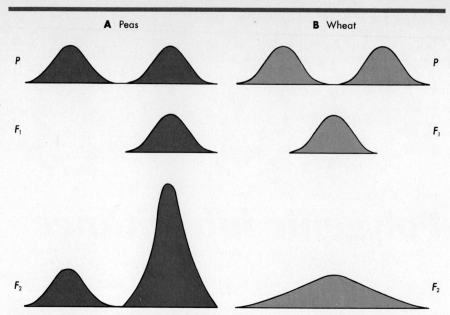

Figure 9-1. *Curves comparing results of crosses involving (A) height, a discontinuous trait in peas and (B) kernel color, a continuous trait in wheat, followed for three generations. Ordinates represent the number of individuals, abscissas the particular quantitative trait.*

Moreover, quantitative inheritance more often deals with a population in which all possible matings occur, and less often with individual matings. This kind of study involves quantification of data and statistical treatment such as means, variances, standard deviations, and others in order to evaluate data properly.

Kernel color in wheat

Among the earliest investigations that gave a significant clue to the mechanism of quantitative inheritance was the work of Nilsson-Ehle (1909) with wheat. One of his crosses consisted of a red-kerneled variety × a white-kerneled strain. Grain from the F_1 was uniformly red, but of a shade intermediate between the red and white of the parental generation. This might suggest incomplete dominance, but by intercrossing members of the F_1, Nilsson-Ehle produced an F_2 in which five phenotypic classes occurred in a 1:4:6:4:1 ratio. Noting that $\frac{1}{16}$ of the F_2 was as extreme in color as either of the parental plants (i.e., as red or as white as ·the P individuals), he theorized that two pairs of genes controlling production of red pigment, were operating in this cross. The key, of course, is in the fraction $(\frac{1}{16})$ that were extreme as either of the original parental strains. Symbolizing the genes for red with the capital letters A and B and their alleles responsible for lack of pigment by a and b, this cross can be diagramed as follows:

$$P \quad AABB \quad \times aabb$$
$$\text{dark red} \quad \text{white}$$
$$F_1 \quad AaBb \quad (\times AaBb)$$
$$F_2 \quad \tfrac{1}{16} AABB + \tfrac{2}{16} AaBB + \tfrac{1}{16} aaBB$$
$$+ \tfrac{2}{16} AABb + \tfrac{4}{16} AaBb + \tfrac{2}{16} aaBb$$
$$+ \tfrac{1}{16} AAbb + \tfrac{2}{16} Aabb + \tfrac{1}{16} aabb$$

Assuming each "dose" of a gene for pigment production increases the depth of color, this F_2 can be sorted phenotypically according to the number of genes for red in this way:

Genotype	Number of genes for red	Phenotype	Fraction of F_2
AABB	4	dark red	$\frac{1}{16}$
AABb, AaBB	3	medium red	$\frac{4}{16}$
AAbb, aaBB, AaBb	2	intermediate red	$\frac{6}{16}$
aaBb, Aabb	1	light red	$\frac{4}{16}$
aabb	0	white	$\frac{1}{16}$

Genes symbolized by capital letters, those "*contributing*" to red color in this case, are termed **contributing alleles.** Those that do not "contribute" to red color (here symbolized by lower case letters) may be designated **noncontributing alleles.** Some geneticists refer to these as *effective* and *noneffective* alleles, respectively. Here, then, is a polygene series of as many as four contributing alleles. The term *polygene* was introduced by Mather (1954), who has summarized the modern interpretation of quantitative inheritance. This term has since found wide usage and is supplanting the older term *multiple gene.*

In an effort to determine whether the mechanism of polygene inheritance is the same as that operating in instances of qualitative characters, several simplifying assumptions must be made. *Assume* that

1. There is no dominance; rather, there exist pairs of contributing and noncontributing alleles.
2. Each contributing allele in the series produces an equal effect.
3. Effects of each contributing allele are cumulative or additive.
4. There is no epistasis among genes at different loci.
5. There is no linkage.
6. Environmental effects are absent or are so controlled that they may be ignored.

Certainly, as the number of pairs of genes increases, the probability of linkage rises, and the effect of environment can be ignored only in the most closely controlled experiments. Furthermore, some geneticists have found evidence suggesting less than universal operation of the first and/or second of these assumptions. Many polygene effects however do appear to operate in a manner consistent with the first four points, and the task is eased considerably if all six *simplifying assumptions* are made.

Calculating the number of polygenes

In another cross in wheat, reported by Nilsson-Ehle, a different red variety was used, with the result that 1/64 of the F_2 was as extreme as either parent, with seven classes in a 1:6:15:20:15:6:1 ratio. A little reflection will serve to suggest the operation here of *three* pairs of genes. From earlier chapters on monohybrid inheritance, recall that, if only one pair of genes were involved, one fourth of the F_2 should be as extreme as either parent. If information for one, two, and three pairs of polygenes is tabulated, a pattern emerges:

Number of pairs of polygenes in which two parents differ	Fraction of F_2 like either parent	Number of genotypic F_2 classes	Number of phenotypic F_2 classes	F_2 phenotypic ratio = coefficients of
1	$\frac{1}{4}$	3	3	$(a + b)^2$
2	$\frac{1}{16}$	9	5	$(a + b)^4$
3	$\frac{1}{64}$	27	7	$(a + b)^6$
n	$(\frac{1}{4})^n$	3^n	$2n + 1$	$(a + b)^{2n}$

Thus, with four pairs of polygenes, 1/256 of the F_2 is as extreme as either parent; with five, only 1/1,024; with 10, the fraction drops to 1/1,048,576; and with 20 pairs only one in 1,099,511,627,776 of the F_2 will have measurements like one parent or the other! The number of genotypic classes increases, of course, with startling rapidity as the number of pairs of polygenes becomes larger: for four pairs of genes there are 81 F_2 classes; five pairs of genes produce 243 F_2 genotypes; 10 pairs, 59,049; and 20 pairs, 3,486,784,401! Thus as the number of polygenes governing a particular trait goes up, the progeny very quickly form a continuum of variation in which class distinctions become virtually impossible to make.

To calculate the number of contributing alleles instead of the number of *pairs* of polygenes, use the expression $(\frac{1}{2})^n$. Dividing the total quantitative difference by the number of contributing alleles, of course, indicates the amount contributed by each effective allele. For example, in the following hypothetical case in pumpkin, how many contributing alleles are operating and how much does each contribute?

P   5-lb fruits $\times$ 21-lb fruits
F_1 13-lb fruits
F_2 $\frac{3}{750}$ 5-lb fruits . . . $\frac{3}{750}$ 21-lb fruits

Note that 3/750, the fraction of the F_2 that has fruits as light or as heavy as the P generation, simplifies to 1/250. In terms of the formula $(\frac{1}{4})^n$, note that $(\frac{1}{4})^n = 1/256$ if $n = 4$ (*pairs* of genes), and 1/250 is close only to this fraction $(\frac{1}{256})$ in the series of $(\frac{1}{4})^n$. Therefore, four pairs of genes must be involved, where the plants producing the heaviest fruits have all eight contributing alleles. Since the total weight difference is 16 pounds $(21 - 5)$, $\frac{16}{8} = 2$ pounds contributed by each effective allele. Alternatively, $(\frac{1}{2})^n$ could be used to determine directly the number of contributing alleles (instead of the number of *pairs*). In this case $(\frac{1}{2})^n = 1/256$ only if $n = 8$. In this example with weight of pumpkin fruits, a weight of 5 lb is termed the *base weight*, and suggests that of all the polygenes that may be involved in fruit weight in this case, the two parental strains were homozygous for all but the four pairs determined by these calculations.

It is also apparent that the number of F_2 phenotypic classes follows a pattern, but one which produces a less dramatic increase as the number of pairs of polygenes becomes larger. Thus with one pair, the F_2 of the cross $AA \times aa$ includes three phenotypic classes that correspond to the number of genotypic classes (AA, Aa, and aa). The two examples from Nilsson-Ehle's work on wheat indicate that the number of F_2 *phenotypic classes* is one more than twice the number of pairs of polygenes, or $2n + 1$.

Note that the *phenotypic ratio* also follows a pattern. One pair gives rise to a 1:2:1 F_2 ratio, two pairs to 1:4:6:4:1, and three to 1:6:15:20:15:6:1. These ratios are the same as the sequence of *coefficients* in binomials of a power equal to *twice* the number of pairs of multiple genes. Thus expansion of $(a + b)^2$ gives a coefficient sequence of 1:2:1, which is the same as the F_2 phenotypic ratio for one pair of polygenes; expanding $(a + b)^4$ produces the coefficient series 1:4:6:4:1, and so on. So an F_2 phenotypic ratio can be determined for any number of pairs of polygenes by thinking of the sequence of coefficients given by expanding a binomial raised to the power $2n$ where, again, n represents the number of pairs of polygenes.

HUMAN EYE COLOR

People clearly differ in eye color, that is, in the amount of melanin pigment in the iris. Except for albinos, no one is without some eye pigmentation. Those with the least pigment have eyes that appear blue, those with the greatest have eyes that appear brown. Blue eyes owe their color to the scattering of white light by the nearly colorless superficial cells of the iris. This effect is greatest in the shorter (blue) wavelengths of the visible spectrum, giving the iris its blue appearance. Close inspection of the apparently nonbrown iris of some persons discloses small to very small flecks of brown or somewhat orange coloration, because the small amount of pigment is most abundant in discrete groups of cells. In other persons the pigment is more evenly distributed, and the eyes appear uniformly blue.

Clearly, though, there is a gradation in eye color, ranging from the lightest blue to the darkest brown ("black"). Human beings simply do not fall into "either-or," blue or brown, categories, but rather, large samples form a continuum of variation, strongly indicative of polygenic inheritance. The number of phenotypic classes recognized is therefore arbitrary and depends in part on the observational techniques and equipment used, and in part on the observer. Although the inheritance of eye color is complex and only incompletely understood, at least nine phenotypic classes[1] may be recognized as a matter of convenience. In order of increasing amounts of melanin pigmentation, these can be designated as light blue, medium blue, dark blue, gray, green, hazel, light brown, medium brown, and dark brown. If one considers that the number of phenotypic classes is one more than twice the number of pairs of polygenes (page 162), nine classes would result from the action of four pairs of genes. Under this hypothesis, a simplified basis for human eye color would be as follows:

Number of contributing alleles in genotype	Eye color
0	Light blue
1	Medium blue
2	Dark blue
3	Gray
4	Green
5	Hazel
6	Light brown
7	Medium brown
8	Dark brown

Regardless of the number of pairs of polygenes postulated (and it must be emphasized that studies to date do not permit a definitive determination of the number of pairs), it is clear that eye color is due to polygenes, some of which may interact in poorly understood ways. No conclusive data are yet available on linkage of genes affecting this trait, although Brues (1946) and others have suggested X-linkage for some.

OTHER HUMAN TRAITS

Skin color in humans also depends on relative amounts of melanin. Studies by Davenport and Davenport (1910) and by later investigators, which attempt to relate sample frequencies of various degrees of pigmentation to models based on different numbers of pairs of polygenes, show best agreement between observation and theoretical expectation for four, five, or six pairs rather than for fewer or for more. As might be expected, some Caucasians in the samples were darker than some blacks and vice versa. Although polygene inheritance is clearly suggested in humans for many quantitative traits, such as height, intelligence, and hair color (except for red versus nonred), no complete hypothesis setting forth the exact number of pairs of genes and measurement of their individual and collective effect has yet been developed for any of these traits.

[1]However, Davenport (1913) designated five phenotypic classes and Hughes (1944) recognized seven.

Transgressive variation

Some progeny may be more extreme than either parent or grandparent. Recall examples in humans where, say, some children are shorter or taller than either parent or any of their more remote ancestors. The same phenomenon sometimes occurs, too, with respect to intelligence, skin color, and eye color. Such examples illustrate **transgressive variation.**

One of the earliest instances of transgressive variation to be reported in the literature was one described in 1914 and 1923 by Punnett and Bailey. They crossed the large Golden Hamburg chicken with the smaller Sebright Bantam. The F_1 was intermediate in weight between the parents and fairly uniform, but a few of the F_2 birds were heavier or lighter than either of the parental individuals. Their results suggested to Punnett and Bailey four pairs of genes, with the Golden Hamburg being of, say, genotype *AABBCCdd* and the Sebright Bantam being of genotype *aabbccDD*. Likewise, children may have darker or lighter eyes than either parents. Some of the problems at the end of this chapter deal with some interesting illustrations of transgressive variation.

Other organisms

Instances of polygene inheritance are known from many other plant and animal species. One of the more suggestive was discovered in tomato by Lindstrom (1924 and 1926). Crosses between the larger-fruited Golden Beauty (average fruit weight 166.5 g) and the smaller-fruited Red Cherry (average fruit weight 7.3 g) varieties produced an F_1 having fruits intermediate in size between the parental types, but distinctly closer to the smaller-fruited variety (average weight 23.9). Such results could be explained by assuming dominance or unequal effect among at least some of the noncontributing alleles.

In cattle both solid color and spotting occur. Solid color is due to a dominant gene, *S*, spotting to its recessive allele, *s*. Studies indicate that the degree of spotting in *ss* individuals depends on a rather large series of polygenes (Fig. 9-2). Those cattle that are *S* − will, of course, not be spotted, regardless of the remainder of the genotype with regard to spotting.

Chai (1970) has shown that the difference between high and low leukocyte count in the mouse is a polygenic trait. He suggests dominance, however, for low count and makes the interesting point that "the evidence is accumulating . . . that a quantitative trait is an aggregate of effects from different biological systems, each of which contributes specific effects that differ in magnitude and biological effect under the control of individual genes. The present results, although not considered as definitive, further indicate this to be the case."

Concluding statement

In summary, the basic mechanisms operative in quantitative inheritance appear to be the same as those for qualitative characters. Such studies as are reported here also further emphasize the fact that many traits are the result of *interaction* of one kind or another among several pairs of genes.

In the case of *quantitative characters* such as those dealt with here, the possible effect of environment must be considered and carefully regulated in any controlled experiment. For example, height in many plants (e.g., corn, tomato, pea, marigold, zinnia) is a genetically controlled character, but it is obvious that such environmental factors as soil fertility, texture, and water, the temperature, the duration and wavelength of incident light, the occurrence of parasites, to name just a few, also affect height. Or, with identical twins (who have identical genotypes) growing up in different kinds of environments, many classical studies

Figure 9-2. *Variation in degree of spotting in a herd of cattle. Amount of spotting depends upon a series of polygenes, but these are hypostatic to* S *(solid color).* (Photo courtesy Ayrshire Breeders' Association.)

indicate sufficiently different intelligence quotients such that environment surely played a significant role. In summary, *genotype* determines the range an individual will occupy with regard to a given quantitative character; environment determines the *point* within the genetically determined range at which an individual's measurement will fall.

Study and evaluation of polygene inheritance requires certain statistical treatments, particularly those that describe populations. These will be examined in the next chapter.

References

Brues, A. M., 1946. A Genetic Analysis of Human Eye Color. *American Journal of Physical Anthropology* (New Series), **4:** 1–36.

Chai, C. K., 1970. Genetic Basis of Leukocyte Production in Mice. *Jour. Hered.*, **61:** 61–71.

Davenport, C. B., 1913. Heredity of Skin Color in Negro-White Crosses. *Carnegie Institution of Washington*, Publication, **188:** 1–106.

Davenport, G. C., and C. B. Davenport, 1910. Heredity of Skin Pigmentation in Man. *Amer. Natur.*, **44:** 641–672.

Hughes, B. O., 1944. The Inheritance of Eye Color—Brown and Nonbrown. *Contributions from the Laboratory of Vertebrate Biology, University of Michigan*, No. **27:** 1–10.

Lindstrom, E. W., 1924. A Genetic Linkage Between Size and Color Factors in the Tomato. *Science*, **60:** 182–183.

Lindstrom, E. W., 1926. Hereditary Correlation of Size and Color Characters in Tomatoes. *Iowa Agr. Exper. Sta. Research Bull.*, **93.**

Lindstrom, E. W., 1929. Linkage of Qualitative and Quantitative Genes in Maize. *Amer. Natur.*, **63:** 317–327.

Mather, K., 1954. The Genetic Units of Continuous Variation. *Proc. IX International Cong. Genet.*, **Part I:** 106–123.

Nilsson-Ehle, H., 1909. *Lunds Univ. Arsskrift N.F. Avd.*, **2:** Bd. 5.

Punnett, R. C., 1923. *Heredity in Poultry.* New York: Macmillan (pp. 44 ff.).

9-1   Which of the following human phenotypes would appear to be based on polygene inheritance: intelligence, absence of incisors, height, **phenylketonuria** (inability to metabolize the amino acid phenylalanine), ability to taste phenylthiocarbamide, skin color, **cryptophthalmos** (failure of eyelids to separate in embryonic development), eye color?

9-2   Show, by means of appropriate genotypes, how parents may have children taller than themselves.

9-3   Suppose another race of wheat is discovered in which kernel color is determined to depend on the action of six pairs of polygenes. From the cross *AABBCCDDEEFF* × *aabbccddeeff*, (a) what fraction of the F_2 would be expected to be like either parent? (b) How many F_2 phenotypic classes result? (c) What fraction of the F_2 will possess any six contributing alleles?

9-4   Two races of corn, averaging 48 and 72 in. in height, respectively, are crossed. The F_1 is quite uniform, averaging 60 in. tall. Of 500 F_2 plants, two are as short as 48 in. and two as tall as 72 in. What is the number of polygenes involved, and how much does each contribute to height?

9-5   Two other varieties of corn, averaging 48 and 72 in. in height, respectively, are crossed. The F_1 is again quite uniform at an average height of 60 in. Out of 3,100 F_2 plants, four are as short as 36 in., and two as tall as 84 in. How many polygenes are involved, and how many inches of height are contributed by each effective allele?

9-6   In problems 6-17 to 6-20 it was pointed out that *Pl* − corn plants are purple, whereas *pl pl* individuals are green. Assume now that the F_1 in problem 9-5 are also *Pl pl*. A breeder wishes to recover a pure-breeding green, 84-in. variety from the cross given in problem 9-5. What fraction of the F_2 will satisfy his requirement?

9-7   Two 30-in. individuals of a hypothetical species of plant are crossed, producing progeny in the following ratio: one 22-in., eight 24-in., twenty-eight 26-in., fifty-six 28-in., seventy 30-in., fifty-six 32-in., twenty-eight 34-in., eight 36-in., and one 38-in. What are the genotypes of the parents? (Start with the first letter of the alphabet and use as many more letters as needed.)

9-8   Two different 30-in. plants of the same species are crossed and produce all 30-in. progeny. What parental genotypes are possible?

9-9   Mr. A has dark brown eyes; his wife's eyes are light blue. Based on the hypothesis that four pairs of polygenes (*A*, *a*; *B*, *b*; *C*, *c*; *D*, *d*) are responsible for human eye color, give the genotype of (a) Mr. A and (b) his wife. (c) Give the phenotype(s) of children they could have.

9-10   A daughter of Mr. and Mrs. A marries a man (Mr. B) having the same genotype as she. What is the probability that they could have a (a) dark-brown-eyed child, (b) dark-blue-eyed child, (c) hazel-eyed child?

9-11   (a) What eye color is most likely to occur in any child of Mr. and Mrs. B? (b) What is the probability of a child having this eye color?

9-12   The children referred to in problem 9-10 illustrate what phenomenon?

9-13   As noted in the text, spotting in certain breeds of cattle is dependent on interaction of *S*— (solid color) or *ss* (spotted) and a number of polygenes for degree of spotting. Assume four pairs of the latter (which is almost certainly too low), designated as *A*, *a*; *B*, *b*; *C*, *c*; *D*, *d*. If data are accumulated from enough *SsAaBbCcDd* × *SsAaBbCcDd* crosses to give a total of, say, 1,024 calves from such matings, how many of these should be unspotted?

9-14   Among *ss* animals, how many different degrees of spotting could be found, using information from problem 9-13?

9-15   Assume height in a particular plant to be determined by two pairs of unlinked polygenes, each effective allele contributing 5 cm to a base height of 5 cm. The cross *AABB* × *aabb* is made. (a) What are the heights of each parent? (b) What height is to be expected in the F_1 if there are no environmental effects? (c) What is the expected phenotypic ratio in the F_2?

9-16   If each pair of alleles in problem 9-15 exhibited complete dominance instead of an additive effect, (a) what are the heights of each parent? (b) What height would be expected in the F_1? (c) What is the expected phenotypic ratio in the F_2?

9-17   If you were dealing with a case of polygenic inheritance in corn, and determined that six pairs of polygenes were involved, what binomial would have to be expanded to arrive at the expected F_2 phenotypic ratio (assume the P generation to consist of two genotypes: (1) all homozygous contributing and (2) all homozygous noncontributing)?

CHAPTER 10

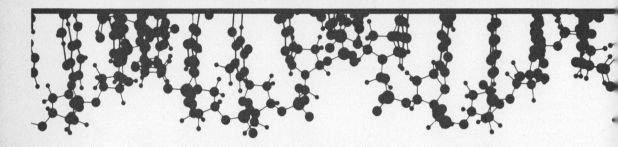

Statistical concepts and tools

In the preceding chapter we saw that polygenes govern continuously variable, quantitative traits. Analysis of this kind of inheritance requires application of certain techniques from the branch of mathematics called statistics. Statistics are useful in two fundamental problems commonly encountered in scientific research: (1) what can be learned about a population from measurements of a sample of it and (2) how much confidence can be placed in judgments about that population.

There is a clear difference between **population** and **sample.** A population consists of an infinite group of individuals, measured for some variable quantitative character. In this context the important aspect of a population is a series of numbers that represent such a variable, quantitative trait. Thus figures showing, for example, gains in weight of adult laboratory rats that have been fed a specific diet for a certain number of days, represent a biological population, as do height measurements of college males or females, or weights of ripe pumpkin fruits. The statistician is concerned with these figures and not with the rats, people and pumpkins themselves. A population is usually infinite in size, and often so large as to be theoretical. For example, height measurements of people or pumpkin fruits might include all individuals that have been or will be alive. Likewise, the rat population might well consist of an infinite number of measurements obtained from an infinite number of experimental rats. Such populations, of course, never actually exist. Even a more discrete problem dealing with a population of an endemic species on an isolated small island includes *all* the individuals of the species on that island. For obvious reasons, it is either impractical or impossible to accumulate measurements for such a group; rarely is the population sufficiently finite for *all* individuals to be measured.

So descriptions of the population must generally be formulated from *samples* of it. To be useful in making estimates of a population, the sample must have been drawn as randomly as possible. In dealing with heights or weights, for example, sample measurements must not be selected more from the taller, or shorter, or heavier, or lighter individuals, but should reflect the same kind and degree of variability as does the population. A large and randomly selected sample is therefore generally representative of the population.

Actual values for populations are constants called **parameters;** estimates of populations based on samples are **statistics.** The statistics are, of course, subject to some degree of chance error resulting from sampling practices, but once a statistic is determined, it is possible to state the range of the corresponding parameter with a particular degree of confidence. The geneticist needs to be able not only to estimate the parameters with which he is concerned, but also to determine how likely it is that he is dealing with individuals from either the same or different populations. In the genetic context, then, statistics may be said to provide

1. A concise description of the quantitative characteristics of the sample.
2. An estimate of
   a. The quantitative characteristics of the population from which the sample was drawn.
   b. How well the sample represents that population.
3. An expression of the probability that two samples differ significantly (or, conversely, only within limits set by chance alone) in terms of a particular theory concerning the reason for observed differences.

Basic statistics

Five principal statistics will provide the estimates and descriptions just cited.

1. THE MEAN

A very elementary statistic, and one with which you are undoubtedly familiar, is the average or **mean.** Calculation of the mean, symbolized by $\bar{x}$ ("x-bar"), may be represented by the formula

$$\bar{x} = \frac{\Sigma x}{n} \tag{1}$$

where Σ, the upper case Greek letter sigma, directs us to sum all following terms, x the individual measurements, and n the number of individuals in the sample. Often, when n is quite large, it becomes convenient to *group* data by *classes*. Thus, in computing the average grade on a quiz in a large class, it might be more practical to tally the number or *frequency* of individuals scoring between 96 and 100 in one class or group, those between 91 and 95 in another, and so on. A "class value" midway between the extremes of each class range is also entered. The tabulated data would then be arranged as follows:

Class range	Class value x	Frequency f	fx
96–100	98	1	98
91–95	93	4	372
86–90	88	8	704
81–85	83	12	996
76–80	78	18	1,404
71–75	73	25	1,825
66–70	68	17	1,156
61–65	63	10	630
etc.	etc.	etc.	etc.

If the data are grouped in this way, the mean will, of course, be given by

$$\bar{x} = \frac{\Sigma fx}{n} \tag{2}$$

Formula (2) loses a little in accuracy over formula (1), but the much simpler arithmetic of its method more than compensates for this.

Although the mean is a necessary statistic, it is a rather uninformative one in that a comparison of means of different samples reflects nothing of their spread, or *variability*. Consider as an example three students; the first has grades of 75, 75, 75; the second, 65, 75, 85; and the third, 50, 75 and 100. Obviously the mean for each is 75, but the distributions reflect quite different spreads. There is no variability in the first student's record and quite a bit in the last. Furthermore, the mean is greatly affected by a few extreme values.

2. VARIANCE

Variability of the population is measured by the **variance,** σ^2:

$$\sigma^2 = \frac{\Sigma(x - \mu)^2}{N} \tag{3}$$

where x represents each individual measurement in the population, μ the population mean, and N the number of individuals comprising the population. Of course, the population mean and the total number of individuals generally are not determinable because it is impractical or impossible to measure every individual in a population of infinite size. These difficulties are avoided by substituting sample values for those of the population:

$$s^2 = \frac{\Sigma(x - \bar{x})^2}{n} \tag{4}$$

But formula (4) is biased in the direction of underestimating σ^2 because, in using the *sample* mean, the number of *independent* measurements is $n - 1$. For example, the series of six values $4 + 3 + 6 + 2 + 1 + 2$ has a mean of 3, which is calculated in such a way that one value is fixed by the sum of the others. That is, given

$$\frac{4 + 3 + 6 + 2 + 1 + x}{6} = 3$$

the value of x can only be 2. To obviate the bias in formula (4), it is multiplied by the correction factor

$$\frac{n}{n - 1},$$

giving

$$s^2 = \frac{\Sigma(x - \bar{x})^2\, n}{n(n - 1)}$$

Cancelling n in numerator and denominator gives an unbiased estimate of the sample variance:

$$s^2 = \frac{\Sigma(x - \bar{x})^2}{n - 1} \tag{5}$$

If data are grouped by classes, as in our calculation of the mean (page 170), formula (5) becomes

$$s^2 = \frac{\Sigma f(x - \bar{x})^2}{n - 1} \tag{6}$$

The sample variance, s^2, provides an unbiased estimate of the population variance (σ^2). But variance is in *squared* units, and we do not express height in square feet, or weight variability in square pounds. This difficulty is resolved by the next statistic to be described.

3. STANDARD DEVIATION

To avoid expressing variability in squared units of measurement, we simply extract the square root of the variance. This statistic is the **standard deviation,** s, which is given by the formula

$$s = \sqrt{\frac{\Sigma f(x - \bar{x})^2}{n - 1}} \tag{7}$$

In effect, the standard deviation reflects the extent to which the mean represents the entire sample. If all individuals had exactly the same value, there would be no variability and the mean would represent the sample perfectly. Examination of formula (7) indicates that the standard deviation would then be *zero*. As the sample becomes more variable, the mean serves progressively less well as an index of the entire sample, and as departures from the mean, class by class, increase, so does the standard deviation. However, extracting the square root of the variance reintroduces bias. Nevertheless, the standard deviation has the necessary advantage of expression in the units of measurement, as well as usefulness in determining other statistics.

To see how this statistic may be calculated and what it discloses regarding the sample, consider some length measurements of 200 hypothetical F_1 plants resulting from a particular cross. Calculation will be facilitated if data are grouped by classes and tabulated as follows:

1	2	3	4	5	6
Class value (cm) x	Fre- quency f	fx	Deviation from mean $(x - \bar{x})$	Squared deviation $(x - \bar{x})^2$	$f(x - \bar{x})^2$
48	8	384	-4.75	22.56	180.50
50	32	1,600	-2.75	7.56	242.00
52	75	3,900	-0.75	0.56	42.19
54	52	2,808	$+1.25$	1.56	81.25
56	28	1.568	$+3.25$	10.56	295.75
58	5	290	$+5.25$	27.56	137.81
	$n = 200$	$\Sigma fx = 10{,}550$			$\Sigma f(x - \bar{x})^2 = 979.50$

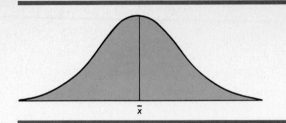

Figure 10-1. *The curve of normal distribution. A perpendicular erected from the abscissa at a value equal to the mean intersects the curve at its highest point and divides the area under the curve into areas of equal size.*

$$\bar{x} = \frac{\Sigma fx}{n} = \frac{10,550}{200} = 52.75 \text{ cm}$$

$$s = \sqrt{\frac{\Sigma f(x - \bar{x})^2}{n - 1}} = \sqrt{\frac{979.5}{199}} = \sqrt{4.92} = 2.218 \cong 2.22$$

This calculation provides a mean, plus or minus a standard deviation; that is, $\bar{x} = 52.75 \pm 2.22$. To understand the meaning of this expression and the information it conveys, "curves of distribution" must be examined.

As data for large samples are plotted with a quantitative measurement, such as length, along the abscissa and numbers of individuals (frequency) along the ordinate, the resulting curve is frequently bell-shaped; the variation is symmetrical about the largest class (or mode), as indicated in Figure 10-1. Such a curve is a **normal curve,** or a curve of normal distribution. If data are carefully plotted and a perpendicular erected from the abscissa at a value equal to the mean, it will intersect such a curve at the latter's highest point and will divide the area under the curve into two equal parts (Fig. 10-1) and, therefore, the sample into two groups of equal size. Now if perpendiculars to the abscissa are erected on it at points having values equal to $\bar{x} + s$ and $\bar{x} - s$, the area under the curve between $\bar{x} - s$ and $\bar{x} + s$ is 68.26 percent of the area under the curve (Fig. 10-2).

Similarly, the area under the curve between $\bar{x} - 2s$ and $\bar{x} + 2s$ is 95.44 percent of the total area; for $\bar{x} \pm 3s$, the area included is 99.74 percent of the total (Fig. 10-3). This means that, *in a normal distribution,* about 68 percent (or roughly two-thirds) of the individuals will have values between $\bar{x} - s$ and $\bar{x} + s$, about 95 percent between $\bar{x} - 2s$ and $\bar{x} + 2s$, and so on. Therefore, if an individual is chosen *at random* from a normally distributed population, the probability is about 0.68 that it will belong to that part of the population lying in the range $\bar{x} \pm s$. Similarly, there is a 0.95 probability that the individual selected will lie within the limits $\bar{x} \pm 2s$ or, only a 0.05 probability that the randomly chosen individual will lie outside those limits. The percentages of the sample determined by the mean plus or minus different multiples of the standard deviation are shown more fully in Table 10-1. Thus the standard deviation is a useful description of the variability of the sample and, if the sample is large and randomly chosen, a good indicator of the variability of the population. As variability of the sample increases, so does its standard deviation.

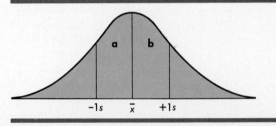

Figure 10-2. *The curve of normal distribution with perpendiculars to the abscissa erected at points showing values of $\bar{x} + s$ and $\bar{x} - s$. Areas a and b each comprise 34.13 percent of the area under the curve.*

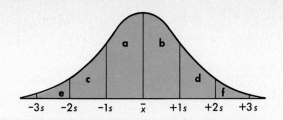

Figure 10-3. *Curve of normal distribution with perpendiculars to the abscissa erected at values* $\bar{x} \pm 1s$, $\bar{x} \pm 2s$, *and* $\bar{x} \pm 3s$. *Areas under the curve are as follows:* a + b = 68.26 *percent;* (a + b) + (c + d) = *95.44 percent;* (a + b) + (c + d) + (e + f) = 99.74 percent.

4. THE STANDARD ERROR OF THE SAMPLE MEAN

If a series of samples were drawn from the same population, their means and standard deviations would probably not be the same. Page 171 carried measurements for a sample of hypothetical plants having a mean of 52.75 cm. Another group of 200 F_2 plants from the same population would be expected, by chance, to have a somewhat different mean. However, if a series of samples of 200 individuals each were drawn at random, we could assume from our knowledge of the laws of probability that a few samples would have relatively low means and a few relatively high, but that most would be intermediate. In fact, if a large number of these successive sample means were plotted, they would be found to form a normal curve, and thus give a fairly clear picture of the population distribution. The population mean and standard deviation could then be calculated. Practical limitations preclude doing exactly this, but the standard deviation of means, or the **standard error of the sample mean,** can be calculated from one representative sample.

Table 10-1. *Percentages of the sample falling within given multiples of the standard deviation from the mean*

Mean ± values of s	Percent of sample included	Mean ± values of s	Percent of sample included
0.1	7.96	2.0	95.44
0.2	15.86	2.1	96.42
0.3	23.58	2.2	97.22
0.4	31.08	2.3	97.86
0.5	38.30	2.4	98.36
0.6	45.14	2.5	98.76
0.675	50.00	2.58	99.00
0.7	51.60	2.6	99.06
0.8	57.62	2.7	99.30
0.9	63.18	2.8	99.48
1.0	68.26	2.9	99.62
1.1	72.86	3.0	99.74
1.2	76.98	3.1	99.80
1.3	80.64	3.2	99.86
1.4	83.84	3.3	99.90
1.5	86.64	3.4	99.94
1.6	89.04	3.5	99.96
1.645	90.00	3.6	99.96
1.7	91.08	3.7	99.98
1.8	92.82	3.8	99.98
1.9	94.26	3.9	99.99
1.96	95.00	4.0	99.99

The standard error of the sample mean, $s_{\bar{x}}$, represents an estimate of the standard deviation of the means of many samples that might be taken, and is a measure of the closeness with which the sample mean, $\bar{x}$, represents the population mean, μ. The size of the sample used and the variability of the population affect the reliability of $\bar{x}$ as an estimate of μ. The greater the variation in the population, the larger the sample needed to provide an adequate representation of the population. Both these factors are taken into account in the formula for calculating the standard error of the sample mean:

$$s_{\bar{x}} = \frac{s}{\sqrt{n}} \tag{8}$$

where s is the standard deviation of the sample and n, of course, is the number of individuals composing the sample.

In the hypothetical sample of 200 F_1 plants (page 171), where $\bar{x} = 52.75$ cm and $s = 2.22$, the standard error of the sample mean becomes

$$s_{\bar{x}} = \frac{2.22}{\sqrt{200}} = \frac{2.22}{14.14} = 0.157, \text{ or about } 0.16$$

Because $s_{\bar{x}}$ represents the standard deviation of a *series* of sample means, and recalling that the areal relationships of a normal curve (Fig. 10-3) indicate that $\bar{x} \pm s$ includes 68.26 percent of the sample, $\bar{x} \pm 2s$ includes 95.44 percent of the sample, etc., the value $s_{\bar{x}} = 0.16$ indicates that there is about a 0.68 probability that μ lies in the range 52.75 cm $\pm$ 0.16, i.e., between 52.59 cm and 52.91 cm. Likewise, the probability that μ is in the range $\bar{x} \pm 2s_{\bar{x}}$, or between 52.43 and 53.07, is about 0.95. Thus $\bar{x} = \mu \pm s_{\bar{x}}$ 68.26 percent of the time by *chance alone*, $\bar{x} = \mu \pm 2s_{\bar{x}}$ in 95.44 percent of the cases, and $\bar{x} = \mu \pm 3s_{\bar{x}}$ 99.74 percent of the time. If the sample is large (> 100), other confidence levels may be determined from Table 10-1. For example, there is a 0.9 probability that μ lies in the range 52.74 $\pm$ 1.654 $s_{\bar{x}}$, or 52.75 $\pm$ 0.263 cm; a 0.5 probability that it is within the range 52.75 $\pm$ 0.675 $s_{\bar{x}}$; and so on. Obviously, the smaller the standard error, the more reliable the estimate of the population mean. As sample size, n, increases, the magnitude of the standard error decreases. Therefore, it is desirable to use samples as large as possible in order to determine characteristics of the population.

In cases involving polygene inheritance, the standard error of the sample mean will give a measure of the population mean. It will also help to fix, for example, a parental mean to which a given fraction of the F_2 may be compared. In other words, in the example of pumpkin fruits (page 162), the question could be raised as to what value between 4 and 6 or 19 and 23 is acceptable as representative of the parental strains so that certain F_2 individuals can be designated as being as extreme as either parent. If the standard error of the mean is calculated for each parental sample, a range is arrived at within which there is 0.68, 0.95, or 0.99 confidence that the population mean lies, and therefore this range provides a valid representation of the parental populations against which we can compare the F_2.

5. STANDARD ERROR OF THE DIFFERENCE IN MEANS

It is often necessary to determine whether the difference in the means of two samples is statistically significant—that is, to determine the likelihood that two sample means represent genetically different populations rather than chance differences in two samples from the same population. A judgmental answer to this problem is provided by a statistic known as the **standard error of the difference in means** (S_d):

$$S_d = \sqrt{(s_{\bar{x}_1})^2 + (s_{\bar{x}_2})^2} \tag{9}$$

For example, consider the statistics developed for the hypothetical group of 200 plants ("sample 1") as compared with like statistics for a second hypothetical sample:

Sample 1	Sample 2
$n_1 = 200$	$n_2 = 200$
$\bar{x}_1 = 52.75$	$\bar{x}_2 = 55.87$
$s_1 = 2.218$	$s_2 = 3.150$
$s_{\bar{x}_1} = 0.16$	$s_{\bar{x}_2} = 0.22$

Substituting in formula (9) to determine S_d for these two samples gives

$$S_d = \sqrt{(0.16)^2 + (0.22)^2}$$

$$= \sqrt{0.026 + 0.048}$$

$$= \sqrt{0.074}$$

$$= 0.272$$

The meaning of a value of $S_d = 0.272$ can be seen by comparing the difference in sample means, here $\bar{x}_2 - \bar{x}_1$, with the standard error of the difference in means, S_d:

$$\frac{\bar{x}_2 - \bar{x}_1}{S_d} \tag{10}$$

Substituting values, formula (10) becomes

$$\frac{3.12}{0.27} = 11.55$$

(Note the smaller mean is always subtracted from the larger in order to have a positive numerator.)

What, then, does the value of S_d and its relation to the difference in sample means signify? Remember that, in a normal curve, the area under the curve equal to $\bar{x} \pm 2s$ comprises about 95 percent of the total area, i.e., 95 percent of the individuals will have a quantitative value between $\bar{x} - 2s$ and $\bar{x} + 2s$. Thus, for a normal distribution, a standard error represents the standard deviation of a series of means, and a standard error of the difference in sample means deals with this same relationship between $\pm 1s$, $\pm 2s$, and so on. Therefore, *if the difference in sample means is greater than twice the standard error of the difference of the sample means, the difference in sample means is considered significant.* Significance begins whenever $\bar{x}_1 - \bar{x}_2 > 2S_d$. Here *significance* means two different populations.

Would a difference in means of, say, *exactly* twice the value of S_d mean that two different populations are represented by the two samples? No! However, the values of $\bar{x}_1 - \bar{x}_2 = 2S_d$ would give 0.95 *confidence* that two populations are involved, a 0.05 probability that the two samples were taken from the same population. *When the probability that two samples have come from the same population falls below 0.05 the difference in means is considered significant.* Therefore, to be significant, $\bar{x}_1 - \bar{x}_2$ must *exceed* $2S_d$. In the example, here, $\bar{x}_1 - \bar{x}_2 = 11.55 \, S_d$; hence, $\bar{x}_1 - \bar{x}_2$ is considered highly significant. The probability that only one population is represented is so very small that it is rejected. Note that a significant difference between two sample means is inherently neither "good" nor "bad."

These five statistics are summarized in Appendix D.

Applications of statistics to genetic problems

Two examples of the application of statistics to genetic problems will demonstrate the usefulness of such analyses.

In the first example, assume a commercial producer of hybrid seed corn wishes
to market grains that will produce plants having ears of very uniform length.
He has two varieties, A and B, both of which produce ears of just under 8 in.,
which is a satisfactory length for his marketing purposes. Although A averages
closer to 8 in., it appears to be more variable than B. The producer grows several
acres of each variety in as uniform an environment as possible, then analyzes
100 ears from each. Variety A has the following statistics:

$$\bar{x} = 7.95 \text{ in.}$$
$$s = 0.52$$
$$s_{\bar{x}} = 0.05$$

Although the sample mean is very close to the desired length, the standard
deviation indicates that two-thirds of the ears of this variety may be expected to
vary up to 0.52 in. from the mean of 7.95 in. Of course, one-third will deviate
more than this.

This is more variability than the grower would prefer, so, for comparison, 100
ears of variety B are similarly analyzed. This variety has these statistics:

$$\bar{x} = 7.88 \text{ in.}$$
$$s = 0.23$$
$$s_{\bar{x}} = 0.02$$

Although the ears average slightly shorter than those of variety A, B has a much
narrower range of variation. Two-thirds of the ears of the latter may be expected
to fall within the range 7.65 to 8.11 in., as compared with 7.43 to 8.47 for A.
Therefore, the breeder elects to use B.

The standard errors of the two samples are useful in indicating to the grower
just how much the mean length of his samples might be expected to vary from
the mean lengths of *all* plants of the two varieties. His samples are sufficiently
large, and the standard errors are quite small. This shows that the samples do
reflect reliably the magnitude of variability in the two varieties.

Another example will show an application of statistics to a more theoretical
type of problem (Table 10-2). Data were accumulated on days to maturity for
two varieties of tomato (Burpeeana Early Hybrid, P_1, and Burpee Big Boy, P_2)
and their hybrids (F_1 and F_2). Maturation time is dependent on both heredity
and environment, so environmental differences must be minimized. This is often
done in randomized plots, whereby different varieties are distributed randomly
in the field. The time elapsing between setting the plants in the field and the
ripening of the first fruit was recorded as "days to maturation." Data were
recorded for samples of 100 plants in each of the four varieties, all of which
were grown in the same season in randomized plots.

The data in Table 10-2 clearly suggest that P_1 and P_2 represent different
populations, and this is confirmed by the standard deviation and standard error
of each, as well as by the standard error of the difference in sample means. In
fact, the difference in sample means is 19.54 days, which is about 108 times the
standard error of the difference in means! Because a difference in sample means
of more than $2S_d$ is considered significant, the difference here is highly significant.

The same statistics for the F_1 and F_2 generations also show that although
differences in maturation times are considerably less than for the paternal strains,
they are significantly different. One therefore has more than 95 percent confidence that the F_1 and F_2 samples represent two genetically different populations.
The difference in means for these two samples is only 0.92 day, but this is about
2.5 times the standard error of the difference in the sample means.

**Table 10-2. Data for two parental and two progeny strains
of tomato based on days required to reach maturity**

Days	P_1	P_2	F_1	F_2
55	1			
56	6			
57	9			1
58	40			1
59	28			2
60	14			3
61	2		3	4
62			8	9
63			20	10
64			31	12
65			19	14
66			10	20
67			8	7
68			1	4
69				3
70				3
71				1
72				2
73				1
74				1
75		4		1
76		12		1
77		20		
78		35		
79		15		
80		10		
81		3		
82		1		
$\bar{x}$	58.38	77.92	64.22	65.14
s	1.14	1.43	1.49	3.43
$s_{\bar{x}}$	0.114	0.143	0.149	0.343
S_d		0.183		0.374

These data clearly suggest polygenes. The F_1 is intermediate between the parents, as is the F_2, but the F_2 has a wider range than the F_1. If P_1 and P_2 are assumed to be completely homozygous, then the variability that each shows must be wholly environmental. The F_1 would then be completely and uniformly heterozygous and its variability environmental. The F_2 would be expected to segregate; therefore, genetic variation is superimposed on environmental.

The F_2 data may be used to furnish a very rough estimate of the number of contributing alleles. Eleven of the 100 F_2 were as extreme as P_1, and 2 as extreme as P_2. Now 11/100 simplifies to about 1/9, and 2/100 to 1/50. The formula for computing numbers of effective alleles from the F_2 data (page 162), shows that 1/16, which indicates four contributing alleles, is between the two extremes of 1/9 and 1/50. This approach is really too simple and probably gives too low an estimate, for it would take many hundreds or thousands of F_2 individuals to provide a reasonable chance of recovering the maximum extremes possible.

Sewall Wright has shown that the number of *pairs* of polygenes (n) can be calculated from the formula

$$n = \frac{R^2}{8(s^2_{F_2} - s^2_{F_1})} \tag{11}$$

where R is the range (greatest difference) between the mean values of extreme phenotypes (whether found in the parental generations or elsewhere), $s^2_{F_2}$ is the *variance* of the F_2, and $s^2_{F_1}$ the variance of the F_1. An application of this formula to the data of Table 10-2 with substitutions shows

$$n = \frac{(77.92 - 58.38)^2}{8(11.76 - 2.22)} = \frac{(19.54)^2}{8(9.54)} = \frac{381.81}{76.32} = 5.003$$

or about five pairs of polygenes.

In using this formula the following assumptions have been made: (1) no environmental effect, (2) no dominance, (3) no epistasis, (4) equal, additive contributions by all loci, (5) no linkage, and (6) complete homozygosity in each parent, with complete heterozygosity in the F_1. With these assumptions, the F_1 mean would be expected to fall midway between the parental means. That it does not suggests that not all of those assumptions are completely valid in this case and/or that the sample is too small. Moreover, the fact that a crude estimate of the number of pairs of genes from the fraction of the F_2 as extreme as either parent is lower than the value given by the Wright formula also illustrates the need for a considerably larger sample.

These are practical contributions of statistical analyses to genetic problems. There are, however, more subtle advantages. In many cases it has been possible to separate genetic mechanisms from sampling errors, and in others to differentiate between genetic and environmental variation. Above all, critical attitudes toward design of experiments and treatment of data have been sharpened.

Problems

The following data were obtained on the weight in pounds of a given sample of pumpkin fruits:

30	26	22	16	24
24	24 :	30	22	14
28	22	16	26	22
24	20	28	14	28
22	24	22	22	26

10-1 What is the mean to the nearest tenth of a pound?

10-2 What is the variance?

10-3 (a) What is the standard deviation to the nearest tenth of a pound? (b) What does this value tell you about the sample?

10-4 If an individual is selected at random from this sample, what should be the probability that it will weigh more than 18 but less than 28 lb. (Assume that the sample is normally distributed.)

10-5 (a) What is the standard error of the sample mean? (b) What information does this give you?

10-6 What is the probability that the population mean lies between about 21 and 25 lb?

Data on a second sample of 25 individuals were collected and the following statistics determined:

$$\bar{x}_2 = 21.8 \text{ lb}$$
$$s_2 = 4.0$$
$$s_{\bar{x}_2} = 0.8$$

10-7 What is the standard error of the difference in sample means?

10-8 What is the approximate probability that samples 1 and 2 represent two different populations?

10-9 In view of your answer to the preceding question, is the difference in sample means significant? Why?

10-10 Two parental strains with mean heights of 64.29 and 135 cm, respectively, were crossed. The F_1 and F_2 that resulted had the following statistics:

	$\bar{x}$	s
F_1	99.64	10.000
F_2	102.11	13.342

If we neglect any possible environmental effects, dominance, linkage, or epistasis, and assume that each effective allele makes the same additive contribution, how many pairs of polygenes are involved?

CHAPTER 11

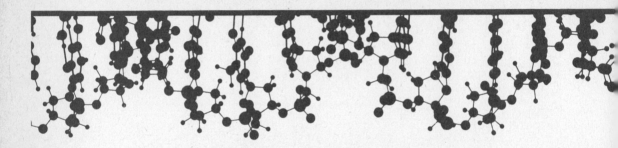

Sex determination

Two types of sexually reproducing animals and plants may be recognized: (1) **monoecious** (from the Greek *monos*, "only," and *oikos*, "house"), in which each individual produces two kinds of gametes, sperm and egg, and (2) **dioecious** (Greek prefix *di-*, "two," and *oikos*), in which a given individual produces only sperms or eggs. In dioecious organisms, the **primary sex difference** concerns the kind of gamete and the primary sex organs by which these are produced. Each sex also exhibits many **secondary sex characters**. In humans these include voice, distribution of body fat and hair, and details of musculature and skeletal structure; in *Drosophila* these include the number of abdominal segments, presence (♂) or absence (♀) of sex combs, and so on (Fig. 11-1). Our immediate problem is to examine the mechanisms that determine the sex of an individual. Two basic types, chromosomal and genic will be discussed, though the distinction is not always a sharp one.

Sex chromosomes

DIPLOID ORGANISMS

XX-XO system. Chromosomal differences between the sexes of several dioecious species were found early in the course of cytological investigations. Henking, a German biologist, in 1891 noted that half the sperms of certain insects contained an extra nuclear structure, the "X body." The significance of this structure was not immediately understood, but in 1902 an American, McClung, reported that the somatic cells of the female grasshopper contained 24 chromosomes, whereas those of the male had only 23. Three years later, Wilson and

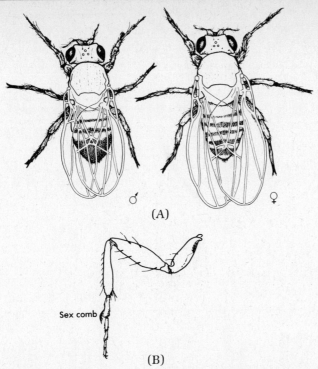

Figure 11-1. Drosophila melanogaster, *(A) male; (B) female. Below, detail of leg showing sex comb of male.*

Stevens succeeded in following both oogenesis and spermatogenesis in several insects. They realized that the X body was a chromosome, so the X body became known as an X chromosome. Thus in many insects, there is a chromosomal difference between the sexes, i.e., females are referred to as XX (having two X chromosomes) and males as XO ("X-oh," having one X chromosome). As a result of meiosis, all the eggs of such species carry an X chromosome. Only half the sperms have one and the other half have none.

XX-XY system. In the same year, 1905, Wilson and Stevens found a different arrangement in other insects. In such cases females were again XX, but males had in addition to one X chromosome, an odd one of a different size, which was called the Y chromosome; thus males were XY. Half the sperms carry an X and half a Y. The so-called XY type occurs in a wide variety of animals, including *Drosophila* and mammals, as well as in at least some plants (e.g., the angiosperm genus *Lychnis*) (Fig. 11-2).

A distinction can thus be made between the X and Y chromosomes associated with sex and those that are alike in both sexes. The X and Y chromosomes are called **sex chromosomes;** the remaining ones of a given complement, which are the same in both sexes, are **autosomes.** In both the XX-XO and XX-XY types described thus far, all the eggs have one X chromosome, whereas the sperms are of two kinds, X and O, or X and Y. In each case the male is the **heterogametic** sex (producing two kinds of sperms), whereas the female is the **homogametic** sex (producing but one kind of egg).

ZZ-ZW system. A final major type of chromosomal difference between the sexes is one in which the female is heterogametic and the male homogametic. The sex chromosomes in this case are often designated as Z and W to avoid

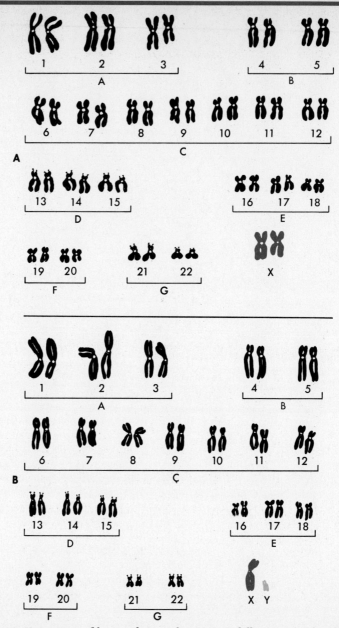

Figure 11-2. *Karyotypes of human beings showing sex differences. (A) female; (B) male. Chromosome pairs 1–22 are autosomes, customarily designated by letter groups. The X and Y are sex chromosomes; structurally, X is similar to group C, and Y to group G.*

confusion with instances in which the female is homogametic. Females are thus ZW and males ZZ. Birds (including the domestic fowl), butterflies and some fishes belong to this group.

MONOPLOID ORGANISMS

Liverworts. Like all sexually reproducing plants, liverworts (division *Bryophtya*) are characterized by a well-marked alternation of generations (see Appendix B), in which a monoploid, sexually reproducing phase or generation (the

gametophyte) alternates in the life history with a diploid, asexually reproducing individual (the **sporophyte**). As early as 1919 the chromosome complement of the sporophyte of the liverwort *Sphaerocarpos* was reported to consist of seven matching pairs, plus an eighth pair in which one of the two chromosomes was much larger than the other. The larger member of this eighth pair has been designated the X chromosome, its smaller partner the Y chromosome. At meiosis, which terminates the diploid sporophyte generation, X and Y chromosomes are segregated, so that of the four meiospores produced from each meiocyte, two receive an X chromosome and two a Y chromosome. Meiospores containing an X chromosome develop into female gametophytes, those with a Y into males. Thus females are X; males, Y; and asexual sporophytes, XY.

Summary of sex chromosome types

The various types of chromosomal differences between the sexes may be summarized as follows:

♀	♂	Examples
XX	XY	*Drosophila*, humans and other mammals, some dioecious angiosperm plants
XX	XO	Grasshopper; many Orthoptera and Hemiptera
ZW	ZZ	Birds, butterflies, and moths
X	Y	Liverworts

Determination of sex under the chromosomal system

Such chromosomal differences as these raise certain fundamental questions. For example, is a *Drosophila* individual a male because of the presence of a Y or because only one X is present? Is an individual female because of the absence of the Y or because of the presence of two X chromosomes? Do the autosomes have anything to do with sex determination? Is the system identical in *Drosophila* and humans, both of which have the XX-XY sex difference? What do genes have to do with the situation? What causes sex reversal, in which an individual of one sex becomes one of the other sex? Or what operates to produce individuals that are part male and part female in species where the sexes are ordinarily separate and distinct?

DROSOPHILA

Primary nondisjunction of X chromosomes. Work on the genetics of *Drosophila melanogaster* showed that sex determination, at least in that animal (where $2n = 8$), was far less simple than the mere XX-XY difference would suggest. From some unusual breeding results in his laboratory, Bridges was able, in a masterpiece of inductive reasoning, to lay the groundwork for a complete understanding of sex determination in *Drosophila* (Bridges, 1916a, 1916b, and 1925).

The gene for wild type red eyes (+) is carried on the X chromosome; a recessive allele (v) produces vermilion eyes in homozygous females and in all males (which, of course, have only one X chromosome). Ordinarily, vermilion-eyed females mated to red-eyed males produce only red-eyed daughters and vermilion-eyed sons:

$$
\begin{array}{ccccc}
P & & vv & \times & +Y \\
P\ gametes: & & \text{\female} & & v \\
& \male & \frac{1}{2}+ & + & \frac{1}{2}Y \\ \hline
F_1 & & \frac{1}{2}+v & + & \frac{1}{2}\,vY \\
& & (\text{red \female}) & & (\text{vermilion \male})
\end{array}
$$

However, in rare instances, crosses of this type produce unexpected vermilion-eyed daughters and red-eyed sons with a frequency of one per 2,000 to 3,000 offspring. Bridges surmised that these unusual progeny are due to a failure of the X chromosomes in an XX female to disjoin during oogenesis. Such *primary nondisjunction* (Bridges, 1916a), he reasoned, would produce three kinds of eggs, the majority of which would contain the normal single X chromosome and a small number, either two X chromosomes or no X at all. If each X chromosome is represented by either X^+ (carrying the dominant gene for red eyes) or X^v (bearing the recessive gene for vermilion eyes), and each *set of three* autosomes is A, Bridge's cross may be represented in this way:

$$
\begin{array}{ll}
P & AAX^vX^v \times AAX^+Y \\
& (\text{vermilion \female})\ (\text{red \male})
\end{array}
$$

$$
\begin{array}{ll}
P\ gametes: \female & AX^v\ (\text{numerous}) + AX^vX^v\ (\text{rare}) + AO\ (\text{rare}) \\
\male & \underline{\qquad AX^+ + AY \qquad}
\end{array}
$$

$$
\begin{array}{lll}
F_1 & AAX^+X^v & \text{red \female (numerous; normal)} \\
& AAX^+X^vX^v & \text{``metafemale'' (rare; die)} \\
& AAX^+O & \text{sterile red \male (rare)} \\
& AAX^vY & \text{vermilion \male (numerous; normal)} \\
& AAX^vX^vY & \text{vermilion \female (rare)} \\
& AAOY & \text{die (rare)}
\end{array}
$$

The metafemales (*AAXXX*) are weak, and seldom live beyond the pupal stage; *AAOY* individuals die in the egg stage. Note that the presence of a Y chromosome does not determine maleness itself, though males without it are sterile.

Secondary nondisjunction. Bridges next mated the exceptional vermilion-eyed females (*AAX^vX^vY*) that arose as a result of primary nondisjunction to normal red-eyed males (*AAX^+Y*), and obtained progeny in these frequencies:

$$
\begin{array}{ll}
0.46 & \text{red \female} \\
0.02 & \text{vermilion \female} \\
0.02 & \text{red \male} \\
0.46 & \text{vermilion \male} \\
0.02 & \text{metafemales} \\
0.02 & OYY
\end{array} \left.\vphantom{\begin{array}{l}a\\b\end{array}}\right\} \text{die}
$$

Occurrence of the vermilion-eyed females and red-eyed males is due to *secondary nondisjunction*. Meiosis in XXY females is expected to be somewhat irregular because of pairing problems and, indeed, Bridges' results confirm this. In oogenesis, synapsis may involve either the two X chromosomes (XX type) with the Y chromosome remaining unsynapsed or one X and the Y (XY type) with the other X remaining free. Bridges found XY synapsis to occur in about 16 percent of the cases and XX synapsis in about 84 percent. After XY synapsis, disjunction segregates the X and Y synaptic partners to opposite poles. The unsynapsed X may go to *either* pole so that XY synapsis produces four kinds of eggs with a frequency of 0.04 each: XX, Y, X, and XY (Fig. 11-3). On the other hand, XX synapsis is followed by disjunction of the two previously synapsed X chromosomes and their movement to opposite poles. The free Y may, of course, go to either pole, but the result is only two kinds of eggs, X and XY, with a frequency of 0.42 each. The overall result of secondary nondisjunction is four kinds of eggs, in these frequencies:

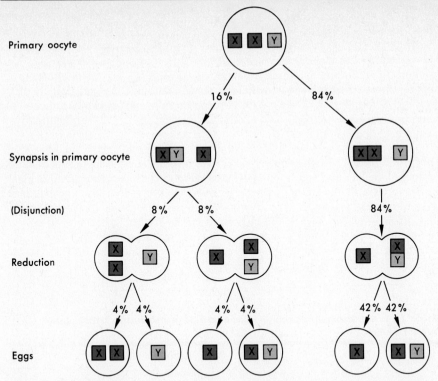

Figure 11-3. *Diagram of secondary nondisjunction in XXY female* Drosophila, *resulting in 46 percent X, 46 percent XY, 4 percent XX, and 4 percent Y eggs. For explanation, see text.* (Based on work of Bridges, 1916a.)

$$0.46\ X^vY + 0.46\ X^v + 0.04\ X^vX^v + 0.04\ Y$$

Fertilization by sperm from a cytologically normal, red-eyed male (X^+Y) produces eight types of zygotes, which may be grouped in six classes:

Frequency	Zygote	Phenotype
0.23	X^+X^vY	red ♀
0.23	X^+X^v	
0.02	$X^+X^vX^v$	metafemale (die)
0.02	X^+Y	red ♂
0.23	X^vYY	vermilion ♂
0.23	X^vY	
0.02	X^vX^vY	vermilion ♀
0.02	YY	die

Bridges verified the chromosomal constitution of each of the viable genotypes.

As Bridges used the terms, *primary nondisjunction* may occur in either XX females or XY males. In the former it leads to the production of XX and O eggs. Occurrence in the first meiotic division of males produces XY and O sperms. Should it take place during the second division, XX, YY, and O sperms are the result. *Second nondisjunction*, on the other hand, occurs in XXY females, where it gives rise to XX, XY, X, and Y eggs. As the term *nondisjunction* implies, these aberrant gametes are produced only as a result of failure of the sex chromosomes to disjoin after synapsis; they are not physically attached.

Attached-X flies. Another strain of flies in which nondisjunction of the X chromosomes occurred in *all* females was discovered by L. V. Morgan (1922). When such females were mated to cytologically normal males carrying a recessive gene on the X chromosome, all viable male progeny were phenotypically like the father (but sterile), whereas all viable female offspring were like the mother. In addition, one-fourth of the total progeny were metafemales and another fourth (*AAOY*) died in the egg stage. Morgan reasoned that in these attached-X females (X̂X̂) the two X chromosomes were physically attached so that only two kinds of eggs, AX̂X̂ and AO, were produced. This explanation was soon confirmed cytologically.

Polyploid flies. Experimentally produced triploid (three whole sets of chromosomes, or 3*n*) and tetraploid (4*n*) flies were next incorporated into Bridges' work, so that many kinds of flies with respect to chromosome complements were ultimately produced. As this work continued, it became increasingly clear that, in *Drosophila* at least, the important key to the sex of the individual was provided by the *ratio of X chromosomes to sets of autosomes*. The Y, then, has nothing to do with sex determination but does govern male fertility. These results are summarized in Table 11-1.

Table 11-1. Summary of chromosomal sex determination in Drosophila (after Bridges)

Number of X chromosomes	Number of sets of autosomes	Number of autosomes	Total chromosome number	X/A ratio	Sex designation
3	2	6	9	1.50	Metafemale
4	3	9	13	1.33	Triploid metafemale
4	4	12	16	1.00	Tetraploid female
2	2	6	8	1.00	Female
3	4	12	15	0.75	Tetraploid intersex
2	3	9	11	0.67	Triploid intersex
1	2	6	7	0.50	Male
1	3	9	10	0.33	Triploid metamale
1	4	12	13	0.25	Tetraploid metamale

Metamales (or supermales) are to the male sex as metafemales (or superfemales) are to the female sex. That is, they are weak, sterile, underdeveloped, and die early. Intersexes are sterile individuals that display secondary sex characters between those of male and female (Fig. 11-4).

From all these results the mechanism of sex determination in *Drosophila* may be summarized:

1. Sex is governed by the ratio of the number of X chromosomes to sets of autosomes. Thus, from Table 11-1, females have an X/A ratio = 1.0, males = 0.5. This relationship applies even to polyploid flies as long as the appropriate X/A ratio is maintained.
2. Genes for maleness per se are apparently carried on the autosomes, those for femaleness on the X chromosome.
3. The Y chromosome governs male *fertility*, rather than sex itself, because AAXY and AAXO flies are both male with regard to secondary sex characters, but only the former produce sperms; it has no effect in AAXXY flies, which have an X/A ratio of 1.0 and are female.

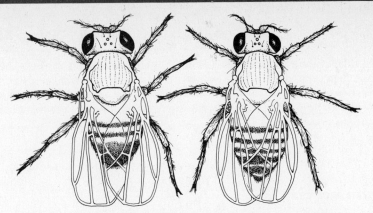

Figure 11-4. *Triploid intersexes (X/A ratio 0.67) in* Drosophila.

4. An X/A ratio > 1.0 or < 0.5 results in certain characteristic malformations (*metafemales and metamales*).

5. An X/A ratio < 1.0 but > 0.5 produces individuals intermediate between females and males (*intersexes*). The degree of femaleness is greater where the X/A ratio is closer to 1.0, and the degree of maleness is greater where that ratio is closer to 0.5.

Other workers have confirmed and extended these observations on intersexes in *Drosophila*. By using X-rays to fragment chromosomes, lines of flies having extra fragments of the X chromosome have been developed. Thus an individual with about 2.67 X chromosomes and 3 sets of autosomes (= 6 chromosomes) has an X/A ratio of 2.67/3 = 0.89, and is an intersex. Its secondary sex characters, however, are more female in nature than a fly with 2.33 X and 3 sets of autosomes and therefore, an X/A ratio of 2.33/3 = 0.78.

The same system in which the X/A ratio is critical is reported for the flowering seed plant (angiosperm) *Rumex acetosa*.

The transformer gene.   One additional complicating factor in sex determination in *Drosophila* is worth examining briefly. A recessive gene, *tra*, on the third chromosome (an autosome), when homozygous, "transforms" normal diploid females (AAXX) into sterile males. The XX *tra tra* flies have many sex characters of males (external genitalia, sex combs, and male-type abdomen) but, as noted, are sterile.

Gynandromorphs.   Concepts of sex determination as developed for *Drosophila* are verified by the occasional occurrence of **gynandromorphs** (or gynanders). These are individuals in which part of the body expresses male characters, whereas other parts express female characters. A bilateral gynandromorph, for example, is male on one side (right or left) and female on the other. The male portions of such flies would be expected to have a male chromosomal composition. Ingenious experiments using known "marker" genes on the X chromosome have provided experimental evidence to confirm this prediction. In *Drosophila*, right and left body halves are determined at the first cleavage of the zygote. Lagging of the X chromosome at this first mitosis can result in two daughter cells of the chromosomal complement AAXX and AAXO when the laggard X fails to be incorporated in a daughter nucleus. The portion of the body developing from the former cell will be normal female, that from the latter (sterile) male. Gynandromorphs represent one kind of *mosaic*, or organisms part of which are composed of cells genetically different from the remaining part.

Sex as a continuum. Instead of an either-or, male or female, XX or XY condition, sex may be viewed as a continuum, which ranges from supermaleness through maleness, intersexes of varying degree, to femaleness and on to superfemaleness. Where an individual places in such a continuum is related to the ratio of individual X chromosomes to sets of autosomes. Of course, it is not the chromosomes as gross structures that are the deciding factors, but rather, *genes* on the chromosomes. Thus genes for maleness are associated with the autosomes, and those for femaleness with the X chromosomes. Yet this entire sex-determining arrangement may, in some cases, be upset by a single pair of recessive autosomal genes (*tra*)!

Human beings

In normal human beings, males are XY and females are XX, just as in *Drosophila.* But is sex here also determined by the X/A ratio? With regard to sex, does the Y chromosome bear genes for male fertility as in *Drosophila* or for male sex per se? To answer these questions it will be helpful to examine (1) the sex chromosomes, (2) the concept of sex differentiation, and (3) human sex anomalies and their chromosomal make-up.

THE SEX CHROMOSOMES

X chromosome. The human X chromosome is medium-length, submetacentric, and intermediate in length between autosomes 7 and 8 (Table 11-2). Its centromere is more nearly central than any of the larger chromosomes of the C group. With the development of fluorescent banding techniques (Chapter 3), visual differentiation of the X chromosomes and morphologically similar autosomes became exact (Fig. 11-5). In mitotic metaphase spreads, the X chromosome measures approximately 5.0 to 5.5 μm, depending on the preparation.

Y chromosome. The Y chromosome in most human males averages roughly 2 μm in length, very slightly longer than the members of the shortest or G group of autosomes (Table 11-2). This chromosome is, however, quite variable in length among different men; in some it is regularly as long as or longer than members of the F group (autosomes 19 and 20), whereas in others it may be less than half the length of the members of the G group. No particular phenotype or syndrome has been consistently associated with either a long or a short Y chromosome. The Y chromosome is acrocentric, even more so than the G autosomes, and its longer arms generally lie close together. Unlike autosomes 21 and 22, the Y has no satellites. With fluorescent banding techniques, the longer arm fluoresces brilliantly in good-quality preparations (Fig. 11-5).

X- and Y-chromatin in interphase nuclei. A clue to the sex chromosomal complement of an individual may be obtained quite simply by examining squamous epithelial cells from scrapings of the lining of the cheek. When stained, most somatic cells of normal females show a characteristic structure, the *Barr body,* so named after its discoverer, Murray Barr, who first described it in 1949 (Fig. 11-6). Females are therefore termed *sex chromatin positive;* normal males, whose cells do not contain a Barr body, are *sex chromatin negative.* Because of various cytological factors, as well as some of the techniques employed, frequency of sex chromatin positive cells in scrapings of oral mucosa from normal females have been reported to range from 36 to 80 percent of cells examined. The Barr body is a small structure (about 1 μm in greatest dimension), hemispherical, disk shaped, rod shaped, or triangular in outline. Ordinarily it lies

Table 11-2. Measurements of relative lengths of human chromosomes in percentage of total monoploid autosome length

Chromosome number	A	B	C
1	9.08	9.11 ± 0.53	8.44 ± 0.433
2	8.45	8.61 ± 0.41	8.02 ± 0.397
3	7.06	6.97 ± 0.36	6.83 ± 0.315
4	6.55	6.49 ± 0.32	6.30 ± 0.284
5	6.13	6.21 ± 0.50	6.08 ± 0.305
6	5.84	6.07 ± 0.44	5.90 ± 0.264
7	5.28	5.43 ± 0.47	5.36 ± 0.271
X	5.80	5.16 ± 0.24	5.12 ± 0.261
8	4.96	4.94 ± 0.28	4.93 ± 0.261
9	4.83	4.78 ± 0.39	4.80 ± 0.244
10	4.68	4.80 ± 0.58	4.59 ± 0.221
11	4.63	4.82 ± 0.30	4.61 ± 0.227
12	4.46	4.50 ± 0.26	4.66 ± 0.212
13	3.64	3.87 ± 0.26	3.74 ± 0.236
14	3.55	3.74 ± 0.23	3.56 ± 0.229
15	3.36	3.30 ± 0.25	3.46 ± 0.214
16	3.23	3.14 ± 0.55	3.36 ± 0.183
17	3.15	2.97 ± 0.30	3.25 ± 0.189
18	2.76	2.78 ± 0.18	2.93 ± 0.164
19	2.52	2.46 ± 0.31	2.67 ± 0.174
20	2.33	2.25 ± 0.24	2.56 ± 0.165
21	1.83	1.70 ± 0.32	1.90 ± 0.170
22	1.68	1.80 ± 0.26	2.04 ± 0.182
Y	1.96	2.21 ± 0.30	2.15 ± 0.137

Column A: Previous Denver-London Conference data.
Column B: Data from 10 cells by Drs. T. Caspersson, M. Hultén, J. Lindsten, and L. Zech. Cells stained with orcein.
Column C: Data from 95 cells provided by Drs. H. Lubs, T. Hostetter, and L. Ewing from 11 normal subjects (6 to 10 cells per person). Average total length of chromosomes (diploid set) per cell: 176 μm. Cells stained with orcein or Giemsa 9 technique.
Percentages for autosomes do not add to 100 because they are averages for a given chromosome in several metaphase spreads.
From *Paris Conference (1971): Standardization in Human Cytogenetics.* In *Birth Defects: Orig. Art. Ser.,* ed. D. Bergsma. Published by The National Foundation–March of Dimes, White Plains, N.Y., Vol. VIII (7), 1972. Used by permission.

appressed to the inner surface of the nuclear membrane or, in nerve cells, may be associated with the nucleolus. In diploid cells of females the number of Barr bodies is one less than the number of X chromosomes.

In somatic cells having two X chromosomes, one replicates later than the other. Mary Lyon (1961, 1962) and others have proposed that the late-replicating X chromosome becomes inactivated early in embryonic life and forms the Barr body. Whether the maternal or paternal X chromosome is inactivated in any given cell depends on chance, but once this has occurred in an embryo cell, indications are that the same chromosome becomes the Barr body in all cells derived therefrom. Thus, females that are heterozygous for genes located on the X chromosomes, are *mosaics;* some patches of tissue express the dominant phenotype, others the recessive.

Staining of XY (or XYY) cells with such fluorescent dyes as quinacrine hydrochloride discloses the brightly fluorescing Y chromosome(s), not only in mitotic metaphase, but also in interphase (Fig. 11-7). The same technique can be used successfully on sperm and on cells in the amniotic fluid.

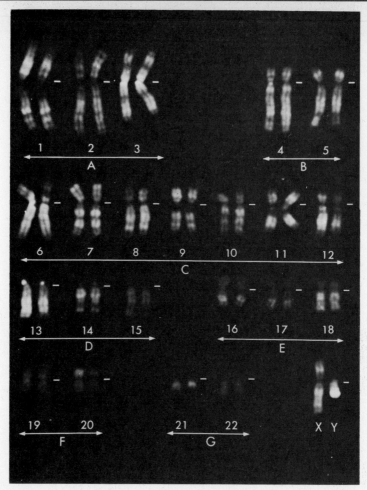

Figure 11-5. *Karyotype of normal human male (46,XY), Q-banding. Banding techniques permit much greater accuracy in matching autosomes and identifying the sex chromosomes. The short bar beside each chromosome pair marks the location of the centromere. Note the brightly fluorescing long arm of the Y chromosome.* (Courtesy Dr. C. C. Lin.)

SEX DIFFERENTIATION

Genetic sex. Normal females ordinarily have two X chromosomes; normal males, one X and one Y. As noted earlier, of course, genes on these chromosomes determine femaleness or maleness. Thus one can speak of females as having the *genetic sex* designation XX and males as having the genetic sex designation XY, although exceptional cases do occur.

Gonadal sex. Chemical substances (inductors) produced by embryonic XX cells act on the *cortical* region of undifferentiated gonads to bring about development of ovarian tissue. In XY embryos, however, inductors stimulate production of testes from the *medulla* of the undifferentiated gonads. Hence the XX genetic sex is ordinarily associated with ovarian *gonadal sex*, and XY with testicular gonadal sex.

Genital sex. The embryonic gonads produce hormones that, in turn, determine the morphology of the external genitalia and the genital ducts. XX embryos

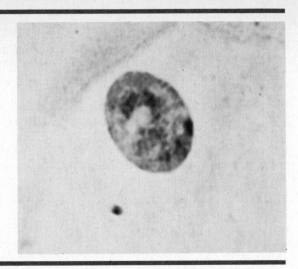

Figure 11-6. Nucleus of normal human female squamous epithelial cell showing prominent, dark Barr body against the nuclear membrane. (Courtesy Carolina Biological Supply Co.)

normally develop ovaries, female external genitalia, and Müllerian ducts. XY embryos, on the other hand, ordinarily develop testes, male external genitalia, and Wolffian ducts. In XX embryos Wolffian ducts are suppressed; in XY embryos the Müllerian ducts remain undeveloped. Thus there is a distinction between male and female *genital sex.*

Somatic sex. Production of gonadal hormones continues to increase until, at puberty, *secondary sex characters* appear. These include amount and distribution of hair (e.g., facial, body, axillary, pubic); pelvis dimensions; general body proportions; subcutaneous fat over hips and thighs, and breast development in the female, as well as increased larynx size and deepening of the voice in the male.

Sociopsychological sex. In most individuals, genetic sex, gonadal sex, genital sex, and somatic sex are consistent; XX persons, for example, develop ovaries, female genitalia, and female secondary sex characters. Ordinarily these persons are raised as females and adopt the feminine gender role under whatever cultural

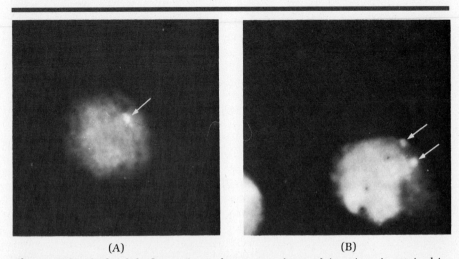

(A) (B)

Figure 11-7. The brightly fluorescing Y chromosome (arrows) *in quinacrine-stained interphase. (A) 46, XY male. (B) 47, XYY male. (Courtesy Dr. C. C. Lin.)*

pattern has been established in the society of which they are members. A similar consistency from genetic sex to sociopsychological sex is seen for XY individuals. On the other hand, some individuals display an inconsistency of some kind or degree among these levels of sexuality. Discordance involving the genetic and anatomical result in intersexuality.

HUMAN SEX ANOMALIES

The Klinefelter syndrome. One in about 500 "male" births produces an individual with a particular set of abnormalities known collectively as the **Klinefelter syndrome** (Fig. 11-8). These persons have a general male phenotype; external genitalia are essentially normal in gross morphology. Although there is some variability in other characteristics, testes are typically small, sperms are usually not produced, and most such men are mentally retarded. Arms are longer than average, some degree of breast development is common, and the voice tends to be higher pitched than in normal males. Klinefelters are sex chromatin positive; the karyotype shows 47 chromosomes, i.e., 47,XXY.

The XXY individual may arise through fertilization of an XX egg by a Y sperm or through fertilization of an X egg by an XY sperm. Although the majority of Klinefelters are born to mothers under the age of 30, this is in large part a reflection of the age group in which most births of all types occur (Fig. 11-9). After a drop in Klinefelter births from ages 27 to 32, there is a small increase again after age 32, whereas total births decrease rapidly in this age group. This fact suggests nondisjunction of the X chromosomes in aging oocytes as a somewhat more important factor than XY nondisjunction during spermatogenesis.

Less frequently, Klinefelters have more than two X chromosomes, even more than one Y (Table 11-3). Generally the greater the number of X chromosomes, the greater the degree of mental retardation.

As a clue to the nature of the sex-determining mechanism in human beings, the important point here is the fact that persons with at least one Y chromosome have the general phenotype of a male, even in the presence of any number of X chromosomes. Thus, "maleness," however defined, is dictated by the presence of at least one Y chromosome. One study has disclosed that in 70 percent of the XYY males studied, both X chromosomes were maternal in origin (the other 30 percent, of course, were cases of paternal origin). There is some suggestion of a maternal age effect, but no evidence at this time for a paternal age effect.

The Turner syndrome. A second major sex chromosome anomaly of interest here is the **Turner syndrome** (Fig. 11-10), in which the individual presents a general female phenotype, but with certain unique departures. These include short stature, "webbing" of the neck, a low hairline on the nape of the neck, and a broadly shield-shaped chest. Slight mental retardation is often noted, but a few Turners achieve high scores on standard IQ tests. Secondary sex characters do not develop; breast development is absent to very slight, pubic hair is reduced or absent, and axillary hair does not develop. Genitalia remain essentially infantile.

Turners are sex chromatin negative (Polani 1961), which suggests presence of only one X chromosome. This was confirmed cytologically in 1969 by Ford and his colleagues. The karyotype is, therefore, 45,X (i.e., AAXO). Studies of X-linked traits (Chapter 12), such as the Xg blood group, reveal that in 72 percent to 76 percent of the Turners so examined, the *paternal* X chromosome is absent. The reason for this kind of disparity is unknown. Mean parental ages (both sexes) are normal; no age effect is noted.

Ordinarily Turners are sterile; however, at least one normal birth and several pregnancies have been reported for presumed Turners. In these cases, though, mosaicism (45,X/46,XX) has not been eliminated with certainty.

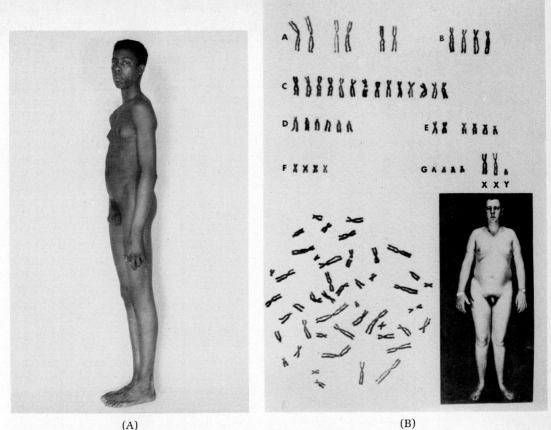

(A)   (B)

Figure 11-8. *The Klinefelter syndrome in man. Such persons are AAXXY (47, XXY). External genitalia are male type, but there is usually some femalelike breast development. (A) General phenotype. (Photo courtesy Dr. Victor A. McKusick.) (B) General phenotype, metaphase chromosomal smear, and karyotype of another Klinefelter individual. (Photo courtesy Ms. Dawn DeLozier.)*

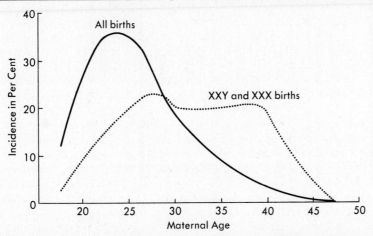

Figure 11-9. *The relationship between maternal age and XXY and XXX births.*

Table 11-3. Sex chromosome anomalies and their estimated frequencies

Designation	Chromosome constitution	Sex chromatin	Somatic chromosome number	General sex phenotype	Usual fertility	Estimated frequency per 1,000	Estimated number in the U.S.*
Klinefelter	AAXXY	+	47	♂	-	2.0	453,000
Turner	AAXO	-	45	♀	(-)†	0.2-0.4	45,300-90,600
Triplo-X	AAXXX	++	47	♀	+	0.75	169,880
Tetra-X	AAXXXX	+++	48	♀	Unknown	Very low	Very low
Triplo-X, Y	AAXXXY	+++	48	♂	-	Very low	Very low
Tetra-X, Y	AAXXXXY	+++	49	♂	-	Very low	Very low
Penta-X	AAXXXXX	++++	49	♀	-	Very low	Very low
XYY	AAXYY	-	47	♂	±‡	0.7-2.0	186,500-453,000
Klinefelter XXXY	AAXXYY	+	48	♂	Not reported	Very low	Very low
Klinefelter XXXYY	AAXXXYY	++	49	♂	Not reported	Very low	Very low

*Based on a 1980 population of 226.5 million with an average birth rate of 2.1 children among women of child-bearing age. (Preliminary 1980 census figures from U.S. Census Bureau.)
†A very few cases of motherhood have been reported for presumed Turners.
‡But generally highly infertile because of low sperm count.

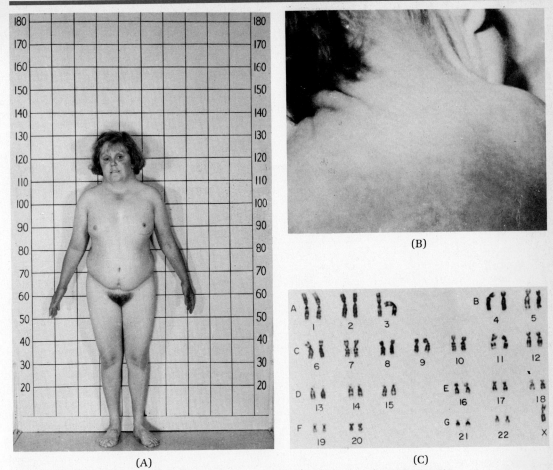

Figure 11-10. *The Turner syndrome in the human female. These persons are AAXO (45,X). Note the female genitalia. (A) General phenotype. (Photo courtesy Dr. Victor A. McKusick.) (B) Detail of webbed neck. (C) Karyotype of a Turner. [(B) and (C) courtesy Ms. Dawn DeLozier.]*

The frequency of live Turner births has been estimated at two to four per 10,000 live births, although a study of 139 cases of gonadal dysgenesis, place the incidence at a mere 1 in 10,000. On the other hand, de Grouchy and Turleau (1977) find the frequency to be 4 in 10,000. This low incidence of Turner births is a reflection of a high rate of intrauterine mortality. Reports of spontaneously aborted 45,X fetuses range from 90 to 97.5 of all Turners conceived. Some 20 to 30 percent of *all* spontaneously aborted fetuses are 45,X.

The conclusion to be drawn from the Turner syndrome is that, in the absence of a Y chromosome, the general sex phenotype is female.

Poly-X females. In 1959 the first known case of a *triplo-X* individual, i.e., 47,XXX, was reported. This was clearly a female in general sex phenotype, but at age 22, she had infantile external genitalia and marked underdevelopment of internal genitalia and breasts. She was somewhat retarded mentally. Since that time many more XXX females have been described, and it is estimated that between 1 in 1,000 and 1 in 2,000 live female births is triplo-X. Some XXX females are essentially normal, but others are retarded and/or show abnormalities of primary and secondary sex characters. Apparently all are fertile, but among more

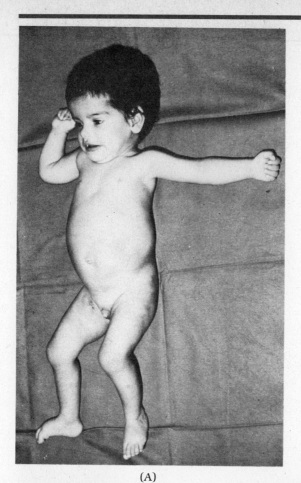

(A)

Figure 11-11. Poly-X individual. This child is *AAXXXXY (49, XXXXY). (A) General phenotype. (B) External genitalia.* (Photos courtesy Ms. Dawn DeLozier.)

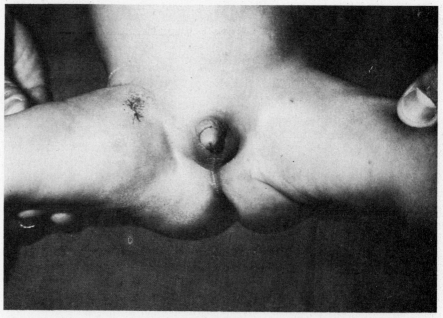

(B)

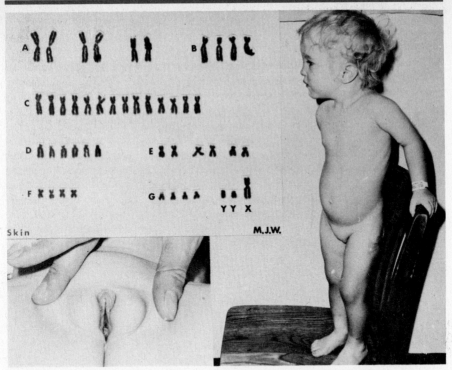

Figure 11-12. *An XYY human who, coincidentally, displays the trait known as testicular feminization which has a genetic basis apart from the XY karyotype. Photo shows general phenotype of a three-year-old, together with karyotype and detail of intermediate genitalia.* (Courtesy Ms. Dawn DeLozier.)

than 30 children of triplo-X mothers, all were XX or XY except for one Klinefelter. A very few tetra-X (48,XXXX) and penta-X (49,XXXXX) persons have been described; manifestations are similar to those of triplo-X individuals but more marked. In general, as the number of X chromosomes increases, intelligence is progressively reduced, as with Klinefelters. Also, it is important to note that multiple-X individuals are female in general phenotype, and show evidence of a maternal age effect (Fig. 11-9). Persons with more than two X chromosomes plus a Y exhibit some degree of testicular abnormality, not greatly unlike those of Klinefelters (Fig. 11-11).

The XYY male. Particular interest has recently focused on behavior patterns in XYY individuals. These are all males, above average in height (>72 in.), with intelligence quotients ranging from about 80 to 118, and a history of severe (facial) acne during adolescence. Abnormalities of both internal and external genitalia have been noted in some of these persons (Fig. 11-12), but no consistent major anomalies occur. This kind of individual was noticed first in 1965 when a high incidence of XYY males (nine of 315 inmates, or a frequency of about 0.029), housed in the maximum security section of a Scottish criminal institution was reported. Many similar reports that followed thereafter in the 1960s,[1] led initially to the notion that XYY males are more aggressive and more likely than the XY male to commit crimes of violence. This conclusion is now viewed as having been premature, for after additional studies of the general population, it

[1]Borgaonkar (1969) has compiled an extensive bibliography on the XYY syndrome.

has become clear that many XYY males do make satisfactory social adjustment; some are indeed of very retiring personality. A 1970 study of 4,366 consecutive births at Yale–New Haven Hospital showed XYY births to occur with a frequency of 0.69 per thousand.

A number of widely publicized criminal cases have served to direct attention to legal aspects of chromosomal aberrations, particularly the XYY male (Table 11-4). Yet the violent crime may be the exception with the 47,XYY individual; offenses predominantly involve theft, arson, and confidence schemes.

In general, courts in the United States have not "excused" criminal behavior because of an XYY karyotype, but some have taken that finding into account in sentencing. Scanty evidence from other countries indicates that some courts and juries do recognize the XYY state as a contributing cause of commission of the crime. In the French case of Daniel Hugon the prosecutor asked for only a 5- to 10-year sentence. Although Hugon was found legally sane, the court considered his XYY karyotype as crime-related and complied with the prosecution's recommendation in passing sentence. The Australian case of Hannell is interesting in that the jury is reported to have deliberated for only eleven minutes before reaching its verdict. Aside from the legal and social questions raised by the 47,XYY individual, it is important to note here that the general sex phenotype is male.

Genetically determined sex reversal. German et al. (1978) report an interesting case of a presumed gene-determined sex reversal in humans. This case involves an extensive family pedigree in which three confirmed 46,XY phenotypic females were intensively studied. Each of these three had vestigial gonads or gonadal tumors in place of normal gonads of either sex. Throughout life, to the time of examination, each appeared to be a normal female and was raised as such. All were somewhat tall for females (approximately 67 to 70 inches), with normal female genitalia, normal uterus and fallopian tubes but, as pointed out, with only rudimentary ovarian tissue.

Based on their study of this family, German and his colleagues conclude that this rare type of XY gonadal dysgenesis is due to a recessive gene, probably X-linked, that prevents testicular differentiation in 46,XY embryos. They conclude that the Y chromosome does not act alone in causing undifferentiated embryonic gonadal tissue to develop into testes, but instead postulate an interaction "between a locus near the centromere of the Y and one somewhere on the X." This interaction, they theorize, must be either (1) induction of a structural gene for testicular development on the X chromosome by a controlling element on the Y chromosome, or (2) a structural gene on the Y chromosome being induced by a control on the X. In the family studied, the conclusion was that the X-linked gene is defective and unable to interact with the Y-borne locus.

Inasmuch as initiation of male gonadal development requires the Y chromosome, implication of some Y-linked product in the process is suggested. Deletions (see Chapter 14) of the short arm of the Y chromosome result in a general female phenotype and rudimentary, nonfunctional gonads. Deletions of the long arm of the Y chromosome, on the other hand, sometimes occur in normal, fertile males. Thus a testis-determining locus would appear to be on or very near the short arm of the Y chromosome.

Other sex chromosome anomalies. A number of other sex chromosome anomalies have been reported, most of which are quite rare (Table 11-3). In every case, though, the presence of even one Y chromosome serves to produce the male sex phenotype, regardless of the number of Xs that may be present. No clear relationship between sex chromosome anomalies in children and any particular parental characteristics, except for some suggestion of a higher risk in mothers with thyroid disorder (hyperthyroid and hypothyroid), has been observed.

Table 11-4. Some criminal cases involving XYY males

Year	Accused	Location	Charge	Plea	Disposition
1968	Daniel Hugon	Paris	Murder	Not guilty by reason of XYY	Seven years' imprisonment
1968	Lawrence Hannell	Melbourne	Murder	Not guilty by reason of insanity	Not guilty; maximum security hospital until "cured"
1968	Ernest Beck	West Germany	Multiple murder	Not guilty by reason of insanity	Guilty; life imprisonment
—	Robert Tait	Melbourne	Murder	Not guilty	Guilty; sentenced to hang, but commuted to life imprisonment
1969	Sean Farley	New York, N.Y.	Murder, rape	Not guilty by reason of XYY	Guilty of first-degree murder
1970	Raymond Tanner	California	Assault with intent to commit murder	Not guilty; insanity by reason of XYY	Guilty; imprisonment

Hermaphroditism. No account of sex chromosome anomalies would be complete without mention of hermaphroditism and pseudohermaphroditism. *True hermaphrodites*, by generally accepted definition, are individuals that possess both ovarian and testicular tissue (Fig. 11-13). The external genitalia are ambiguous, but often more or less masculinized; secondary sex characters vary from more or less male to more or less female. Some are reared as males, some as females. Ordinarily, true hermaphrodites are sterile because of rudimentary ovatestes. However, at a meeting of The American Society of Human Genetics (Vancouver, B.C., October 1978), Brazzel, Tegenkamp, and Ashcom reported a pregnancy (unknown to the individual!) that terminated in delivery of a stillborn child to a 25-year-old true hermaphrodite after about 30 weeks' gestation. Even more remarkable was the fact that this individual engaged in male sexual activity in her early years but gradually shifted to a preference for the female role. Surgical procedures disclosed a tumor containing muscle plus ovarian tissue on the left side, with an ovatestis in a sac in the right groin. As these researchers put it, "True hermaphroditism . . . is rare; however, pregnancy and bisexual activity in such an individual must indeed be considered more rare."

The incompletely known, apparently multiple, causes of the various degrees of hermaphroditism require a discussion beyond the scope of an introductory text. However, some cytological and morphological characteristics are interesting in that they demonstrate that sex development can go awry in a variety of ways. In a study of 108 cases of true hermaphroditism, for example, 59 were 46,XX in karyotype; 21 were 46,XY; and 28 were mosaics. All but two of the mosaics were found to possess some Y chromosome cell lines, and the same condition may be suspected in the two apparent nonmosaics. Accounting for the development of testicular tissue in the absence of a Y chromosome presents some difficulties, of course. Apparent 46,XX true hermaphrodites could be accounted for without conflicting with the apparent necessity of a Y chromosome for testicular development by assuming loss of the Y chromosome from a 47,XXY fetus at an appropriate stage of embryo development. The most suggestive evidence may well come from the 46,XX/46,XY mosaics, a condition that seems to indicate that many true hermaphrodites may originally have had both XX and XY cell lines.

Pseudohermaphrodites have either testicular or ovarian tissue, generally rudimentary, but not both. On the basis of the usual chromosomal constitution (genetic sex), two major classes, *male* and *female*, are distinguished. The former are most often 46,XY or 46,XY/45,X mosaics. Very rarely is the karyotype 46,XX or some type of sex chromosome mosaic. External genitalia are ambiguous; if the balance is shifted somewhat toward male, the individual is usually assigned to the male sex. A penislike organ of variable size is present, as is often a urogenital sinus. In adolescence pubic and axillary hair develops and the voice deepens. Often some breast development, as in females, occurs. This condition is referred to as *masculinizing male pseudohermaphroditism* (Fig. 11-14). Male pseudohermaphroditism is distinguished from another sex aberration, the testicular feminization syndrome, which is caused by mutation of a gene that de Grouchy and Turleau (1977) feel is very probably located on the X chromosome. These individuals are invariably 46,XY, but they present a feminine phenotype (See also page 198.)

Feminizing male pseudohermaphrodites (Fig 11-15) are, like the masculinizing variety, usually 46,XY (hence, *male*) or, less often, a 46,XY/45,X (or other) mosaic. These individuals present a general female sex phenotype, but often (not always) have poorly developed secondary sex characters. Some lead a normal *female* sex life, though they cannot conceive.

Female pseudohermaphrodites are 46,XX (hence, *female*) but present a more or less masculine phenotype (Fig. 11-16). The external genitalia are ambiguous; ovaries are present but are immature or rudimentary. This condition arises most

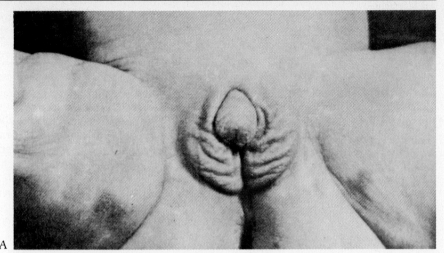

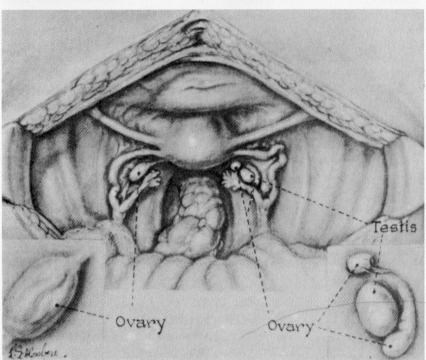

Figure 11-13. *True hermaphroditism in a five-month-old child. (A) Ambiguous external genitalia. The clitoris is enlarged, resembling a penis; the labio-scrotal folds are fused. (B) Drawing of operative findings, showing ovary (left) and an ova-testis (right).* (From M. Bartalos and T. A. Baramki, 1967. *Medical Cytogenetics.* Baltimore, The Williams & Wilkins Company. Used by permission of authors and publisher.)

frequently by virilization of a female fetus because of either an inherited proliferation of the adrenal glands or a prenatal hormone imbalance in the mother.

Pseudohermaphroditism is summarized in Table 11-5.

Mechanism of sex determination in human beings. Individuals that have at least one Y chromosome are, with the exception of unusual cases such as

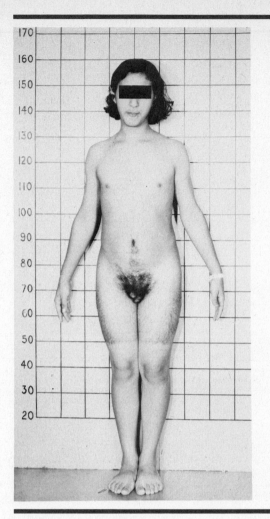

Figure 11-14. *A 16-year-old male pseudohermaphrodite of the masculinizing variety (46, XY), reared as a female. Note body hair, lack of breast development, and ambiguous external genitalia. (Courtesy Dr. T. A. Baramki.)*

Table 11-5. Summary of human pseudohermaphrodites

Type	Variety	Gonads	External genitalia	Genetic sex	General sex phenotype	Sex of rearing
♂	Masculinizing	Testes ± dystrophic	Ambiguous but ± ♂	XY	♂ or ± ♀	♂ (or ♀)
	Feminizing	Testes ± dystrophic; often inguinal	± ♀	XY	♀	♀
♀	—	Ambiguous; often immature ovaries	Ambiguous but ± ♂	XX	± ♂	♂ (or ♀)

± means more or less.
Parentheses around a sex designation indicate a minority of cases.
For details see text.

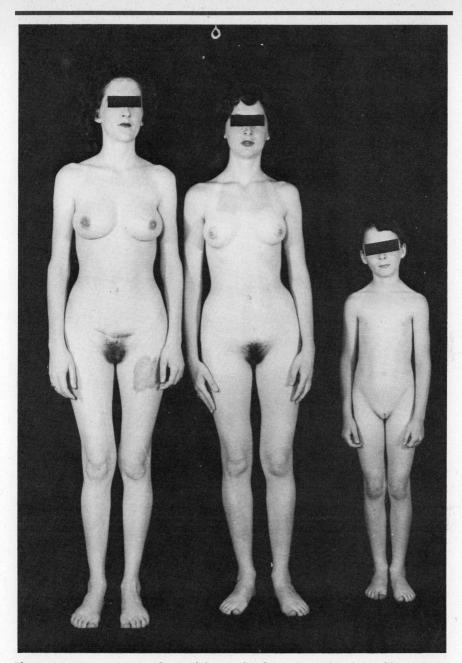

Figure 11-15. *Feminizing male pseudohermaphroditism (46, XY) in three siblings.* (From M. Bartalos and T. A. Baramki, 1967. *Medical Cytogenetics.* Baltimore, The Williams & Wilkins Company. Used by permission of authors and publisher.)

those described by German et al. (1978), ordinarily male with regard to external genitalia and general phenotype, though they may be sterile (Table 11-3). In contrast, persons with one or more X chromosomes are *ordinarily* phenotypically female as long as no Y is present, though, again, infertility sometimes occurs.

Recent chromosome mapping techniques appear to have established the locus for a testis-determining factor on the nonfluorescent segment of the Y chromosome (Table 6-2). Recall that, in the presence of the Y chromosome, but (fol-

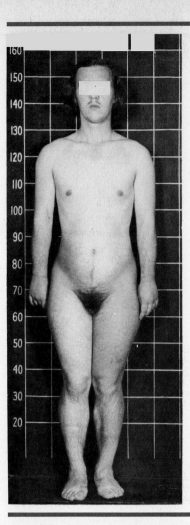

Figure 11-16. *A 16-year-old female pseudo-hermaphrodite (46, XX), showing short, stocky build, body hair, and lack of breast development. External genitalia are ambiguous.* (From M. Bartalos and T. A. Baramki, 1967. *Medical Cytogenetics.* Baltimore, The Williams & Wilkins Company. Used by permission of authors and publisher.)

lowing German's second hypothesis) with a normally functioning X-linked controlling element, inductors are produced in the embryo, thus stimulating the medulla of the undifferentiated gonads to produce testicular tissue (p. 190). In the absence of Y, but in the presence of one or more X chromosomes, the cortical region of the embryonic gonads begins to differentiate into ovarian tissue. Persons with one or more X chromosomes, but without a Y, ordinarily have the general female phenotype. On the other hand, even 49,XXXXY persons are male in general phenotype, though sterile. The rare type of case described by German and his colleagues stands as an exception to this generalization. In any event, sex in human beings is determined by genes on the X and Y chromosomes.

Do the autosomes play any part in sex determination in humans? Persons with exceptional numbers of autosomes are still male or female in general phenotype, which is consistent with genetic sex.

The *usual* situation in human beings may be summarized as follows:

1. Autosomes play no part in determining sex.
2. Genes on the Y chromosome determine maleness (provided a controlling——X-linked?——gene permits testicular differentiation).
3. Genes on the X chromosome determine femaleness in the absence of any Ys.

Sex determination in humans, then, although related to the X-Y chromosome make-up, operates differently than in *Drosophila*. Mammals generally appear to follow the mechanism outlined for human beings.

Plants

One flowering plant, the wild campion (*Lychnis dioica*, formerly *Melandrium*) of the pink family (Caryophyllaceae), has been extensively investigated. This plant illustrates still another variation of the XX-XY system. In angiosperms, the plant recognized by name is the asexual, diploid sporophyte generation. The sexual phase is microscopic and contained largely within the tissues of the sporophyte, on which it is parasitic. Flowers contain either or both of two essential organs, stamens and/or pistils. The former produce microspores that develop into male-gamete-bearing plants (at one stage these are the well-known pollen grains). Pistils produce and contain the egg-bearing sexual plant. Many plants have so-called perfect flowers, which contain both stamens and one or more pistils. *Lychnis* has imperfect flowers, which bear either stamens *or* pistils. The species is, moreover, dioecious, so that there are staminate (male-producing) individuals and pistillate (female-producing) ones. Though the nomenclature is not accurate, staminate plants are often referred to as male plants and pistillate as female.

In *Lychnis*, staminate plants are XY and pistillate plants are XX. The X/A ratio bears no relation to "sex" but, through studies of plants with multiple sets of chromosomes, the X/Y ratio is found to be critical. X/Y ratios of 0.5, 1.0 and 1.5 are found in plants having only staminate flowers; in plants whose X/Y ratio is 2.0 or 3.0, occasional perfect flowers occur among otherwise all staminate flowers. In plants having four sets of autosomes, four X chromosomes and a Y, flowers are perfect but with an occasional staminate one. This situation is summarized in Table 11-6 and illustrated in Figure 11-17.

Extensive investigations of the cytology of *Lychnis* show that, where portions of the X or Y chromosome are deleted, the two chromosomes compare as shown in Figure 11-18. The sex chromosomes are quite dissimilar in size, i.e., the Y is larger than the X, and each is larger than any autosome. Only a small portion of the X is homologous with a similar small bit of the Y, as suggested by their synaptic figures in meiosis. When region I is deleted, perfect-flowered plants are

Table 11-6. "Sex" and X/Y ratios in Lychnis

Chromosome constitution	X/Y ratio	"Sex"
2 AXYY	0.5	♂
2 AXY		
3 AXY	1.0	♂
4 AXY		
4 AXXXYY	1.5	♂
2 AXXY		
3 AXXY		
4 AXXY	2.0	♂ with occasional ⚥ flower
4 AXXXXYY		
3 AXXXY	3.0	♂ with occasional ⚥ flower
4 AXXXY		
4 AXXXXY	4.0	with occasional ♂ flower

♂ = staminate; ♀ = pistillate; ⚥ = perfect.
Each A = one set of 11 autosomes, each X = an X chromosome, each Y = a Y chromosome.

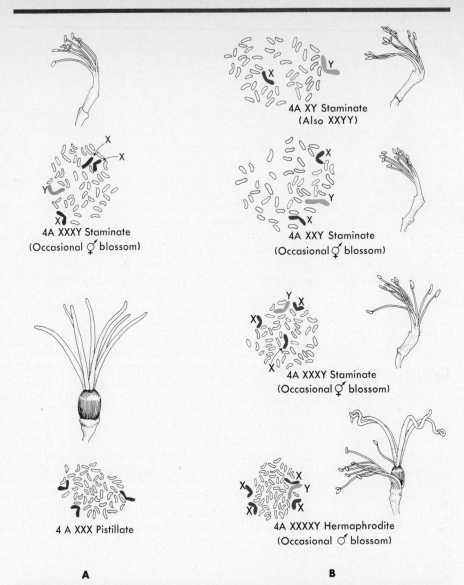

Figure 11-17. (A) Reproductive parts and camera lucida drawings of the somatic chromosomes from 4A XXXY and 4A XXX plants of Lychnis (Melandrium). Note that a single Y chromosome produces a staminate plant having occasional perfect flowers; in the absence of the Y chromosomes, plants are pistillate. (B) Reproductive parts and camera lucida drawings of the somatic chromosomes of Lychnis in a series showing pistillate-determining tendency of the X chromosome. All plants shown are tetraploid with respect to autosomes, but differ in the number of X chromosomes present. Tendency to produce pistillate flowers increases as the number of X chromosomes rises. (Redrawn from H. W. Warmke, 1946, "Sex Determination and Sex Balance in Melandrium." American Journal of Botany, **33:** 648–660, by permission of the author and the Botanical Society of America, Inc., Washington.)

produced, whereas loss of region II produces pistillate plants (that is, whereas AAXY plants are staminate, AAXY^{-II} plants are pistillate), and deletion of region III produces sterile staminate plants with aborted stamens.

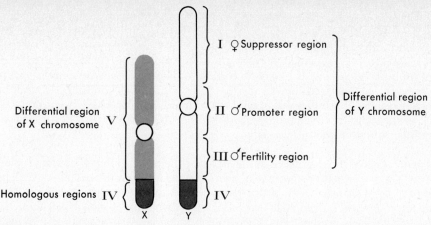

Figure 11-18. *Comparison of X and Y chromosomes of the plant* Lychnis. *Regions I, II, and III bear holandric genes, and region V, sex-linked genes. Genes in region IV are termed incompletely sex-linked.* Lychnis *is unusual in that its Y chromosome is the larger of the two sex chromosomes.*

Genic determination of sex

Sex does not appear to be controlled in all dioecious organisms by numerous genes on two or more chromosomes, however. Several examples will serve to illustrate some variations on the so-called chromosomal method.

Asparagus. Rick and Hanna (1943) presented evidence to show that sex in asparagus is determined by a single pair of genes, with "maleness" (i.e., formation of staminate flowers) the dominant character. Asparagus is normally dioecious; some plants produce only staminate flowers, others only pistillate. Rudimentary pistils occur in staminate flowers, and abortive stamens in pistillate blossoms. These rudimentary organs are ordinarily nonfunctional. However, the pistils in staminate flowers do very rarely function to produce viable seeds. Flower structure in this species is such that these seeds most likely result from self-pollination. In effect, this is a "male × male" cross from the genetic standpoint. The basis for the occasional functional pistils in staminate plants is not known, but both genetic and environmental factors have been suggested.

Rick and Hanna germinated 198 seeds from these uncommon functional pistils produced on staminate plants, and found a progeny ratio of 155 staminate to 43 pistillate. This is a close approximation to a 3:1 ratio, as a chi-square test will indicate. One-third of the staminate progeny, when crossed with normal pistillate plants, yielded only staminate offspring. The rest proved to be heterozygous, and produced staminate and pistillate offspring in a ratio close to 1:1.

If "maleness" is represented as $A-$ and "femaleness" as aa, the original "male × male" cross would be

$$P \; Aa \times Aa$$
$$F_1 \; \tfrac{3}{4} A- \; + \tfrac{1}{4} aa$$

On the basis of monohybrid inheritance, one-third of the $A-$ (staminate) plants should be AA and two-thirds Aa. Crosses between these genotypes and normal pistillate (aa) individuals would produce the results reported by Rick and Hanna.

Honeybee. Worker and queen bees are diploid females, with 32 chromosomes. Drones, on the other hand, are males that have but 16 chromosomes. Through extensive studies on the parasitic wasp *Habrobracon* it was discovered that femaleness in such hymenopterans is determined by heterozygosity at a number of different loci on several chromosomes. The situation is analogous to a series of multiple alleles but is based on chromosomal segments that contain several genes each. Monoploid individuals that hatch from unfertilized eggs cannot be heterozygous; hence they are male. Diploid males however, have been produced experimentally by developing individuals that are heterozygous in a sufficient number of chromosome segments.

Corn. Unlike the organisms so far described, corn (*Zea mays*) is monoecious. The "tassel" consists of staminate flowers and the ear of pistillate flowers. Among several controls over "sex" in this plant are two interesting pairs of genes. The genotype *bs bs* ("barren stalk") results in plants having no ears at all, though a normal tassel is present (Fig. 11-19). Such individuals are staminate ("male"). Gene *ts* ("tassel seed"), when homozygous, converts the tassel to pistillate flowers so that ears develop at the top of the plant (Fig. 11-20). Therefore *ts ts* individuals are pistillate ("female"). If ♂ represents staminate flowers, and ♀ pistillate ones, the various genotypes and phenotypes are as follows:

Genotype	Phenotype	
Bs − Ts −	Normal monoecious	♂ ♀
bs bs Ts −	Staminate	♂
Bs − ts ts	Pistillate; ears terminal *and* lateral	♀ ♀
bs bs ts ts	Pistillate; ears terminal only	♀

Plants of genotype *bs bs Ts −* plus either *Bs − ts ts* or *bs bs ts ts* comprise a dioecious race of corn. Such a race may easily be produced by making the cross *bs bs ts ts* × *bs bs Ts ts*, which will segregate pistillate and staminate in a 1:1 ratio. Perhaps a similar sort of development has occurred in the evolution of dioecism. The practical advantage of the dioecious state to the breeder is obvious.

Chlamydomonas. The common microscopic, unicellular, monoploid green alga *Chlamydomonas* reproduces sexually by **isogametes.** These are gametes in which no morphological differences can be observed. Syngamy is selective, in that the gametes of all individuals are not equally capable of fusing. Mating strains, designated as + and −, rather than male and female, are recognized. Plus strains conjugate only with minus. Mating type appears to be related to physiological differences within the cell and is, in turn, known to be genetically determined. Thus, zygotes produce four cells by meiosis; two become plus adults and two become minus. Diploid adults have been developed by mitosis of zygotes. All of such unusual 2*n* individuals are of the minus mating strain, which indicates dominance of this trait. Progeny ratios resulting from crosses involving diploids confirm this. So, in this simple alga where evolution of sex has not progressed very far, sex appears to be controlled by a single pair of genes that are responsible for physiological rather than morphological differences.

Figure 11-19. Barren stalk corn, bs bs. Note the absence of ears. (From *The Ten Chromosomes of Maize*, DeKalb Agricultural Association, DeKalb, Illinois. Reproduced by permisison.)

Figure 11-20. Tassel seed corn, ts ts. Note silks of developing ear at top of stem in place of tassel. (From *The Ten Chromosomes of Maize*, DeKalb Agricultural Association, DeKalb, Illinois. Reproduced by permission.)

Conclusion

Sex represents something of a continuum in many organisms, but it is basically gene-determined. Species differ with respect to (1) the number of genes that appear to play a part and (2) the locations of those genes (X chromosome, Y chromosome, autosomes, cytoplasm). The genes an individual receives at the moment of syngamy determine which kind of gonads, and therefore of gametes, will be formed. Later processes of sex differentiation are quite distinct from the initial event of sex determination. In insects the critical factors in sex differentiation appear to be intracellular. In humans and other mammals, however, sex differentiation is hormonal. Sex hormones, produced by the gonads, interact with endocrine glands elsewhere in the body to affect the multiplicity of secondary sex characters. These characters may be markedly affected by hormone injections; therefore, the potential of each individual for either sex is suggested.

References

Barr, M. L., 1966. The Sex Chromosomes in Evolution and Medicine. *Can. Med. Assoc. Jour.*, **95**: 1137–1148.

Barr, M. L., F. R. Sergovich, D. H. Carr, and E. L. Shaver, 1969. The Triplo-X Female: An Appraisal Based on a Study of 12 Cases and a Review of the Literature. *Can. Med. Assoc. Jour.*, **101**: 247–258.

Borgaonkar, D. S., 1969. 47,XYY Bibliography. *Ann. Genet.*, **12**: 67–70.

Borgaonkar, D. S., H. M. Herr, and J. Nissim, 1970. DNA Replication Pattern of the Y Chromosome in XYY and XXYY Males. *Jour. Hered.*, **61**: 35–36.

Brazzel, J., T. Tegenkamp, and R. Ashcom, 1978. Pregnancy in a Bisexually Active True Hermaphrodite, 46,XX. *Program and Abstracts*, The American Society of Human Genetics, 29th Annual Meeting. Chicago: The University of Chicago Press.

Bridges, C. B., 1916a. Nondisjunction as Proof of the Chromosome Theory of Heredity. *Genetics*, **1**: 1–52.

Bridges, C. B., 1916b. Nondisjunction as Proof of the Chromosome Theory of Heredity (concluded). *Genetics*, **1**: 107–163.

Bridges, C. B., 1925. Sex in Relation to Chromosomes and Genes. *Amer. Natur.*, **59**: 127–137.

Cohen, M. M., and M. W. Shaw, 1965. Two XY Siblings with Gonadal Dysgenesis and a Female Phenotype. *New England Jour. Med.*, **272**: 1083–1088.

de Grouchy, J., and C. Turleau, 1977. *Clinical Atlas of Human Chromosomes*. New York, John Wiley & Sons.

Ferguson-Smith, M. A., 1970. Chromosomal Abnormalities II: Sex Chromosome Defects. In V. A. McKusick and R. Claiborne, eds. *Medical Genetics*. New York: HP Publishing Co.

Fichman, K. R., B. R. Migeon, and C. J. Migeon, 1980. Genetic Disorders of Male Sexual Differentiation. In H. Harris and K. Hirschhorn, eds., *Advances in Human Genetics*, volume 10. New York: Plenum Press.

German, J., 1967. Autoradiographic Studies of Human Chromosomes. In J. F. Crow and J. V. Neel, eds. *Proc. Third Int. Cong. Human Genetics*. Baltimore: The Johns Hopkins Press.

German, J., 1970. Studying Human Chromosomes Today. *Amer. Scientist*, **58**: 182–201.

German, J., J. L. Simpson, R. S. K. Chaganti, R. L. Summit, L. B. Reid, and I. R. Merkatz, 1978. Genetically Determined Sex-Reversal in 46,XY Humans, *Science*, **202**: 53–56.

Jacobs, P. A., A. G. Baikie, W. M. Court Brown, T. N. MacGregor, N. Maclean, D. G. Harnden, 1959. Evidence for the Existence of the Human "Superfemale." *Lancet*, **2**: 423.

Jacobs, P. A., M. Brunton, M. M. Melville, R. P. Brittain, and W. F. McClermont, 1965. Aggressive Behavior, Mental Subnormality and the XYY Male. *Nature*, **208**: 1351–1352.

Lederberg, J., and E. L. Tatum, 1946. Gene Recombination in *Escherichia coli*. *Nature*, **158**: 558.

Lubs, H. A., and F. H. Ruddle, 1970. Chromosomal Abnormalities in the Human Population: Estimation of Rates Based on New Haven Newborn Study. *Science*, **169**: 495–497.

Lyon, M. F., 1961. Gene Action in the X-Chromosome of the Mouse (*Mus musculus* L.). *Nature*, 190: 372–373.

Lyon, M. F., 1962. Sex Chromatin and Gene Action in Mammalian X-Chromosomes. *Amer. Jour. Human Genetics*, 14: 135–148.

McKusick, V. A., and R. Claiborne, eds., 1973. *Medical Genetics*. New York, HP Publishing Co.

Morgan, L. V., 1922. Non-Criss-Cross Inheritance in *Drosophila melanogaster*. *Biol. Bull.*, 42: 267–274.

Polani, P. E., 1961. Turner's Syndrome and Allied Conditions. *Brit. Med. Bull.*, 17: 200–205.

Price, W. H., J. A. Strong, P. B. Whatmore, and W. F. McClermont, 1966. Criminal Patients with XYY Sex-Chromosome Complement. *Lancet*, 1: 565–566.

Rick, C. M., and G. C. Hanna, 1943. Determination of Sex in *Asparagus officinalis* L. *Amer. Jour. Bot.*, 33: 711–714.

Rook, A., L. Y. Hsu, M. Gertner, and K. Hirschhorn, 1971. Identification of Y and X Chromosomes in Amniotic Fluid Cells. *Nature*, 230: 53.

Sternberg, W. H., and D. L. Barclay, 1967. Women with Male Sex Chromosomes: The Syndromes of Testicular Feminization and XY Gonadal Dysgenesis. *Jour. Amer. Med. Women's Assn.*, 22.

Summit, R. L., and D. Bergsma, eds. 1978. *Sex Differentiation and Chromosomal Abnormalities*. Annual Review of Birth Defects, 1977. Proceedings of the 1977 Memphis Birth Defects Conference, Part C. New York: Alan R. Liss, Inc.

Sumner, A. T., J. A. Robinson, and H. J. Evans, 1971. Distinguishing Between X, Y, and YY-Bearing Spermatozoa by Fluorescence and DNA Content. *Nature New Biol.*, 229: 231–233.

Turner, C. D., 1964. Special Mechanisms in Anomalies of Sex Differentiation. *Amer. Jour. Obstet, and Gyn.* 90(No. 7, pt. 2): 1208–1226.

Turner, H. H., 1938. A Syndrome of Infantilism, Congenital Webbed Neck, and Cubitus Valgus. *Endocrinology*, 23: 566–574.

Zeuthen, E., J. Nielsen, and H. Yde, 1973. XYY Males Found in a General Male Population: Cytogenetic and Physical Examination. *Hereditas*, 74(2): 283–290.

Problems

11-1 What is the sex designation for each of the following fruit flies (each A = one set of autosomes, each X = one X chromosome): (a) AAXXXX, (b) AAAAAXX, (c) AAXXXXXX, (d) AAAAAXXX, (e) AAAAXXXX, (f) AAAXY?

11-2 What is the sex designation for the following human beings: (a) AAXXX, (b) AAXXXYYY, (c) AAXO, (d) AAXXXXXY, (e) AAXYYY?

11-3 What phenotypic ratio results in corn from selfing *Bs bs Ts ts* plants?

11-4 In *Drosophila*, what fraction of the progeny of the cross *Tra tra* XX × *tra tra* XY is "transformed"?

11-5 What is the sex ratio in the progeny of the cross given in problem 11-4?

11-6 In poultry, the dominant gene, *B*, for barred feather pattern is located on the Z chromosome. Its recessive allele, *b*, produces nonbarred feathers. What is the genotype of a (a) nonbarred female, (b) barred male, (c) barred female, (d) nonbarred male?

11-7 Give the phenotypic and sex ratios in the progeny of the following crosses in poultry: (a) nonbarred ♀ × heterozygous barred ♂, (b) barred ♀ × heterozygous barred ♂.

11-8 In poultry, removal of the ovary results in the development of the testes. Thus a female can become converted to a male, producing sperm and developing male secondary sex characters. If such a "male" is mated to a normal female, what sex ratio occurs in the progeny?

11-9 An XXY *Drosophila* female is mated to a normal male. The progeny include 5 percent metafemales. If secondary nondisjunction occurred, what was the frequency of XX eggs?

11-10 Based on your answer to 11-9, and the assumption that Y eggs occur with a frequency of 0.1, what percentage of the progeny of the cross of problem 11-9 will be phenotypically normal females?

11-11 An attached-X female *Drosophila* is mated to a normal male. (a) What fraction of the zygotes become viable females? (b) What fraction of the zygotes are metafemales in chromosomal constitution? (c) What fraction of the viable, mature progeny are sterile males?

11-12 If the female parent of problem 11-11 is $AA\widehat{X^+X^w}$ and the male AAX^wY, what will be the eye color of (a) viable, mature male progeny and (b) viable, mature female progeny?

11-13 How can the human Y chromosome be distinguished in suitably stained preparations?

11-14 If you assume parents to be AAXX and AAXY, how could you account in humans for children of each of the following types: (a) AAXYY, (b) AAXXY, (c) AAXO?

11-15 Numerous cases of human mosaics are reported in the literature. How could you account cytologically for an AAXX-AAXO mosaic?

11-16 No individuals, either live births or spontaneously aborted fetuses (abortuses), completely lacking any X chromosomes (e.g., AAOY) have ever been discovered. Why is this to be expected?

11-17 From the standpoint of survival of the species, which system of sex determination, that of *Asparagus* or of humans, seems to offer the greater advantage? Why?

11-18 For humans, give the genetic sex of (a) most true hermaphrodites, (b) masculinizing male pseudohermaphrodites, (c) feminizing male pseudohermaphrodites, (d) female pseudohermaphrodites.

11-19 Can you suggest any reason why such persons as those having the karyotype 48,XXXY are highly sterile?

11-20 A human blood antigen, Xg^a, is due to a dominant gene (symbolized also as Xg^a) on the short arm of the X chromosome. Its recessive allele may be symbolized Xg. Phenotypes are designated as either $Xg^a +$ or $Xg^a -$. Examine the following pedigree of a Turner individual and determine the source of the single X chromosome in the Turner child (was it maternal or paternal?).

$Xg^a -$ $Xg^a +$

$Xg^a -$

11-21 In terms of the gene symbols given in the preceding problem, what is the genotype of each of the three persons in 11-20 (a)?

11-22 Examine the following pedigrees of a Klinefelter individual; which parent was the source of the extra X chromosome?

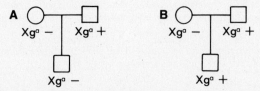

A $Xg^a -$ $Xg^a +$ **B** $Xg^a -$ $Xg^a +$

$Xg^a -$ $Xg^a +$

11-23 In terms of the gene symbols given in problem 11-20, what is the genotype of each of the three persons in pedigree 11-22 (a)?

CHAPTER 12

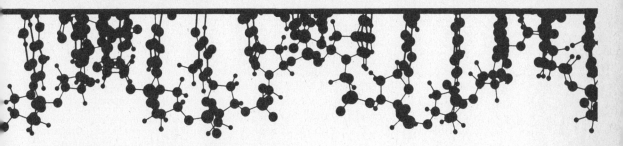

Inheritance related to sex

The fact that males in *Drosophila* and in humans have an X and Y chromosome, whereas females have two Xs and no Y, raises some interesting genetic possibilities. This is especially true in view of the fact that the sex chromosomes are not entirely homologous and therefore, inheritance patterns related to the sex of the individual should be expected to be somewhat different from those examined previously. For example, genes that occur only on the X chromosome will be represented twice in females, once in males; recessives of this type might be expected to show up phenotypically more often in males. Genes located exclusively on the X chromosome are called **sex-linked,** or **X-linked, genes.** On the other hand, genes that occur only on the Y chromosome can produce their effects only in males; these are **holandric genes** (Greek, *holos,* "whole," and *andros,* "man"). Still other mechanisms are known whereby a given trait is limited to one sex (**sex-limited genes**), or even in which dominance of a given allele depends on the sex of the bearer (**sex-influenced genes**). Genes that occur on homologous portions of the X and Y chromosomes are called **incompletely sex-linked.** The first four of these types of sex-related inheritance will be examined in this chapter.

Sex linkage (X-linkage)

Drosophila.   From 1904 to 1928 T. H. Morgan taught at Columbia University, and attracted many students who were later to become brilliant geneticists. Most of these people worked in what become known fondly as "the flyroom." This must have been a remarkable and stimulating group; as one of them (Sturtevant, 1965) described it, "There was an atmosphere of excitement in the laboratory, and a great deal of discussion and argument about each new result as the work rapidly developed."

For example, in a long line of wild-type, red-eyed flies, an exceptional white-eyed male was discovered. To the Columbia group, this was a new character, apparently produced through mutation or change of the gene for red eyes (Fig. 12-1). Morgan and his students crossed this new male to his wild-type sisters; all the offspring had red eyes, which indicated that white was recessive. An F_2 of 4,252 individuals was obtained; 3,470 were red-eyed and 782 white-eyed. This is not a good representation of the expected 3:1 ratio, but later evidence indicated that white-eyed flies do not survive as well as their wild-type sibs and are therefore less likely to be counted. However, the important point here is that all 782 white-eyed F_2 flies were males! About an equal number of males had red eyes.

From Chapter 11 you will recall that these genes for eye color are on the X chromosome. Again using Y to represent the Y chromosome (which does not carry a gene for eye color), $+$ for red eyes, and w for white eyes (alleles on the X chromosome), Morgan's crosses may be represented as follows:

$$
\begin{array}{ccc}
\text{P} & +\ + & \times & wY \\
& red\ \female & & white\ \male \\
\text{F}_1 & \tfrac{1}{2}\ +\ w & \text{and} & \tfrac{1}{2}\ +\ Y \\
& red\ \female & & red\ \male \\
\text{F}_2 & \tfrac{1}{4}\ +\ +, \tfrac{1}{4}\ +\ w, \tfrac{1}{4}\ +\ Y, \tfrac{1}{4}\ wY \\
& red\ \female\quad red\ \female\quad red\ \male\quad white\ \male
\end{array}
$$

Morgan correctly predicted that white-eyed females would be produced by the cross $+\ w \times wY$. A perpetual stock of white-eyed flies of both sexes was then established by mating white-eyed males and females.

An important characteristic of sex-linked inheritance emerges from an examination of the original Morgan cross. Note that the F_2 white-eyed males have received their recessive gene from their F_1 mothers, and these, in turn, have received gene w from their own white-eyed fathers. This "crisscross" pattern from father to heterozygous daughter (often termed a *carrier*) to son is typical for a recessive sex-linked gene. Moreover, in this particular pedigree, about half the sons in the F_2 show the trait, whereas none of the daughters do. Of course, in the cross $+\ w \times wY$ half the daughters as well as half the sons have the recessive phenotype. Notice that normal females (AAXX) carry two of these genes and thus may be either homozygous ($+\ +$ or ww) or heterozygous ($+\ w$), but normal males (AAXY) can only be **hemizygous** ($+\ Y$ or wY).

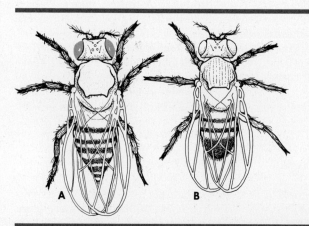

Figure 12-1. *Eye color mutation in* Drosophila melanogaster. *(A) Red-eyed female; (B) white-eyed male.*

traits are known in humans; most of these are recessive. The first to be described
in the literature, a type of red-green color blindness in which the green-sensitive
cones are defective (*deutan* color blindness), is due to an X-linked recessive gene.
It affects about 8 percent of human males, but only about 0.7 percent of females.
Females, of course, may be homozygous normal, heterozygous, or (rarely) homozygous for the defective gene, inasmuch as females have two X chromosomes
and, therefore, a greater chance of receiving a gene for normal vision from at
least one parent. Heterozygous women have completely normal red-green vision.
Males, on the other hand, receive either a dominant gene for normal or a recessive gene for defective red-green color vision (from the *mother*, who contributes
their X chromosome). Deutan color blindness appears to be the most commonly
encountered sex-linked trait in human beings. A different kind of red-green color
blindness, the *protan* type, produces a defect in the red-sensitive cones, but is
much less common than the deutan type, occurring in only some 2 percent of
males, and in but 4 women out of 10,000. Still other forms of color blindness,
some X-linked and some autosomal, are also known in humans.

Hemophilia, a well-known disorder in which blood clotting is deficient because
of a lack of the necessary substrate thromboplastin, is likewise a sex-linked
recessive condition. Two types of sex-linked hemophilia are recognized:

1. *Hemophilia A*, characterized by lack of antihemophilic globulin (Factor
   VIII). About four-fifths of the cases of hemophilia are of this type.
2. *Hemophilia B*, or "Christmas disease," (after the family in which it was
   first described in detail), which results from a defect in plasma throm-
   boplastic component (PTC, or Factor IX). This is a milder form of the
   condition.

From an unusual family in which both types of hemophilia were segregating,
Woodliff and Jackson (1966) concluded that the two loci involved were far apart
on the X chromosome. Incomplete evidence suggests a distance of more than 40
map units. Hemophilia is well known in the royal families of Europe, where it
is traceable to Queen Victoria, who must have been heterozygous (Fig. 12-2).
No hemophilia is known in her ancestry, hence it is surmised that she arose
from a mutant gamete.

Male hemophiliacs occur with a frequency of about one in 10,000 male births
(0.0001) and heterozygous females may be expected in about twice that frequency. (In Chapter 15 methods of calculating such probabilities will be explored.) Under a system of random mating, hemophilic females would be expected
to occur once in $10,000^2$, or 100 million births. But this probability is reduced
by the likelihood that male hemophiliacs will die before reaching reproductive
age (unless medically treated in order to extend life expectancy). Moreover, a
hemophilic girl would be likely to die by adolescence. Consequently, few cases
of female hemophiliacs are known, though they have been reported in some
pedigrees involving first-cousin marriages. Because clotting time in different hemophiliacs varies somewhat, it has been suggested that the condition is affected
by a number of modifying genes. Heterozygous women can be detected by a
small increase in clotting time as well as lower levels of Factor VIII. Whissell et
al. (1965) find that Factor VIII levels in heterozygous women are distributed on
a normal curve whose mean is lower than that for homozygous normal females.

Deficiency of the enzyme glucose-6-phosphate dehydrogenase (G6PD) is another important trait that results from action of a recessive X-linked gene, which
is estimated to affect up to 100 million persons. The enzyme is directly involved
in a minor glycolysis pathway in red blood cells. G6PD deficiency is common in
blacks (about 10 percent of American black males) and in persons in Mediterranean areas (10 to 40 percent of Sardinian males in malarial areas, but less
than 1 percent in Sardinian males of nonmalarial areas). The disorder is rare in

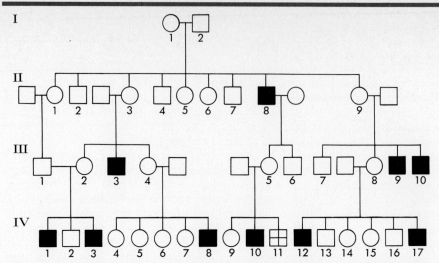

Figure 12-2. *Pedigree of some of the descendants of Queen Victoria showing incidence of hemophilia (shaded symbols). (I-1) Queen Victoria; (II-8) Leopold, Duke of Albany; (III-3) Frederick of Hesse; (III-8) Victoria Eugenie, who married Alfonso XIII of Spain; (III-9) Lord Leopold of Battenberg; (III-10) Prince Maurice of Battenberg; (IV-1) Waldemar of Prussia; (IV-3) Henry of Prussia; (IV-8) Tsarevitch Alexis of Russia; (IV-10) Rupert, Viscount Trematon; (IV-12) Alfonso of Spain; (IV-17) Gonzolo of Spain; (IV-11) died in childhood; (II-2) represents Edward VII of England, great grandfather of Elizabeth II.*

whites in areas where malaria is not indigenous. Aside from the apparent advantage of G6PD deficiency in affording some protection against malaria, the condition is noteworthy for the destruction of erythrocytes, with consequent severe hemolytic anemia, when certain drugs are administered. These drugs include para-amino salicylic acid, the sulfonamides, naphthalene, phenacetin, and primaquine (primaquine is an antimalarial agent). Inhalation of the pollen or ingestion of seeds of the broad bean (*Vicia faba*) produces the same result. Anemia caused by the broad bean is known as **favism.** In the absence of these incitants, neither hemizygous recessive males nor heterozygous and homozygous recessive females suffer ill effects.

Many variants of this recessive gene are known, each of which affects a different portion of the polypeptide chain of the enzyme (*polymorphism*). These variants range in effect from severe through mild to no enzyme deficiency; some even produce increased enzyme activity.

Some of the other sex-linked traits in humans include two forms of diabetes insipidus, one form of anhidrotic ectodermal dysplasia (absence of sweat glands and teeth), absence of central incisors, certain forms of deafness, spastic paraplegia, uncontrollable rolling of the eyeballs (nystagmus), a form of cataract, night blindness, optic atrophy, juvenile glaucoma, juvenile muscular dystrophy, and white forelock (a patch of light frontal hair on the head). Most of these are fairly clearly due to recessive genes. On the other hand, hereditary enamel hypoplasia (*hypoplastic amelogenesis imperfecta*), in which tooth enamel is abnormally thin so that teeth appear small and wear rapidly down to the gums, is due to a dominant sex-linked gene.

Mapping of X-linked genes.   Once two or more genes are clearly shown to be X-linked (by a consistent "crosscross" pattern of inheritance) some pedigree data extending over at least three generations can be used to determine at least preliminary map distances. Necessary information includes: (1) appearance of the traits in grandfather and grandson, but not in the mother of the latter, (2) identification of grandsons having a doubly heterozygous mother, and (3) determination of *cis* or *trans* linkage in the mother.

An illustration will clarify this. Assume the following symbols and phenotypes for two known X-linked genes:

Xm + production of blood antigen Xm   *R* normal deutan color vision
Xm − nonproduction of antigen Xm   *r* deutan color blindness

A hypothetical family pedigree might look like this:

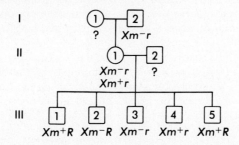

In this pedigree III-2 and III-4 are recombinants, which suggests (on the basis of this very small sample) a crossover frequency of 2/5, or 40 percent. If enough data are accumulated to form a larger sample, the map distances between these two loci can be more accurately determined. Actual data do suggest a much smaller map distance than the 40 cM arrived at here. This method of X-chromosome mapping is often referred to as the *grandfather method*. Note that it is not necessary to have even the phenotypes of the father (II-2) or of the maternal grandmother (I-1).

Sex linkage in other organisms.   Sex linkage in XX-XO species is, of course, just as it is in *Drosophila* and humans, because sex-linked genes are, by definition, those on the X chromosome. Here again only females can be heterozygous.

In birds, where the female is the heterogametic sex, the situation is reversed, although the mechanism is unchanged. Sex-linked genes follow the "crisscross" pattern, but from mother through heterozygous sons to granddaughters.

Although the barred feather pattern, which results from action of a dominant sex-linked gene in Plymouth Rock chickens (Fig. 12-3), is a frequently cited illustration, a pair of alleles governing speed of feather growth is a more inter-

Figure 12-3. *Barred feather pattern, caused by a dominant sex-linked gene, in Plymouth Rock female (left) and male (right).*

esting one. Both traits have been used for identifying chick sex, but feather color patterns are sometimes modified by other genes. Gene k, for slow feather growth, eliminates such complications and can be detected within hours after hatching. For example, since males are homogametic, the cross $+ W(♀) × kk(♂)$ results in two kinds of progeny, $+k$ (males with normal feather growth) and kW (females with slower feather growth). Gene k has no effect on other characters of commercial value.

Sex-linked lethals. The gene for hemophilia is actually a recessive, **sex-linked lethal,** for it may often cause death. Slight scratches, accidental injuries, or even bruises, which would not be serious in normal persons, often result in fatal bleeding for the hemophiliac. Sex-linked lethals, then, by bringing about death will alter the sex ratio in a progeny.

Duchenne (or progressive pseudohypertrophic) muscular dystrophy is a disorder in which the affected individual, though apparently normal in early childhood, exhibits progressive wasting away of the muscles, resulting in confinement to a wheelchair by about age 12, and death in the teen years (Fig. 12-4). Like hemophilia, this disorder is due to a recessive sex-linked gene. At present, no

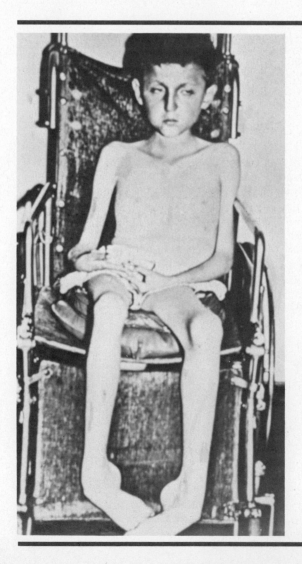

Figure 12-4. Duchenne (pseudohypertropic) muscular dystrophy, *caused by a recessive X-linked gene. This case will terminate in death.* (Courtesy Muscular Dystrophy Association of America, Inc., New York.)

means of arresting or preventing this condition is known; a given genotype dooms the bearer at conception to death in adolescence. The gene responsible is a lethal, and will change the sex ratio in a given group of offspring over time. If we let + represent the normal (dominant) gene and d represent the gene for muscular dystrophy and consider children of many marriages between heterozygous women and normal men, $+d \times +Y$, the offspring would be expected in a ratio of $\frac{1}{4}$ $++$, $\frac{1}{4}$ $+d$, $\frac{1}{4}$ $+Y$, and $\frac{1}{4}$ dY at birth. Individuals of the last genotype however die before age 20. So an initial approximately 1:1 female-to-male ratio will later become 2:1 female-to-male. If a recessive sex-linked lethal kills before birth, the ratio of female to male live births is changed from nearly 1:1 to 2:1. A 2:1 female-to-male ratio is always a strong indication of a sex-linked recessive lethal.

HOLANDRIC GENES

Holandric genes are those that occur normally on the Y chromosome only and, therefore, are not expressed in females. As noted earlier, the human Y chromosome does bear a gene responsible for a testis-determining factor, probably

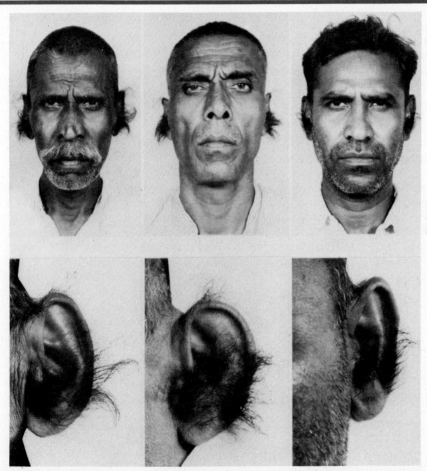

Figure 12-5. "Hairy ears," a trait common in certain parts of the world. Once thought to be due to a holandric gene, this trait is now considered to be either sex-limited or sex-influenced. (From C. Stern, W. R. Centerwall, and S. S. Sarker, "New Data on the Problem of Y-linkage of Hairy Pinnae," *Amer. Jour. Hum. Gen.*, **16:** 455–471, 1964. By permission of Grune and Stratton, Inc. Photo courtesy Dr. Curt Stern.)

on the short arm (Table 6-2). Histocompatibility antigen genes have been located on the Y chromosome in humans, mouse, rat, and guinea pig. It has been tentatively suggested that gene(s) controlling spermatogenesis are on the long arm of Y. Some of the genes that contribute to height are thought also to be on this chromosome, as well as genes that determine slower maturation of the individual. As yet no other *unequivocal* Y locus has been demonstrated for humans. It was once thought that a gene that determined excessive hair development on the ears (**hypertrichosis**) is holandric. This trait occurs rather frequently among East Indians, but also appears in Caucasions, Australian aborigines, and Japanese. Pedigrees have been published showing hypertrichosis in every male descendant from an affected man, but present thinking is that this trait is either sex-limited or sex-influenced. These latter types of inheritance are described in the next sections.

Other traits have been suggested as having Y chromosome loci, but Stern (1973) argues convincingly against their holandric nature. The best known of these is the so-called porcupine man, born in England in 1716. In this male, the skin turned yellow at the age of a few weeks, gradually blackened and became covered with rough scales with bristly outgrowths. This abnormal skin, which occurred over the whole body except for face, palms, and soles, is reported in some accounts to have been sloughed at intervals, only to have the condition recur. Original reports indicate that the trait appeared in all six sons of this man, as well as in all their male descendants, but in no females. This pattern of inheritance is certainly explainable on the basis of a holandric gene. However, careful investigation by Stern (1973) of the pedigree showed some affected females and some unaffected males, which suggests an autosomal dominant.

Few other genes appear to have a holandric pattern. Either the chromosome is largely genetically inert or Y-based genes include some alleles of such low frequency that they have not yet come to light.

Sex-limited genes

Sex-limited genes are those whose phenotypic expression is determined by the presence or absence of one of the sex hormones. Their phenotypic effect is thus *limited* to one sex or the other.

Perhaps the most familiar example occurs in the domestic fowl where, as in many species of birds, males and females may exhibit pronounced differences in plumage. In the Leghorn breed, males have long, pointed, curved, fringed feathers on tail and neck, but feathers of females are shorter, rounded, straighter, and without the fringe (Fig. 12-6). Thus males are cock-feathered and females hen-feathered. In such breeds as the Sebright bantam, birds of both sexes are hen-feathered. However, in others (Hamburg or Wyandotte) both hen- and cock-feathered males occur, but all females are hen-feathered. Leghorn females are hen-feathered, males all cock-feathered.

It has been shown that feathering type depends on a single pair of autosomal genes, H and h, in the following manner:

Genotype	♀	♂
HH	hen-feathered	hen-feathered
Hh	hen-feathered	hen-feathered
hh	hen-feathered	cock-feathered

Thus Sebright bantams are all *HH*, Hamburgs and Wyandottes may be *H −* or *hh*, but Leghorns are all *hh*. Cockfeathering, where it occurs, is *limited* to the male sex.

Figure 12-6. *Hen-feathering (left) and cock-feathering (right) in domestic fowl.* **Cock-feathering** *is characterized by long, pointed, curving neck and tail feathers.*

A trait in humans of interest here is dizygotic (two egg) twinning. Obviously a trait limited in expression to females, it is transmitted by members of both sexes. Occurrence of multiple births after administration of pituitary gonadotropins suggests the physiological basis, but whether this is a monogenic or polygenic trait is difficult to determine. Family pedigrees indicate recessiveness of the trait. Insufficient data on monozygotic twinning precludes a conclusion on its genetic basis.

Sex-limited inheritance patterns are quite different from those of sex-linked genes. The latter may be expressed in either sex, though with differential frequency. Sex-limited genes express their effects in only one sex or the other, and their action is clearly related to sex hormones. They are principally responsible

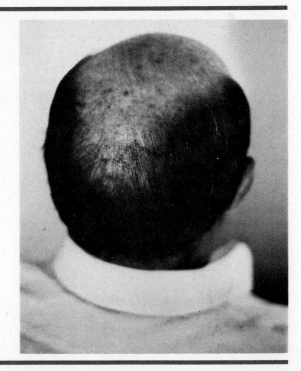

Figure 12-7. *Pattern baldness in a man, a sex-influenced trait dominant in males and recessive in females.*

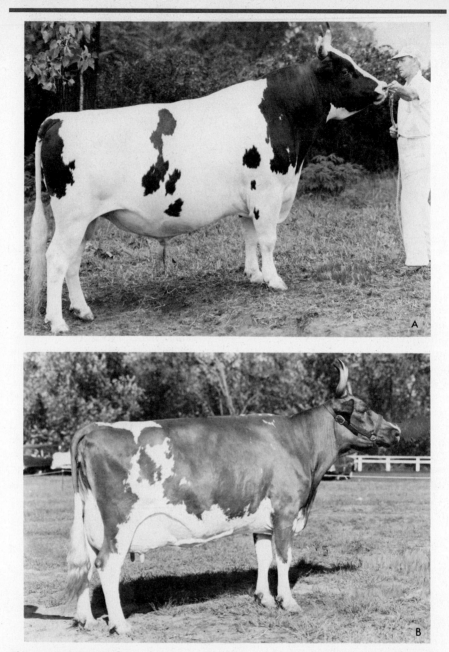

Figure 12-8. A sex-influenced trait in cattle. (A) Mahogany and white, dominant in males, recessive in females; (B) red and white, dominant in females, recessive in males. (Courtesy Ayshire Breeders' Association, Brandon, Vermont.)

for secondary sex characters. Beard development in human beings is such a sex-limited character as men normally have beards, whereas women normally do not. Yet studies indicate no significant difference between the sexes in number of hairs per unit area of skin surface, only in their development. This appears to depend on sex hormone production, changes in which may result in a genuine bearded lady.

Sex-influenced genes

In contrast to sex-limited genes, where one expression of a trait is limited to one sex, **sex-influenced** genes are those whose dominance is *influenced* by the sex of the bearer.

Although baldness may arise through any of several causes (e.g., disease, radiation, thyroid defects), "pattern" baldness exhibits a definite genetic pattern. In this condition, hair gradually thins on top, leaving ultimately a fringe of hair low on the head (Fig. 12-7). Pattern baldness is more prevalent in males than in females, where it is rare and usually involves marked thinning rather than total loss of hair on the top of the head. Numerous pedigrees clearly show that pattern baldness is due to a pair of autosomal genes operating in this fashion:

Genotype	♂	♀
BB	Bald	Bald
Bb	Bald	Not bald
bb	Not bald	Not bald

Gene *B* behaves as a dominant in males and as a recessive in females; it appears to exert its effect in the heterozygous state only in the presence of male hormone. A number of reports in the literature dealing with abnormalities that lead to hormone imbalance or with administration of hormone support this view (Hamilton, 1942, 1951). Some authors prefer to classify pattern baldness as sex-limited because the trait is generally less completely manifested in females, but it seems best to restrict the concept of sex-limited genes to those cases where one of the phenotypic expressions is *limited entirely to one sex* or the other because of anatomy.

A few well-known cases of sex-influenced genes occur in lower animals—e.g., horns in sheep (dominant in males) and spotting in cattle (mahogany and white dominant in males, red and white dominant in females, Fig. 12-8). Some hypothetical crosses involving these characters are explored in problems at the end of this chapter.

References

Hamilton, J. B., 1942. Male Hormone Stimulation Is Prerequisite and an Incitant in Common Baldness. *Amer. Jour. Anat.*, **71:** 451–480.

Hamilton, J. B., 1951. Patterned Loss of Hair in Man: Types and Incidence. *Ann. N.Y. Acad. Sci.*, **53:** 708–728.

McKusick, V. A., 1978. *Mendelian Inheritance in Man*, 5th ed. Baltimore: The Johns Hopkins University Press.

Milham, S., Jr., 1964. Pituitary Gonadotropin and Dizygotic Twinning. *Lancet II:* 566.

Morgan, T. H., 1910. Sex Limited Inheritance in *Drosophila. Science*, **32:** 120–122. Reprinted in J. A. Peters, ed., 1959. *Classic Papers in Genetics.* Englewood Cliffs, N.J.: Prentice-Hall.

Morton, N. E., C. S. Chung, and M. P. Mi, 1967. Genetics of Interracial Crosses in Hawaii. *Monographs in Human Genetics*, vol. 3. Basel, S. Karger.

Stern, C., 1973. *Principles of Human Genetics*, 3rd ed. San Francisco, W. H. Freeman.

Sturtevant, A. H., 1965. *A History of Genetics.* New York: Harper & Row.

Summitt, R. L., and D. Bergsma, eds., 1978. *Birth Defects: Original Article Series Volume XIV, Number 6C, Proceedings of the 1977 Memphis Birth Defects Conference, Sex Differentiation and Chromosomal Abnormalities.* New York: Alan R. Liss, Inc.

Whissell, D. Y., M. S. Hoag, P. M. Aggeler, M. Kropatkin, and E. Garner, 1965. Hemophilia in a Woman. *Amer. Jour. Med.*, **38**(1): 119–129.

Woodliff, H. J., and J. M. Jackson, 1966. Combined Haemophilia and Christmas Disease. A Genetic Study of a Patient and His Relatives. *Med. Jour. Aust.*, **53:** 658–661.

Wyshak, G. and C. White, 1965. Genealogical Study of Human Twinning. *Am. Jour. Pub. Health*, **55:** 1586–1593.

Problems

12-1 "Bent," a dominant sex-linked gene, B, in the mouse, results in a short, crooked tail; its recessive allele, b, produces normal tails. If a normal-tailed female is mated to a bent-tailed male, what phenotypic ratio should occur in the F_1?

12-2 Nystagmus is a condition in humans characterized by involuntary rolling of the eyeballs. The gene for this condition is incompletely dominant and sex-linked. Three phenotypes are possible: normal, slight rolling, severe rolling. A woman who exhibits slight nystagmus and a normal man are considering marriage and ask a geneticist what the chance is that their children will be affected. What will he tell them?

12-3 "Deranged" is a phenotype in *Drosophila* in which the thoracic bristles characteristically disarranged and the wings are vertically upheld. Crosses between deranged females and normal males result in a 1:1 ratio of normal females to deranged males in the progeny. What is the mode of inheritance and how would you describe the dominance of this gene?

12-4 In poultry, sex-linked gene B, which produces barred feather pattern, is completely dominant to its allele, b, for nonbarred pattern. Autosomal gene R produces rose-comb; its recessive allele, r, produces single comb in the homozygous state. A barred female, homozygous for rose-comb, is mated to a nonbarred, single-comb male. What is the F_1 phenotypic ratio?

12-5 Members of the F_1 from problem 12-4 are then crossed with each other. What fraction of the F_2 is barred rose, and are these male or female?

12-6 In what ratio does (a) barred, nonbarred and (b) rose, single segregate in the F_2 of problem 12-5?

12-7 A family has five children, three girls and two boys. One of the latter died of muscular dystrophy at age 15. The others have graduated from college and are concerned over the probability that their children may develop the disease. What would you tell them?

12-8 "Jimpy" is a trait in the mouse characterized by muscular incoordination which results in death at an age of three to four weeks. Crosses between heterozygous females and normal males produce litters in which half the males are jimpy. What type of gene is jimpy?

12-9 At locus 0.3 on the X chromosome of *Drosophila* there occurs a recessive gene, l, which is lethal in the larval stage. A heterozygous female is crossed to a normal male; what F_1 adult sex phenotypic ratio results?

12-10 A woman with defective tooth enamel and normal red-green color vision, who had a red-green blind father with normal tooth enamel and a mother who had defective tooth enamel and normal red-green vision, marries a red-green blind first cousin with normal tooth enamel. What is the probability of their having a child with normal tooth enamel and red-green color blindness if crossing-over does not occur?

12-11 What is the probability, if they have three children, that these will be two red-green blind girls with normal teeth and one boy with defective teeth but normal red-green color vision?

12-12 The bald phenotype can sometimes be distinguished at a relatively early age. A nonbald, red-green blind man marries a nonbald, normal-visioned woman whose mother was bald and whose father was red-green blind. What is the probability that they will have each of the following children: (a) bald girls, (b) bald normal-visioned boys, (c) nonbald boys, (d) red-green blind children of either sex?

12-13 In addition to the allelic pair determining pattern baldness (B, b) described in this chapter, consider early baldness to be due to another autosomal gene (E) on a different pair of chromosomes and also dominant in males. The phenotype for ee may be either late baldness or nonbaldness, depending on sex and the genotype for the B, b alleles. Two doubly heterozygous persons ($BbEe$) marry. (a) What is or will be the phenotype of the male parent? (b) What is or will be the phenotype of the female parent? (c) What is the phenotypic ratio among *male* children of couples such as this one? (d) What is the phenotypic ratio among *female* children of couples such as this one?

12-14 A mahogany-and-white cow has a red-and-white calf. What is the sex of the calf?

12-15 The cross of a horned ewe ($♀$) and a hornless ram produces a horned offspring. What is the sex of the F_1 individual?

12-16 In the clover butterfly, males are always yellow, but females may be either yellow or white. What kind of inheritance is operating?

12-17 In the clover butterfly, white is dominant in females. Crossing of two heterozygotes produces what F_1 phenotypic ratio?

12-18 Milk yield in cattle is governed by several pairs of autosomal genes. Of the types of sex-related inheritance described in this chapter, which fits milk yield?

12-19 Length of index finger relative to that of the fourth (ring) finger in human beings has been ascribed to a pair of sex-influenced genes, with short index finger dominant in males. Give the expected phenotypic ratios to be expected in children of the following marriages: (a) homozygous short $♂$ × short $♀$, (b) homozygous short $♂$ × homozygous long $♀$, (c) heterozygous short $♂$ × heterozygous long $♀$, (d) heterozygous short $♂$ × short $♀$.

12-20 How could you differentiate between a sex-linked recessive trait and a holandric one?

12-21 Why is nothing known regarding dominance of holandric genes?

12-22 How could you differentiate between a sex-linked dominant gene and a holandric one?

12-23 If hairy ears (hypertrichosis) were a holandric trait, what kind of children would be produced by a hairy-eared man?

12-24 In poultry, a particular cross produced an F_1 ratio of 3 hen feathering:1 cock feathering. One-third of the hen-feathered progeny were males. What were the parental genotypes?

12-25 Which of the individuals in Figure 12-2 of the text was heterozygous for hemophilia?

12-26 What is the probability that any woman shown in generation IV of Figure 12-2 was heterozygous?

12-27 King Edward VII (Fig. 12-2) married Princess Alexandra of Denmark; their son was George V, who married Princess Mary of Teck. George VI, son of George V, married Lady Elizabeth Bowes-Lyon; one of their children is the present Queen Elizabeth II, whose husband is Prince Philip. None of these men were hemophilic, and the disease did not occur in the families of their wives. Would it have been possible for Queen Elizabeth II to have had any hemophilic children?

12-28 A woman, heterozygous for long index finger and for hemophilia A, marries a nonhemophilic man who is heterozygous for short index finger. What is the probability that their child will be a hemophiliac with (a) short index finger or (b) long index finger? (c) Of what sex(es) will such children be?

12-29 Recall from Chapter 11 the XX-XY chromosomal situation in the flowering plant *Lychnis*. A dominant X-linked gene, *N*, produces normal broad leaves; its recessive allele, *n*, produces narrow, almost grasslike leaves. A heterozygous broad-leaved pistillate plant is pollinated by a broad-leaved staminate plant. (a) Give the genotypes of the two parents. (b) What fraction of the progeny should have narrow leaves? (c) Will these narrow-leaved plants be pistillate, staminate, or both?

12-30 A woman, heterozygous for hemophilia A but homozygous normal at the B locus, marries a first cousin who suffers from hemophilia B only. (a) What type(s) of hemophilia could occur in their children? (b) Would this be in daughters or in sons? (c) Would any of the daughters be carriers for either type of hemophilia? Explain. (d) What is the probability of their having a normal child who will not transmit either disorder to future generations?

CHAPTER 13

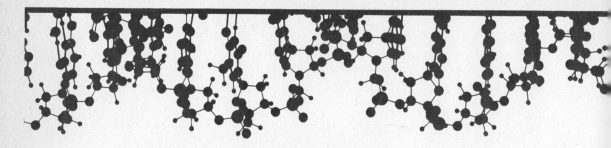

Chromosomal aberrations: changes in number

In earlier chapters it was pointed out that each species of plant and animal is characterized by a particular chromosome complement or set, represented once in monoploid cells (e.g., gametes and spores) and twice in diploid cells. Possession of such sets of chromosomes, or **genomes,** gives to each species a specific chromosome number (see Table 3-1). But irregularities sometimes occur in nuclear division, or "accidents" (as from radiation) may befall interphase chromosomes so that cells or entire organisms with aberrant genomes may be formed. Such chromosomal aberrations may include whole genomes, entire single chromosomes, or just parts of chromosomes. Thus cytologists recognize (1) **changes in number of whole chromosomes (heteroploidy)** and (2) **structural modifications.** Heteroploidy may involve either entire sets of chromosomes (**euploidy**), or loss or addition of single whole chromosomes (**aneuploidy**). Each may produce phenotypic changes, modifications of phenotypic ratios, or alteration of linkage groups. Many are of some evolutionary significance.

Changes in chromosome number

EUPLOIDY

Monoploids.   Euploids are characterized by possession of entire sets of chromosomes; monoploids carry one genome (n); diploids, two ($2n$); and so on. Monoploidy is rare in animals (the male honeybee is an outstanding exception), but common in plants. In most sexually reproducing algae and fungi and in all bryophytes (liverworts and mosses), the monoploid phase represents the dominant part of the life cycle and is the plant recognized by species name. In vascular plants this stage is short-lived and microscopic. Occasionally adult monoploid vascular plants may be recognized in natural populations, but they are ordinarily weak, small, and highly sterile. The first such case appears to have been reported in 1924, in the Jimson weed (*Datura stramonium*); there is good evidence that these developed from unfertilized eggs. Sterility in monoploids is due to extreme irregularity of meiosis because of the impossibility of chromosomal pairing and the very low probability of chromosome distribution in complete sets to daughter nuclei. For example, consider a hypothetical monoploid corn plant ($n = 10$). Only one set of 10 chromosomes is present in the sporocytes of such a plant. Each of these unsynapsed chromosomes has one chance in two of going to a given pole of the spindle (if normal and essentially simultaneous anaphase-I movement is assumed); i.e., $(\frac{1}{2})^1$ chance of going to, say, the "upper" pole. Thus, in this kind of plant the probability of a normal monoploid nucleus at the end of the first meiotic division is $(\frac{1}{2})^{10} = 1/1024$. The same hurdle would exist for the second meiotic division, and thus there is a very low probability of a normal monoploid gamete from a monoploid sporocyte. The odds against a normal monoploid gamete, of course, increase dramatically with increase in number of chromosomes in a set. These comments apply to the development of sperms from a microsporocyte because all four microspores are normally functional, but in events leading to the production of an egg, only one of the four megaspores functions in the production of a female gametophyte. This results in an even lower probability of a viable egg.

Polyploidy. Euploids with three or more complete sets of chromosomes are called **polyploids.** This condition is rather common in the plant kingdom but rare in animals. Whereas one 4*n* plant, for example, would produce 2*n* gametes and could, in many species, be self-fertilized to produce more 4*n* progeny, the probability of *two* such *animals* (one of each sex) occurring and mating is extremely low. Furthermore, the imbalance of sex-determining mechanisms that would result from polyploidy would be expected to result in sterility because of aberrant meiosis, or to produce individuals that, for many morphological reasons, might be at a considerable disadvantage in mating. But this is not true in many *plants*. Many plant genera include species whose chromosome numbers constitute a euploid series. The rose genus, *Rosa*, includes species with the somatic numbers 14, 21, 28, 35, 42, and 56. Notice that each of these numbers is a multiple of 7. Thus this is a euploid series of the basic monoploid number 7, which gives diploid, triploid, tetraploid, pentaploid, hexaploid, and octoploid series. All but the diploids may be collectively referred to as polyploids. One authority has estimated that at least two-thirds of all grass species are polyploids.

In many instances in plants, morphological differences between diploids and their related polyploids are not great enough for taxonomists to give the latter forms species rank, although the polyploids can usually be distinguished visually. It was discovered that a large amount of morphological variation within a single species of swamp saxifrage (*Saxifraga pensylvanica*) is associated with polypoidy and that diploids, triploids, and tetraploids differ consistently in several respects. The greater chromosome number of the polyploids was found to be reflected in larger cell size (Fig. 13-1). Measurement of lower leaf epidermal cell size, for example, of randomly selected diploids and tetraploids showed mean cell area of diploids to be 1,608 μm^2 for tetraploids, 2,739 μm^2. The greater cell size of tetraploids was associated with larger size of plant and plant parts, and a lower length-to-width ratio of leaves, which gives 2*n* and 4*n* plants distinctly different appearances in the field. Leaves of tetraploids are noticeably wider for their length than those of the 2*n* plants (Fig. 13-2). Such differences between diploid and tetraploid were found to be significant; statistical tests provided a very high

228

CHAPTER 13
CHROMOSOMAL
ABERRATIONS:
CHANGES IN
NUMBER

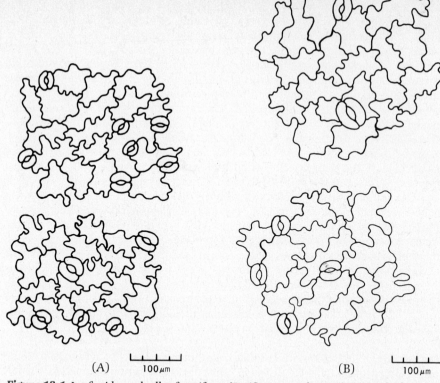

Figure 13-1. *Leaf epidermal cells of saxifrage* (Saxifraga pensylvanica). *(A) Diploids; mean cell area ranges from 1,147 μm² to 1,897 μm². (B) Tetraploids; mean cell area ranges from 2,378 μm² to 3,408 μm².*

degree of confidence in the premise that two different populations exist, as seen in the following comparison of length-to-width ratios of leaves:

	Diploids	Tetraploids
$\bar{x} =$	4.67	3.44
$s =$	1.109	0.837
$s_{\bar{x}} =$	0.158	0.119
$S_d =$	0.194	

The difference in sample means is $4.67 - 3.44 = 1.23$, which is 6.34 times greater than the standard error of the difference in sample means. Chromosomes of the various members of the series are illustrated in Figure 13-3.

In general, tetraploids are often hardier, more vigorous in growth, and able to occupy less favorable habitats. The tetraploid swamp saxifrage extends a little farther west into drier habitats than the diploid in Minnesota. The same habitat difference occurs in tetraploids and diploids in Ohio for *Tradescantia ohioensis*. Tetraploids are often hardier, more vigorous in growth, and/or have larger flowers and fruits.

In many cultivated plants tetraploid varieties are commercially more desirable than their diploid counterparts and are commonly available from seed or plant suppliers (Fig. 13-4). Many important crop plants are polyploids or include polyploid (usually tetraploid) forms: alfalfa, apple, banana, barley, coffee, cotton,

(A)   (B)

Figure 13-2. *Herbarium specimens of* Saxifraga pensylvanica. *(A) Diploid; (B) tetraploid. Note differences if leaf length-width ratio. Other characteristic dissimilarities occur in shapes and sizes of flower parts, fruits, and seeds.*

peanut, potato, strawberry, sugar cane, tobacco, and wheat are a few examples. As might be expected, polyploid varieties with an even number of genomes (e.g., tetraploids) are often fully fertile, whereas those with an odd number (e.g., triploids) are highly sterile. This latter fact is applied in marketing seeds that contain triploid embryos. Triploid watermelons, for example, are nearly seedless and are listed in a number of seed catalogs (Fig. 13-5). Triploids are commonly created by crossing the normal diploid, whose gametes are n, with a tetraploid, gametes of which are $2n$.

Production of polyploids. Polyploids may arise naturally or be artificially induced. In plants it appears that diploidy is more primitive and that polyploids have evolved from diploid ancestors. In natural populations this may arise as the result of interference with cytokinesis once chromosome replication has occurred and may occur either (1) in somatic tissue, which gives tetraploid branches, or (2) during meiosis, which produces unreduced gametes. It has been found that chilling may accomplish this in natural populations.

Application of the alkaloid colchicine, derived from the autumn crocus (*Colchicum autumnale*), either as a liquid or in a lanolin paste, induces polyploidy. Although chromosome replication is not interfered with, normal spindle formation is prevented and the double number of chromosomes becomes incor-

230

CHAPTER 13
CHROMOSOMAL
ABERRATIONS:
CHANGES IN
NUMBER

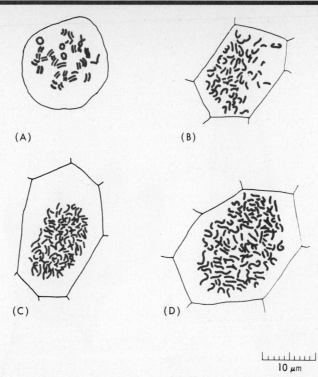

(A)

(B)

(C)

(D)

10 μm

Figure 13-3. *Camera lucida drawings of the chromosomes of Saxifraga pensylvanica. (A) microsporocyte at synapsis; (B) root cell, diploid; (C) root cell, triploid; (D) root cell, tetraploid.*

porated within a common nuclear membrane. Subsequent nuclear divisions are normal, so that the polyploid cell line, once initiated, is maintained. Polyploidy may also be induced by other chemicals (acenaphthene and veratrine) or by exposure to heat or cold.

Autopolyploidy and allopolyploidy. If a tetraploid is developed by the colchicine treatment, for example, its cells contain four genomes, all of the same species. Such a polyploid is an **autotetraploid.** Autopolyploids may, of course, exist with any number of genomes. The same situation results if an individual is formed from the fusion of two diploid gametes; the four sets of chromosomes all belong to the same species.

Synapsis in autotetraploids usually involves synapsis in *pairs* of homologs (i.e., two pairs of bivalents, Fig. 13-6A); this is the case in the previously mentioned autotetraploid of *Saxifraga pensylvanica.* However, quadrivalents, or a trivalent and a univalent sometimes occur (Fig. 13-6B,C). Particularly where bivalents are formed, meiosis and the subsequent gamete formation are essentially normal. Random pairing and disjunction in an autotetraploid of genotype *AAaa*, where bivalents are formed, will produce a gamete genotypic ratio of 1 *AA*:4 *Aa*:1 *aa* as may be seen in the following diagram. Lines connecting genes show possible combinations:

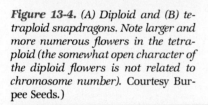

Figure 13-4. *(A) Diploid and (B) tetraploid snapdragons. Note larger and more numerous flowers in the tetraploid (the somewhat open character of the diploid flowers is not related to chromosome number). Courtesy Burpee Seeds.)*

Selfing such an autotetraploid produces some unusual progeny phenotypic ratios as will be seen shortly.

Crossing-over between linked genes in an autotetraploid complicates gamete genotypic ratios. Progeny ratios produced by, say, an *AABB/aabb* individual, depend on (1) distance between the *A* and *B* loci, and (2) whether synapsis involves two bivalents, one quadrivalent, or a univalent plus a trivalent. Fertility is generally not materially reduced if bivalents regularly form, but the other two possibilities (quadrivalent or trivalent plus a univalent) lead to degree of sterility commensurate with the frequencies of each type of configuration.

On the other hand, polyploids may develop (and be developed) from hybrids between different species. These are **allopolyploids;** the most commonly encountered type is the allotetraploid, which has two genomes from each of the two ancestral species. The Russian cytologist Karpechenko (1928) synthesized a new genus from crosses between vegetables belonging to different genera, the radish (*Raphanus*) and the cabbage (*Brassica*). These plants are fairly closely related and belong to the mustard family (*Cruciferae*). Each has a somatic chromosome number of 18, but those of radish have many genes that do not occur in cabbage chromosomes, and vice versa. Karpechenko's hybrid had in each of its cells 18 chromosomes, nine from radish and nine from cabbage. Members of the very unlike genomes failed to pair, and the hybrid was largely sterile. A few 18-chromosome gametes were formed, however, and a few allotetraploids were thereby produced as an F_2. These were completely fertile, because two sets each of radish and cabbage chromosomes were present and pairing between homologs occurred normally. The allotetraploid, or **amphidiploid,** was named

232

CHAPTER 13
CHROMOSOMAL
ABERRATIONS:
CHANGES IN
NUMBER

(A)

(B)

Figure 13-5. *(A) Diploid watermelon with numerous seeds; (B) triploid variety with few and imperfect seeds. (Courtesy Burpee Seeds.)*

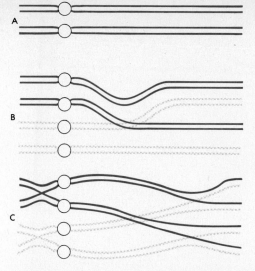

Figure 13-6. *Diagram of synapsis in autotetraploids. (A) Bivalents; (B) one trivalent and one univalent; (C) tetravalent. (B) usually leads to high frequencies of chromosomally abnormal gametes and a high degree of sterility.*

Raphanobrassica. Unfortunately, the resulting plant had the root of cabbage and leaves of radish and is of no direct economic importance. The method does, however, offer a means of producing fertile interspecific or intergeneric hybrids.

Polyploidy in humans.   Complete polyploid human beings are, as might be expected, quite rare and the few cases known are either spontaneously aborted fetuses (**abortuses**) or stillborn. A few live for a matter of hours. Gross and multiple malformations, which reflect the extreme genic imbalance of these individuals, occur in all cases. Table 13-1 presents a sampling of human polyploids reported in the literature. In a 1963 paper Carr described the chromosome complement of 227 abortuses. Some chromosome anomaly was found in 50 of these; two were triploids and one was tetraploid. None of the latter were mosaics so far as could be determined. Stern (1973) estimates that about 15 percent of all spontaneously aborted fetuses are either triploids or tetraploids. A large number of abortuses and live-born children that die very early are mosaics for diploid-polyploid cell lines. An unusual live-born triploid (69,XXX) is shown in Figure 13-7.

Table 13-1. Polyploidy in human abortuses and live births

Reference	69,XXY	69,XXX	69,XYY	92,XXXX	92,XXYY
Penrose and Delhanty (1961)	2				
Makino et al. (1964)	3				
Szulman (1965)	4	1			
Carr (1965)	6	2	1	1	1
Patau et al. (1963)		2			
Schindler and Mikamo (1970)	1*				
Butler et al. (1969)		1*			

*Born alive; died within hours.

234

CHAPTER 13
CHROMOSOMAL
ABERRATIONS:
CHANGES IN
NUMBER

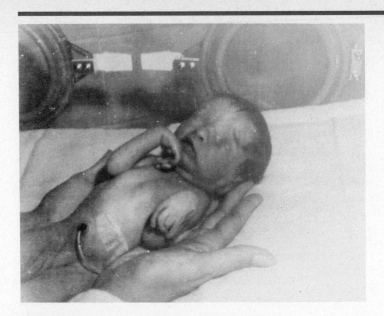

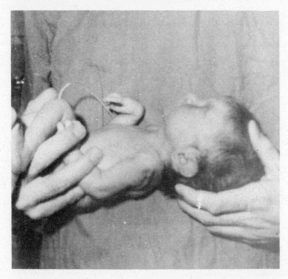

Figure 13-7. A liveborn human triploid (69, XXX). (Photos courtesy Ms. Dawn DeLozier.)

Origin of polyploid human embryos is difficult to explain satisfactorily. Although fusion of a normal monoploid gamete with a diploid one would, of course, produce a triploid zygote that might even progress to some stage of embryo development, little concrete evidence exists for occurrence of unreduced gametes in mammals. These could be produced in theory, however, by incorporation of the chromosomes of the first or second polar body within the egg nucleus, or by an analogous abnormality during spermatogenesis, or by meiosis of an exceptional tetraploid oocyte or spermatocyte. Fertilization of a normal monoploid egg by more than one sperm has also been suggested, because polyspermy has been found to occur occasionally in rabbits and rats, but there is no clear evidence for this in humans. In summary, polyploidy in humans, whether complete or as a mosaic, leads to gross abnormalities and death.

Evolution through polyploidy. Interspecific hybridization combined with polyploidy offers a mechanism whereby new species may arise suddenly in natural populations. Cytogenetic investigations of such instances of speciation in many cases have involved real "detective work" and even culminated in the artificial production of the new species.

An excellent example is furnished by the work of Huskins (1930) and of Marchant (1963) on the saltmarsh grass *Spartina*. A European species, *S. maritima*, occurs along the Atlantic coast of Europe and adjacent Africa. *S. alterniflora*, an eastern North American species (Fig. 13-8), was introduced accidentally into Great Britain in the eighteenth century. The two species are morphologically quite different from each other and have different chromosome numbers, $2n = 60$ for *S. maritima* and 62 for *S. alterniflora*. The American species gradually spread after introduction in western Europe, and the two species grew intermixed. In the 1870s a sterile putative F_1 hybrid (now called *S. × Townsendii*) was collected. This latter form has a somatic chromosome number of 62, rather than the expected 61 ($= 60/2 + 62/2$), apparently because of meiotic abnormality in one of its parents. Chromosomes in this genus are small and not easily distinguished from each other on the basis of morphology. Thus, it is nearly impossible to determine the species origin of the chromosome complement of *S × Townsendii* with reference to *S. alterniflora* and *S. maritima*. In any event, the hybrid is sterile because of the lack of meiotic pairing in its mixture of *alterniflora* and *maritima* chromosomes. In the 1890s a new, vigorous, fertile *Spartina*, which quickly spread over the British coasts and into France, was discovered. This new form, named *S. anglica*, has a somatic chromosome number of 124 in most cases, although some with 122 and 120, respectively, have also been reported. From its morphology and cytology, *S. anglica* appeared to be an amphidiploid of the cross *S. alterniflora × S. maritima*. This did prove to be the case, for *S. anglica* has been created (or re-created) in experimental plots, a triumph of cytological research (Fig. 13-9).

Similarly, other workers investigated the origin of New World cotton (*Gossipium*). Briefly, Old World cotton has 26 rather large chromosomes, whereas a Central and South American species has 26 much smaller ones. The cultivated cotton has 52, of which 26 are large and 26 smaller, and was suspected of being an allotetraploid of a cross between Old and New World species. Eventually the cultivated species was reconstituted by crossing the two putative parents and using colchicine to double its chromosome number.

Caution must be exercised in considering polyploidy as a mechanism of evolution. Basically, polyploidy adds no new genes to a gene pool but, rather, results in new combinations, especially in allopolyploids. Phenotypic effects of autopolyploidy generally represent merely exaggerations of existing characters of the species. Possession of multiple genomes reduces the likelihood that a recessive mutation will be expressed until and unless its frequency becomes quite high in the population. Polyploidy, therefore has the potential for actually decreasing genetic variation. So although entities recognized as new species do arise through allopolyploidy, and although vigor and often geographic range may be increased by autopolyploidy, polyploidy must be viewed in its proper perspective as both a positive and negative force in evolution.

ANEUPLOIDY

Trisomy. The Jimson weed (*Datura stramonium*) shows a considerable amount of morphological variation in many traits but particularly in fruit characters. The normal chromosome number for this plant is $2n = 24$, but in a now classical study, Blakeslee and Belling (1924) showed that each of several particular morphological variants had 25 chromosomes. One of the 12 kinds of chromosomes was found to be present in triplicate; that is, the somatic cells were $2n + 1$.

236

CHAPTER 13
CHROMOSOMAL
ABERRATIONS:
CHANGES IN
NUMBER

(A)

(B)

Figure 13-8. Spartina alterniflora. *(A) Habitat at edge of salt marsh. The 50-foot-wide border directly adjacent to the water is a pure growth of* S. alterniflora. *(B) Closer view*

Such a **trisomic** plant has three of each of the genes of the extra chromosome. Because the Jimson weed has 12 pairs of chromosomes, 12 recognizable trisomics should be possible, and Blakeslee and his colleagues succeeded in producing all of them (Fig. 13-10). Trisomics usually arise through nondisjunction so that some gametes contain two of a given chromosome.

When trisomics are crossed, ordinary Mendelian ratios do not result. For ex-

(C)

of pure stand of S. alterniflora *in bloom bordering the salt water. (C) Single flowering spike of* S. alterniflora *(All photos by the author in Cape Cod National Seashore Park, Eastham, Mass.)*

ample, the trisomic "poinsettia" (Fig. 13-10) has an extra ninth chromosome (i.e., triplo-9), which carries gene P (purple flowers) or its allele p (white flowers). One possible genotype, therefore for a purple-flowering "poinsettia" is PPp. Crossing two such plants produces a 17:1 phenotypic ratio. Pollen (and hence sperms) in *Datura* carrying either more or less than 12 chromosomes is nonfunctional. Megaspores (and subsequently developed eggs), however, are not so affected.

238

CHAPTER 13
CHROMOSOMAL
ABERRATIONS:
CHANGES IN
NUMBER

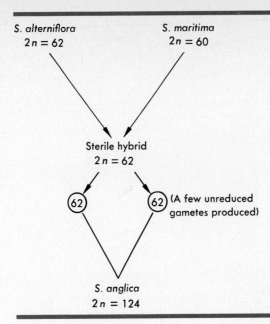

Figure 13-9. Development of Spartina anglica, an allotetraploid (amphidiploid). The cross S. alterniflora × S. maritima produced a sterile hybrid (named S. × townsendii by Marchant, 1963) having 62 chromosomes instead of the expected 61. Production of infrequent gametes having all 62 chromosomes, followed by their fusion, results in the fertile hybrid S. anglica. Some amphidiploid plants with 120 and 122 chromosomes have also been reported.

Meiosis in trisomics ordinarily results in two of the three homologs going to one pole, and one to the other. As a result, some gametes carry various combinations of two homologs and others carry but one. With this in mind, the cross $PPp \times PPp$ may be represented in this way:

P $\qquad$ PPp ♀ $\qquad \times \qquad$ PPp ♂
"purple poinsettia" $\qquad$ "purple poinsettia"

P gametes:

♀ $\qquad \frac{1}{6}$ ea.: $P + P + Pp + Pp + PP + p$
♂ $\qquad \frac{1}{3}$ ea.: $P + P + p$

F¹ $\frac{4}{18}PP$ $\qquad$ homozygous purple diploid
$\frac{4}{18}Pp$ $\qquad$ heterozygous purple diploid
$\frac{5}{18}PPp$ $\qquad$ heterozygous purple trisomic ("poinsettia")
$\frac{2}{18}Ppp$ $\qquad$ heterozygous purple trisomic ("poinsettia")
$\frac{2}{18}PPP$ $\qquad$ homozygous purple trisomic ("poinsettia")
$\frac{1}{18}pp$ $\qquad$ homozygous white diploid

Purple and white segregate in a 17:1 ratio. Other ratios are, of course, possible with different parental genotypes.

Other aneuploids. Aneuploids other than trisomics are reported in the literature, but the best-known ones occur in the Jimson weed because of the extensive work of Blakeslee and his associates and because most aneuploids are lethal in many organisms. Types of aneuploids are summarized in Table 13-2 along with the other modifications of chromosomes considered thus far.

TRISOMY IN HUMANS

Down's syndrome. An important and tragic instance of trisomy in humans involves Down's syndrome, named for the nineteenth-century British physician J. Langdon Down, who first described the syndrome in 1866 (Fig. 13-11). The incidence in the general population is roughly 1 in 600 live births. Manifestations of Down's syndrome include (1) mental retardation (I.Q. in the range 25 to 75, with most in the lower 40s), (2) below-average height, (3) a specific peculiarity

Normal
DIPLOID

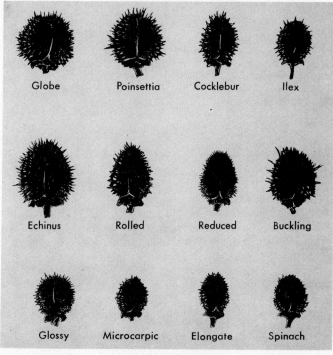

Globe   Poinsettia   Cocklebur   Ilex

Echinus   Rolled   Reduced   Buckling

Glossy   Microcarpic   Elongate   Spinach

TRISOMICS

Figure 13-10. *Fruits of normal diploid Jimson weed (top) and its 12 possible trisomics (below). Each of the latter was produced experimentally by Blakeslee and Belling. (Re-drawn from A. B. Blakeslee and J. Belling, "Chromosomal Mutations in the Jimson Weed,* Datura stramonium." Journal of Heredity, **15:** *195–206, 1924, by permission of the American Genetic Association, Washington, D.C.)*

of the upper eyelid (epicanthal fold) that suggests Oriental eyes, (4) somewhat sloping forehead, (5) low or flattened nose bridge, (6) low-set ears, (7) short, broad hands with characteristic abnormalities of the palm prints, and often (8) cardiac malformations. The mouth is usually open, with the tongue protruding. Life expectancy is reduced. Sexual maturity is generally not attained; males appear to be sterile, but a few females are reported to have borne children.

Individuals with Down's syndrome are trisomic for one of the G group of autosomes[1] (Fig. 13-12), originally identified as number 21, the second smallest of the autosomes. Recent work with fluorescence microscopy, however, shows that in actuality, the smallest (formerly designated 22) chromosome is present in triplicate. Because the notation *triplo-21* has become so entrenched in the literature, it has been agreed to renumber the two members of the G group, with 22 becoming 21 and vice versa. Although this is contrary to the otherwise standard system of assigning successively higher numbers to successively shorter

[1]Discovered in 1959 by Lejeune et al.

240

CHAPTER 13
CHROMOSOMAL
ABERRATIONS:
CHANGES IN
NUMBER

Table 13-2. Summary of variations in chromosome number (heteroploidy)

Type	Designation	Chromosome complement (where one set consists of four chromosomes, numbered 1, 2, 3, and 4)
Euploids		
Monoploid	n	1-2-3-4
Diploid	$2n$	1-2-3-4 1-2-3-4
Triploid	$3n$	1-2-3-4 1-2-3-4 1-2-3-4
Autotetraploid	$4n$	1-2-3-4 1-2-3-4 1-2-3-4 1-2-3-4
Allotetraploid	$4n$	1-2-3-4 1-2-3-4 1'-2'-3'-4' 1'-2'-3'-4'
etc.		
Aneuploids		
Trisomic	$2n + 1$	1-2-3-4 1-2-3-4 1
Triploid tetrasome	$3n + 1$	1-2-3-4 1-2-3-4 1-2-3-4 1
Tetrasomic	$2n + 2$	1-2-3-4 1-2-3-4 1-1
Double trisomic	$2n + 1 + 1$	1-2-3-4 1-2-3-4 1-2
Monosomic	$2n - 1$	1-2-3-4 2-3-4
Nullisomic	$2n - 2$	2-3-4 2-3-4

autosomes, confusion between older and newer literature on Down's syndrome is prevented. The important fact is the presence of an extra one of the shortest of the autosomes. The determinants of the Down phenotype have been localized to trisomy of a segment of the longer arm of chromosome 21 (the proximal part of band q21).

From the cytological viewpoint, two types of Down's syndrome may be recognized: (1) *triplo-21*, in which the affected individual is trisomic for autosome 21, with a total somatic complement of 47 chromosomes,[2] and (2) *translocation*, in which the extra twenty-first chromosome has become attached to another

[2]About 92.5 percent of persons with Down's syndrome are free 21 trisomics.

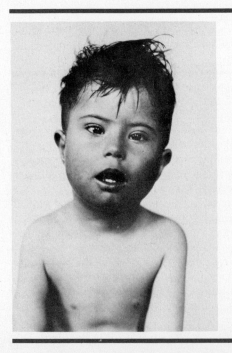

Figure 13-11. *Down's syndrome in a child.* (Courtesy National Foundation, New York.)

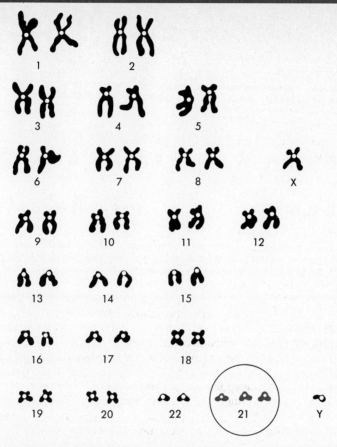

Figure 13-12. *Karyotype of trisomic Down's syndrome male having 47 chromosomes (triplo-21).* (Courtesy National Foundation, New York.)

autosome, most frequently one of the D group, now generally agreed to be number 14. More rarely the translocation involves 15 and 21, or even one of the G group, probably another 21. A karyotype of a translocation Down's syndrome, therefore, shows 46 chromosomes, one of which is 14-21 (Fig. 13-13). Both types involve an extra number 21; in one case it is a separate entity; in the other it has become attached to another chromosome. Phenotypes of both triplo-21 and translocation mongoloids are identical.

Incidence of Down's syndrome is closely related to maternal age (Fig. 13-14). No consistent correlation with paternal age has been discerned. Triplo-21s arise through nondisjunction, chiefly in oogenesis. A female is born with all the oocytes she will ever produce—nearly 7 million. These remain in an arrested prophase-I of meiosis from before birth until ovulation, generally at the rate of one per menstrual cycle (about 28 days) after puberty. A given oocyte may remain in this state of suspended development for some 12 to 45 years or so. On the other hand, the testes produce about 200 million sperms a day, and meiosis in a spermatocyte requires 48 hours or less to complete.

Nondisjunction during oogenesis is thus a function of senescence of the oocytes, whereby separation of homologs is interfered with in some way. This may be through the breakdown of chromosomal fibers or through deterioration of the centromere. Presence of a virus as well as radiational damage (whose cumulative effect would be greatest in older individuals) have been suggested as causative factors. Cells are especially susceptible to damage by viruses and by radiation

242

CHAPTER 13
CHROMOSOMAL
ABERRATIONS:
CHANGES IN
NUMBER

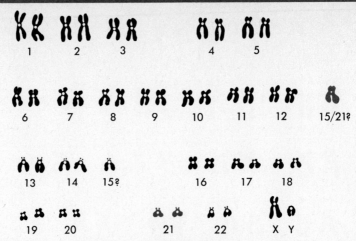

Figure 13-13. *Karyotype of Down's syndrome male. This individual has 46 chrommo-somes, but only one normal 15 and a large one formed by union of a 21 and the other 15. Two normal 21s are also present, giving the individual three "doses" of chromosome 21.*

during division. Of course, the older the woman, the longer these agents will be able to act on her oocytes. Up to three times greater incidence of Down's syndrome has been reported in children born to mothers (age group for age group) who have had infectious hepatitis prior to pregnancy.

Some family studies suggest a specific gene that interferes with disjunction; such genes are known in some organisms. The probability that autoimmune diseases may play a part in certain cases is evidenced by the fact that there is about a threefold increase in risk of Down's syndrome child at any given maternal age where high thyroid antibody levels are present. In one study, 30 of 177

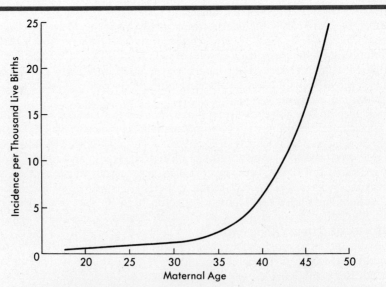

Figure 13-14. *Relationship between maternal age and birth of children with Down's syndrome.*

mothers of Down's children had had some form of clinical thyroid disorder, compared with only 11 of 177 control mothers (i.e., those without Down's syndrome children).

Another explanation may be based on the fact that an egg will degenerate in the fallopian tube within about one day if not fertilized. If fertilization should occur toward the close of the period of viability, degenerative changes in the egg will lead to chromosomal abnormalities in zygotene in some animal species, probably including humans. The correlation of trisomy-21 with increased maternal age under this explanation probably is a reflection of less-frequent intercourse in older couples. That this is not a *principal* factor is seen in the shape of the curve in Figure 13-14. Delayed fertilization, acting alone, would be expected to produce a linear slope, but, as shown in Figure 13-14, the increased incidence of Down's syndrome in the later maternal years is exponential. In any event, although the causes are not fully clear, the maternal age effect is well established.

Mosaics with Down's syndrome, who exhibit the classical symptoms in varying degree, probably arise through failure of the 21s to segregate at mitosis during embryogeny, which produces a fetus with cells of three types: (1) normal 21, 21, (2) triplo-21, and (3) monosomic-21. The latter group's cells appear unable to survive, and thus lead to a disomic-trisomic mosaic. Should mitotic segregation of the 21s fail at the first division of the zygote, however, two daughter cells, one trisomic and the other monosomic, would result. Death of the latter cell would then lead to production of a wholly trisomic embryo.

Examination of parents of translocation mongoloids almost always discloses that one of them has only 45 chromosomes, including one 21, one 14, and a fused 14-21. Such "carriers" are phenotypically normal, inasmuch as their genetic material is present in the proper amount. More often, the *mother* of a translocation Down's syndrome child is the carrier. The reason why carrier fathers rarely produce translocation children is unknown. If the centromere of the maternal translocation chromosome is that of number 14, four kinds of eggs are produced, and this leads apparently to three kinds of children:

Egg	Sperm	Zygote	Progeny
14-21, 21	14, 21	14, 14-21, 21, 21	Translocation Down's syndrome
14-21	14, 21	14, 14-21, 21	Translocation carrier
14, 21	14, 21	14, 14, 21, 21	Normal
14	14, 21	14, 14, 21	Usually lethal

However, among cell cultures available to research workers are two 45,XY, − 21 (i.e., monosomy 21) cell lines, one from a five-and-a-half-year-old patient, the other from a 14-month-old child. Some partial 21 monosomics have been reported.

Edwards' syndrome. First described in 1960 by Edwards and his colleagues, trisomy-18 (Fig. 13-15) is now well known in more than 50 published cases.[3] Incidence is about 0.3 per 1,000 births. It is characterized by multiple malformations, chief among which are malformed, low-set ears; small, receding lower jaw; flexed, clenched fingers; cardiac malformations; and various deformities of skull, face, and feet. Harelip and cleft palate often occur. Death takes place generally around three to four months of age, but may be delayed for nearly two years. One mentally retarded female, age 15, was reported in 1965. Triplo-18 children show evidence of severe mental retardation. The defect is about three times more frequent in females; the reasons for this are not clear.

[3]Edwards et al. first described this condition as trisomy-17.

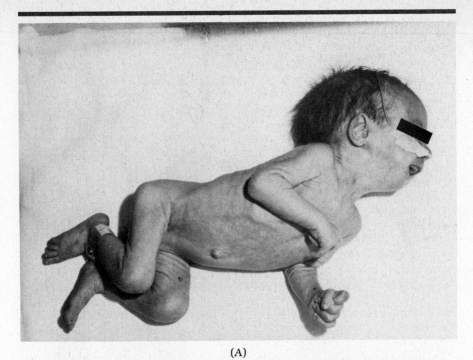

(A)

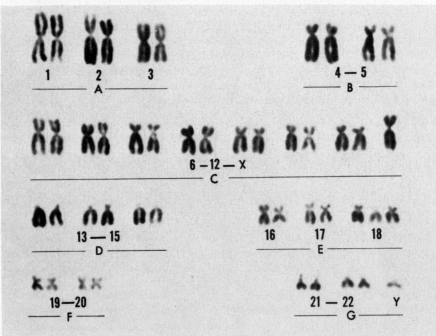

(B)

Figure 13-15. *Edwards' syndrome (trisomy-18). (A) Young triplo-18 child, showing low-set ears, small, receding lower jaw, and flexed, clenched fingers. (B) Karyotype of same child. Note the seven E-group autosomes. (Photos courtesy Dr. Richard C. Juberg.)*

As in Down's syndrome, there is a pronounced maternal age effect, although plotting frequency against maternal age produces a bimodal curve (Fig. 13-16). The secondary peak in the early 20s reflects the normal maternal age group of maximum births, whereas the pronounced peak from 35 to 45 years is clearly related to greater age of the mother.

Trisomy-13 (Patau's syndrome). In 1960 Patau and associates described a case of multiple malformations in a newborn child, established as a trisomic for one of the D-group of autosomes, now generally agreed upon as number 13. Individuals appear to be markedly mentally retarded and very frequently have sloping forehead, harelip, and cleft palate; the last two malformations are so strongly developed that the face has a severely deformed appearance. Polydactyly (both hands and feet) is almost always present; the hands and feet are also characterisically deformed (Fig. 13-17). Cardiac and various internal defects (kidneys, colon, small intestine) are common. Death usually occurs within hours or days, but may abort spontaneously (Fig. 13-18). In an extreme case of longevity, a living triplo-13, aged five years, was reported. There is a slight excess of affected females. Age of the mother is not as clearly a factor as in Edwards' syndrome, although tabulated cases do form a bimodal curve. However, in one study 40 percent of the triplo-13s were born to mothers over 35, whereas this age group accounted for less than 12 percent of all births. Incidence in the general population is of the order of 0.2 per 1,000 births.

Trisomy and spontaneous abortion. It appears that trisomy for autosomes other than 13, 18, and 21 may well be lethal. Boué and Boué (1973) examined 1,457 abortuses, of which 892 (61 percent) had some chromosomal abnormality, including XO, trisomy D, trisomy E, and triploidy. McCreanor et al. (1973) reported spontaneous abortion of an empty embryo sac, wall cells of which were trisomic for autosome 7, as revealed by Giemsa banding. Kajii and associates

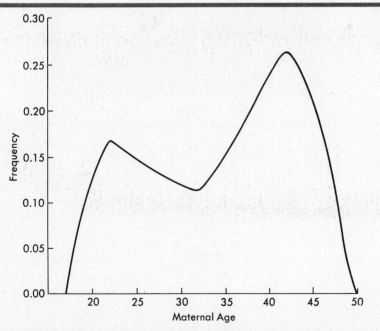

Figure 13-16. *Maternal age effect in Edwards' syndrome. The peak between maternal ages 35 and 45 is the significant one here. See text for details.*

246

CHAPTER 13
CHROMOSOMAL
ABERRATIONS:
CHANGES IN
NUMBER

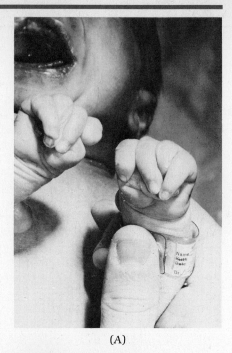

Figure 13-17. *Patau's syndrome (tri-somy-13). (A) Clenched fists. (B) Deform-ities of feet. See text for details.* (Photos courtesy Dr. Richard C. Juberg.)

(A)

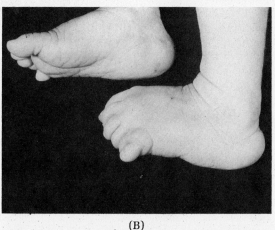

(B)

(1973) reported that of the 22 possible human autosomal trisomics, they found all but those for numbers 1, 5, 12, 17, and 19 in abortuses. By 1977 (de Grouchy and Turleau) complete trisomies only for chromosomes 5, 12, and 17 had not been reported, even in abortuses; for autosomes 3 and 19 only partial trisomies appear to have been reported. A trisomic for one of the C-group of autosomes is shown in Figure 13-19. Inasmuch as the karyotype shown in that figure antedates development of banding techniques, no method was available for more precise identification of the extra chromosome.

Trisomy for even a part of a chromosome usually produces some degree of deformity. A child, trisomic for the short arm of autosome 2, is shown in Figure 13-20.

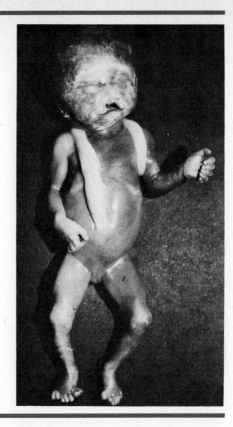

Figure 13-18. *A trisomy-13 abortus.*
(Photo courtesty Ms. Dawn DeLozier.)

NONHUMAN TRISOMICS

That Down's syndrome may also occur in other animals is clear from the work of McClure et al. (1969), which described a trisomic young female chimpanzee. This animal displayed clinical and behavioral features similar to those of human mongolism with scores below normal in a variety of behavioral and postural tests. Chimpanzees have a diploid chromosome number of 48 (Table 3-1) but in the animal in question most of the blood cells examined had 49, with an extra small acrocentric chromosome that matched pair 22 (the second smallest of the autosomal pairs). This same chromosome was identifiable in triplicate even in the few cells where only a total of 46, 47, or 48 chromosomes could be counted. Both parents were cytologically and behaviorally normal. McClure and his colleagues note that "a comparable condition has not been reported in nonhuman primates. The occurrence of this condition in a lower primate again emphasizes the close phylogenetic relation between man and the great apes and may provide a model for studying this relatively frequent human syndrome." Trisomy-21 has also been reported in the gorilla (de Grouchy et al., 1973).

References

Blakeslee, A. F., and J. Belling, 1924. Chromosomal Mutations in the Jimson Weed, *Datura stramonium. Jour. Hered.*, **15:** 195–206.
Boué, A., and R. Boué, 1973. Évaluation des Erreurs Chromosomiques au Moment de la Conception. *Biomedicine*, **18:** 372–374.

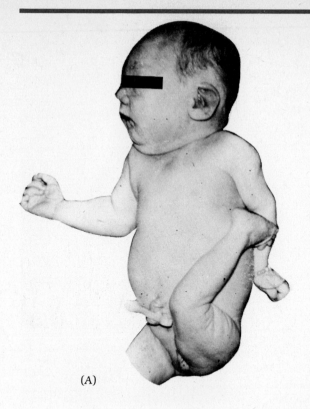

(A)

Figure 13-19 *C-trisomy in a female. (A) External features; note ears, jaw, hands, and leg position. (B) Karyotype of same child showing 17 chromosomes in the C, X group instead of the normal 16, for a total of 47. (From R. C. Juberg, E. F. Gilbert, and R. S. Salisbury, 1970, "Trisomy-C in an Infant with Polycystic Kidneys and Other Malformations." Jour. Pediatr., 76: 598–603. Used by permission of Dr. Juberg, who supplied the photos, and the C. V. Mosby Co.)*

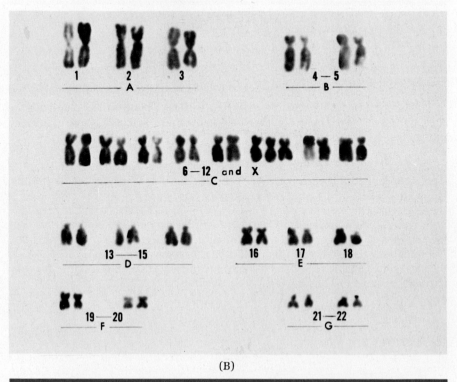

(B)

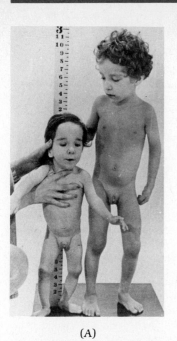
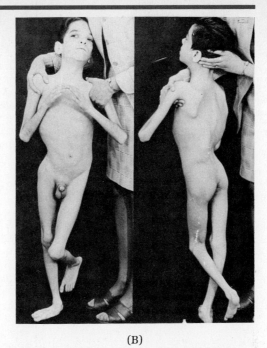

(A)   (B)

Figure 13-20. *Trisomy for the short arm of autosome 2. (A) with normal sibling. (B) Two views of the same subject at an older age.* (Photos courtesy Ms. Dawn DeLozier.)

Burns, G. W., 1942. The Taxonomy and Cytology of *Saxifraga pensylvanica L.* and Related Forms. *Amer. Midl. Nat.*, **28:** 127–160.

Butler, L. J., C. Chantler, N. E. France, and C. G. Keith, 1969. A Liveborn Infant with Complete Triploidy (69, XXX). *Jour. Med. Genet.*, **6:** 413–421.

Carr, D. H., 1963. Chromosome Studies in Abortuses and Stillborn Infants. *Lancet*, **2:** 603–606.

Carr, D. H., 1965. Chromosome Studies in Spontaneous Abortions. *Obstet. Gynecol.*, **26:** 308–326.

Carter, C. O., and K. A. Evans, 1961. Risk of Parents Who Have Had One Child with Down's Syndrome (Mongolism) Having Another Child Similarly Affected. *Lancet*, **2:** 785–787.

Conen, P. E., and B. Erkman-Balis, 1966. Frequency and Occurrence of Chromosomal Syndromes. 1. D-Trisomy, 2. E-Trisomy. *Amer. Jour. Hum. Genet.*, **18:** 374–398.

Day, R. W., 1966. The Epidemiology of Chromosome Aberrations. *Amer. Jour. Hum. Genet.*, **18:** 70–80.

Doxiadis, S., S. Pantelakis, and T. Valaes, 1970. Down's Syndrome and Infectious Hepatitis. *Lancet*, **1:** 897.

Edwards, J. H., D. G. Harnden, A. H. Cameron, V. M. Grosse, and O. H. Wolff, 1960. A New Trisomic Syndrome. *Lancet*, **1:** 787–789.

German, J., 1970. Studying Human Chromosomes Today. *Amer. Scientist*, **58:** 182–201.

de Grouchy, J., M. Lamy, S. Thieffry, M. Arthuis, and C. Salmon, 1963. Dysmorphie Complexe avec Oligophrénie: Délétion des Bras Courts d'un Chromosome 17–18. *Compt. Rendus Acad. Sci. Paris*, **256:** 1028–1029.

de Grouchy, J., and C. Turleau, 1977. *Clinical Atlas of Human Chromosomes.* New York: John Wiley & Sons.

de Grouchy, J., C. Turleau, M. Roubin, and F. Chavin-Colin, 1973. Chromosomal Evolution of Man and the Primates (*Pan troglodytes, Gorilla gorilla, Pongo pygmaeus*). *Nobel Symposium 23: Chromosome Identification.* New York: Academic Press.

Hook, E. B., R. Lehrke, A. Roesner, and J. J. Yunis, 1965. Trisomy-18 in a 15-Year-Old Female. *Lancet*, **2:** 910–911.

Huskins, C. L., 1930. The Origin of *Spartina townsendii*. *Genetica*, **12:** 531–538.

250

CHAPTER 13
CHROMOSOMAL
ABERRATIONS:
CHANGES IN
NUMBER

Juberg, R. C., E. F. Gilbert, and R. S. Salisbury, 1970. Trisomy C in an Infant with Polycystic Kidneys and Other Malformations. *Jour. Pediatrics*, **76:** 598–603.

Kajii, T., N. Niikawa, A. Ferrier, and H. Takahara, 1973. Trisomy in Abortion Material. *Lancet*, **2:** 1214.

Karpechenko, G. D., 1928. Polyploid Hybrids of *Raphanus sativus L. X. Brassica oleracea L. Ztschr. ind. Abst. Vererb.*, **48:** 1–83.

Lejeune, J., M. Gautier, and R. Turpin, 1959. Les Chromosomes Humains en Culture de Tissus. *Compt. Rendus Acad. Sci. Paris*, **248:** 602–603.

Makino, S., M. S. Sasaki, and T. Fukuschima, 1964. Triploid Chromosome Constitution in Human Chorionic Lesions. *Lancet*, **1:** 1273.

Marchant, C. J., 1963. Corrected Chromosome Numbers for *Spartina x townsendii* and Its Parent Species. *Nature*, **199:** 929.

McClure, H. M., K. H. Belden, W. A. Pieper, and C. B. Jacobson, 1969. Autosomal Trisomy in a Chimpanzee: Resemblance to Down's Syndrome. *Science*, **165:** 1010–1011.

McCreanor, H. R., F. M. O'Malley, and R. A. Reid, 1973. Trisomy in Abortion Material. *Lancet*, **2:** 972–973.

Pantelakis, S. N., O. M. Chryssostomidou, D. Alexiou, T. Valaes, and S. A. Doxiadis, 1970. Sex Chromatin and Chromosome Abnormalities Among 10,412 Liveborn Babies. *Arch. Dis. Childhood*, **45:** 87–92.

Patau, K., S. L. Inhorn, and E. Therman, 1963. Lethal Chromosome Constitutions in Man. In S. J. Geerts, Ed. *Genetics Today: Proceedings of the XI International Congress on Genetics*, volume **1.** New York: Macmillan.

Patau, K., D. W. Smith, E. Therman, S. L. Inhorn, and H. P. Wagner, 1960. Multiple Congenital Anomaly Caused by an Extra Autosome. *Lancet*, **1:** 790–793.

Penrose, L. S., and J. D. A. Delhanty, 1961. Triploid Cell Cultures from a Macerated Foetus. *Lancet*, **1:** 1261–1262.

Schindler, A. -M., and K. Mikamo, 1970. Triploidy in Man: Report of a Case and a Discussion on Etiology. *Cytogenetics*, **9:** 116–130.

Szulman, A. E., 1965. Chromosomal Aberrations in Spontaneous Human Abortions. *New Eng. Jour. Med.*, **272:** 811–818.

Problems

13-1 Application of colchicine to a vegetative bud of a homozygous tall diploid tomato plant (*DD*) causes development of a tetraploid branch. What is the genotype of the somatic cells of this branch?

13-2 Flowers are produced on the tetraploid branch of the plant in problem 13-1. What is the genotype of the gametes?

13-3 Pollinating one of the flowers of problem 13-2 with pollen from a diploid dwarf plant produces embryos of what genotype?

13-4 If the plant in problem 13-1 had been heterozygous tall (*Dd*), what would be the genotype of the somatic cells of the tetraploid branch?

13-5 Give the gamete genotypes produced on the tetraploid branch of problem 13-4. In what ratio are they produced?

13-6 (a) Self-pollinating flowers on one of the tetraploid branches referred to in problem 13-4 would produce embryos with what specific kind of ploidy? (b) What is the probability of a dwarf plant in the progeny?

13-7 If an autotetraploid plant heterozygous for two pairs of genes, e.g., of genotype *AAaaBBbb,* is self-pollinated, what is the probability of an *aaaabbbb* plant in the progeny?

13-8 If an autotetraploid plant of genotype *AAaaBBbbCCcc* is self-pollinated, what is the probability of an *aaaabbbbcccc* progeny individual?

13-9 If an autotetraploid plant, heterozygous for n pairs of genes (i.e., *AAaaBBbbCCcc* ... and so on for n pairs of alleles) is self-pollinated, what mathematical expression would you solve to determine the probability of a completely recessive progeny individual?

13-10 If only pairs of alleles that show complete dominance are considered, what is the effect of tetraploidy on phenotypic variability?

13-11 A number of species of the birch tree have a somatic chromosome number of 28. The paper birch (*Betula papyrifera*) is reported to occur with several different chromosome numbers. Individuals with the somatic numbers 56, 70, and 84 are known. With regard to chromosome number, how should the 28, 56, 70, and 84 chromosome individuals be designated?

13-12 The sugar maple (*Acer saccharum*) and the box elder (*Acer negundo*) each have diploid chromosome numbers of 26. Note that they are different species of the same genus. However, hybrids between the two are sterile. What explanation can you offer?

13-13 Should it be possible to secure a fertile hybrid of the cross sugar maple × box elder? How?

13-14 Different species of rhododendron have somatic chromosome numbers of 26, 39, 52, 78, 104, and 156. By what means does evolution appear to be taking place in this genus?

13-15 What appears to be the basic monoploid chromosome number in rhododendrons?

13-16 How many sets are represented in the species with 156 chromosomes?

13-17 Referring back to Table 13-1, what was the sex phenotype of the triploid child reported by (a) Schindler and Mikamo, (b) Butler et al.?

13-18 If polyspermy were demonstrated to occur in human beings, what would be the chromosomal complement of an embryo resulting from polyspermy involving an X and a Y sperm?

13-19 Suppose the first mitosis in a normal 46,XY zygote were abnormal so that it became a tetraploid cell. (a) What would be the chromosomal complement of that tetraploid cell? (b) Do you find any evidence in this chapter or its references to suggest that a viable child would result?

13-20 In the Jimson weed what gamete ratios are produced by (a) *Ppp* ♀, (b) *Ppp* ♂, (c) *PPP* ♀, (d) *PP* ♂, (e) *PPp* ♂?

13-21 What is the F_1 phenotypic ratio produced in Jimson weed by crossing (a) purple *Ppp* ♀ × purple *PPp* ♂, (b) *PPp* ♀ × *Pp* ♂, (c) *Ppp* ♀ × *Ppp* ♂?

13-22 "Eyeless" (eyes small or absent) is a recessive character whose gene is located on the small chromosome IV of *Drosophila melanogaster*. Both triplo-IV eggs and triplo-IV sperms are functional. The dominant gene for normal eyes is designated +, the recessive for eyeless *ey*. What is the F_1 phenotypic ratio produced by each of the following crosses: (a) + *ey ey* × + *ey ey*, (b) + + *ey* × + *ey ey*, (c) + + *ey* × + + *ey*?

13-23 The Jimson weed (*Datura*) has 12 pairs of chromosomes, and 12 different viable trisomics are known. How would you explain the fact that the only trisomies reported for human *live* births involve chromosomes 8, 9, 13, 14, 18, 21, 11, X, and Y and that only trisomy 8, X, and Y appear not to reduce life expectancy measurably?

13-24 What explanation could you offer for the fact that although human trisomics are known for a number of different autosomes, no autosomal monosomics are reported among live births?

CHAPTER 14

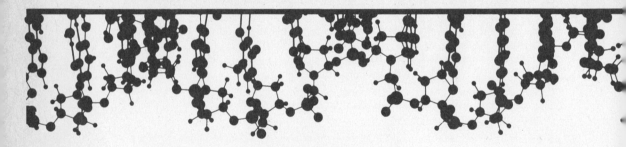

Chromosomal aberrations: structural changes

For all their complexities of structural organization, chromosomes are far from indestructible. Through such agents as radiation and chemicals breakage can occur, which may result in genetic damage to subsequent generations. Suitable precautions, of course, must be taken in X-ray diagnoses, and nuclear weaponry constitutes a definite hazard. A study of 43 Marshall Islanders 10 years after accidental exposure to radioactive fallout following testing of a high-yield nuclear device at Bikini disclosed chromosomal breaks in 23 of 43. A wide variety of chemical substances, some formerly or presently in common use by humans, have been implicated or suspected in inducing chromosomal damage. Breaks may also occur "naturally," for no assignable cause. Any of several cytological and genetic consequences may result. Breaks may be detected in almost any dividing cell, though certain kinds offer special advantage. Particularly useful in this regard are the salivary glands of dipteran insects such as *Drosophila*. Genetic effects of chromosomal damage or aberrant behavior are detectable in unexpected phenotypes, gross physical malformations, altered enzyme production, altered linkage relationships, reduced fertility (or complete infertility), and an increase in spontaneous abortions.

Salivary gland chromosomes. Structural changes in chromosomes that result from breakage are most profitably studied in (1) salivary gland chromosomes of dipteran insects (such as *Drosophila*) and (2) meiocytes during the process of meiosis. Because of their large size and banded structure, salivary gland chromosomes are particularly appropriate for such studies.

Nuclei of the salivary gland chromosomes of the larvae of dipterans like *Drosophila* have unusually long and wide chromosomes, 100 or 200 times the size of the chromosomes in meiosis or mitosis of the same species (Fig. 14-1). This is particularly surprising, because the salivary gland cells do not divide after the glands are formed, yet their chromosomes replicate several times and become unusually long as well. Moreover, these chromosomes are marked by numerous cross bands that are apparent even in unstained nuclei. These bands are quite constant in size and spacing for a given normal chromosome. These multiple chromosomes appear to be in a perpetual prophase and are synapsed. Thus any difference in banding between homologs can be easily compared.

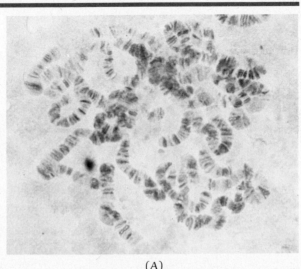

Figure 14-1. Salivary gland chromosomes of Drosophila melanogaster *showing the banding that permits rather accurate cytological mapping. (A) Entire complement. (B) Partial complement, enlarged.* [(A) Courtesy Miss Chris Arn. (B) Courtesy Mr. John Derr.]

(A)

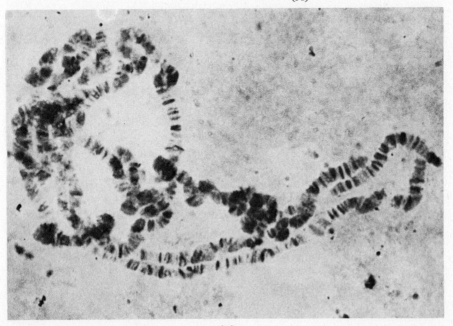

(B)

254

CHAPTER 14
CHROMOSOMAL
ABERRATIONS:
STRUCTURAL
CHANGES

TYPES OF STRUCTURAL CHANGES

A chromosomal break may or may not be followed by a "repair." If segments are rejoined in the original configuration, the break normally passes undetected. Should such repair not be effected, one or more of the aberrations listed in Table 14-1 occur. Note that a deletion in one chromosome may be accompanied by a translocation or a duplication in another. Reciprocal translocations, in which two nonhomologs *exchange* segments, often unequal in length, may also occur.

Table 14-1. *Types of aberrations produced by chromosomal breaks*

Type	Description	Gene changes
Normal	(*ABCDEFGH*)	None
Deletion	No rejoining; chromosomal segment lost	*ABFGH, CDEFGH*, etc.
Inversion	Broken segment reattached to original chromosome in reverse order	*ABFEDCGH*, etc.
Duplication	Broken segment becomes attached to homolog that has experienced a break; homolog then bears one block of genes in duplicate	*ABCDEFGEFGH*
Translocation	Broken segment becomes attached to a nonhomolog resulting in new linkage relations	*LMNOPQRCDEFGH*, etc.

DELETIONS

Deletions and cytological mapping. Perhaps the simplest result of breakage is the loss of a part of a chromosome. Portions of chromosomes without a centromere (*acentric fragments*) lag in anaphasic movement and are lost from reorganizing nuclei. Such loss of a portion of a chromosome is called a **deletion** or a deficiency.

Deletions often make possible *cytological mapping* of chromosomes. For example, if large numbers of fruit flies of autosomal genotype *ABCDEF . . ./ABCDEF . . .* (i.e., homozygous dominant for several traits) are subjected to X-irradiation, a few of the individuals may suffer a break and deletion in one of the chromosomes bearing these genes. If a deletion for genes *C* and *D* occurs in one primary spermatocyte, then two of the four sperms produced from it would have an *intercalary deletion* in this particular autosome:

Primary spermatocyte (irradiated)	$\dfrac{AB--EF \ldots}{ABCDEF \ldots}$
Secondary spermatocytes	*ABEF . . ., ABCDEF . . .*
Spermatids and sperms	*ABEF . . ., ABEF . . ., ABCDEF . . ., ABCDEF . . .*

Mating such males to recessive females, *abcdef . . ./abcdef . . . will produce offspring, some of which will have the genotype abcdef . . ./ABEF.* Therefore these flies will express the recessive phenotype for genes *c* and *d*, whereas those receiving an *ABCDEF . . .* sperm from the male parent will express the dominant phenotype for all six genes. The expression of genes *c* and *d*, which would have been obscured by genes *C* and *D* had these been present, is called **pseudodominance.** This is really not a good term, but it is a useful one and is firmly established in the literature.

Next, flies showing pseudodominance are mated with normal *abcdef . . ./abcdef . . .* individuals. Half the offspring will now exhibit pseudodominance for the same genes. Examination of the salivary gland chromosomes of larvae of this cross will quickly disclose those individuals that are *abcdef . . ./ABEF*. Remember, these chromosomes are banded and pair exactly, band for band. The segment that includes genes *C* and *D* will also include some of the bands, and the giant chromosomes of *abcdef . . ./ABEF . . .* will appear somewhat as shown in Figure 14-2, where the normal, unaltered one shows a characteristic **deletion loop.**

Another strain of flies with an *overlapping deficiency* such as *ABC− −F . . ./ ABCDEF . . .* might be developed and mated as in the preceding case. Giant chromosomes of *abcdef . . ./ABCDF . . .* individuals will again display the characteristic deficiency loop. Comparison with the loops produced by the *abcdef . . ./ ABEF . . .* individuals permits easy detection of the precise chromosomal region that bears gene *D*; its locus can be seen to be associated with a particular band (Fig. 14-2).

Detailed *cytological maps* of *Drosophila* have been prepared in just this way. The geneticist now has visible bands within which genes appear to be located, and their distances can be measured in ordinary units. Comparison of cytological and genetic maps confirms the linear sequence given by the latter. Cytological maps, however, show known genes to be more uniformly distributed over the chromosome than is suggested by the genetic map. This results from the greater degree of interference near the centromere and termini of the chromosome. For example, in chromosome II of *Drosophila* the centromere is at locus 55.0. Painter's map (Fig. 6-4) shows gene *pr* (purple eyes) to be at locus 54.5, just 0.5 unit to the left of the centromere. This is about 1 percent of the 55 map units between the centromere and the end of the left arm. Yet on the salivary gland, map gene *pr* is about 12 genetic map units to the left of the centromere, or nearly 22 percent of the total distance involved. Figure 14-3 compares genetic and cytological map locations for several genes of chromosome II.

Deletions in humans. Probably the best-known disorder to be associated definitely with a deletion in humans is the *cri du chat,* or cat-cry, syndrome

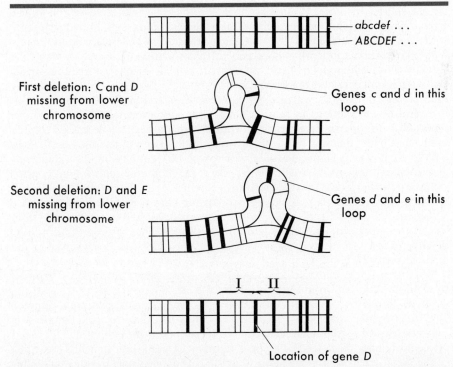

Figure 14-2. *Determining which band of salivary gland chromosomes bears a given gene is done by a series of overlapping deficiencies. See text for details.*

256

CHAPTER 14
CHROMOSOMAL
ABERRATIONS:
STRUCTURAL
CHANGES

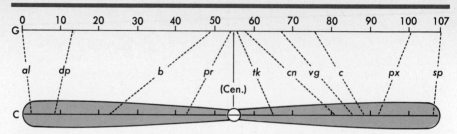

Figure 14-3. *Diagrammatic comparison of genetic map (G) and cytological map (C) of chromosome II of* Drosophila melanogaster. *Note particularly the much closer spacing of genes near the centromere on the genetic map, as opposed to the distances between them on the cytological map. This reflects the interference effect of the centromere. (Cen = centromere.)*

described by Lejeune and colleagues in 1963 and based on three cases. Manifestations include a characteristic high-pitched, plaintive cry, very similar to that of a kitten in distress. De Grouchy and Turleau (1977) report that acoustic recordings of *cri du chat* infants and of a kitten yield identical tracings. This symptom is associated with a malformation of the larynx. Other characteristic malformations involve the head and face. Mental retardation is severe, with the intelligence quotient usually below 20. Other nonspecific malformations and dysfunctions involving the brain, heart, eyes, kidneys, and skeleton may also occur. At least 120 cases have been studied since the initial description. The condition appears not to be significantly life-threatening; many affected individuals attain adulthood.

The karyotype is unvarying; in each case a large part of the smaller arm of one of the number 5 autosomes is deleted (Fig. 14-4). The standard designation for the shorter arm of a nonmetacentric chromosome is p; that for the longer arm is q. A deletion is indicated by a superscript minus sign, added segments by a superscript plus sign. Hence the karyotype of a *cri du chat* patient is $5p^-$. Cytological studies indicate association of major symptoms with deletion of a small portion of band 5p14p15. In most, if not all cases, the deletion arises during gametogenesis in one parent or the other; both parents of affected children present normal karyotypes. No parental age effect is discernible.

Several other deletions in human beings are well known, and new ones are reported in the literature from time to time. German (1970) described a male infant that had a deletion of the short arm of a number 4 autosome ($4p^-$). Such a child is shown in Figure 14-5A, its karyotype in Figure 14-5B. The child was described as "unusually small, (with) severe psychomotor retardation, convulsions, a wide flat nasal bridge, prominent forehead, cleft palate, and congenital heart disease." Some mentally defective patients have been found to have other chromosomal deletions; de Grouchy, for example, reported one such child who lacked the short arm of one of its number 18 autosomes. Notice that, in *most* departures from the normal human karyotype, the individual is abnormal in one or more respects.

Patients with chronic myelocytic leukemia carry a short chromosome, (named the Philadelphia chromosome (Ph^1), after the city where the first case was discovered), in all bone marrow cells and cells originating in the bone marrow (i.e., neutrophils, eosinophils, normoblasts, and megakaryocytes). Ph^1 persists in bone marrow, and often in cells derived therefrom, during remission. Detailed cytological study discloses Ph^1 to be a number 22 autosome that has lost most of the distal part of its longer arm ($22q^-$). This deleted part of autosome 22 is often seen to be translocated to one of the other autosomes, preferentially to the distal end of 9q. Some early confusion over whether number 21 or 22 is involved,

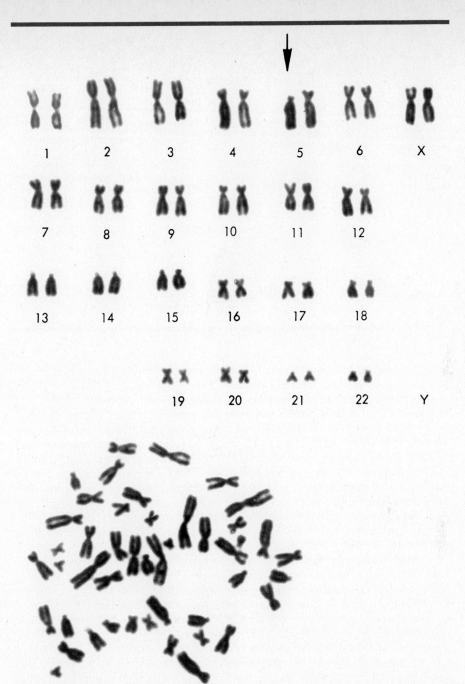

Figure 14-4. *Karyotype of child with* cri du chat *syndrome. Note delection of the short arm of one of the number 5 autosomes.* (Photo courtesy Dr. James German, New York Blood Center.)

derived from the morphological similarity between numbers 21 and 22. However, banding techniques have clearly revealed that Ph^1 is $22q^-$, that is, the *weakly* fluorescent G chromosome. Although it is not yet possible to state with certainty whether Ph^1 is a cause or an effect of chronic myelocytic leukemia (if a "cause," then what is the cause of the deletion?), it is interesting to note that

258

CHAPTER 14
CHROMOSOMAL
ABERRATIONS:
STRUCTURAL
CHANGES

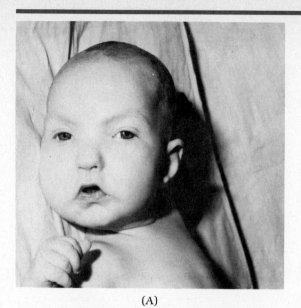

Figure 14-5. *(A) Infant with a deletion of the short arm of one of the number 4 autosomes. The syndrome is described in the text. (B) Karyotype of the child in Figure (A). Note missing segment of the short arm of one of the number 4 autosomes (arrow). (Photos courtesy Dr. James German, New York Blood Center. Copyright 1970 by The Society of the Sigma Xi. Used by permisison.)*

(A)

several investigators have reported that not only is Ph^1 itself not lethal, but it actually accelerates cell multiplication.

Juberg and Jones (1970) reported an 18-month-old male with Down's syndrome who died of acute myelocytic leukemia. Cytological investigation disclosed two cell lines in the bone marrow and in the peripheral blood, one of 47 chromosomes, the other of 51 (Fig. 14-6B). The 47-chromosome karyotype shows five G-group chromosomes, two with deletions of all of the short arm (Gp^-). In the 51-chromosome line, the G group consisted of seven chromosomes, three with no detectable short arms (Fig. 14-6A). In addition, this cell line showed two extra chromosomes in other groups, one in the C group and one in the F. The Gp^- chromosome was first reported in 1962 in two siblings suffering from chronic lymphocytic leukemia and was designated the Christchurch chromosome (Ch^1), from the New Zealand city where the case was found.

A number of phenotypic abnormalities were present in the case reported by Juberg and Jones, but abnormalities, when present, appear not to be constant from one case to another. In 45 persons described to 1970, only 17 had some abnormality, and only 2 of 30 carriers displayed any phenotypic abnormality. Further study is necessary before the nature of the inconstant effect of the Ch^1 chromosome can be clarified.

Deletion of part of the short arm of one of the X chromosomes produces a typical Turner syndrome. Deletion of part of the long arm of an X, on the other hand, results in an atypical, Turnerlike syndrome. The latter individuals are of normal height, but have vestigial gonads and are sterile. No other anomalies are consistently associated with this cytological condition. By contrast, apparent deletions in the Y chromosome are not known to produce any serious consequences, probably because of the relative genetic inertness of the Y.

Increasing use is being made of cultured amniotic fluid cells obtained by amniocentesis[1] to determine fetal karyotypes. Amniotic cells appear to show a rather high frequency of tetraploidy, unaccompanied by any cytological abnormality of the fetus, but structural aberrations, trisomy, and the like are fairly

---

[1]Amniocentesis involves the withdrawal of fluid, containing sloughed fetal epithelial cells, from the amniotic cavity. Chromosomal analysis as well as enzyme studies can then be made from subcultured cells.

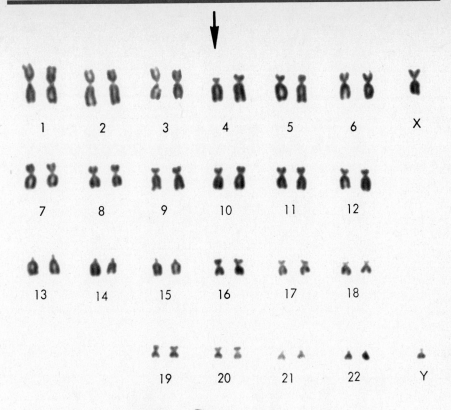

1 2 3 4 5 6 X

7 8 9 10 11 12

13 14 15 16 17 18

19 20 21 22 Y

(B)

260

CHAPTER 14
CHROMOSOMAL
ABERRATIONS:
STRUCTURAL
CHANGES

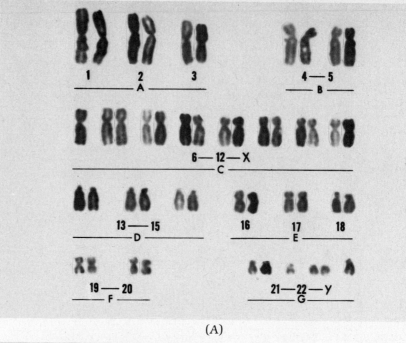

(A)

(B)

Figure 14-6. *Karyotype showing abnormal G chromosome (Ch¹, or Christchurch) from a case of acute myelocytic leukemia. (A) From peripheral blood cell of patient; 47 chromosomes, with 5 members of the G group, two of which have short-arm deletions (Gp⁻). (B) From bone marrow cell of patient, showing 51 chromosomes.* [From R. C. Juberg and B. Jones, 1970. "The Christchurch Chromosome (Gp⁻): Mongolism, Erythroleukemia, and an Inherited Gp⁻ Chromosome (Christchurch.) *New Eng. Jour. Med.,* **282**:292–297. Photos courtesy Dr. Richard C. Juberg. Used by permission of authors and publisher.]

easy to detect from these cells around the sixteenth week of gestation. This information can, of course, be extremely useful to parents and physician in considering a possible termination of the pregnancy.

INVERSIONS

Inversions develop most likely when breaks occur at a place where a chromosome forms a tight loop during synapsis. An inversion heterozygote, in which only one of two homologs has a particular inversion, forms a characteristic **inversion loop** in prophase-I of meiosis (Fig. 14-7). With a little care differences between deletion loops and inversion loops can be distinguished easily.

An inversion requires two breaks in a chromosome, followed by reinsertion of the segment in the opposite direction. A particular block of genes thus occurs in reverse sequence (Table 14-1). Inversions may either include the centromere (*pericentric inversions*) or not include the centromere (*paracentric inversions*). Homologous chromosomes, with identical inversions in each member, pair and undergo normal distribution in meiosis. On the other hand, should only one member of the chromosome pair experience a given inversion, synapsis produces a characteristic inversion loop (Fig. 14-8B). If a chiasma forms within a paracentric inversion, for example, a dicentric bridge is produced at anaphase-I, which results in the loss of an acentric fragment (Fig. 14-8C). The bridge itself breaks as anaphase-I progresses, and this results in additional aberrations such as deletions or duplications. Although inversions have been referred to as "crossover suppressors," such a term is inaccurate because crossing-over itself is not suppressed but, rather, crossover *products* are eliminated. Comparison of banded karyotypes of humans and the apes reveals numerous paracentric as well as pericentric inversions in humans that are not paralleled in apes.

TRANSLOCATIONS

As indicated on page 254, translocations involve the shift of part of one chromosome to another, nonhomologous chromosome. Two principal types of translocations are recognized:

1. *Simple*, in which, following breaks, a segment of one chromosome is transferred to another, nonhomologous chromosome, where it occupies an intercalary location.
2. *Reciprocal* (interchange), in which segments, which need not be of the same size, are exchanged between nonhomologous chromosomes.

Reciprocal translocations are well known in animals and plants that have been studied extensively. They have been produced frequently in *Drosophila* (as well as in other organisms) by radiation and have occurred widely in natural populations of the evening primrose (*Oenothera*), the plant whose resulting phenotypic variability led De Vries to formulate his mutation theory of evolution. Simple translocations have been found in humans.

Figure 14-7. *Diagrammed representation of synapsis of a pair of homolous chromosomes, the lower of which has sustained an inversion.*

262

CHAPTER 14
CHROMOSOMAL
ABERRATIONS:
STRUCTURAL
CHANGES

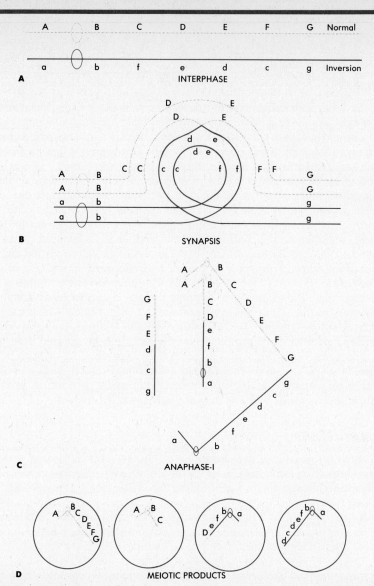

Figure 14-8. *Crossing-over within the inversion loop of a paracentric inversion decreases the frequency of functional, chromosomally normal gametes. (A) one pair of chromosomes, one member of which has previously sustained an inversion involving loci c, d, e, and f. (B) synapsis of homologous chromosomes showing chromatids and a crossover between nonsister chromatids within the inversion loop. (C) Anaphase-I. Two normal daughter chromosomes (A B C D E F G and (a b c d e f g), one dicentric (A B C D e f b a) and one acentric (G F E d c g). The acentric is lost, either in meiosis or in an early subsequent mitosis; the dicentric often breaks (if the centromeres move in opposite directions) randomly. (D) The four meiotic products (e.g., meiospores or pollen mother cells in flowering plants or primary spermatocytes or primary oocyte and polar bodies) with two normal nuclei (left and right) plus two abnormal nuclei (center).*

Two reciprocal translocation types are recognized: homozygotes and heterozygotes (Fig. 14-9). The former may have normal meiosis and, in fact, be difficult to detect cytologically unless morphologically dissimilar chromosomes are involved, or banding patterns differ markedly. Reciprocal translocations are characterized genetically by altered linkage groups and by the fact that a gene with

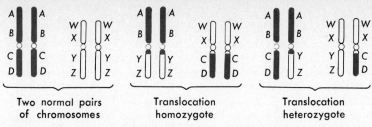

Figure 14-9. *Translocation homozygotes and heterozygotes compared.*

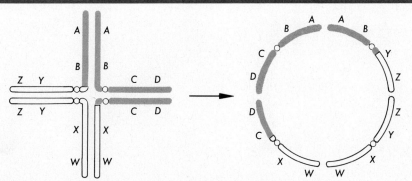

Figure 14-10. *Cross-shaped figure occurring in late prophase-I because of translocation heterozygosity. Such a figure often opens out into a ring.*

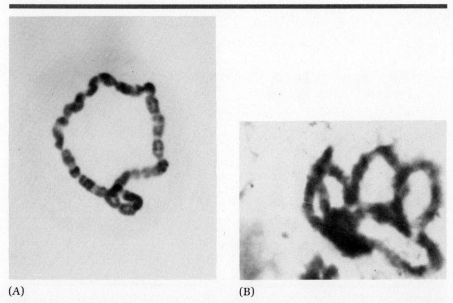

(A)   (B)

Figure 14-11. *Ring chromosomes in the tropical plant* Rhoeo discolor. *(A) Ring of 12 chromosomes. (B) Ring of several chromosomes.* [(A) Courtesy Dr. L. F. LaCour, John Innes Institute; (B) Courtesy Mr. Gary Bock.]

264

CHAPTER 14
CHROMOSOMAL
ABERRATIONS:
STRUCTURAL
CHANGES

"new neighbors" may produce a somewhat different effect in its new location. Such a **position effect** will be examined shortly.

Translocation heterozygotes, however, are marked by a considerable degree of meiotic irregularity. Peculiar and characteristic formations occur at synapsis because of the difficulty of attaining a pairing of homologous parts. Typically, a cross-shaped formation is seen in prophase-I; this often opens out into a ring. (Figs. 14-10 to 14-12).

In addition to altered linkage groups, translocation heterozygotes are frequently partially sterile because between half and two-thirds of gametes (in

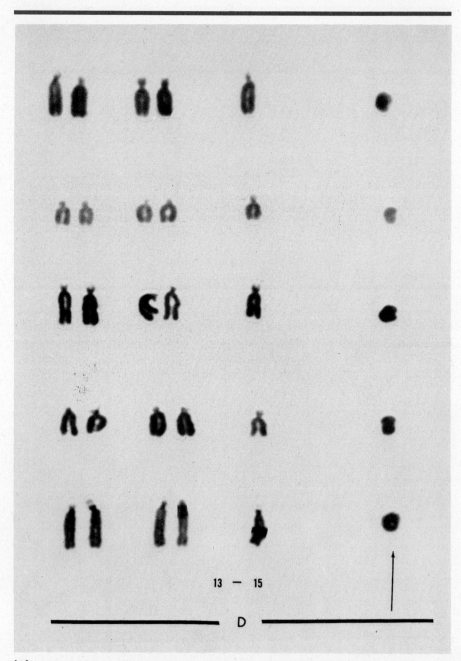

13 — 15

D

(A)

animals) or meiospores (in plants) fail to receive the full complement of genes required for normal development. Semisterility resulting from reciprocal translocations is easily observed in such plants as corn. Ears lack about half the kernels, and these are arranged irregularly (Fig. 14-13). Abortive pollen is reduced in size.

Translocations in humans. Translocations in addition to the 14-21 associated with Down's syndrome have been reported in a few instances for human beings. Those most likely to receive attention involve chromosomal segments present in triplicate. Among others, D/D, G/G, B/D, and C/E have been detected. When members of the several groups of autosomes are not easily differentiated, translocations in humans are designated by two letters separated by a slash; the first letter indicates the chromosome group that supplied the translocated segment and the second, the group that received the transferred segment.

In most cases there are numerous phenotypic abnormalities, and spontaneous abortion, or death within a few months after birth, usually ensues. Those who do live longer almost always exhibit mental retardation. A probable unbalanced translocation has been reported in which a large part of one of the D-group chromosomes was present in triplicate and a small part of one of the B chromosomes was lacking. There were numerous anatomical abnormalities, and death of the subject occurred at seven months. Forty-six chromosomes were

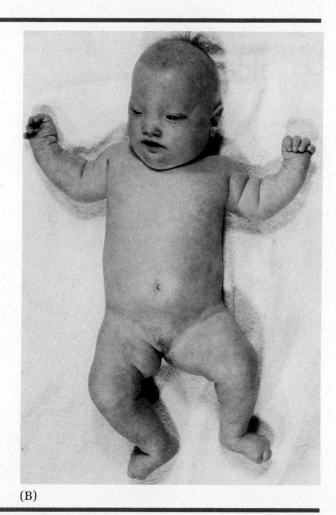

Figure 14-12. Ring-D chromosome in a human female. (A) Partial karyotype of the D group showing ring chromosomes at the right. (B) Phenotypic abnormalities associated with the ring-D chromosome. (Courtesy Dr. Richard C. Juberg. From R. C. Juberg, M. S. Adams, W. J. Venema, and M. G. Hart, 1969. "Multiple Anomalies Associated with a Ring-D Chromosome." Jour. Med. Gen. **6:** *314–321. Used by permisison of senior author and publisher.)*

(B)

266

CHAPTER 14
CHROMOSOMAL
ABERRATIONS:
STRUCTURAL
CHANGES

Figure 14-13. *Partial sterility results from a reduction in functional gametes caused by multiple translocations.* (Courtesy DeKalb Agricultural Association, Inc., DeKalb, Illionis.)

present, including an unusually long one believed to be a B + D. Both parents were cytologically normal.

A C/E translocation with effects resembling Down's syndrome was apparently first detected in 1969 at a Boston hospital. The child had one extra long chromosome in its complement of 46, interpreted as an E with added material from a C-group member, which material was thus present in triplicate. This diagnosis was made on the basis of the mother's karyotype, which showed one shortened member of group C (believed to be number 8) and a longer than normal E chromosome (probably number 18). The mother was phenotypically normal, as would be expected, because she had the normal amount of chromosomal material.

DUPLICATIONS AND THE POSITION EFFECT

Duplications occur when a portion of a chromosome is represented more than twice in a normally diploid cell. The extra segment may be attached either to the chromosome whose loci are repeated or to a different linkage group, or it may even be present as a separate fragment. Duplications are useful in studying the quantitative effects of genes normally present only in pairs in diploid cells.

The first duplication to receive critical study was the *bar eye* variant in *Drosophila*. The wild-type eye is essentially oval in shape; the bar eye phenotype is characterized by a narrower, oblong, bar-shaped eye with fewer facets. The now classical studies of Bridges (1936) showed this trait to be associated with the duplication of a segment of the X chromosome called section 16A, as observed in salivary gland chromosomes. Each added section 16A intensifies the bar phenotype. However, the narrowing effect is greater if the duplicated segments are on the same chromosome. If we let A represent one section 16A in a given X chromosome, we can recognize the genotypes and phenotypes listed in Table 14-2. Other arrangements are of course also possible, but these show clearly that the bar effect of a given number of duplicated 16A sections is intensified if the duplications occur in one X chromosome rather than being divided between the two of the female. Compare heterozygous ultrabar and homozygous bar eyes, for example.

Table 14-2. Comparison of genotypes and phenotypes for bar eye in Drosophila females

X chromosomes	Phenotype		Mean number of facets
A/A	Normal		779
AA/A	Heterozygous bar eye		358
AA/AA	Homozygous bar eye		68
AAA/A	Heterozygous Ultrabar		45
AAA/AAA	Homozygous Ultrabar		25

A = One section 16A of the X chromosome.

Such a change in the effect of the genes or genes in a chromosomal segment is known as the **position effect.** In bar eye, each added segment narrows the eye still farther, and this effect is enhanced as more duplications occur in one chromosome. Other duplications are known that produce the opposite effect, counteracting the effect of mutant genes. Moreover, duplications need not always be immediately adjacent to exert this position effect.

268

CHAPTER 14
CHROMOSOMAL
ABERRATIONS:
STRUCTURAL
CHANGES

POSSIBLE CAUSATIVE CHEMICAL AGENTS
OF CHROMOSOMAL ABERRATIONS IN HUMANS

Many chemical substances have been suspected as causal agents of chromosomal damage in the human species. We shall review the sometimes contradictory evidence relating to LSD, marijuana, nicotine, cyclamates, DDT, caffeine, asbestos fibers, and two (former) hair dye components.

LSD and chromosome damage. Because of the therapeutic use of **LSD** (lysergic acid diethylamide) in treatment of some mental disorders and its frequent illicit use, many investigators have recently turned their attention to the question of whether this drug produces chromosomal and genetic damage. Studies have been carried on in vivo and in vitro with humans, other animals, and plants. Conclusions are contradictory, as summed up by Maugh (1973) in these words: "Many studies have shown, for example, that LSD can cause chromosome damage in cultured human cells, that it can cause birth defects in laboratory animals and humans, and that it is carcinogenic. An equally large number of studies have shown that there is no evidence for such effects, particularly at LSD concentrations that might be encountered in drug abuse." Recent studies, however, indicate that in at least some individuals of some species an increased frequency of chromosome breaks is found.

It has been found that addition of LSD to cultured human leukocytes results in a considerable increase in chromosomal abnormalities, principally a significantly larger number of breaks, most of which involve number 1, the largest of the autosomes. This team also observed a similar increase in a paranoid schizophrenic who had been extensively treated for four years with the drug. Subsequently, numerous studies on plants and animals, including humans, have appeared; about half provide evidence for chromosome damage; the remainder either do not or are inconclusive.

A significant increase in frequency of chromosome breaks has been found in a large number of illicit users of LSD and other drugs. A high incidence of breaks in root tip cells of barley, about half of which are near the centromere, has also been reported. A study of effects of LSD in rye disclosed an increase in breaks, fragments, and univalents in meiotic material. Similarly, meiotic chromosomes that had previously been treated with heavy doses of LSD showed a considerable increase in breaks and fragments over control animals. Yet other studies in rabbits, hamsters, and rats yielded no increase in chromosomal breaks in either mothers or embryos. Long (1972), in a review of the literature involving a total of 161 children of parents who took LSD, concluded that LSD does not cause chromosome breakage in humans.

Some of the disagreement among these and other studies may arise because of the possibility that some persons may be more susceptible than others to chromosomal damage by LSD. Few reports can easily make use of data taken before and after LSD use in the same individuals. Moreover, many of the studies in the literature are not, and cannot be, based on standardized doses because they involve illicit users of probably quite variable drug concentration and purity. Furthermore, some of the subjects admit to use of other drugs as well. Much remains to be learned from in vivo studies of various kinds of cells, especially gametes, under rigidly controlled experimental conditions, and more data are needed on children born to LSD users. In such a controlled experiment, Hungerford et al. (1968) first established aberration frequencies in cultured leukocytes from a control group, and also from an experimental group whose members were later to receive LSD therapy. Aberrations scored included acentric fragments, dicentrics, deletions, breaks and gaps. These researchers report that administration of three intravenous doses (usually of 200 µg each) was followed by some increase in aberration frequency, as well as the appearance of new abnormalities in the experimental group. However, they found a return to the lower control levels ensued within one to six months after the last dose.

Birth of a child with several congenital abnormalities to parents who had both used LSD prior to conception, but not during pregnancy, has been reported by Hsu and others (1970). Although the infant was found to have 46 chromosomes, she was trisomic for autosome 13 and exhibited a translocation involving members of the D group. Hsu and his colleagues suspect drug-induced damage to maternal germ cells; the child, they theorize, resulted from fertilization of an egg having an unbalanced chromosome complement.

At the present time, then, risk of chromosomal damage in users and of damage to their descendants cannot be eliminated from consideration, although the period of susceptibility may be limited. But Dishotsky and colleagues (1971), after surveying the literature, conclude that, in humans, "pure LSD in moderate doses does not damage chromosomes in vivo (and) does not cause detectable genetic damage." Certainly human beings would not necessarily react to the same amount of LSD in the same way as would barley, for example, and different persons very likely react differently to the same dosage. In view of these conflicting findings, it is difficult to reach a definitive conclusion.

Marijuana. The active ingredient in the smoke of leaves of marijuana (*Cannabis sativa*) is Δ^9-tetrahydrocannabinol (Δ^9-THC or THC). There is considerable interest in the question of possible cytological and genetic effects of marijuana because of its frequent illicit use. In the work previously reported in connection with LSD, hamsters were given daily subcutaneous injections of marijuana extract distillate containing 17.1 percent Δ^9-THC for 10 consecutive days in amounts of 1, 10, and 100 mg per kilogram of body weight. Another group of hamsters was given the same dosages of Δ^9-THC in combination with 1,000 μg LSD per kilogram of body weight. No significant elevation of leukocyte chromosome breaks over controls was found. Exposure of cultured leukocytes from four healthy human subjects to 0.1, 10, and 100 μg Δ^9-THC per milliliter of culture solution produced no increase in chromosome breaks over controls. However, the 100 μg culture cells did not grow and showed no mitoses at all.

Leuchtenberger et al. (1973a, 1973b) exposed human adult lung fibroblast cells to four puffs of 25 ml each per day of smoke from cigarettes, which contained 1.8 g of marijuana. Another set of cultured cells was exposed to two puffs per day of smoke from tobacco cigarettes. Various cytotoxic effects, including death and a considerable decrease in mitoses were found in both types of cultures, but these were more marked in cells exposed to tobacco smoke than to marijuana smoke! Cells of both cultures showed various mitotic abnormalities up to 45 or more days after exposure. Leuchtenberger and his associates conclude that smoke from both marijuana and tobacco cigarettes does produce disturbances in the mitotic process, in the chromosomal complement and, therefore, in the genetic equilibrium.

Asbestos fibers. The apparent relationship between some types of lung cancer and inhalation of asbestos fibers has been widely reported by the various news media. Cancerous tissue often shows a wide variety of chromosomal aberrations, from polyploidy (often aneuploidy) to structural defects. Sincock and Seabright (1975) exposed Chinese hamster cell lines to asbestos fibers in the culture medium for periods of 48 hours and 5 days. Cells in both the 48-hour and 5-day cultures showed a significant increase in polyploidy (mostly tetraploidy), occurrence of fragments (paired isodiametric fragments, chromatid breaks), and chromosomal breaks and rearrangements. The number of cells that showed chromosomal aberrations was not significantly different in the two time-period cultures; that is, five-day exposure produced no significant increase in number of cells with such damage over those exposed for 48 hours.

270

CHAPTER 14
CHROMOSOMAL
ABERRATIONS:
STRUCTURAL
CHANGES

Hair dye components. Until the late 1970s, widely used ingredients in commercial hair dyes included, among others, 2-nitro-*p*-phenylenediamine (2-NPPD) and 4-nitro-*o*-phenylenediamine (4-NOPD). Benedict (1976) found that both substances were highly mutagenic and both induced chromosomal changes in hamster cell cultures. These included many chromosomal breaks, as well as lesser numbers of quadriradials, triradials, dicentrics, and ring chromosomes. In regard to cells with chromosomal fragments, Benedict points out "This type of abnormal metaphase may be highly significant in the production of malignant transformation since at this nontoxic dose (10^{-5} molar 2-NPPD), the surviving cell could easily be aneuploid, with chromosome imbalances. We believe such chromosome changes may be related to malignant transformation and (tumor production)." He concludes that such hair dye ingredients as 2-NPPD and 4-NOPD could constitute potential hazards in human use because of their higher concentration (10^{-2} *M*) in hair dyes.

Other substances. A great variety of other substances that are in wide use have been implicated or suspected in the induction of chromosomal damage. In addition to the work of the Leuchtenberger group on nicotine, other research disclosed a variety of cytotoxic effects of nicotine on human leukocytes in vitro, but no chromosomal aberrations. Gross abnormalities, however, do occur in mouse cells.

The once widely used artificial sweetener sodium cyclamate is reported to produce chromosomal breaks and gaps in cultured human leukocytes, and calcium cyclamate produces both aneuploidy and breaks in bone marrow cells of the gerbil. No chromosome anomalies, however, have been found in spermatocytes of mice that had received sodium cyclamate in drinking water.

The insecticide DDT [1,1,1-trichloro-2,2-bis(p-chlorophenyl)ethane] has been found not to produce a significant increase of chromosome breaks in hamster cells in vitro. On the other hand, a significant increase in the incidence of chromosome aberrations has been found in mice—one of the few records of DDT-induced chromosome damage.

Caffeine has been suspected of causing chromosome damage; however, in another study, no significant increase in damage levels of cultured leukocytes was found in human volunteers who had ingested caffeine in four 200 mg lots per day for a month. In yeast treated with 0.1 percent caffeine in the nutrient medium, both a decrease in crossover frequency and an increase in chromosome breaks has been reported. It should be noted that the amount of caffeine administered in these tests is not especially large and, in fact, is much lower than the level consumed by many heavy coffee drinkers.

CHROMOSOMAL ABERRATIONS AND EVOLUTION

Speciation, the evolutionary divergence of segments of a population to the point that they are no longer able to combine their genes through sexual reproduction, is a complex process with no single causative mechanism. The diversification on which evolution is built does, in every case, however, require alterations in the genetic material itself (mutation, Chapter 19), and/or changes in its arrangement (chromosomal aberrations), both of which lead to reproductive incompatibility. The effectiveness of both of these factors is increased if the population is broken into two or more isolated groups. In this way frequencies of mutant genes may often be spread more rapidly, probability of *different* mutations that arise in separated groups is increased, and differing groups are often prevented from crossing by various isolating mechanisms.

Each of the chromosomal aberrations described in Chapters 13 and 14 is, if not lethal or disabling, related to the development of new taxonomic entities, though each rarely acts alone over time. Thus, trisomy in the Jimson weed leads to morphological differences that are recognizable and forms the basis for an

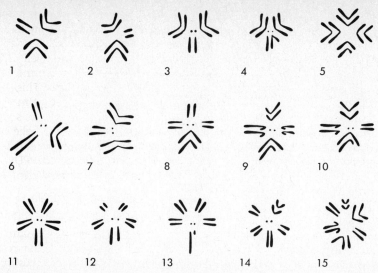

Figure 14-14. *Male karyotypes in several secies of* Drosophila. *The X and Y chromosomes are at the bottom of each drawing. (1)* D. willistoni; *(2)* D. prosaltans; *(3)* D. putrida; *(4)* D. melanogaster; *(5)* D. ananassae; *(6)* D. spinofemora; *(7)* D. americana; *(8)* D. pseudoobscura; *(9)* D. azteca; *(10)* D. affinis; *(11)* D. virilis; *(12)* D. funebris; *(13)* D. repleta; *(14)* D. montana; *(15)* D. colorata.

aneuploid series. Certainly aneuploid series, which may arise in a variety of ways, do exist in many plant families and genera and are associated with interspecific morphological differences.

But structural changes too are important in a gradual, step-by-step change. Many inversions are known to differentiate the several species of *Drosophila*. Thus *D. pseudoobscura* and *D. persimilis*, which are morphologically very similar, produce sterile male but fertile female hybrids and differ in four major inversions. On the other hand, *D. pseudoobscura* and *D miranda* produce completely sterile offspring when crossed. The very complex pairing arrangements assumed by giant chromosomes in hybrid larvae of this interspecific cross indicate repeated and extensive inversions. Translocations in the evening primrose, *Oenothera*, were largely responsible for the differences on which DeVries based his mutation theory. A series of many reciprocal translocations in isolated populations may lead to complete reproductive incompatibility, as well as definitive morphological differences, so that speciation may be said to have occurred. Comparison of the chromosome complements of several species of *Drosophila* (Fig. 14-14) suggests all manner of chromosomal aberrations, including translocations (even unions of whole chromosomes) and deficiencies, at least. Note the aneuploid series 3, 4, 5, 6.

On the other hand, the disabling and lethal effects of many types of chromosomal aberrations, especially in higher animals such as humans, remind us that aneuploidy, polyploidy, and structural changes in chromosomes may carry a high negative selection value.

References

Benedict, W. F., 1976. Morphological Transformation and Chromosome Aberrations Produced by Two Hair Dye Components. *Nature,* **260:** 368–369.

272

CHAPTER 14
CHROMOSOMAL
ABERRATIONS:
STRUCTURAL
CHANGES

Bridges, C. B., 1936. The Bar "Gene," a Duplication. *Science*, **83:** 210–211. Reprinted in J. A. Peters, ed., 1959. *Classic Papers in Genetics*, Englewood Cliffs, N.J.: Prentice-Hall.

Dishotsky, N. I., W. D. Loughman, R. E. Mogar, and W. R. Lipscomb, 1971. LSD and Genetic Damage. *Science* **172:** 431–440.

de Grouchy, J., M. Lamy, S. Thieffry, M. Arthuis, and C. Salmon, 1963. Dysmorphie Complexe avec. Oligophrénie: Délétion des Bras Courts d'un Chromosome 17-18. *Compt. Rendus Acad. Sci. Paris*, **256:** 1028–1029.

de Grouchy, J., and C. Turleau, 1977. *Clinical Atlas of Human Chromosomes*. New York: John Wiley & Sons.

de Grouchy, J., C. Turleau, M. Roubin, and F. Chavin-Colin, 1973. Chromosomal Evolution of Man and the Primates (*Pan troglodytes, Gorilla gorilla, Pongo pygmaeus*). *Nobel Symposium 23: Chromosome Identification*. New York: Academic Press.

Hsu, L. Y., L. Strauss, and K. Hirschorn, 1970. Chromosome Abnormality in Offspring of LSD User. *Jour. Amer. Med. Assn.*, **211:** 987–990.

Hungerford, D. A., K. M. Taylor, C. Shagass, G. U. LaBadie, G. B. Balaban, and G. R. Paton, 1968. Cytogenetic Effects of LSD 25 Therapy in Man. *Jour. Amer. Med. Assn.*, **206:** 2287–2291.

Johnson, G. A., and S. M. Jalal, 1973. DDT-Induced Chromosome Damage in Mice. *Jour. Hered.*, **64:** 7–8.

Lejeune, J., 1963. Trois Cas de Délétion du Bras Court d'un Chromosome 5. *Compt. Rendus Acad. Sci. Paris*, **257:** 3098–3102.

Leuchtenberger, C., R. Leuchtenberger, U. Ritter, and N. Innui, 1973a. Effects of Marijuana and Tobacco Smoke on DNA and Chromosomal Complement in Human Lung Explants. *Nature*, **242:** 403–404.

Leuchtenberger, C., R. Leuchtenberger, and A. Schneider, 1973b. Effects of Marijuana and Tobacco Smoke on Human Lung Physiology. *Nature*, **241:** 137–139.

Long, S. Y., 1972. Does LSD Induce Chromosomal Damage and Malformations? A Review of the Literature. *Teratology*, **6:** 75–90.

Sincock, A., and M. Seabright, 1975. Induction of Chromosome Changes in Chinese Hamster Cells by Exposure to Asbestos Fibers. *Nature*, **257:** 56–58.

Problems

14-1   Suggest a meiotic configuration at synapsis for a situation where six chromosomes have undergone reciprocal translocations of about the same length.

14-2   Below are diagrams of a pair of autosomes (for simplicity chromatids are not shown). Loci are indicated by Arabic numerals along the chromosome. For each of the following structural changes in "autosome 7" (maternal and paternal homologs are designated A and B, respectively) construct a simple diagram showing synaptic figures assumed in meiosis in each of the cases given below the chromosome diagrams; chromatids need be shown only for part (c):

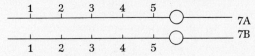

(a) a deletion in 7B involving loci 2 and 3;
(b) an inversion in 7B involving loci 2, 3, and 4;
(c) a reciprocal translocation with autosome 8B involving loci 1, 2, and 3 of autosome 7B, and loci F, G, and H of autosome 8B (all of the latter on the long arm of 8B, which has the sequence (centromere), D, E, F, G, H, (terminus).
(d) a duplication involving loci 2, 3, and 4 in 7B.

14-3   Let the diagrams immediately below represent a pair of human autosomes 6 (with A and B representing maternal and paternal homologs, respectively), each of which consists of two sister chromatids. Letters along the chromatids represent several hypothetical loci.

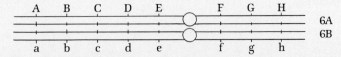

(a) Diagram the synaptic figure assumed by the two homologs (showing chromatids) if chromosome 6 sustains one nonsister chromatid crossover between loci F and G.

(b) Diagram chromosome 6, with loci, as it would appear in each of four sperms derived from the primary spermatocyte.

14-4 Differentiate between paracentric and pericentric inversions.

14-5 Assume a chromosome with the following gene sequence (the o represents the centromere):

$$A\ B\ C\ D\ o\ E\ F\ G\ H$$

You find the following aberrations in this chromosome; for each identify the specific kind of aberration:

(a) $A\ B\ C\ D\ o\ E\ F\ H$

(b) $A\ D\ C\ B\ o\ E\ F\ G\ H$

(c) $A\ B\ C\ D\ C\ D\ o\ E\ F\ G\ H$

14-6 In relation to the information in Table 14-2, where would AAA/AA and $AAAA/AAA$ be properly placed? Note that progressively narrower eyes are shown from the top to the bottom of Table 14-2.

14-7 In *Drosophila*, e is a gene at locus 70.7 on chromosome III; ee flies have ebony bodies, much darker than wild-type flies of genotype $+ +$ or $+ e$. If the cross $+ + \times e\ e$ yields a small percentage of ebony flies, but greater than could be accounted for by the known mutation frequency of this gene, (a) what would you suspect as the cause? (b) What would you look for as confirmation?

14-8 Evaluate the probable cytological and genetic effect on human beings of (a) LSD, (b) marijuana.

14-9 In a hypothetical organism, gene a maps genetically three map units from the centromere. Would a cytological map be expected to show this gene farther from, closer to, or the same distance from the centromere? Why?

CHAPTER 15

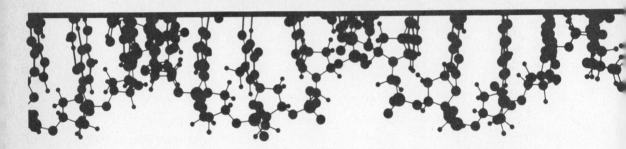

Population genetics

Up to this point our concern has been largely with results of experimental breeding programs, and family pedigree studies, from which certain now familiar genotypic and phenotypic ratios emerge. In producing such F_2 ratios as 1:2:1 and 9:3:3:1, the parental generation was usually of such genotypes as $AA \times aa$, or $AABB$ and $aabb$; that is, the alleles A and a as well as B and b were introduced *in equal frequency*. Similarly, in all the other experiments which have been dealt with in this text, gene frequency was intentionally included among the controlled factors. But, in natural populations the frequencies of alleles may vary considerably over both space and time. For example, the gene for polydactyly is dominant to its allele, yet the polydactylous phenotype is fairly infrequent among newborn infants, even though the trait is not known to play any part in survival and no more than a very minor role in choosing a mate. Apparently the *frequency of the dominant gene here is lower than that of its recessive allele;* this pair obviously does not exist in the *population* in the 1:1 ratio so commonly encountered in the laboratory. But the frequencies of these and other genes might well be expected to be unequal in different populations and to differ in the same population at different times. Population genetics deals with the analysis of changes in gene frequencies in a population over time.

Just as gene frequency is controlled in the genetics laboratory, so is the mating pattern, generation after generation. But outside the laboratory, mating is often largely a chance or random affair. In natural populations, as a result, an ultimate **equilibrium frequency** is attained by alleles, governed by such factors as

1. **Breakdown of isolating mechanisms,** whereby additional numbers of one allele or the other are introduced from outside the group.

2. Frequency of mutation (change in the nature of the gene).
3. Selection (environmental and reproductive).
4. Random genetic drift.

The ultimate equilibrium frequency, which *approximates the binomial distribution*, is then maintained, subject only to random genetic drift, until and unless admixture from outside the population occurs, mutation frequency changes, and/ or selection forces change.

In 1908, G. H. Hardy, a British mathematician, and the German physician W. Weinberg independently developed a relatively simple mathematical concept, now referred to as the Hardy-Weinberg theorem, to describe this genetic equilibrium. This principle is the foundation of population genetics and, in essence, states that *in the absence of migration, mutation, and selection*, gene and genotypic frequencies remain constant within certain narrow, determinable limits, generation after generation, in a large, randomly mating population. The Hardy-Weinberg theorem may be used to determine the frequency of each allele of a pair or of a series, as well as frequencies of homozygotes and heterozygotes in the population. This chapter will examine applications of this theorem and the forces that alter gene frequencies.

Calculating gene frequency

Codominance. The M-N blood type furnishes a useful example of a series of phenotypes due to a pair of codominant genes. None of the three possible phenotypes, M, MN, and N, appears to have any selection value; in fact, most persons do not know their M-N type. Frequencies of the two alleles involved will be calculated for samples from two different groups.

On page 144, a study of M-N types for 6,129 white Americans living in New York City, Boston, and Columbus, Ohio was cited. Because the closely related alleles *S* and *s* were not included in the data, these individuals can be grouped according to genotypes on the M-N system alone:

Genotype	Number
MM	1,787
MN	3,039
NN	1,303
	6,129

To calculate frequencies of the two alleles *M* and *N*, remember that these 6,129 persons possess a total of $6,129 \times 2 = 12,258$ alleles. The number of *M* alleles, for example, is $1,787 + 1,787 + 3,039$. Thus, calculation of the frequencies for *M* and *N* may be worked out in this way:

$$(1)\ M = \frac{1,787 + 1,787 + 3,039}{12,258} = \frac{6,613}{12,258} = 0.5395$$

$$(2)\ N = \frac{1,303 + 1,303 + 3,039}{12,258} = \frac{5,645}{12,258} = 0.4605$$

So frequencies of the two alleles in *this* sample are almost equal, and this is reflected in the close approximation to a 1:2:1 ratio so often seen in laboratory results.

Gene frequencies expressed as decimals may be used directly to state probabilities. Assume this sample to be representative of the population; then there is a probability of 0.5395 that, of the chromosomes bearing this pair of alleles, any one selected randomly will bear gene M, and 0.4605 that any one will bear N. Thus, the frequencies of the three phenotypes to be expected in the population are:

Genotype	Phenotype	Phenotypic frequency
MM	M	$0.5395 \times 0.5395 = 0.2911$
MN } NM	MN	$2(0.5395 \times 0.4605) = 0.4968$
NN	N	$0.4605 \times 0.4605 = \underline{0.2121}$
		1.0000

Notice that these genotypic frequencies follow a binomial distribution. If we let p represent the frequency of M, and q that of N, the total array of probabilities of the three genotypes (MM, MN, and NN) is

$$P^2 + 2pq + q^2$$

which is the expansion of $(p + q)^2$. Note also that $p + q = 1$ and, of course, so does $(p + q)^2$. Thus the probability of a type-M individual, for example, under a system of random mating is given by the expression p^2 where the value of p has already been calculated to be 0.5395. Similarly, probability of MN persons (the heterozygotes) is given by $2pq$, and of N persons by q^2.

Alternatively, the frequencies of M and N could have been calculated by another approach. Recall that an individual of genotype NN, for example, represents the simultaneous occurrence of two events of equal probability, namely the fusion of two gametes, each with the genotype N. Thus q^2 has the value 1,303/6,129 = 0.2126, and $q = \sqrt{0.2126}$, or 0.46. Because $p + q = 1$, $p = 1 - q$, or $1 - 0.46 = 0.54$.

For comparison, look briefly at the following sample of 361 Navaho Indians from New Mexico:

Phenotype	Number
M	305
MN	52
N	$\underline{4}$
	361

This sample is far from a laboratory-type 1:2:1 ratio. Do such frequencies mean that the Navahos do not conform to the same genetic laws as the previous sample? The answer is "no" as soon as it is recalled that the 1:2:1 proportion is based on equal frequencies of the alleles in the population. The raw data clearly suggest that M is considerably more frequent in Navaho Indians than is the N allele. Using the same method as was employed for the earlier sample to calculate gene frequencies produces this result:

$$\text{let } p = \text{frequency of } M = \frac{305 + 305 + 52}{722} = 0.9169$$

$$\text{let } q = \text{frequency of } N = \frac{52 + 4 + 4}{722} = \frac{0.0831}{1.0000}$$

Applying these gene frequencies to the sample produces these genotypic frequencies:

Genotype	Gene frequencies	Genotype probability	Genotypes in sample	
			Expected	Observed
MM	$p^2 = (0.9169)^2$	0.8407	303.5	305
MN	$2pq = 2(0.9169 \times 0.0831)$	0.1524	55.0	52
NN	$q^2 = (0.0831)^2$	0.0069	2.5	4
		1.0000	361.0	361

A chi-square test of the data for the Navaho sample shows $\chi^2 = 1.071$; for the earlier sample of 6,129 persons, $\chi^2 = 0.0237$. So deviation from expectancies based on the calculated gene frequencies is well below the level of significance in each case. Note that although there are three phenotypic classes, there is but *one* degree of freedom, because only two alleles, M and N, are involved. Therefore, only one phenotypic class can be set at random. For example, in a sample of 400 persons having just 200 M alleles and therefore 600 N alleles, any number of individuals *up to* 100 may be of genotype MM. If there are, say, 60 persons of type M, the other classes are thereby automatically determined:

Phenotype	Number of persons	Number of alleles	
		M	N
M	60	120	—
MN	80	80	80
N	260	—	520
Totals	400	200	600

Complete dominance. An interesting phenotypic trait with no known selection value is the ability or inability to taste the chemical phenylthiocarbamide ("PTC," $C_7H_8N_2S$), also called phenylthiourea. This was reported by Fox in 1932, who found a similar situation for several other thiocarbamides. The test is a simple one that can easily be performed by any genetics class. The usual procedure is to impregnate filter paper with a dilute aqueous solution of PTC (about 0.5 to 1 g per liter), allow it to dry, then place a bit of the treated paper on the tip of the tongue. About 70 percent of the white American population can taste this substance, generally as very bitter, rarely as sweetish. Although the physiological basis is unknown, tasting ability does depend on a completely dominant gene, which we will designate as T. Thus tasters are $T-$ (i.e., TT or Tt); nontasters are tt.[1]

Of 280 genetics students in the author's classes in one year, 198 were tasters and 82 were nontasters. From such data the frequencies of genes T and t in the sample may be readily calculated. The 82 nontasters (29.29 percent of the sample) are persons of genotype tt and in the Hardy-Weinberg theorem may be represented by q^2. Therefore,

$$q^2 = 0.2929 \text{ and}$$

$$q = \sqrt{0.2929} = 0.5412$$

which is the frequency of gene t. Because only a pair of alleles are involved, frequencies of the two genes must total 1, $p + q = 1$, and $p = 1 - q$. Therefore,

[1]Blakeslee and Salmon (1936) report the lowest concentration detectable by tasters to vary from 1:500,000 to 1:5,000, with a mean around 1:80,000. Sensitivity among tasters decreases with age, and females are slightly more sensitive than males.

in this example p, the frequency of gene T, equals $1 - 0.5412$, or 0.4588. Frequencies of homozygous and heterozygous tasters may now be computed. The binomial expansion $p^2 + 2pq + q^2$ shows the distribution of the three possible genotypes:

$$
\begin{aligned}
p^2 &= TT &&= (0.4588)^2 &&= 0.2105 \\
2pq &= Tt &&= 2(0.4588 \times 0.5412) &&= 0.4966 \\
q^2 &= tt &&= (0.5412)^2 &&= \underline{0.2929} \\
& && && \ 1.0000
\end{aligned}
$$

By testing representative samples of different populations, the frequencies of T and t in those groups may be similarly calculated.

Multiple alleles. The binomial $(p + q)^2 = 1$ can be used only when two alleles occur at a particular locus. For cases of multiple alleles, one simply adds more terms to the expression. Recall that the A-B-O blood groups are determined by a series of three multiple alleles, I^A, I^B, and i, if the various subtypes are neglected. Hence in a gene-frequency analysis, let

$$p = \text{frequency of } I^A$$

$$q = \text{frequency of } I^B$$

$$r = \text{frequency of } i$$

$$\text{and } p + q + r = 1$$

Thus genotypes in a population under random mating will be given by $(p + q + r)^2$.

To see how this trinomial is applied, consider the following sample of 23,787 persons from Rochester, New York:

Phenotype	Number	Frequency
A	9,943	0.418
B	2,379	0.100
AB	904	0.038
O	10,561	0.444
	23,787	1.000

The frequency of each allele may now be calculated from these data, where p, q, and r represent the frequencies of genes I^A, I^B, and i, respectively. The value of r, that is, the frequency of gene i, is immediately evident from the figures given:

$$r^2 = 0.444, \text{ hence}$$

$$r = \sqrt{0.444} = 0.6663 \ (= \text{frequency of } i)$$

The sum of A and O phenotypes is given by $(p + r)^2 = 0.418 + 0.444 = 0.862$; therefore,

$$p + r = \sqrt{0.862} = 0.9284$$

so $p = (p + r) - r = 0.9284 - 0.6663 = 0.2621 \ (= \text{frequency of } I^A)$. Because $p + q + r = 1$, $q = 1 - (p + r) = 1 - 0.9284 = 0.0716 \ (= \text{frequency of } I^B)$; genotypic frequencies, as shown in Table 15-1, can now be calculated. The probability figures arrived at for this large sample check quite closely with those arrived at for other samples of the general United States population.

Samples taken from other races and/or nationalities, however, may show quite different frequencies for these alleles. In a sample of Navaho Indians, the following gene frequencies were obtained in one study:

$$I^A = 0.1448$$

$$I^B = 0.0020$$

$$i = 0.8532$$

If this sample is representative of the Navaho population, the percentage of individuals of each genotype and phenotype may be calculated from these frequencies, as shown in Table 15-2. Thus, one may either calculate gene frequencies from numbers of each phenotype or compute percentages of the population that have each phenotype if gene frequencies are known.

Table 15-1. Calculation of genotypic frequencies for a sample of 23,787 persons living in Rochester, New York

Phenotypes	Genotypes	Genotypic frequencies	Population probability based on sample	
O	ii	r^2		0.4440
A	$I^A I^A$	p^2	0.0687 ⎱	0.4180
	$I^A i$	$2pr$	0.3493 ⎰	
B	$I^B I^B$	q^2	0.0051 ⎱	0.1005
	$I^B i$	$2qr$	0.0954 ⎰	
AB	$I^A I^B$	$2pq$		0.0375
				1.0000

Table 15-2. Probabilities of genotypes and phenotypes for a sample of Navaho Indians

Phenotypes	Genotypes	Genotypic frequencies	Population probability based on sample	
O	ii	r^2	$(0.8532)^2 =$	0.7280
A	$I^A I^A$	p^2	$(0.1448)^2 = 0.020967$	0.2680
	$I^A i$	$2pr$	$2(0.1448 \times 0.8532) = 0.247087$ ⎱	
B	$I^B I^B$	q^2	$(0.002)^2 = 0.000004$	0.0034
	$I^B i$	$2qr$	$2(0.002 \times 0.8532) = 0.003413$ ⎱	
AB	$I^A I^B$	$2pq$	$2(0.1448 \times 0.002) =$	0.0006
				1.0000

The two preceding examples clearly demonstrate that different frequencies of genes I^A, I^B, and i may occur in different populations. That this is true, especially for I^A and I^B, is shown in Figures 15-1 and 15-2.

Sex linkage. Thus far, in considering gene frequencies, only autosomal genes have been discussed. The same techniques, with one modification, may be used in treating sex-linked genes, however. Because human males have only one X chromosome, they cannot reflect a binomial distribution for random combination of pairs of sex-linked genes as do females. Equilibrium distribution of genotypes for a sex-linked trait, where $p + q = 1$, is given by

$$(\male)\ p + q$$

$$(\female)\ p^2 + 2pq + q^2$$

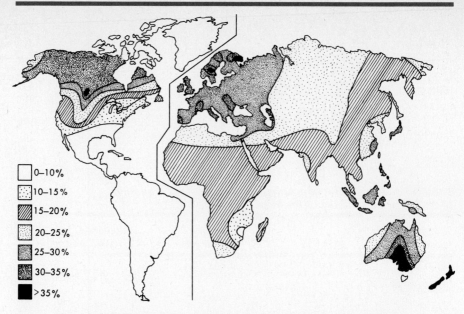

Figure 15-1. *Generalized world distribution of I^A for native populations.*

Consider, for example, red-green color blindness. This trait is due to a sex-linked recessive, as noted in Chapter 12, and may be designated r. About 8 percent of all males suffer from this deutan color-blindness. This shows at once that q, the frequency of gene r, is 0.08 and p, the frequency of its normal allele, R, 0.92. Thus the frequency of color-blind females is expected to be $q^2 = 0.0064$. This is about what is found. Sex-linked dominants may be handled in the same way; in the case of normal color vision, with value of $p = 0.92$, the incidence of normal women is $p^2 + 2pq = 0.9936$.

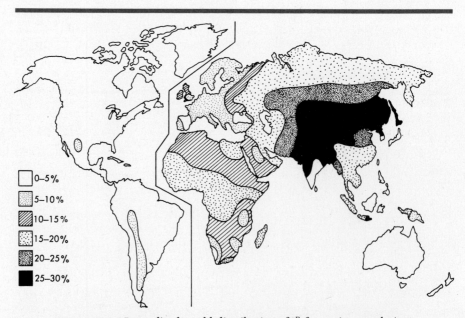

Figure 15-2. *Generalized world distribution of I^B for native populations.*

Factors affecting gene frequency

Isolating mechanisms. Any mechanism that prevents gene exchange is termed an *isolating mechanism*. Broadly considered, these may be either (1) geographic or physical, such as great distances or mountain or oceanic barriers that keep populations apart, or (2) other mechanisms that effectively prevent gene exchange between populations in the same area. Because geographic separation keeps populations physically isolated, there is a question of whether in all cases such groups would remain reproductively isolated if brought together. In fact, it is clear in many cases that physically isolated populations do exchange genes when the isolation is ended.

Table 15-3. Percentage of PTC tasters in various human populations

Population	Place	Sample size	Percent tasters	Gene frequencies	
				T	t
Welsh	Five towns	237	58.7	0.36	0.64
Eskimo (unmixed)	Labrador and Baffin	130	59.2	0.36	0.64
Arab	Syria	400	63.5	0.40	0.60
American white	Montana	291	64.6	0.41	0.59
Eskimo (mixed)	Labrador and Baffin	49	69.4	0.45	0.55
American white	Columbus, Ohio	3,643	70.2	0.45	0.55
American students	Delaware, Ohio	280	70.7	0.46	0.54
American black	Alabama	533	76.5	0.52	0.48
Flathead Indians (mixed)	Montana	442	82.6	0.58	0.42
Flathead Indians (unmixed)	Montana	30	90.0	0.68	0.32
American black	Ohio	3,156	90.8	0.70	0.30
African black	Kenya	110	91.9	0.72	0.28
African black	Sudan	805	95.8	0.80	0.20
Navaho Indians	New Mexico	269	98.2	0.87	0.13

Comparison of frequencies for mixed and unmixed Eskimo and Flathead Indian samples (Table 15-3) shows that gene frequencies in a population may be changed by admixture of genes from other populations. Presumably the "mixing" in these cases is from the western European-American complex, where percentage of tasters averages roughly 62–72 percent with corresponding frequencies of T of 0.384 to 0.471. Note how the frequency of this gene has been increased by admixture in the Eskimo. The same sort of change, though in the opposite direction, has occurred by outbreeding of Flathead Indians where the frequency of T is high (0.683) in the unmixed group and somewhat lower (0.583) in the mixed population. The lower frequency of T in the American black as compared to African blacks has undoubtedly come about in similar fashion. Data for the multiple allele series I^A, I^B, and i, for the Rh alleles D and d, and for the M-N pair show similar population differences (Table 15-4).

Because of the actual or possible exchange of genes between previously separated populations once they are permitted to intermingle, it is often preferred to confine the concept of isolating mechanisms to those that prevent gene exchange between populations that occupy the same area. These include one or more of the following: restriction to quite different habitats (especially in plants), reproductive maturation at different times, mating behavioral differences and/ or physical incompatibility of genitalia (animals), destruction of sperm (principally in animals) or lack of development of pollen tubes (plants), and/or death of the zygote or of the embryo. In addition, some populations produce sterile

Table 15-4. Gene frequencies for three blood group loci for selected populations

Population	I^A	I^B	i	D	d	M	N
U.S. white	0.28	0.08	0.64	0.61	0.39	0.54	0.46
U.S. students*	0.26	0.07	0.67	0.65	0.35	—	—
U.S. black	0.17	0.14	0.69	0.71	0.29	0.48	0.52
West African	0.18	0.16	0.66	0.74	0.26	0.51	0.49
American Indian	0.10	0.00	0.90	1.00	0.00	0.76	0.24

*1,698 Ohio Wesleyan University genetics students.

hybrids, as in the mule (donkey × horse) and in many horticultural varieties of plants, effectively preventing establishment of new self-perpetuating genetic lines.

In summary, different populations are often characterized by particular gene frequencies, which produce phenotypic frequencies that may be expected to fluctuate narrowly around a mean until something occurs to alter gene frequencies. Figure 15-3 indicates how frequencies of homozygotes and heterozygotes are changed by a shift in gene frequencies.

On page 275 it was noted that in large populations changes in gene frequency may come about not only through alteration of isolating mechanisms but also through mutation and selection. But in populations of finite size, an additional factor comes into play. This is random fluctuation in gene frequency, or **genetic drift.** Each of these latter three forces as mechanisms of change in gene frequency will now be examined.

Mutation. Basically, a mutation is a sudden, random alteration in the genotype of an individual. Strictly speaking, it is a change in the genetic material itself, but the term is often loosely extended to include chromosomal aberrations such as those already considered (Chapters 13 and 14). Its importance in the genetics of populations is to provide new material on which selection can operate as well as to alter gene frequencies.

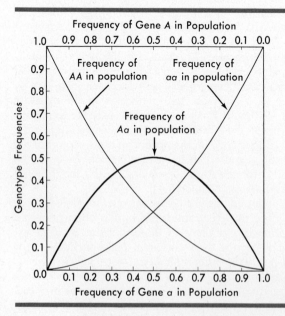

Figure 15-3. The effect of changes in gene frequencies on frequencies of genotypes in a population.

If, for example, gene T mutates to t, the relative frequencies of the two alleles are changed. If the mutation $T \rightarrow t$ recurs consistently, T could disappear from the population. But not only is mutation recurrent, it is also reversible, with a known frequency in many cases. These *back mutations* will at least slow the otherwise inexorable shift of T to t and perhaps prevent the total disappearance of T. But notice that the mutation $T \rightarrow t$ inevitably would shift gene frequencies over time (1) unless the rate of back mutation ($T \leftarrow t$) equals the rate of forward mutation ($T \rightarrow t$), and (2) only if possession of gene t either gives its bearer an advantage or confers no disadvantage. Thus mutation as a force in altering gene frequencies can scarcely be considered apart from the factor of selection, and their *combined* effect will be considered later in this chapter.

Equilibrium resulting from mutation is easily derived algebraically. Because the rate of mutation $T \rightarrow t$ generally does not equal the rate of change $T \leftarrow t$, we can show this relationship as

$$T \underset{v}{\overset{u}{\rightleftarrows}} t$$

where u is the rate of mutation $T \rightarrow t$, and the back mutation $T \leftarrow t$ occurs at rate v, with $u \neq v$. Generally $u > v$; the reasons for this will be clear when the molecular structure of the genetic material and its operation have been examined.

If q represents the frequency of mutating gene t in any one generation and p, the frequency[2] of its mutating allele T, the change in frequency of t resulting from mutation will be governed by

1. **The addition of t, as determined by**
   a. the rate of forward mutation (u), and
   b. the frequency of T (p); that is,
   c. to the limit set by up.
2. **The loss of t, as determined by**
   a. the rate of back mutation (v), and
   b. the frequency of t (q); that is,
   c. to the limit set by vq.

Frequencies of T and t will be in equilibrium under mutation when additions and losses balance, that is, when $up = vq$. The *rate* of change in q under mutation, or Δq_m, can be stated as

$$\Delta q_m = up - vq \tag{1}$$

Thus, $\hat{q}_m$, the equilibrium frequency of t under mutation, will equal the rate $T \rightarrow t$, divided by the sum of the rates $T \rightarrow t$ and $T \leftarrow t$:

$$\hat{q}_m = \frac{u}{u + v} \tag{2}$$

If, for example, T mutates to t three times as often as t mutates to T, then

$$u = 3v$$

and

$$\hat{q}_m = \frac{3v}{3v + v} = \tfrac{3}{4}$$

So the population will reach equilibrium with regard to the frequencies of the mutating alleles T and t when q, the frequency of t, is 0.75, and p, the frequency of T, is $1 - 0.75$, or 0.25.

[2]Some authors prefer to designate the frequency of the dominant allele as $1 - q$.

Complete selection. The common colon bacillus (*Escherichia coli*) is usually killed by the antibiotic streptomycin; the normal, wild-type phenotype is thus "streptomycin-sensitive." The alleles for this trait are designated str^+ (streptomycin-sensitive) and str (streptomycin-resistant). Because vegetative cells of *E. coli* are monoploid, any given cell has the genotype str^+ or str. In a freely growing culture of this organism, about one in 10 million cells can be shown by appropriate culture techniques to be streptomycin-resistant. Gene str^+ mutates to str with a frequency of about 1×10^{-7}. Now in its normal streptomycin-free environment, neither str^+ nor str confers any advantage or disadvantage. Hence the frequency of str would reach equilibrium after a number of generations. However, let streptomycin be introduced into the environment, and possession of str becomes a distinct advantage; without it an individual does not survive. So with a specific environmental change, str individuals survive, str^+ cells do not. One might say an ordinarily inconsequential gene (str) has thus suddenly assumed a very high positive selection value and, at the same time, str^+ has acquired a strong negative selection value.

Sickle-cell anemia affords an interesting and simple case of selection as a force in maintaining, in certain human populations, a surprisingly high frequency of an otherwise deleterious gene, Hb^S. Briefly, this gene is most frequent in almost precisely those regions where falciparum malaria is prevalent. These areas include principally the Mediterranean basin, central Africa from roughly 10° north latitude to 25° south latitude, the valley of the Nile, Madagascar, and portions of the Arabian peninsula. There is evidence that S-hemoglobin produces conditions unfavorable to the growth of the malarial parasite in the blood. Hence an environment that includes this disease confers some selective advantage on individuals that have S-hemoglobin. Homozygous "normal" persons, $Hb^A Hb^A$, do not suffer from sickle-cell anemia, of course, but are more susceptible to malaria; $Hb^S Hb^S$ individuals are resistant to malaria but develop the anemia and die before reaching reproductive maturity. Heterozygotes, $Hb^A Hb^S$, have some advantage with respect to both malaria and sickle-cell anemia.

The impact of selection may also be seen in an extreme example from tomato. In this plant several linkage groups contain dominant genes that confer resistance to different strains of the leaf mold fungus (*Cladosporium*). The disease is not ordinarily a problem in outdoor culture but frequently becomes severe in commercial greenhouses, where temperature and humidity may be consistently high enough to cause severe infection of susceptible plants. In such cases, homozygous recessives, which are susceptible, are quickly killed and may not live to reproductive age. For simplicity, consider this situation as the result of a single pair of alleles, C and c, with cc plants being susceptible to the fungus. A population of plants in which the frequencies of C and c are each 0.5 at the outset, as well as continuous greenhouse culture in which seed for new plantings is obtained from the crop being grown, will be assumed.

Initially, then the parental generation may be represented as follows:

$$\text{Frequency of } C = p = 0.5; \text{ frequency of } c = q = 0.5$$

$$
\begin{array}{cccc}
\text{P Genotypes:} & CC & Cc & cc \\
\text{Genotypic frequencies:} & p^2 + & 2pq + & q^2 \\
= & 0.25 + & 0.50 + & 0.25
\end{array}
$$

If cc individuals are unable to reproduce in this environment, the breeding population is reduced to CC and Cc plants, which occur now in the ratio of 1:2 as follows:

$$CC: \frac{0.25}{0.25 + 0.50} = 0.33 \text{ (new genotypic frequency)}$$

$$Cc: \frac{0.50}{0.25 + 0.50} = 0.67 \text{ (new genotypic frequency)}$$

This leaves only $C- \times C-$ crosses possible, with consequent changes in phenotypic and genotypic frequencies in the next generation (Table 15-5).

Table 15-5

Crosses	Frequencies	F_1 genotypic frequencies		
		CC	Cc	cc
$CC \times CC$	$(0.33)^2$ = 0.11	0.11		
$CC \times Cc$	$2(0.33 \times 0.67)$ = 0.44	0.22	0.22	
$Cc \times Cc$	$(0.67)^2$ = 0.45	0.11	0.23	0.11
		0.44	0.45	0.11

The frequency of c has declined in a single generation from 0.5 to 0.33 $(= \sqrt{0.11})$, and C's frequency has risen from 0.5 to 0.67.

In effect, we have been considering *complete selection* against a recessive lethal. We can represent the frequency of such a gene after any given number of generations as

$$q_n = \frac{q_0}{1 + nq_0} \tag{3}$$

where q_n = the frequency of the recessive lethal after n additional generations, and q_0 = the initial frequency of that gene. Thus, in our tomato example, where $q_0 = 0.5$ and $n = 1$,

$$q_1 = \frac{0.5}{1 + (1 \times 0.5)} = \frac{0.5}{1.5} = 0.33$$

In a closed population, with no admixture from other groups, only mutation can prevent the frequency of a gene so radically selected against from dropping quickly toward zero. The curve for complete selection against a recessive gene, however, is hyperbolic, with the final slope being determined less by mutation than by the fact that the deleterious gene persists in relatively rare heterozygotes that more frequently mate with homozygous dominant individuals. The number of additional generations (n) required to reduce q from a value of q_0 to some desired value, q_n, is given by the equation

$$n = \frac{1}{q_n} - \frac{1}{q_0} \tag{4}$$

Thus, to reduce q from 0.5 $(= q_0)$ to 0.33 $(= q_n)$, one additional generation is required:

$$n = \frac{1}{0.33} - \frac{1}{0.5} = 3 - 2 = 1$$

With this equation the frequency of a recessive lethal for any number of additional generations can be calculated readily, as shown in Table 15-6.

Note that the frequency of the lethal is reduced by half in generation two, but four more are required to reduce it by half again, and so on. Furthermore, application of equation (4) shows that the number of generations required to halve the frequency of such a gene is very large when the initial frequency is quite low. On the other hand, had the lethal been a dominant instead of a recessive, it would have been eliminated in one generation in the absence of complicating influences.

Table 15-6. Reduction in frequency (q) of a recessive lethal by generations

Generation	q
0	0.500
1	0.333
2	0.250
3	0.200
4	0.167
5	0.143
6	0.125
7	0.111
8	0.100
9	0.091
10	0.083
50	0.019
100	0.010
1,000	0.001

Partial selection. More often the genotypes involved differ much less in their relative advantage and disadvantage, and the homozygous recessive, for example, will be eliminated much more slowly than in the illustration from tomato. For instance, assume genotype $A-$ produces 100 offspring, all of which reach reproductive maturity in a given environment, whereas genotype aa produces only 80 that do so. The proportion of the progeny of one genotype surviving to maturity (relative to that of another genotype) may be designated as its fitness or *adaptive value*, W. Thus, in this case W_{A-} may be set arbitrarily at 1; W_{aa} is then 0.8. The measure of reduced fitness of a given genotype is referred to as its **selection coefficient, s.** The relationship between adaptive value (W) and the selection coefficient (s) can be represented as

$$W = 1 - s, \text{ or } s = 1 - W$$

In the example being considered here, $s = 0$ for genotype $A-$, and $1 - 0.8 = 0.2$ for aa.

If p = the frequency of gene A, and q = the frequency of its recessive allele a, selection against the latter is as shown in Table 15-7. By the same method used to calculate frequencies of genes T and t, for example, in the Hardy-Weinberg equilibrium (page 277), the information in Table 15-7 gives an equation for determining the frequency of gene a when the value of W_{aa} is known:

$$q_1 = \frac{q_0 - sq_0^2}{1 - sq_0^2} \tag{5}$$

where q_0 again is the frequency of a in a given generation, q_1 its frequency one generation later under selection, and s the selection coefficient.

If, for instance, $q = 0.5$ and $s = 0.2$, then substituting in equation (5), $q_1 =$

Table 15-7. Selection against a recessive allele

	AA	Aa	aa	Total
Initial frequency	p^2	$2pq$	q^2	1
Adaptive value (W)	1	1	$1 - s$	
Frequency after selection	p^2	$2pq$	$q^2(1 - s)$	$p^2 + 2pq + q^2 - sq^2 = 1 - sq^2$

0.4737. So with *partial selection* against a fully recessive gene, the decrease in its frequency is much less per generation than for complete selection against a recessive lethal. The change in frequency of gene a under selection (Δq_s, or $q_1 - q_0$) can be represented by the equation

$$\Delta q_s = \frac{-sq_0^2 p}{1 - sq_0^2} \tag{6}$$

Upon substituting here, $\Delta q_s = -0.0263$. If q_0 were very small, as it would ordinarily be in the case of deleterious mutations, the quantity Δq_s becomes almost equal to $-sq_0^2$.

In time a large and randomly mating population attains a genetic equilibrium that is the resultant between selection forces and rate of mutation. This is apparent in many different species of animals, including humans, as well as plants. But once a population ceases to be isolated, gene frequencies may begin to change if genetic material is contributed by other populations.

Assortative mating. Not all populations are randomly mating. In populations of self-fertilizing plants, for example, given genotypes, in effect, mate only with like genotypes. This is **complete positive genotypic assortative mating** and results in rapid increase in frequencies of homozygotes. For instance, assume a population where genotypic frequencies in generation 0 are 0.25 AA + 0.5 Aa + 0.25 aa, that is, $p_0 = q_0 = 0.5$. With complete positive genotypic assortative mating, progeny frequencies in generation 1 can be calculated as follows:

Mating	Proportions of matings	Progeny genotypes AA	Aa	aa
$AA \times AA$	0.25	0.25		
$Aa \times Aa$	0.5	0.125	0.25	0.125
$aa \times aa$	0.25			0.25
Total	1.00	0.375	0.25	0.375

Similarly, progeny in successive generations would occur with the frequencies shown in Table 15-8. With continued inbreeding, frequencies of homozygotes approach the values of p_0 and q_0, respectively. Heterozygotes become so low in frequency that they may easily be eliminated—for example, by an unseasonal freeze (plants) or through predators (animals). Should this happen, the fre-

Table 15-8. Progeny genotypic frequencies under complete positive genotypic assortative mating where $p_0 = q_0 = 0.5$

Generation	Progeny genotypes and frequencies AA	Aa	aa
0		1	
1	$\frac{1}{4} = 0.25$	$\frac{2}{4} = 0.5$	$\frac{1}{4} = 0.25$
2	$\frac{3}{8} = 0.375$	$\frac{2}{8} = 0.25$	$\frac{3}{8} = 0.375$
3	$\frac{7}{16} = 0.4375$	$\frac{2}{16} = 0.125$	$\frac{7}{16} = 0.4375$
4	$\frac{15}{32} = 0.46875$	$\frac{2}{32} = 0.0625$	$\frac{15}{32} = 0.46875$
5	$\frac{31}{64} = 0.484375$	$\frac{2}{64} = 0.03125$	$\frac{31}{64} = 0.484375$
n	$\dfrac{2^n - 1}{2^{n+1}}$	$\dfrac{2}{2^{n+1}}$	$\dfrac{2^n - 1}{2^{n+1}}$

quencies of *AA* individuals would then equal p_0, and the frequency of *aa* individuals would equal q_0. A similar, but less rapid, decrease in frequency of heterozygotes will occur in populations where like genotypes, although not "forced" to mate with each other, may mate preferentially (partial positive genotypic assortative mating).

On the other hand, mating may occur preferentially (or even only) between *unlike* genotypes (outbreeding). This is **negative genotypic assortative mating,** and may, of course, be either complete or partial. Negative genotypic assortative mating results in an increase in the frequency of heterozygotes.

Similarly, *phenotypic* assortative mating may occur. **Positive phenotypic assortative mating** consists in mating of like phenotypes, i.e., $A- \times A-$ and *aa* $\times$ *aa;* **negative phenotypic assortative mating** involves mating of unlike phenotypes, that is, $A- \times aa$. The effect of the former is to increase homozygosity, of the latter to increase heterozygosity.

Combined effect of mutation and selection. Infants normally exhibit a "startle response," in which they stiffen their arms and legs upon hearing a sudden noise. Continuation of this reaction beyond about the age of four to six months may be symptomatic of Tay-Sachs disorder (infantile amaurotic idiocy), which is due to a recessive autosomal gene. Homozygous recessives fail to produce the enzyme hexosaminidase A (hex A), with the result that a lipid, ganglioside G_{M2}, accumulates in the brain. Hex A is one of two enzymes in the sequence of lipid metabolism reactions (the other is hexosaminidase B) and consists of α and β polypeptide chains. Tay-Sachs disorder results from a mutation in the gene responsible for the synthesis of the α polypeptide chain, with the result that functional hex A is not produced.[3] Reference to Table 6-2 will show you that the gene controlling hex A production is on autosome 7; that for hex B formation is on chromosome 5. Both defective genes are recessive.

Mental and motor deterioration rapidly follow the onset of symptoms in the Tay-Sachs disorder, accompanied by paralysis and degeneration of the retina, which leads to blindness; the culmination is death, generally before the age of four. This gene has a much higher frequency in Jewish people of middle and northern Europe (Ashkenazi Jews) than in other groups. Its frequency in the Jewish population of New York City is about 0.015, but approximately 0.0015 in non-Jewish individuals.

The selection coefficient for the Tay-Sachs gene is, of course, 1, inasmuch as homozygotes die in infancy or very early childhood. A reasonable estimate of forward mutation (u) appears to be 1×10^{-6}, although some reports range as high as 1.1×10^{-5}. Because of its lethality early in life, the rate of back mutation (v) is effectively zero. In summary, then, for the Jewish population of New York City

$$q = 0.015 = 1.5 \times 10^{-2}$$

$$p = 0.985 = 9.85 \times 10^{-1}$$

$$u = 0.000001 = 1 \times 10^{-6}$$

$$v = 0$$

$$s = 1$$

What, then, is the equilibrium frequency of this recessive lethal in the Jewish population of New York City under the *combined* effect of mutation and selection?

---

[3]Failure to produce *both* enzymes results in Sandhoff disease, whose symptoms are similar to those seen in the Tay-Sachs syndrome. Hex B is a polymer of only β chains, so Sandhoff disease involves defects in both the α and β chains.

Recall (from page 283) that the rate of change in frequency of a recessive gene under mutation is given by

$$\Delta q_m = up - vq$$

Substituting,

$$\Delta q_m = (1 \times 10^{-6} \times 9.85 \times 10^{-1}) - (0 \times 1.5 \times 10^{-2})$$

$$= (1 \times 10^{-6} \times 9.85 \times 10^{-1}) - 0$$

$$= up = 9.85 \times 10^{-7}$$

This last figure is approximately equal to u, because $0.000000985 \cong 0.000001$. Hence

$$\Delta q_m \cong u \tag{7}$$

Turning for a moment to frequency change under selection, we use equation (6) from page 287

$$\Delta q_s = \frac{-sq_0^2 p}{1 - sq_0^2}$$

Substituting in this equation,

$$\Delta q_s = \frac{-(0.015)^2 \times 1 \times 0.985}{1 - 1(0.015)^2}$$

$$= \frac{-2.25 \times 10^{-4} \times 9.85 \times 10^{-1}}{1 - 2.25 \times 10^{-4}}$$

$$= \frac{-2.216 \times 10^{-4}}{9.998 \times 10^{-1}}$$

$$= -2.22 \times 10^{-4}$$

So

$$\Delta q_s \cong -sq_0^2 \tag{8}$$

since $-2.22 \times 10^{-4} \cong -2.25 \times 10^{-4}$.

Thus $\Delta q_m \cong u$, $\Delta q_s \cong -sq^2$, and $\Delta q = \Delta q_s + \Delta q_m = -sq^2 + u$. But, by definition, at equilibrium, $\Delta q = 0$. Therefore at equilibrium,

$$u - sq^2 = 0$$

and

$$u = sq^2$$

Rewriting this latter equation to solve for q^2,

$$q^2 = \frac{u}{s}$$

and therefore

$$\hat{q} = \sqrt{\frac{u}{s}} \tag{9}$$

Substituting in equation (9) for the Tay-Sachs problem,

$$\hat{q} = \sqrt{\frac{1 \times 10^{-6}}{1}} = 1 \times 10^{-3}$$

Hence in the population under consideration here, the equilibrium frequency of the Tay-Sachs gene under the combined effect of mutation and selection, is 0.001. The frequency of Tay-Sachs births in this group, then, at equilibrium would be 1×10^{-6}, that is, $(1 \times 10^{-3})^2$, as compared with the present frequency of 2.25×10^{-4}, or $(0.015)^2$.

Random genetic drift. From generation to generation the number of individuals carrying a particular allele, either in the homozygous or heterozygous state, may be expected to vary somewhat, so that gene frequencies will fluctuate about a mean. The amplitude of this fluctuation is *random genetic drift* and is due to the vagaries of chance mating and to the fact that, even in cases where $p = q = 0.5$, theoretical ratios (e.g., 3:1, 1:1, 1:2:1) are certainly not always produced. If a population is large, drift is nondirectional and of small magnitude; it may be expected to vary within rather narrow limits above and below the mean. But in *small* populations all the progeny might, by chance alone, be of the same genotype with respect to a particular pair of alleles, for example, Aa. Their progeny (the F_1) would be expected in the genotypic ratio 1 AA:2 Aa:1 aa. The F_2 is assumed to arise from random mating among members of the F_1. The probability of any of the possible F_1 matings is then a function of the frequency of each F_1 genotype. Under these conditions, **fixation** will occur in one-eighth of the progeny (F_2) per generation:

F_1 mating	Probability of mating	Values of p and q in mating population		Progeny (F_2)
		p	q	
$aa \times aa$	$\frac{1}{4} \times \frac{1}{4} = \frac{1}{16}$	0	1	all aa
$Aa \times aa$	$2(\frac{2}{4} \times \frac{1}{4}) = \frac{4}{16}$	0.25	0.75	1 Aa:1 aa
$Aa \times Aa$	$\frac{2}{4} \times \frac{2}{4} = \frac{4}{16}$	0.5	0.5	1 AA:2 Aa:1 aa
$AA \times aa$	$2(\frac{1}{4} \times \frac{1}{4}) = \frac{2}{16}$	0.5	0.5	all Aa
$AA \times Aa$	$2(\frac{1}{4} \times \frac{2}{4}) = \frac{4}{16}$	0.75	0.25	1 AA:1 Aa
$AA \times AA$	$\frac{1}{4} \times \frac{1}{4} = \frac{1}{16}$	1	0	all AA

The formation of a new population by migration of a sample of individuals may likewise lead to different gene frequencies. Imagine a large population in which $p = 0.4$ and $q = 0.6$. The most probable values of p and q, therefore, in a *sample* from this population are also 0.4 and 0.6, respectively. Expected deviation from these values is, of course, provided by the standard deviation. The standard deviation for a simple proportionality, such as heads versus tails, is given by

$$s = \sqrt{\frac{pq}{n}} \tag{10}$$

where n is the number of observations. But when gene frequencies are calculated from the frequency of homozygous recessive phenotypes, as has been done here, the formula for standard deviation becomes

$$s = \sqrt{\frac{pq}{2N}} \tag{11}$$

where N is the number of (diploid) individuals in the sample.

If, in the large population under consideration, a sample of 50,000 persons is taken,

$$s = \sqrt{\frac{0.24}{100,000}} = 0.00155$$

That is, in any sample of 50,000 individuals from this population with $p = 0.4$ and $q = 0.6$, 68 percent of the time p will lie between 0.39845 and 0.40155, i.e., 0.4 ± 0.00155; 95 percent of the time it will fall within the range 0.4 ± 0.0031, or 0.3969 to 0.4031. But if the sample were to consist of only 50 persons,

$$s = \sqrt{\frac{0.24}{100}} = 0.049$$

In 68 percent of samples of this size, p would be expected to fall within the range 0.4 ± 0.049, or between 0.351 and 0.449. Similarly, in 95 percent of the cases of samples of 50, p would be within the range of 0.302 to 0.498. Formation of a new population by emigration of a sample as small as 50 might be expected to lead purely by chance to a very different gene frequency in the next generation. But, had the sample consisted of only two randomly selected individuals, $s = \sqrt{\frac{0.24}{4}} = \sqrt{0.06} = 0.245$. In 68 percent of such samples p would be expected by chance to fall in the range 0.4 ± 0.245, or from 0.155 to 0.645, but in 95 percent of the cases $p = 0.4 \pm 0.49$ or a range of -0.09 (effectively, of course, zero) to 0.89. Likewise, q in any sample of two individuals would equal 0.6 ± 0.245 with a probability of approximately 0.68, and 0.6 ± 0.49, or a range of 0.11 to 1.09 (the latter is effectively 1) with a probability of 0.95.

A presumed example of such a case of genetic drift has been reported for the Dunkers of Pennsylvania. These are members of a religious sect who migrated from Germany in the early eighteenth century and have remained relatively isolated. The frequency of blood group A in this small group is almost 0.6, whereas it is between 0.40 and 0.45 in German and American populations, and the I^B allele is nearly absent in the Dunkers, whereas group B persons comprise 10 to 15 percent of German and American populations.

The mechanism of evolution

The principles thus far examined, especially in this and the two preceding chapters, constitute the basic mechanisms whereby new species evolve, sometimes slowly, sometimes suddenly, from preexisting ones. A species is more a taxonomic concept than a concrete entity, and its parameters necessarily differ somewhat among various groups of animals and plants. Thus a species in bacteria is quite a different concept from one in birds. Species "boundaries" in viruses and vertebrates have different emphases, as they do even in more closely related units likely freely hybridizing, genetically plastic willows versus the more stable maples. However, in very general terms, speciation, or the origin of new species, depends largely on such cytological and genetic mechanisms as these that have been described:

1. Chromosomal aberrations, especially translocations and allopolyploidy (in plants).
2. Addition of new genetic material by mutation (although most are either neutral or disadvantageous in an existing environment).
3. Changes in gene frequencies:
   a. By mutation.
   b. Through random drift.
   c. By migration or the breaking down of isolating barriers.
   d. By environmental selection.

These agencies may sooner or later break up larger populations into smaller units that (1) develop genetically determined, distinctive morphological and/or physiological characters that differ from those of other such units, and (2) become reproductively isolated from related groups, and develop thereby into "Mendelian populations" that have their own gene pools. How, in what direction(s) and to what extent, evolution thus directed will develop depends on an intricate interaction among a variety of influences. But basically only those changes in the genetic material that can be passed on to succeeding generations and do not disappear from the gene pool can serve as agents of evolution in living organisms.

References

Blakeslee, A. F., and T. N. Salmon, 1935. Genetics of Sensory Thresholds: Individual Taste Reactions for Different Substances. *Proc. Nat. Acad. Sci. (U.S.)*, **21:** 84–90.

Fox, A. L., 1932. The Relationship Between Chemical Composition and Taste. *Proc. Nat. Acad. Sci. (U.S.)*, **18:** 115–120.

Glass, H. B., M. S. Sacks, E. F. Jahn, and C. Hess, 1952. Genetic Drift in a Religious Isolate: An Analysis of the Causes of Variation in Blood Group and Other Gene Frequencies in a Small Population. *American Naturalist*, **86:** 145–159.

Kaback, M. M., ed., 1977. *Tay-Sachs Disease: Screening and Prevention*. New York: Alan R. Liss, Inc.

Lewontin, R. C., 1967. Population Genetics. In H. L. Roman, ed. *Annual Review of Genetics*, vol. 1. Palo Alto, Calif.: Annual Reviews, Inc.

Lewontin, R. C., 1973. Population Genetics. In H. L. Roman, ed. *Annual Review of Genetics*, vol. 7. Palo Alto, Calif.: Annual Reviews, Inc.

Morton, N. E., 1969. Human Population Structure. In H. L. Roman, ed. *Annual Review of Genetics*, vol. 3. Palo Alto, Calif.: Annual Reviews, Inc.

Reed, T. E., 1969. Caucasian Genes in American Negroes. *Science*, **165:** 762–768. Reprinted in L. Levine, ed. *Papers on Genetics*. St. Louis: C. V. Mosby.

Stern, C., 1943. The Hardy-Weinberg Law. *Science*, **97:** 137–138. Reprinted in L. Levine, ed., 1971. *Papers on Genetics*. St. Louis, C. V. Mosby.

White, M. J. D., 1969. Chromosomal Rearrangements and Speciation in Animals. In H. L. Roman, ed. *Annual Review of Genetics*, vol. 3. Palo Alto, Calif.: Annual Reviews, Inc.

Problems

15-1 If you consider the data set forth in this chapter for M, MN, and N persons in this country and assume that you do not know the genotype of either yourself or your parents: (a) What genotype are you most likely to have? (b) What genotype are you least likely to have?

15-2 A sample of 1,000 persons tested for M-N blood antigens was found to be distributed: M, 360; MN, 480; N, 160. What is the frequency of genes *M* and *N*?

15-3 A sample of 1,522 persons living in London disclosed 464 of type M, 733 of type MN, and 325 of type N. Calculate gene frequencies of *M* and *N*.

15-4 A sample of 200 persons from Papua (southeast New Guinea) showed 14 M, 48 MN, and 138 N. Calculate gene frequencies for *M* and *N*.

15-5 A sample of 100 persons disclosed 84 PTC tasters. Calculate gene frequencies for *T* and *t*.

15-6 (a) How many heterozygotes should there be in the sample of problem 15-5? (b) How many *TT* persons?

15-7 Albinism is the phenotypic expression of a homozygous recessive genotype. One source estimates the frequency of albinos in the American population as 1 in 20,000. What percentage of the population is heterozygous for this gene?

15-8 Alcaptonuria, which results from the homozygous expression of a recessive auto-somal gene, occurs in about 1 in 1 million perons. What is the proportion of heterozygous "carriers" in the population?

15-9 A sample of 1,000 hypothetical persons in the United States showed the following distribution of blood groups: A, 450; B, 130; AB, 60; O, 360. Calculate the frequencies of genes I^A, I^B, and i.

15-10 Another sample of 1,000 hypothetical persons had these blood groups: A, 320; B, 150; AB, 40; O, 490. What is the frequency in this sample of each of the following genotypes: $I^A I^A$, $I^A i$, $I^B I^B$, $I^B i$, $I^A I^B$, ii?

15-11 Assume the data of Table 15-1 to represent accurately the distribution of A, B, AB, and O blood groups in the United States population. Using the chi-square test, determine whether the sample of problem 15-9 represents a significant deviation. (Round the frequencies of Table 15-1 to the whole numbers 440, 420, 100, and 40, respectively, to obtain calculated values for computing chi-square.)

15-12 A sample of 429 Puerto Ricans showed the following gene frequencies: I^A, 0.24; I^B, 0.06; i, 0.70. Calculate the percentage of persons in this sample with A, B, AB, and O blood.

15-13 What percentage of the sample in the preceding problem is (a) homozygous A, (b) heterozygous B?

15-14 If one man in 25,000 suffers from hemophilia A, what is the frequency of gene h in the population?

15-15 One man in 100 exhibits a trait that results from a certain sex-linked recessive gene. What is the frequency of (a) heterozygous women, (b) homozygous recessive women?

15-16 Data represented in Table 15-4 show a frequency of 0.64 for gene i and 0.61 for the Rh allele D. If only these two genes are considered, what would be the frequency of $O+$ persons in the population?

15-17 Considerable progress is being made in the eradication of malaria. If malaria is eventually eliminated, what effect will this be likely to have on the frequency of gene Hb^S?

15-18 Great effort is being made to eliminate the muscular dystrophy caused by gene d. If we are someday successful in eliminating the *effect* of this gene in *treated* individuals, will this result in genetic improvement or deterioration in the world population?

15-19 Use the tabulated data (text) concerning frequency of gene c in tomato to determine the frequency of this gene after one more generation of inbreeding.

15-20 (a) If we consider the tabulated data (text) concerning frequency of gene c in tomato, what would be the frequency of this gene in the twentieth generation of inbreeding? (b) If the parental generation is considered as generation 1, in what generation would the frequency of c be reduced to exactly 0.005?

15-21 Gene f is an autosomal recessive lethal that kills ff individuals before they reach reproductive age. If, in an isolated population, this gene had a frequency of 0.4 in the P generation, what would be its frequency in the F_2?

15-22 For a certain pair of alleles, completely dominant gene A has an initial frequency of 0.7 and an adaptive value of 1. Its recessive allele has a frequency of 0.3 and an adaptive value of 0.5. What is the frequency of a in the next generation?

15-23 Gene A mutates to its recessive allele a four times more frequently than a mutates to A. What will be the equilibrium frequency of a under mutation?

15-24 If gene B has a forward mutation rate of 5×10^{-6} and a back mutation rate of 1×10^{-6}, what is the equilibrium frequency of gene b under mutation?

15-25 In a given population gene C has a frequency of 0.2. If 50 individuals from that population are sampled, there is a 0.68 probability that the frequency of C will be no more than how much above or below 0.2 in that sample?

15-26 Persons with cystic fibrosis of the pancreas occur with a frequency of about 0.0004.

This condition is due to a recessive autosomal gene and is fatal in childhood. (a) Give the frequency of this gene in the present population. (b) If the rate of forward mutation is assumed to be 4×10^{-6}, what is the equilibrium frequency of the gene for cystic fibrosis under the combined effect of mutation and selection?

15-27 Based on your answer to 15-26(a), and assuming a human generation to be 30 years, (a) how many years would be required to reduce this frequency to 0.01? (b) Is this likely to occur? Why?

15-28 Let p represent the frequency of dominant autosomal gene A and q the frequency of its recessive allele a. For a randomly mating population, give a mathematical expression for (a) the frequency of an AA individual, (b) the probability of an $AA \times AA$ mating, (c) the frequency of an Aa individual, (d) the probability of an $Aa \times Aa$ mating, (e) the probability of an $Aa \times aa$ mating, (f) the total of all possible matings.

15-29 Give a mathematical expression for the frequency of (a) dominant and (b) recessive progeny phenotypes from a *single* $Aa \times Aa$ mating.

15-30 For the *entire* randomly mating population, with all possible matings equally free to occur, give (a) a mathematical expression for the frequency of recessive progeny in the population resulting from such random matings and (b) the numerical value of this expression if $p = q = 0.5$.

15-31 Using the information in Table 15-8, calculate the frequency of heterozygotes in generation 10.

15-32 In view of your answer to the preceding question, what would be the frequency of each of the two homozygous genotypes in generation 10?

15-33 Phenylketonuria (PKU) is a condition characterized by such low intelligence that persons not diagnosed at birth and put on immediate treatment never reproduce. This condition is produced by an inherited inability to metabolize the essential amino acid phenylalanine and is caused by a completely recessive autosomal gene. PKUs have an incidence of about 3.6×10^{-6}. (a) Assuming lack of diagnosis in time to treat the condition effectively, what is the frequency of the PKU gene at present? (b) What is the incidence of homozygous normal persons? (c) What is the incidence of heterozygotes? (d) If u is assumed to be 4×10^{-6}, what is the gene's equilibrium frequency under the combined effect of mutation and selection? (e) At this value of u, what is the probability of two homozygous normal persons producing a PKU child?

15-34 Assume a certain dominant gene to have a forward mutation rate of 2×10^{-6} and its recessive allele to have a frequency now of 0.015 and a selection coefficient of 0.5. What is the equilibrium frequency of the recessive gene under the combined effect of mutation and selection?

15-35 As noted in the text, some reports of the rate of forward mutation for the recessive Tay-Sachs gene are as high as 1.1×10^{-5}. Calculate the equilibrium frequency of this gene under the combined effect of mutation and selection using this higher rate of forward mutation.

15-36 Recall from problem 12-19 that length of index finger relative to that of the fourth finger is thought to be due to a sex-influenced autosomal gene, with gene F (shorter index finger) dominant in males and recessive in females. A class of 85 genetics students consisted of 32 males with short index finger, 8 males with long index finger, 27 females with short index finger, and 18 females with long index finger. Based on the hypothesis of pair of sex-influenced genes, calculate the frequency of (a) gene F in the females and (b) gene f in the males. (c) How many of the males in this sample should be of genotype FF?

CHAPTER 16

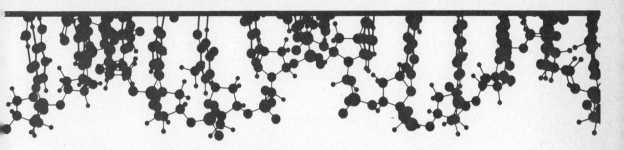

The identification of the genetic material

Up to this point, a variety of genetic observations has been made in a wide assortment of organisms. It has been seen that all these observations can be explained by theorizing that genes occur at specific loci, and in linear order on chromosomes. The following questions, however, must now be considered:

1. What is *the* genetic material? Is it the same in all organisms?
2. How does this genetic material operate to produce detectable phenotypic traits, and can its operation explain such phenomena as dominance and recessiveness?
3. In terms of *the* genetic material, what actually is a gene? Can its molecular configuration be determined?
4. In terms of the nature of the genetic material and of the gene, what is mutation? That is, when a gene changes and produces a different effect, what happens to it at the level of its molecular structure?
5. Differentiation in multicellular organisms, as well as many intermittent biochemical reactions in both eukaryotes and prokaryotes, suggest that not all genes function all the time. If this is so, how are genes regulated?
6. Is this genetic material (are these genes) all located in the chromosomes of eukaryotes, or does the cytoplasm play any part in inheritance? How does the answer to this question relate to prokaryotes?
7. Can answers to such questions as these be used in any way to mitigate or offset the effect of disadvantageous or lethal genes, or even to replace them?

296

CHAPTER 16
THE
IDENTIFICATION
OF THE GENETIC
MATERIAL

These questions will be pursued in the remaining chapters; in this one the problem of identifying the genetic material itself will be examined. Basically, there are three approaches which deal primarily with microorganisms. It is much easier to find answers to these questions in bacteria and viruses than it is in rather complex individuals such as ourselves. Interestingly enough, knowledge gained from studies on these simpler forms is found to be perfectly applicable to all the more highly evolved ones. These three approaches are **transformation, transduction,** and **conjugation.**[1]

Bacterial transformation

The "Griffith effect." In 1928 Frederick Griffith published a paper in which he cited a number of what were, at that time, remarkable results for which he had no explanation. His observations involved a particular bacterium, *Diplococcus pneumoniae*, which is associated with certain types of pneumonia. This organism occurs in two major forms: (1) *smooth* (S), whose cells secrete a covering capsule of polysaccharide materials, which causes its colonies on agar to be smooth and rather shiny, and which is virulent in that it produces bacterial pneumonia in suitable experimental animals; and (2) *rough* (R), cells of which lack a capsule, whose agar colonies have a rough, rather dull surface, and which is nonvirulent. Smooth and rough characters are directly related to the presence or absence of the capsule, and this trait is known to be genetically determined.

The smooth types (S) can be distinguished by their possession of different capsular polysaccharides (designated as I, II, III, IV); the specific polysaccharide is antigenic and genetically controlled. Mutations from smooth to rough occur spontaneously with a frequency of about one cell in 10^7, though the reverse is much less frequent. Mutation of, say, an S-II to rough may occur, and these may rarely revert to smooth. When that happens the smooth revertants are again S-II.

In the course of his work Griffith injected laboratory mice with live R pneumococci derived from an S-II culture; the mice suffered no ill effects. Injection of mice with a living S-III culture (or smooth cultures of any other antigenic type) was fatal, but logically enough, use of heat-killed suspensions of either S or R bacteria did not produce pneumonia. Inoculating the animals with live R-II bacteria (i.e., a rough form derived from S-II) *plus* dead S-III, however, resulted in a high mortality. This surely was an unexpected turn of events. The heat-killed S-III were checked; all indeed were dead. Yet both living R-II and S-III organisms were subcultured following autopsy of the dead mice!

An explanation was not immediately forthcoming. It was as though the killed S-III individuals were somehow restored to life, but this was patently absurd. The then rather recently understood phenomenon of mutation might be implicated, but the frequency of occurrence in the Griffith experiments was far too high to be compatible with known mutation rates. Scientists were left only with the idea that in some way the heat-killed cells conferred virulence on the previously nonvirulent strain; in short, the living cells were somehow *transformed.* So the Griffith effect gradually became known as *transformation* and turned out to be the first major step in the identification of *the* genetic material.

Identification of the "transforming substance." Sixteen years after Griffith's work, Avery, MacLeod, and McCarty reported successful repetition of the earlier work, but in vitro, and were able to identify the "transforming substance." They tested fractions of heat-killed cells for transforming ability. Completely negative results were obtained with fractions containing only the polysaccharide

[1]Transformation and transduction are described in this chapter; conjugation is covered in Appendix B-1.

capsule, various cell proteins, or ribonucleic acid (RNA); only extracts containing deoxyribonucleic acid (DNA) were effective. Even highly purified fractions containing DNA and less than 2 parts protein per 10,000, for example, retained the transforming ability. DNA, plus even minute amounts of protein, plus proteolytic enzymes, was fully effective. But DNA, plus DNase, an enzyme that destroys DNA, lost its transforming capability.

Therefore it began to appear that *beyond any reasonable doubt, DNA must be the genetic material.* Moreover, this landmark of genetic research indicated also that *the genetic material, DNA, differs from the end product it determines.* Evidence has been since accumulated to show that transformation is of fairly wide occurrence in bacteria and probably occurs in the bluegreen algae as well.

Transduction

The clear implication of DNA as the genetic material that was furnished by transformation experiments was extended and confirmed by a series of experiments begun by Zinder and Lederberg in 1952 on the mouse typhoid bacterium (*Salmonella typhimurium*). Their experiments involved the process of **transduction,** in which a bacterium-infecting virus (**phage**) serves as the vector transferring DNA from one bacterial cell to another. To understand this process fully we must first examine the phage structure and "life cycle."

Structure of T-even phages. There is no typical virus structure, but there are several modifications of a basic structure. In general terms, viruses consist of an outer, inert, nongenetic protein "shell" and an inner "core" of genetic material. In many cases this genetic material is DNA, but in some cases it is RNA. Perhaps the best-known viruses from a structural standpoint are the so-called T-even phages (e.g., T2, T4) that infect the colon bacillus (*Escherichia coli*).

The T phages are of a general "tadpole" shape, differentiated into a *head* and a *tail* region. The former is an elongate, bipyramidal, six-sided structure composed of several proteins (Figs. 16-1 and 16-2). Its diameter is approximately 0.065 μm, its length about 0.1 μm. Within the head is a closed, no-end DNA molecule some 68 μm in length. The dimensions of the head are such that the DNA molecule must be packed tightly within it.

The tail is a hollow cylinder, most of which can contract lengthwise. The approximate uncontracted dimensions are: length, 0.08 μm, and diameter, 0.0165 μm. Contraction changes these dimensions to 0.035 μm (length) and 0.025 μm (diameter). The tail can be visualized in electron microscopy as bearing some 24 helical striations. Contraction involves compression of this striated sheath region. The core diameter is approximately 0.007 μm, and encloses a longitudinal canal 0.0025 μm in diameter. Six spikes and six tail fibers, the latter bent at an angle about midway in their length, arise from a hexagonal plate at the distal end of the tail. Tail fibers are not visible in the mature phage until it is adsorbed to the surface of a bacterial cell. Phage T4 is shown in electron micrograph view in Figure 16-1 and in diagrammatic sectional view in Figure 16-2.

Other phages are quite different in morphology. Phage X174, for example, appears in electron micrographs as a cluster of 12 identical, adherent, but morphologically distinct spherical subparticles. Two general types of phage replication cycles can be recognized: (1) *virulent* phages, in which infection is followed by *lysis* (bursting) of the host cell and the release of new, infective phages, and (2) *temperate* phages, in which infection only rarely causes lysis.

Life cycle of a virulent phage. Infection begins with a chance collision between phage and bacterial cell, followed by attachment of phage to one of the numerous receptor sites on the bacterial cell (Fig. 16-3); the sheath then con-

298

CHAPTER 16
THE
IDENTIFICATION
OF THE GENETIC
MATERIAL

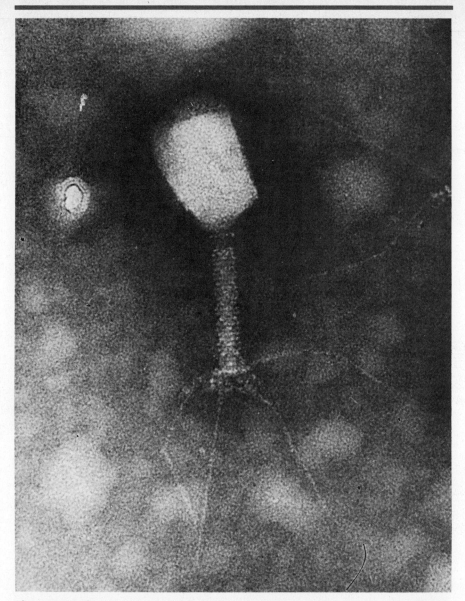

Figure 16-1. *Electron micrograph of a T4 bacteriophage,* × *630,000.* (Courtesy Dr. Thomas F. Anderson, Institute for Cancer Research, Philadelphia.)

tracts, driving the core through the host cell wall. The phage DNA and a small amount of protein then enter the bacterial cell.

Determination of immediately succeeding events is facilitated by the fact that the protein outer shell contains sulfur but no phosphorus, whereas DNA contains phosphorus but no sulfur. Using two samples of phage, one of which contains radioactive ^{35}S and the other radioactive ^{32}P, Hershey and Chase were able to show in 1952 that all the phage DNA enters the host cell following attachment; most of the protein remains outside. Details of the Hershey and Chase experiment are diagramed in Figure 16-4. An *eclipse period* ensues, during which the phage DNA replicates numerous times within the bacterial cell. Toward the end of the eclipse period the phage DNA directs the production of protein coats and assembly of some 50 to 200 new, infective viral particles. Within a short time (e.g., 13

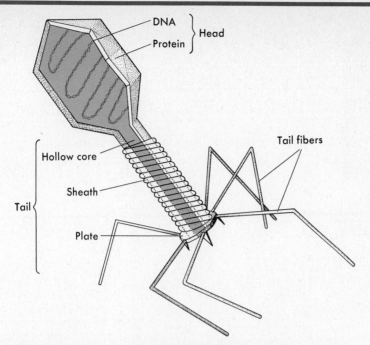

Figure 16-2. *Diagrammatic longitudinal section of a T4 bacteriophage. Only a small portion of the total DNA content is shown.*

minutes for phage T1, and 22 minutes for phage T2), a phage-produced enzyme (lysozyme) brings about lysis of the host cell and release of the mature phages (Fig. 16-3). The following steps in the process can be recognized:

1. *Attachment* of phage tail to specific receptor sites on the bacterial cell wall.
2. *Injection* of phage DNA.
3. *Eclipse period*, in which no infective phage is recoverable if the bacterial cell is artificially lysed and during which synthesis of new phage DNA and protein shells is taking place.
4. *Assembly* of phage DNA into new protein shells.
5. *Lysis* of host cell and release of several hundred infective phage particles.

Life cycle of a temperate phage. Temperate phages do not ordinarily lyse their host; in such cases the phage DNA becomes integrated into the bacterial DNA as *prophage.* Prophages replicate synchronously with host DNA; progeny bacterial cells then contain this bacterial-plus-phage DNA. Under these conditions both infection by a virulent phage and maturation of infective virus particles are prevented. Infrequently, however, the association between host and phage DNA may be terminated (deintegration); the lytic cycle then follows as in the case of a virulent phage. Because the bacterial hosts in this case are potentially subject to lysis, they are termed *lysogenic.* Both phages lambda (λ) and P22, for example, behave as temperate phages.

Generalized transduction. In the early 1950s Zinder and Lederberg looked for indications that the then recently described process of conjugation might occur in the mouse typhoid bacterium *Salmonella typhimurium.* In one experiment they cultured two different strains of this bacterium, *met⁻ his⁻* and *phe⁻ trp⁻ tyr⁻* (which signify defects in the ability to synthesize the amino acids

300

CHAPTER 16
THE
IDENTIFICATION
OF THE GENETIC
MATERIAL

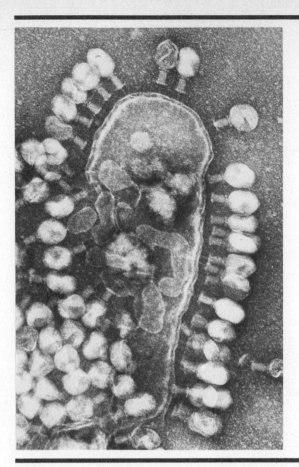

Figure 16-3. *Electron micrograph of T4 phage attacking* Escherichia coli. *Not only can phage particles be seen attached by their tail fibers to the cell wall of the bacterium* (top and right), *but new phage particles are shown being released from the lysed bacterial cell* (left). (Photo by Dr. L. D. Simon, courtesy Dr. Thomas F. Anderson, Institute for Cancer Research, Philadelphia.)

methionine, histidine, phenylalanine, tryptophan, and tyrosine, respectively). Neither strain would grow on a medium that lacked the amino acids that the strain could not synthesize. In one of their experiments, the two strains were grown in liquid media in a U-tube, but separated by a membrane whose pores were too small to permit passage of the bacterial cells. Thus, cell-to-cell contact (and, therefore, conjugation) could not occur. When, however, filters of a pore size that permitted passage of the temperate phage P22 (but too small for the bacteria) were used, wild-type cells were recovered.

Additional work demonstrated clearly that P22 was, in fact, the agent of recombination. Transformation was excluded as a possibility by the simple expedient of treating *cell-free* extracts of one strain with DNase to destroy any naked DNA. Again recombination occurred. So, rather than demonstrating either conjugation or transformation, Zinder and Lederberg had determined a then new type of recombination, one mediated by a virus. To this process they gave the name **transduction.** Its distinguishing characteristic is that the vector of recombination is a phage.

A temperate phage usually exists as a *prophage*, i.e., integrated into the bacterial DNA. In an occasional cell the prophage becomes detached and, like a virulent phage, lyses the host cell. With phage P22, for example, a segment of host DNA, which happens to be about the same length as the normal phage genome, becomes incorporated in a phage shell instead of phage DNA. *Any* portion of host DNA of the appropriate length may be so incorporated. In the new host, the transducing DNA may be integrated into the newly infected cell's genome. If transduced genes and those of the new host are alleles, the transduced

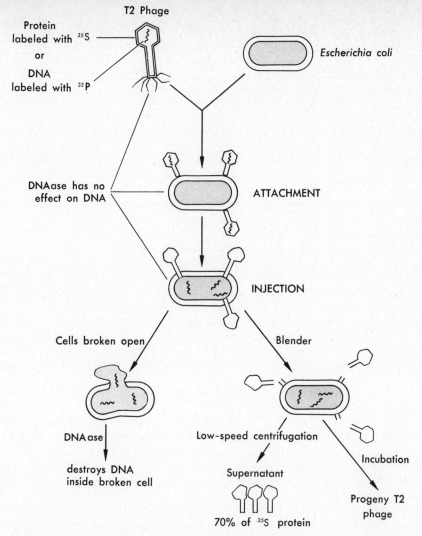

Figure 16-4. *Diagram of life cycle of a virulent phage showing details of the Hershey-Chase experiment. Phage particles and bacterial cells are not to scale. (From Fraser,* Viruses and Molecular Biology, *Macmillan, 1967. By permission.)*

gene(s), upon inclusion in the host DNA, produce recombinant progeny. This type of transduction, in which the transducing DNA may involve any of the bacterial genes, is referred to as **generalized transduction.**

Generalized transducing phages carry *only* bacterial genes, and their DNA may, by crossing-over, replace a corresponding segment of DNA in the newly infected cell. For example, if the donor DNA includes genes tyr^+ trp^+ (which are close together on the *E. coli* chromosome), and the recipient is tyr^- trp^-, any tyr^+ trp^- and tyr^- trp^+ progeny must be recombinants. The distance between the two loci may be calculated from the ratio of recombinants to total transduced cells. Recombinant frequency varies inversely with distance between loci, just as in eukaryotes.

Specialized transduction. Some phages are able to integrate into the host cells DNA at only certain positions. Phage lambda (λ), for example, infects

302

CHAPTER 16
THE
IDENTIFICATION
OF THE GENETIC
MATERIAL

Escherichia coli and can occupy only a site between the *gal* (galactose) and *bio* (biotin) genes. Occasionally the phage DNA is excised incorrectly with the result that the excised DNA carries some bacterial and some viral genes (Fig. 16-5). This "hybrid" DNA is then incorporated into phage shells as usual. Upon infecting a new host cell, the phage may integrate (at its normal site), and create a bacterial cell that is partially diploid. Such *merozygotes* can be used in (1) mapping of some of the bacterial genes on either side of the phage insertion point, and (2) a determination of which host allele, normal or defective, is dominant. The pieces of DNA that may exist either free in the bacterial cytoplasm or integrated into its DNA are designated *episomes*. These are discussed further in Chapter 21.

Note the distinction between generalized and **specialized transduction.** Generalized transducing phages carry *only bacterial* genes, and these may *replace* those of the newly infected host; specialized transducing phages, by contrast, carry *both bacterial and viral* genes. Whereas genes carried by generalized transducing phages may replace those of the bacterial cell, genes borne by specialized transducing phages may be *added* to the genome of the newly infected cell.

When it is recognized that the material transferred during conjugation (Chapter 6 and Appendix B) is also DNA, it is clear that the three processes—conjugation, transformation, and transduction—all implicate DNA as *the* genetic material. Experimental evidence rapidly accumulated to show that, in all but the RNA-containing viruses,[2] DNA has this important property, as it does also in all eukaryotes, such as humans. The essential feature of heredity, then, is the transmission of this "information tape" unchanged from one generation to the next. If we are to identify genes, we must understand the structure of this all-important molecule that has been appropriately called the "thread of life."

Deoxyribonucleic acid

History.   Interestingly enough, recognition of deoxyribonucleic acid as the genetic material was slow in coming. By 1869 Friedrich Miescher, then a 22-year-old Swiss physician, had isolated (by remarkably advanced techniques) from nuclei of pus cells obtained from discarded bandages in the Franco-Prussian War, and from salmon sperm, a previously unidentified macromolecular substance, to which he gave the name *nuclein*. Although he was unaware of the structure and function of nuclein, he submitted his findings for publication. The editor who received the paper was dubious about some aspects of the report and delayed publication for two years while he tried repeating some of what were to him the more questionable aspects of Miescher's work. Finally, in 1871, Miescher's report was published, but it made little immediate impact. He continued his careful work up to his death in 1895, recognizing (with the help of his student, Altmann, in 1889) that nuclein was of high molecular weight and was associated in some way with a basic protein, to which he gave the name *protamine*. By 1895 the pioneering cytologist E. B. Wilson suspected that "inheritance . . . may be effected by the physical transmission of a particular chemical compound from parent to offspring."

Nuclein was later renamed *nucleic acid*, the name still used, and work on it continued slowly in several laboratories. Early in this century the biochemist Kossel identified the constituent nitrogenous bases of nucleic acid, as well as its 5-carbon sugar, and phosphoric acid. Kossel's work and the later investigations of Ascoli, Levene, and Jones during the first quarter of this century disclosed the two kinds of nucleic acid, deoxyribonucleic acid and ribonucleic acid. Development of DNA-specific staining techniques by Feulgen and Rossenbeck in 1924 enabled Feulgen to demonstrate in 1937 that *most* of the DNA content of a cell

---

[2]For example, influenza, poliomyelitis, and tobacco mosaic viruses.

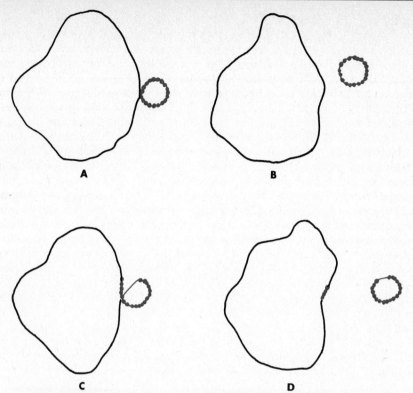

Figure 16-5. *(A) Normal association of prophage* DNA *(beaded) with the bacterial chromosome, and (B) its separation. (C) "Faulty" attachment between prophage and bacterial DNA, resulting in (D) an exchange of* DNA *between prophage and bacterium. As a consequence of this latter, aberrant association, the phage* DNA *includes a bit of bacterial DNA and vice versa.* (Redrawn from Werner Braun, *Bacterial Genetics*, W. B. Saunders and Company, 2nd ed., 1965. By permission of author and publisher.)

is located in the nucleus. Portugal and Cohen have written a highly readable account of a century of work on DNA, from Miescher's discovery to the 1970s.

Structure. In 1953 James Watson and Francis Crick proposed a molecular model for DNA after about a year and a half of joint work at Cambridge University. So completely was their model substantiated by subsequent investigations that this team shared a Nobel Prize in 1962.

Publication of their proposal was the outcome of intensive work that involved not only Watson and Crick, but also Maurice Wilkins, Rosalind Franklin, and Linus Pauling. Both Wilkins and Franklin prepared X-ray diffraction pictures on which the ultimate model was based. They and Pauling considerably sharpened Watson's and Crick's notion of the structure of the molecule in discussions and by correspondence. Though Watson, in his engrossing account of those years (*The Double Helix*, 1968), emphasized chiefly his and Crick's contribution (which was considerable), it appears that Franklin not only came to some of the correct conclusions earlier than did Watson and Crick, but indeed might well have at least shared in the Nobel Prize had not her untimely death at age 37 cut short a brilliant career.

The basic structure deduced is that of a very long molecule of high molecular weight, composed of two sugar-phosphate strands on the outside of the mole-

304

CHAPTER 16
THE
IDENTIFICATION
OF THE GENETIC
MATERIAL

cule. These strands are oriented in opposite directions, and together form a double helix, a complete right-handed (clockwise) turn every 34 angstrom units (i.e., 3.4 nm). These sugar-phosphate "backbones" are crossconnected internally by nitrogen-containing bases, two of which are in the class of chemicals known as **purines** and two **pyrimidines.** The whole, then, may be likened to a rope ladder that is helically twisted, as suggested in Figure 16-6.

The sugar is the 5-carbon **deoxyribose** whose molecular structure is shown in Figure 16-7; these are linked together by **phosphoric acid** bonded to the 3' and 5' carbons of the sugar.[3] The purine and pyrimidine bases are spaced 3.4 angstrom units (0.34 nm or 3.4×10^{-4} μm) apart, with the result that there are 10 base pairs per complete turn of the sugar-phosphate strands. Each pair of bases is, therefore, oriented 36° clockwise from the preceding pair. The diameter of the molecule is about 20 angstrom units.

Typically, only four different nitrogenous bases occur, the purines **adenine** and **guanine** and the pyrimidines **thymine** and **cytosine.** Their molecular structure, shown in Figure 16-8, is such that adenine is connected by two hydrogen bonds only to thymine, compared to three hydrogen bonds linking cytosine only to guanine:

[3]It is customary to prime the positions of carbons in the sugar to distinguish them from carbon positions in the organic bases.

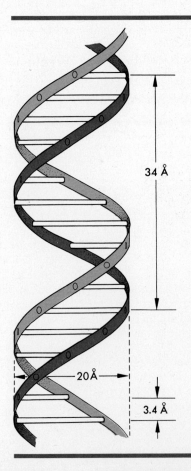

34 Å

20Å

3.4 Å

Figure 16-6. The general structure of double-stranded deoxyribonucleic acid. (1Å = 0.0001 μm or 0.1 nm)

Purines		Pyrimidines
Adenine	=	Thymine
Guanine	≡	Cytosine

Because pairing occurs in this way, the amounts of adenine and thymine should be equal to each other, as should the amounts of cytosine and guanine. Indeed, this has repeatedly been shown to be the case (within the limits of experimental error). Notice in Table 16-1 that for all sources listed (except phage X174) the ratios A/T and G/C are near 1.

The DNA of phage X174 is single-stranded (except during its replicative phase), hence its values for A/T and G/C depart considerably from unity. In fact, whenever this kind of deviation is found it is indicative of singlestrandedness. Note also that values for A + T/C + G vary widely from well below to well above 1; that is, although the relationships A = T and C = G are valid, it is also true that A + T ≠ C + G in most cases. Certainly the structure of DNA does not

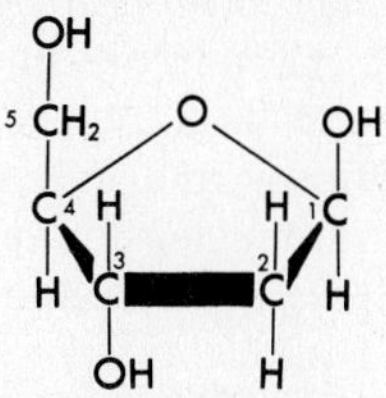

*Figure 16-7. Molecular structure of **deoxyribose**, the sugar of DNA.*

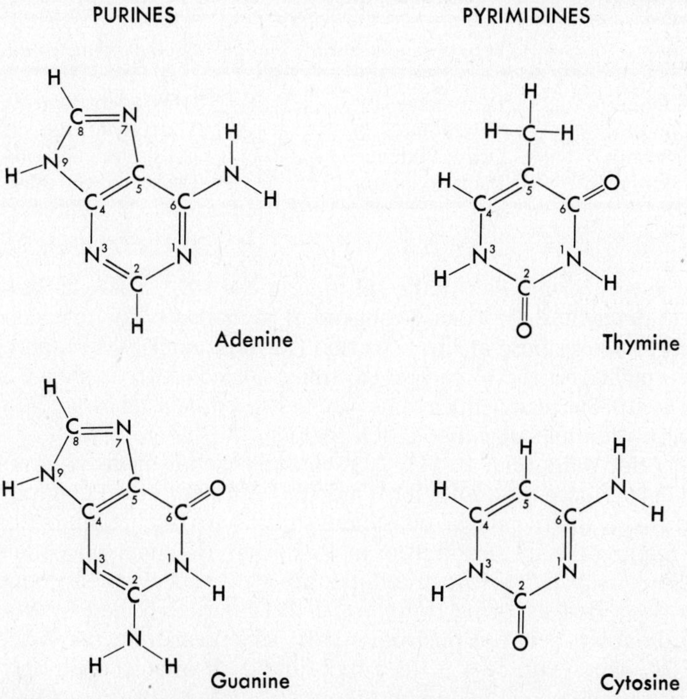

Figure 16-8. Molecular structure of the four nitrogenous bases of DNA.

306

CHAPTER 16
THE
IDENTIFICATION
OF THE GENETIC
MATERIAL

Table 16-1. Comparison of nucleotide composition of DNA

Source	A	T	G	C	$\dfrac{A}{T}$	$\dfrac{G}{C}$	$\dfrac{A + T}{G + C}$
Human sperm	31.0	31.5	19.1	18.4	0.98	1.03	1.67
Salmon sperm	29.7	29.1	20.8	20.4	1.02	1.02	1.43
Euglena nucleus	22.6	24.4	27.7	25.8	0.93	1.07	0.88
Euglena chloroplast	38.2	38.1	12.3	11.3	1.00	1.09	3.23
Escherichia coli	26.1	23.9	24.9	25.1	1.09	0.99	1.00
Mycobacterium tuberculosis	15.1	14.6	34.9	35.4	1.03	0.98	0.42
Phage T2	32.6	32.6	18.2	16.6*	1.00	1.09	1.87
Phage X174	24.7	32.7	24.1	18.5	0.75	1.30	1.35
Drosophila	27.3	27.6	22.5	22.5	0.99	1.00	1.22
Corn (*Zea*)	25.6	25.3	24.5	24.6	1.01	1.00	1.04

*5-hydroxymethyl cytosine.

require equality in that relationship. In fact, *DNA of different species is distinguished in large measure by the relative numbers of AT and CG pairs, their sequence, whether these occur as AT or TA and as CG or GC, and the number of such base pairs* (hence the length of the DNA molecules).

Certain useful terms are applied to parts of the DNA molecule. One phosphate and its attached sugar are referred to as **deoxyribose phosphate;** one deoxyribose molecule plus its attached purine or pyrimidine is a **nucleoside** (deoxyribonucleoside or deoxynucleoside), as shown in Figure 16-9; and a purine or a pyrimidine plus one deoxyribose phosphate constitutes a **nucleotide** (deoxyribonucleotide or deoxynucleotide), as shown in Figure 16-10. Deoxyribonucleosides and deoxyribonucleotides have specific names, as shown in Table 16-2.

Table 16-2. Deoxyribonucleosides and deoxyribonucleotides

Base	Deoxyribonucleoside	Deoxyribonucleotide
Adenine	Deoxyadenosine	Deoxyadenylic acid
Thymine	Thymidine	Thymidylic acid
Cytosine	Deoxycytidine	Deoxycytidylic acid
Guanine	Deoxyguanosine	Deoxyguanylic acid

Techniques involving enzymatic splitting of the DNA molecule have made it possible to determine the exact arrangement of its parts. The nitrogenous bases are linked to deoxyribose at the 1′ carbon (forming nucleosides), and the phosphate is attached to the 5′ carbon (forming nucleotides), as shown in Figures 16-9 and 16-10. Successive nucleotides are linked by 3′, 5′ phosphodiester bonds, as shown in the molecular model of a segment of DNA consisting of four deoxyribonucleotide pairs (Fig. 16-11). That is, the 3′ and 5′ hydroxyl groups of two different deoxyribose molecules form a double ester with the PO_4 group. As seen in Figure 16-11, the two sugar-phosphate strands are oriented in opposite directions. In reading the *left* strand from top to bottom the sugar-phosphate linkages are 5′, 3′, whereas in the *right* strand they are 3′, 5′. The two "backbone" strands are thus described as being *antiparallel*. It was determined by Watson that hydrogen bonding between nucleotide pairs give measurements very similar to AT and GC pairs (Fig. 16-12). In a molecule such as this, with only the four "usual" bases, the possible kind and number of sequences are virtually infinite. The DNA of different species is, in fact, distinctive through such qualitative and quantitative differences in nucleotide pairs, although not all DNA is built exactly

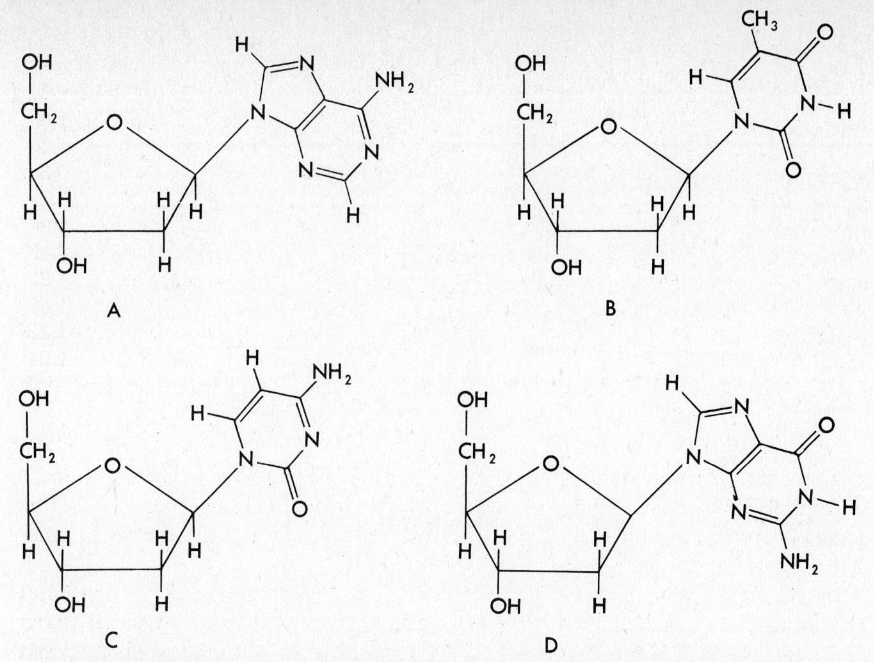

Figure 16-9. *The four deoxyribonucleosides: (A) Deoxyadenosine. (B) Thymidine. (C) Deoxycytidine. (D) Deoxyguanosine.*

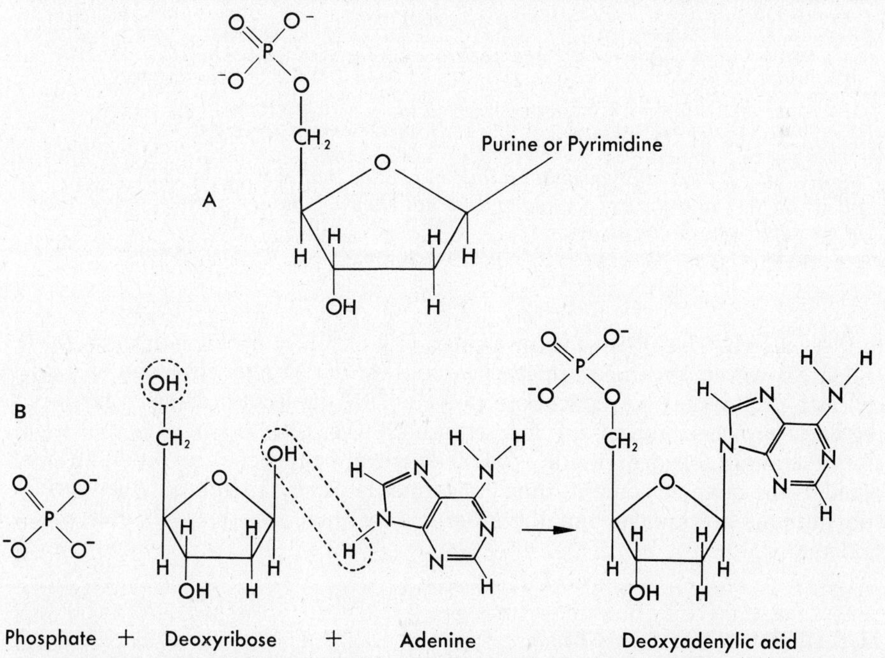

Phosphate + Deoxyribose + Adenine → Deoxyadenylic acid

Figure 16-10. *(A) Structural model for a deoxyribonucleotide. A purine is attached to deoxyribose by a bond connecting a ring-nitrogen at the number 9 position to the 1' position of the sugar; a pyrimidine is attached to deoxyribose by a bond connecting a ring-nitrogen at the number 3 position to the 1' position of the sugar. (B) Formation of the deoxyribonucleotide deoxyadenylic acid. The other three deoxyribonucleotides are formed in the same way.*

308

CHAPTER 16
THE
IDENTIFICATION
OF THE GENETIC
MATERIAL

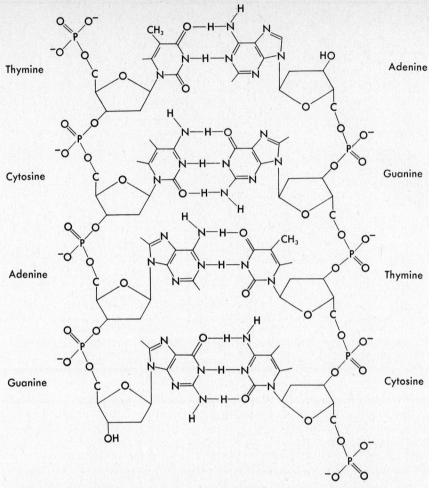

Figure 16-11. *Molecular model of a four nucleotide pair segment of DNA. For simplicity the helical nature of the molecule is not shown. Note antiparallel orientation of the sugar-phosphate strands.*

to this pattern. The DNA of some phages (e.g., φX174) is only single-stranded. It does, however, become temporarily double-stranded after infection of a host cell as a preliminary to replication. In some DNA molecules, other, rarer bases regularly replace some of the four common ones. The T-even phages, for instance, contain 5-hydroxymethyl cytosine instead of cytosine, but in an amount equal to the guanine content, thus indicating its pairing qualities. Other similar substitutions are known, but the exact role of these rare bases is only partly understood.

LOCATION OF DNA IN CELLS

As was pointed out earlier, the Feulgen stain technique is specific for DNA; any cellular structure containing DNA retains a purple color. Not only is this process specific for DNA, but the intensity of staining is directly proportional to the amount of deoxyribonucleic acid present. Measurements of light absorption by structures colored by the Feulgen process can be used to determine the relative amounts of this nucleic acid present. Such techniques, applied to many different

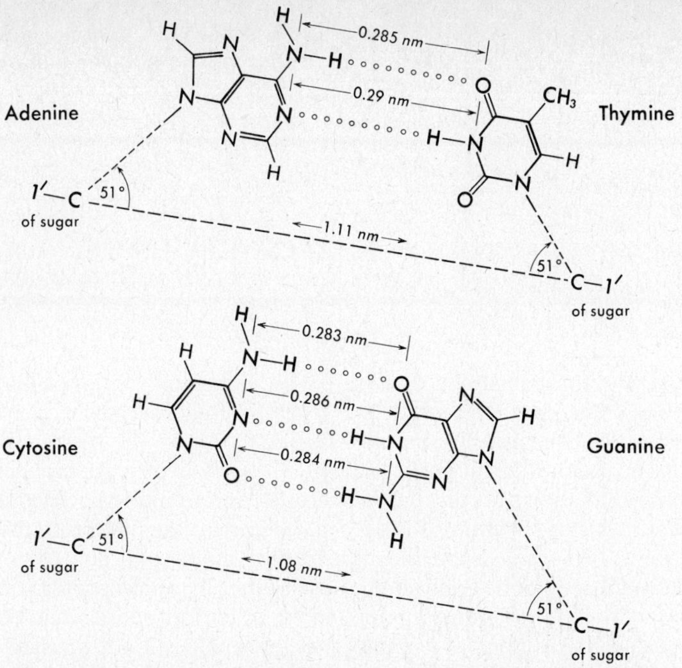

Figure 16-12. *Details of hydrogen bonding between deoxyribonucleotide pairs. Note the close similarity of measurements of AT and
GC pairs. (0.1 nm = 1 Å.)*

kinds of eukaryotic cells, show DNA to be almost entirely restricted to the chromosomes.[4] Such microspectrophotometric techniques also employ ultraviolet light, the peak absorption of which is near 260 nm. It is significant that although ultraviolet radiation is a relatively weak mutagenic agent, its most effective wavelengths as such are also near 260 nm.

The physical arrangement of DNA in the chromosomes of eukaryotes has receive intensive attention in recent years, with the result that our knowledge has been considerably enhanced. It has been known for some time that the amount of DNA per cell nucleus, and per bacterial cell, is many times the volume of the containing structures (Table 16-3 and Figure 16-13).

Obviously, DNA, which behaves as a linear molecule in crossing-over and which bears genes in apparent linear sequence, cannot simply extend in a straight line from one end of a eukaryotic chromosome or bacterial cell to the other. From a physical standpoint, the great length of DNA molecules relative to that of the structures containing them imposes stringent requirements on "packaging." Yet, on genetic bases, each DNA molecule must behave as if it did run in a straight line from end to end of the containing structure. Electron microscopy, particularly scanning electron microscopy, has shown clearly the physical arrangement of DNA in chromosomes of eukaryotes. Figure 16-14 shows such a micrograph of a metaphase chromosome of the Chinese hamster, in which can be seen the highly coiled chromatin (nucleoprotein) fibers, or *microconvules*. In general, these fibers range from about 50 to 500 Å, depending somewhat on the species, but largely on the technique used in preparation. Interestingly enough, the centromeric region appears as a constriction in each chromatid but displays

---

[4]The fact that small amounts of DNA are also found in the cytoplasmic organelles chloroplasts and mitochondria only reinforces the cytological, chemical, and genetic evidence that DNA is genetic material. See also Chapter 21.

310

CHAPTER 16
THE
IDENTIFICATION
OF THE GENETIC
MATERIAL

Table 16-3. Approximate quantitative characteristics of the DNA of some organisms and a phage

Organism	Approximate molecular weight, DNA	Number of deoxyribonucleotide pairs	Length, μm	Length, inches
Phage T2	1.3×10^8	2×10^5	68	0.003
Escherichia coli	2.8×10^9	4.3×10^6	1,465	0.06
Drosophila melanogaster (diploid cell)	2.2×10^{11}	3.38×10^8	1.15×10^5	4.5
Human (2n)	3.9×10^{12}	6×10^9	2.04×10^6	80.3

no obvious structural element that might be correlated with the kinetochore (compare with Fig. 4-9). Figure 16-15 is a transmission electron micrograph of one of the small G-group human chromosomes. Again the seemingly tangled network of nucleoprotein fibers is evident. A portion of a metaphase human chromosome that shows accumulations of nucleoprotein fibers folded into loops and connecting sister chromatids is seen in the transmission electron micrograph of Figure 16-15. In some electron micrographs, the several chromosomes of a complement are seen to be connected to each other by nucleoprotein fibers. This suggests to some that an entire monoploid set of chromosomes may be foldings of one long chromatin fiber. It is hardly necessary to point out that this idea has met with less than unanimous acceptance. Yet, in spite of certain problems that remain in the interpretation of attachment of spindle fibers to kinetochores in nuclear division, and in the observed occurrences of sister chromatid exchanges, Bahr wrote in 1977, "In view of the very suggestive electron micrographs and numerous light microscope observations, the concept of large circles of DNA in

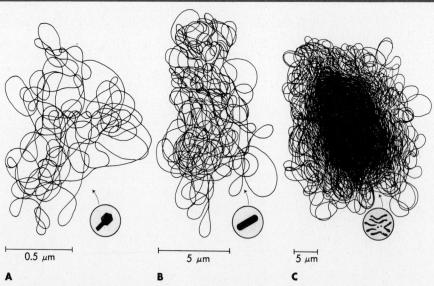

0.5 μm	5 μm	5 μm
A	**B**	**C**

Figure 16-13. Total length of DNA of (A) bacteriophage T-4 (about 68 μμ); (B) the bacterium, E. coli (about 1,100 μm); (C) the fruit fly. Drosophila melanogaster (about 16,000 μm). Compare these lengths to the size of the structures into which the DNA is packed. (Redrawn from F. W. Stahl, The Mechanics of Inheritance, © 1964, Prentice-Hall, Inc., Englewood Cliffs, N.J., by permission of author and publisher.)

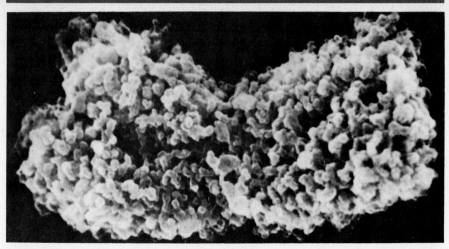

Figure 16-14. *Scanning electron micrograph of an isolated metaphase chromosome showing the highly coiled chromatin fibers (microconvules) that comprise the chromatids,* × *28,000. The bar represents 0.5* μm. (Micrograph kindly supplied by Dr. Wayne Wray. From M. L. Mace, Jr., Y. Daskal, H. Busch, V. P. Wray, and W. Wray, 1977. "Isolated Metaphase Chromosomes: Scanning Electron Microscopic Appearance of Salt-Extracted Chromosomes." *Cytobios,* **19:** 27–40. Used by permission of Dr. Wayne Wray and The Faculty Press, Cambridge, England.)

which chromosomes represent local foldings deserves serious consideration." (See Fig. 16-16.) A different, more traditional view was expressed by Lewin in 1980 in these words, "The DNA of a eucaryotic cell is not, of course, a single duplex molecule, for it is organized into many individual chromosomes." Although the structural configuration within individual chromosomes has now been clearly and amply demonstrated, it has not been as easy to relate this structure to the sequential arrangement of the genes.

On the other hand, the bacterial "chromosome" is unequivocally a single long DNA molecule in the form of a "naked" (i.e., not histone-complexed), continuous, closed, no-end structure. That it is highly folded and coiled can be determined from electron microscopy, as well as from the quantitative relationship of the bacterial DNA molecule and the cell that contains it. For example, an *Escherichia coli* cell is about 2 μm in length, whereas its DNA molecule is some 700 times longer. Moreover, a bacterial cell may contain more than one of these DNA molecules, depending on its stage in the life cycle.

Cyclic quantitative changes in DNA content.   By appropriate quantitative measurements of DNA content of cells, a number of striking parallels with chromosome behavior and our earlier postulates regarding genes can be seen:

1. The quantity of DNA detected in gametes is, within limits of experimental error, half that of diploid meiocytes.
2. Zygotes contain twice the amount of DNA in gametes.
3. The amount of DNA increases during interphase by a factor of 2. Telophase nuclei have half the DNA content of late-interphase (G_2) or early-prophase nuclei.
4. DNA content of polyploid nuclei is proportional to the number of sets of chromosomes present; that is, tetraploid nuclei can be shown to have twice as much DNA as diploid cells.

312

CHAPTER 16
THE
IDENTIFICATION
OF THE GENETIC
MATERIAL

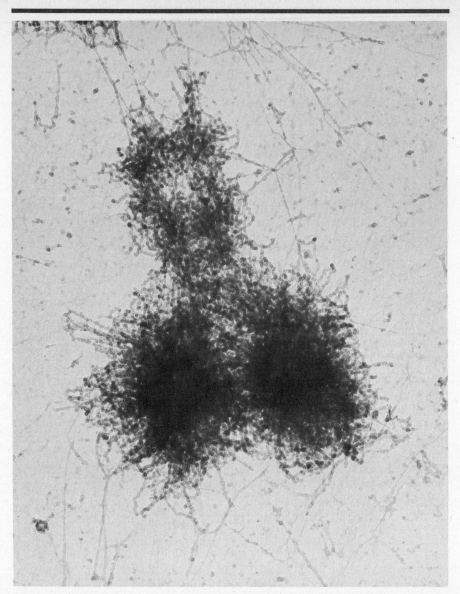

Figure 16-15. *One of the G group of human chromosomes. Satellites and their extensions, appearing like horns, are at the upper portion of the chromosome, just above the centromeric region. Transcription, translocation, and so forth, take place with great precision in this tangled mass of nucleoprotein fibers.* (Scanning electron micrograph courtesy of Dr. G. F. Bahr. From G. F. Bahr, 1977. Chromosomes and Chromatin Structure. In J. J. Yunis, *Molecular Structure of Human Chromosomes,* published by Academic Press, New York. Used by permission of author and publisher.)

Replication of DNA

Because DNA is the genetic material in all but some viruses, its mode of increase from one cell cycle to the next should be such as to *replicate* exactly. Although some details are yet to be clarified, recent work of several laboratories has disclosed the major steps in the process, particularly in *Escherichia coli* and some of the DNA phages. As might be expected, the replicative process in eukaryotes

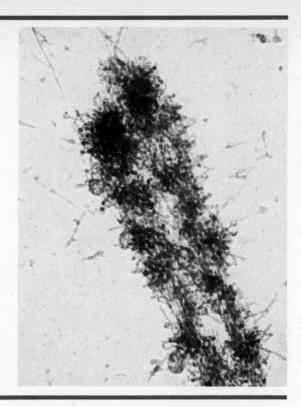

Figure 16-16. Nucleoprotein strands, in portions of two sister chromatids, folded into tandem loops (chromomeres). (Electron micrograph courtesy Dr. G. F. Bahr. From G. F. Bahr, 1977. Chromosomes and Chromatin Structure. In J. J. Yunis, ed., *Molecular Structure of Human Chromosomes,* published by Academic Press, New York. Used by permission of Dr. Bahr and the publisher.)

is less well understood than it is for prokaryotes. The process will be examined first in *E. coli*, then in eukaryotes.

Initiation of replication in E. coli. In *E. coli* the continuous, no-end DNA molecule, or "chromosome," retains its circularity during replication. Replication involves an initial unwinding of the double helix to form two single-strand templates (Fig. 16-17A). This is accomplished by an **endonuclease,** an enzyme that cuts ("nicks") one of the sugar-phosphate backbone strands at specific points. Next the weak hydrogen bonds that link purine-pyrimidine bases are broken, a process that is often referred to as "unzipping" of double-stranded DNA. The result is the formation of a *replication bubble*, which subsequently extends as a Y-shaped **replication fork** (Fig. 16-17B). In bacteria and many DNA phages this extending is *bidirectional*.

Unwinding of template DNA. Before the template strands of DNA can be replicated, an unwinding must occur in order to produce greater lengths of single strands. This is accomplished by **unwinding proteins,**[5] some 200 of which bind to single-stranded DNA in the vicinity of the replication fork, destabilizing the double helical portion of the molecule (Fig. 16-17B). Each molecule of unwinding protein complexes with about 10 deoxyribonucleotides. Some 2,000 unpaired nucleotides are thus available on each single-stranded arm of the replication fork for complementary pairing of the bases of the newly synthesized strand.

Priming. The DNA polymerases, which are responsible for adding new complementary deoxyribonucleotides to the template strand, cannot function with-

---

[5]*Unwinding protein* is one of several "relaxing enzymes" responsible for reducing the number of superhelical turns in circular DNA molecules.

314

CHAPTER 16
THE
IDENTIFICATION
OF THE GENETIC
MATERIAL

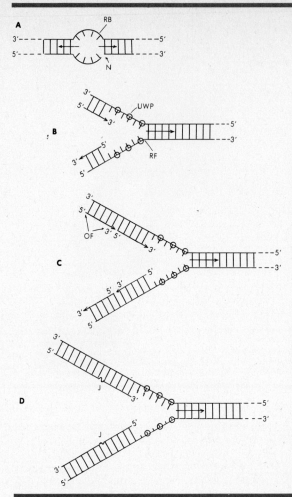

Figure 16-17. *Replication of DNA. (A) Replication bubble(s) (RB) forms at unique site(s) determined by a nick (N) produced in one sugar-phosphate strand by endonuclease. In many organisms replication bubbles "open up" replication forks (RF) bidirectionally (horizontal arrows). (B) Unwinding of template strands through action of unwinding proteins (UWP). Only one of the replication forks is shown here. (C) Discontinuous synthesis of both daughter strands in their 5' → 3' direction through addition of new nucleotides at their 3' ends, producing Okazaki fragments (OF). (D) Continued discontinuous synthesis of daughter strands; the Okazaki fragments are joined (at J) through action of DNA ligase. For greater simplicity, the RNA primers (see text) are not shown here.*

out available 3'OH groups. Generation of 3'OH groups to which the DNA polymerases add new nucleotides is provided by synthesis of a short segment of primer *RNA*, catalyzed by an RNA polymerase. The primer is attached to the 3' end of the template strand. One of the DNA polymerases then adds DNA monomers to the RNA primer, which is later removed enzymatically, leaving a gap that is filled in later.

The DNA polymerases. Three principal DNA polymerases, I, II, and III, are responsible for the template-directed condensation of new deoxyribonucleoside triphosphates (Fig. 16-17). Two inorganic phosphates are then split off (Fig. 16-18), leaving deoxyribonucleotide monophosphates to make up the growing strand. Synthesis of the complementary strands proceeds from the 3'OH terminus of the RNA primer, and extends the newly synthesized DNA strand in its 5' → 3' direction.

DNA polymerase I is a single polypeptide chain of molecular weight 109,000; about 400 molecules are present in each *E. coli* cell. This enzyme binds to single-stranded DNA and is largely responsible for filling in gaps between small precursor oligonucleotides. It also exerts a "proofreading" function in that, although its polymerizing activity is in the 5' → 3' direction (by adding deoxyribinucleotide monomers in antiparallel fashion on the daughter strand), it also "edits"

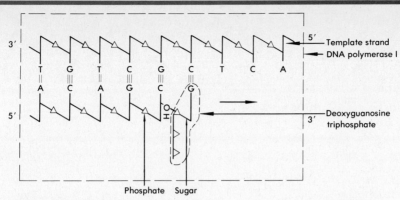

Figure 16-18. *Template-directed condensation of new deoxyribonucleoside tri-phosphates, as catalyzed by* DNA *polymerase I in* Escherichia coli. *See text for details.*

the daughter strand by removing incorrect, nonpairing nucleotides, due to a 3′ exonuclease capability. In addition, DNA polymerase I excises thymine dimers in the 5′ → 3′ direction. In the same fashion this exonuclease activity is able to digest the RNA primer in the 3′ → 5′ direction and, at the same time, fill in the gap with deoxyribonucleotides. Recent work, however, indicates that DNA polymerase I, while necessary for cell viability, is not indispensable for DNA *replication.*

DNA polymerase II is composed of a single polypeptide chain of molecular weight 90,000 to 120,000. It has the same polymerization capability as DNA polymerase I, as well as 3′ → 5′ exonuclease activity, but lacks a 5′ → 3′ exonuclease function. Moreover, it cannot utilize a nicked primer or one that is extensively single-stranded. Its functions are but imperfectly known, although it does appear to bring about repair of ultraviolet lesions. There are about 100 molecules of this enzyme per *E. coli* cell.

DNA polymerase III has a molecular weight of 180,000 and is composed of two polypeptide chains of about 140,000 and 40,000 daltons, respectively. About 10 molecules occur in each *E. coli* cell. It is the principal DNA polymerizing or replicating enzyme in the cell in that it catalyzes addition of DNA monomers to the RNA primer. It is distinguished from the other two DNA polymerases not only by its molecular structure but also by (1) an inability to utilize single-stranded DNA, (2) lack of a 5′ → 3′ exonuclease activity, (3) greater instability, and (4) a lower affinity for deoxyribonucleoside triphosphates.

A second form of DNA polymerase III, designated DNA polymerase III* ("**DNA polymerase III-star**"), probably is the physiologically active state of DNA polymerase III. It is a product of the same *E. coli* gene as is DNA polymerase III, has similar sensitivity to salt and ethanol, and is converted to DNA polymerase III on heating. It is believed to be a tetramer of four identical 90,000 dalton subunits and may be a dimeric form of DNA polymerase III. Its activity and function are not yet clearly understood, but it initially forms a complex with the template primer.

In summary, DNA polymerase III closes gaps of 10 to 50 nucleotides on double-stranded DNA in which small single-stranded gaps occur. It also acts as a 3′ → 5′ exonuclease, hydrolysing single-stranded DNA into 5′-mononucleotides. DNA polymerase I by itself can elongate DNA in the 5′ → 3′ direction; DNA polymerase II can elongate primed DNA that is complexed with DNA binding protein. When accompanied by certain other prerequisite conditions, DNA polymerase II or III will elongate primed DNA that is combined with elongation factors I and

316

CHAPTER 16
THE
IDENTIFICATION
OF THE GENETIC
MATERIAL

II. It has been suggested that DNA polymerases I and II do not perform *essential* functions, inasmuch as mutants lacking these enzymes are viable. Some characteristics of the DNA polymerases are given in Table 16-3.

Table 16-3. Some characteristics of the DNA polymerases of E. coli*

	DNA polymerases		
	I	*II*	*III*
Molecular weight	109,000	120,000	180,000
Structure	single chain	single chain	double chain
Molecules/cell	400	100	10
Product of gene†	pol A	pol B	dna E
Polymerization $5' \rightarrow 3'$	+	+	+
Exonuclease $5' \rightarrow 3'$	+	−	−
Exonuclease $3' \rightarrow 5'$	+	+	+
Lability at 37°C	−	−	+
Nucleotides polymerized per min at 37°C	~1,000	~50	~15,000
Template-primer:			
Primed single strands‡	+	−	+
Nicked duplex	+	−	−
Polymer synthesis de novo	+	−	−
Affinity for triphosphate precursors	low	low	high

*Data from Kornberg (1980).
†See Figure 6-7 for map locations in *E. coli*, strain K-12.
‡A long single strand with a short complementary strand annealed with it.

Discontinuous synthesis and Okazaki fragments. The DNA polymerases all add new deoxyribonucleotides in the $5' \rightarrow 3'$ direction by addition at the $3'$ end of the nascent complementary strand; that is, *both* daughter strands are extended only in the $5' \rightarrow 3'$ direction (Figs. 16-17C and D). Furthermore, daughter strands are added in increments of 1,000 to 2,000 nucleotides, rather than singly. These facts require the construction of daughter strands in segments, which then pair complementarily, extending double-strandedness *discontinuously*. Although it was thought initially that one strand was extended continuously and the other discontinuously, most molecular geneticists now agree that daughter strands are extended *discontinuously on both template strands*.

The major evidence for discontinuous synthesis was developed by Okazaki and his colleagues, who showed that daughter strands in *E. coli* were added as the 1,000 to 2,000 nucleotide segments just noted and discontinuously on *both* sides of the replication fork (Fig. 16-17C).

Visualization of replicating bacterial DNA. Countless electron micrographs of replicating bacterial DNA have been published, giving objective evidence for the series of steps just outlined. In both *E. coli* and phage lambda (λ), a single replication bubble is formed, near gene O in λ and near the *ilv* locus on the *E. coli* K-12 map (Fig. 6-7). From the replication bubble the replication fork proceeds bidirectionally in both of these cases. In *E. coli*, for instance, these facts result in the formation of a θ-*like* (thetalike) configuration as replication advances (Fig. 16-19).

In bacterial conjugation and in intracellular multiplication of many viruses, a *rolling circle* model has been invoked. Here again, replication begins with a nick in one of the template strands which thereby has a $5'$ end and a $3'$ end. Synthesis of a daughter strand begins by addition of complementary deoxyribonucleotides on the nicked template strand. As replication proceeds, the "unraveling" of the

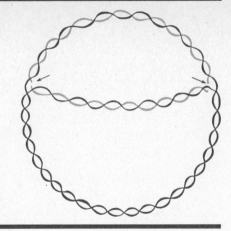

Figure 16-19. *Thetalike configuration assumed by replicating, circular, no-end, double helical DNA molecule of* Escherichia coli. *Arrows indicate the two replication forks. These have progressed bidirectionally from a single replication bubble.*

single linear strand is extended by the rotation of the closed (unnicked) strand about an imaginary axis, hence the appellation "rolling circle model" (Fig. 16-20).

Replication in eukaryotes.

In general terms, replication of eukaryotic DNA proceeds somewhat along the lines of the process in *E. coli*. At least three DNA polymerases are involved, designated as α, β, and γ.

The first of these, DNA polymerase α, in humans reportedly has a molecular weight of ~175,000 and consists of a subunit of 87,000 daltons; that of other sources (e.g., calf thymus) has somewhat different characteristics. Its action resembles that of *E. coli* DNA polymerase I, and its level begins to rise in the G_1 stage of interphase.

DNA polymerase β has a molecular weight of 30,000 to 45,000, depending on the animal source. It is widely distributed in multicellular animal species but has not been found in bacteria, other plants, or protozoa. Its role in DNA synthesis and repair remains unresolved, according to Chang, although its conservation through some 500 million years of biological evolution (from sponges to human beings) suggests an indispensable role, perhaps particularly in DNA repair.

The widely distributed DNA polymerase γ can copy DNA templates but preferentially copies synthetic ribohomopolymers (e.g., poly-A). Natural RNA templates are apparently not copied. It has not yet been sufficiently purified to determine its structure, but the molecular weight of the human enzyme appears to be somewhere between 110,000 and 300,000.

The replicon.

The DNA of each eukaryotic chromosome is replicated by formation of numerous replication bubbles, with replication proceeding bidirectionally from each. The sequentially replicating segments thus formed each constitute a **replicon,** or replication unit. In each chromosome there are numerous replicons in series, each controlled by an operatorlike segment (see Chapter 20) called a *replicator.* (By contrast, *E. coli* has but one replicon.)

Accuracy and speed of the replicative process.

Whatever the precise details of the operation of the replicative mechanism, its accuracy and speed are quite high. Kornberg and his associates were able to synthesize biologically active ϕX174 DNA of approximately 5,400 nucleotides (recall that the DNA of this phage is single-stranded except in its replicative form). Infection of susceptible *E. coli* cells with this synthetic DNA resulted in replication of ϕX174 particles identical with natural phage, as well as lysis of host cells in the usual fashion.

318

CHAPTER 16
THE
IDENTIFICATION
OF THE GENETIC
MATERIAL

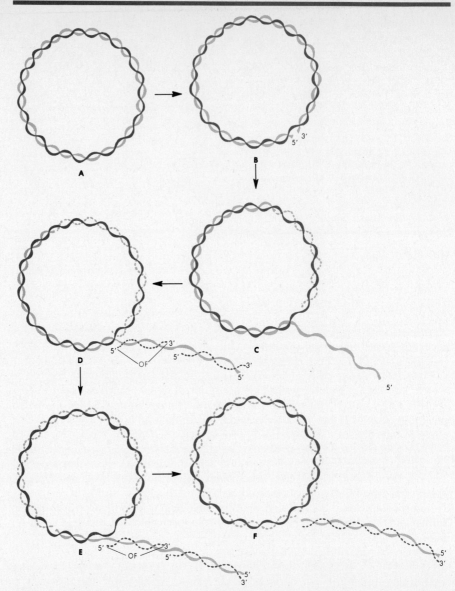

Figure 16-20. *The rolling circle model of* DNA *replication. (A) Closed circle of* DNA *before replication. (B) Nicking of one strand by an endonuclease, providing free 5′ and 3′ ends. (C-F) Successive stages in rolling of "old," continuous strand and "peeling off" of the complementary old strand with synthesis of new complementary strands. In (E) Okazaki fragments (OF) are seen on the template old strand. These have been joined by action of* DNA *ligase in (F). Circularization follows after the stage shown at (F).*

Speed of synthesis in vitro has been reported as 500 to 1,000 nucleotides per minute, but in vivo speeds as high as 100,000 per minute have been calculated.

Single-stranded viruses such as φX174 cannot, of course, follow exactly the procedure outlined for double-stranded bacterial DNA. The single strand present in the infective phage serves as a template in the host cell for synthesis of a complementary strand, and forms a double-stranded molecule (the *replicative form*) that replicates many times. Finally, however, the replicative form begins to produce only single (infective) strands, which are then assembled within protein shells as mature phage and released.

However, certain RNA tumor viruses (e.g., Rous sarcoma virus and Rauscher mouse leukemia virus) either produce or induce an enzyme, RNA-dependent DNA polymerase, that catalyzes synthesis of DNA from an RNA template. Replication of these and some other RNA viruses takes place through a DNA intermediate rather than through an RNA intermediate as described for other RNA viruses. Furthermore, this discovery is providing more information on carcinogenesis by RNA viruses.

Semiconservative nature of DNA replication. Under the strand separation hypothesis, each newly replicated double helix must consist of one "old" strand and one "new" one. By a series of ingenious experiments, Meselson and Stahl showed that this does, indeed, occur. Bacteria were cultured for some time in a medium containing only the heavy isotope of nitrogen, ^{15}N, until all the DNA should be labeled with heavy nitrogen. These cells were next removed from the ^{15}N medium, washed, and transferred to a medium all of whose nitrogen was the usual isotope, ^{14}N. After one bacterial generation, a sample of cells was removed and the DNA extracted. All DNA that replicated once in the ^{14}N medium should be "hybrid"; that is, they should consist of one strand whose nitrogen is all ^{15}N ("old") and one in which the nitrogen should be all ^{14}N ("new"), as suggested in Figure 16-21.

Differentiation between ^{14}N-DNA and ^{15}N-DNA is readily made by *density gradient centrifugation*. In the now generally employed technique, particles of different densities (such as ^{14}N-DNA and ^{15}N-DNA) are suspended in a concentrated solution of the salt of a highly soluble heavy metal such as cesium chloride and subjected to intense centrifugation in an ultracentrifuge. The centrifugal field produces a density gradient in the tube of both the salt and the suspended particles. After several hours the gradient reaches an equilibrium in which densities of the solution and suspended material match at a level in the tube where all the suspended particles of like density collect in a narrow band (Fig. 16-22). Here the particles will remain, subject only to their diffusibility. It is possible to separate particles differing in density by extremely small amounts with this technique.

When the "hybrid" DNA of Meselson and Stahl was ultracentrifuged in a cesium chloride solution, after some 20 to 50 hours all this DNA was found by ultraviolet absorption (260 nm) to have taken a position exactly intermediate between that previously established for ^{15}N-DNA and ^{14}N-DNA. After two cell generations, in which DNA has replicated a second time, half would be expected to contain only ^{14}N and half should be "hybrid." Therefore there should be two bands of DNA in the ultracentrifuge tube, one at the intermediate position, and one at the "all ^{14}N" position, and this was precisely what was observed. This pattern of replication, in which each of the "old" strands serves as one of the two in each new DNA molecule, is referred to as *semiconservative replication*.

The results of these experiments are entirely consistent with the strand-separation theory (Fig. 16-23), although they do not eliminate such other possibilities as lengthwise extension of the unseparated double strand. But there is good experimental evidence from other sources that this latter type of behavior does not occur. In fact, autoradiographs of DNA from *E. coli* that had been allowed to replicate in radioactive thymidine show that replicating DNA does indeed have a Y-shaped configuration. In *E. coli*, autoradiography reveals two such Y-shaped regions in the replicating circular DNA molecule. Although the question of how the strands unwind to permit this type of replication is only partially answered, the nature of the triggering mechanism does seem to relate to certain initiator nucleotide sequences. Considerable progress has been made in the last few years in elucidating the process of replication, and a virtually complete

320

CHAPTER 16
THE
IDENTIFICATION
OF THE GENETIC
MATERIAL

Original DNA molecule
(all ¹⁵N labeled)

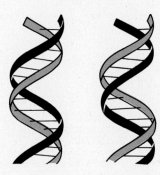

Once-replicated "hybrid" DNA
(each molecule contains one ¹⁵N
and one ¹⁴N strand)

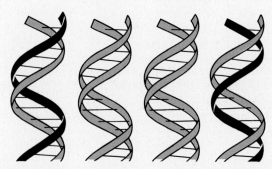

Twice-replicated DNA
(Half the molecules "hybrid"
and half containing only ¹⁴N)

Figure 16-21. *The Meselson and Stahl experiment demonstrating the semiconservative replication of DNA. See text for details. Strands labeled with* ¹⁵N *are darkly shaded;* ¹⁴N-*labeled strands are lightly shaded.*

understanding may be forthcoming soon. Incidentally, the possibility of errors in the replication process (including "proofreading") mistakes suggest a mechanism for mutation. This will be explored in Chapter 19.

Behavior of DNA in transformation and transduction. In both transformation and transduction DNA from another bacterium enters the cell that is about to be "converted." In the former process, naked DNA derives from dead cells, and quite a bit less than its DNA complement penetrates the living cell. Cells can be transformed only when they are in a state of "competence." In transduction, on the other hand, DNA is "injected" from a virus. The transducing DNA may be either all bacterial or bacterial-plus-viral DNA.

If transforming DNA is to bring about transformation, it must be integrated

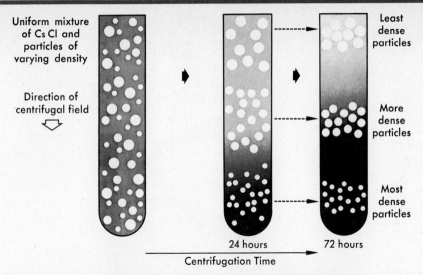

Uniform mixture
of Cs Cl and
particles of
varying density

Direction of
centrifugal field

Least
dense
particles

More
dense
particles

Most
dense
particles

24 hours   72 hours

Centrifugation Time

EQUILIBRIUM DENSITY GRADIENT CENTRIFUGATION

Figure 16-22. *The mechanics of density gradient centrifugation in a cesium chloride solution.* (Adapted from D. Fraser, *Viruses and Molecular Biology*, Macmillan, 1967. By permission.)

into the host DNA. Each of the two complementary strands of *Bacillus subtilis*, for instance, has the same transforming capability, but only one is physically integrated into the recipient's genome. Both original (*endogenous*) and transforming (*exogenous*) DNA molecules replicate synchronously. Double breaks occur, followed by rejoining such that a double crossover is formed. Breakage is likely to be an enzymatically controlled process, though the exact nature of the enzyme action is still unclear. There is, for example, no detectable twisting of bacterial chromosomes or parts thereof around each other in transformation, transduction, or conjugation. Repair of the broken sugar-phosphate strands involves DNA ligase.

It has been clearly shown by appropriate labeling techniques that such a break-and-exchange mechanism is substantially correct. The sequence of events that follows entry of "foreign" DNA in both transformation and transduction now appears to be

1. Association or pairing of transforming or transducing DNA with corresponding lengths of the recipient's DNA.
2. Synchronous replication of both "foreign" (exogenous) and recipient's (endogenous) DNA.
3. Enzymatic breaking of both exogenous and endogenous DNA.
4. Exchange between exogenous and endogenous DNA by enzymatic repair, resulting in
5. Formation of a recombinant DNA molecule that "breeds true" for the traits introduced by transformation or transduction.

Conjugation (described in Appendix B-1), transformation, and transduction provide solid evidence that the genetic material, in all but the RNA viruses, is deoxyribonucleic acid. Our next concern will be to relate the gene, both now and back into remote reaches of geological time, to this remarkable substance that is so widely distributed in plant and animal species.

322

CHAPTER 16
THE
IDENTIFICATION
OF THE GENETIC
MATERIAL

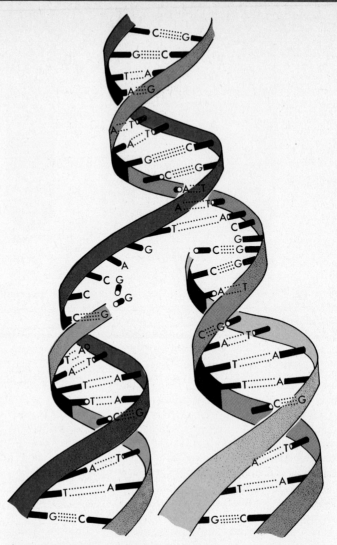

Figure 16-23. DNA replication by the strand separation theory where "old" and "new" duplexes are assumed to unwind and wind simultaneously as nucleotides are assimilated into newly developing strands. (Redrawn from D. Fraser, *Viruses and Molecular Biology,* Macmillan, 1967. By permission.)

References

Avery, O. T., C. M. MacLeod, and M. McCarty, 1944. Studies on the Chemical Nature of the Substance Inducing Transformation of Pneumococcal Types. *Jour. Exp. Med.,* **79:** 137–158. Reprinted in J. A. Peters, ed., 1959. *Classic Papers in Genetics.* Englewood Cliffs, N.J., Prentice-Hall.

Bahr, G. F., 1977. Chromosomes and Chromatin Structure. In J. J. Yunis, ed., *Molecular Structure of Human Chromosomes.* New York: Academic Press.

DuPraw, E. J., 1970. *DNA and Chromosomes.* New York: Holt-Rinehart-Winston.

Edenberg, H. J., and J. A. Huberman, 1975. Eukaryotic Chromosome Replication. In H. L. Roman, A. Campbell, and L. M. Sandler, eds. *Annual Review of Genetics,* vol. **9.** Palo Alto, Calif.: Annual Reviews, Inc.

Goulian, M., 1971. Biosynthesis of DNA. In E. E. Snell, P. D. Boyer, A. Meister, and R. L. Sinsheimer, eds. *Annual Review of Biochemistry*, vol. **40.** Palo Alto, Calif.: Annual Reviews, Inc.

Goulian, M., A. Kornberg, and R. L. Sinsheimer, 1967. Enzymatic Synthesis of DNA, XXIV. Synthesis of Infectious Phage φX-174 DNA. *Proc. Nat Acad. Sci. (U.S.),* **58**(6): 2321–2328.

Griffith, F., 1928. The Significance of Pneumococcal Types. *Jour. Hygiene,* **27:** 113–156.

Hershey, A. D., and M. Chase, 1952. Independent Functions of Viral Protein and Nucleic Acid in Growth of Bacteriophage. *Jour. Gen. Physiol.,* **36:** 39–56. Reprinted in G. S. Stent, ed., 1965, 2nd ed., *Papers on Bacterial Viruses.* Boston: Little, Brown.

Hotchkiss, R. D., and M. Gabor, 1970. Bacterial Transformation, with Special Reference to Recombination Process. In H. L. Roman, ed. *Annual Review of Genetics*, vol. **4:** Palo Alto, Calif.: Annual Reviews, Inc.

Kornberg, A., 1980. *DNA Replication.* San Francisco: W. H. Freeman and Company.

Lark, K. G., 1969. Initiation and Control of DNA Synthesis. In E. E. Snell, P. D. Boyer, A. Meister, and R. L. Sinsheimer, eds. *Annual Review of Biochemistry*, vol. **38.** Palo Alto, Calif.: Annual Reviews, Inc.

Lehman, I. R., 1974. DNA Ligase: Structure, Mechanism, and Function. *Science,* **186:** 790–797.

Lewin, B., 1974. *Gene Expression*, vol. **1,** *Bacterial Genomes.* New York: John Wiley & Sons.

Lewin, B., 1980. *Gene Expression*, vol. **2,** *Eucaryotic Chromosomes.* New York: John Wiley & Sons.

Lewin, B., 1977. *Gene Expression*, vol. **3,** *Plasmids and Phages.* New York: John Wiley & Sons.

Lewis, B. J., J. W. Abrell, R. G. Smith, and R. C. Gallo, 1974. Human DNA Polymerase III (R-DNA Polymerase): Distinction from DNA Polymerase I and Reverse Transcriptase. *Science,* **183:** 867–869.

Meselson, M. S., and F. W. Stahl, 1958. The Replication of DNA in *Escherichia coli. Proc. Nat. Acad. Sci. (U.S.),* **44:** 671–682.

Ogawa, T., and T. Okazaki, 1980. Discontinuous DNA Replication. *Ann. Rev. Biochem.,* **49:** 421–457. Palo Alto, Calif.: Annual Reviews, Inc.

Okazaki, R., T. Okazaki, K. Sakabe, K. Suzimoto, and A. Sugino, 1968. Mechanism of DNA Chain Growth. I. Possible Discontinuity and Unusual Secondary Structure of Newly Synthesized Chains. *Proc. Natl. Acad. Sci. (U.S.),* **59:** 598–605.

Okazaki, T., and R. Okazaki, 1969. Mechanism of DNA Chain Growth. **IV.** Direction of Synthesis of T4 Short DNA Chains as Revealed by Exonucleolytic Degradation. *Proc. Natl. Acad. Sci. (U.S.),* **64:** 1242–1248.

Ozeki, H., and H. Ikeda, 1968. Transduction Mechanisms. In H. L. Roman, ed., 1968. *Annual Review of Genetics*, vol. **2.** Palo Alto, Calif.: Annual Reviews, Inc.

Sayre, A., 1975. *Rosalind Franklin and DNA.* New York: W. W. Norton & Co., Inc.

Sheinin, R., J. Humbert, and R. E. Pearlman, 1978. Some Aspects of Eukaryotic DNA Replication. In E. E. Snell, P. D. Boyer, A. Meister, and C. C. Richardson, eds. *Annual Review of Biochemistry*, vol. **47.** Palo Alto, Calif.: Annual Reviews, Inc.

Tomasz, A., 1969. Some Aspects of the Competent State in Genetic Transformation. In H. L. Roman, ed. *Annual Review of Genetics*, vol. **3.** Palo Alto, Calif.: Annual Reviews, Inc.

Watson, J. D., 1971. The Regulation of DNA Synthesis in Eukaryotes. In D. M. Prescott, L. Goldstein, and E. McConkey, eds. *Advances in Cell Biology.* New York: Appleton-Century-Crofts.

Watson, J. D., 1976, 3rd ed. *Molecular Biology of the Gene.* New York: W. A. Benjamin.

Watson, J. D., and F. H. C. Crick, 1953. Molecular Structure of Nucleic Acids. A Structure for Deoxyribonucleic Acid. *Nature,* **171:** 737–738. Reprinted in J. H. Taylor, ed., 1965. *Selected Papers on Molecular Genetics.* New York: Academic Press.

Wickner, S. H., 1978. DNA Replication Proteins of *Escherichia coli.* In E. E. Snell, P. D. Boyer, A. Meister, and C. C. Richardson, eds. *Annual Review of Biochemistry*, vol. **47.** Palo Alto, Calif.: Annual Reviews, Inc.

Yunis, J. J., 1977. *Molecular Structure of Human Chromosomes.* New York: Academic Press, Inc.

Zinder, N. D., and J. Lederberg, 1952. Genetic Exchange in *Salmonella. Jour. Bact.,* **64:** 679–699.

324

CHAPTER 16
THE
IDENTIFICATION
OF THE GENETIC
MATERIAL

Problems

16-1 What are some of the genetic phenomena that can be investigated in corn, fruit flies, peas, and human beings that cannot readily be studied in *Neurospora*, bacteria, and viruses?

16-2 If one strand of DNA is found to have the sequence 5′ AACGTACTGC 3′, what is the sequence of nucleotides on the 3′, 5′ strand?

16-3 For a molecule of *n* deoxyribonucleotide pairs, give a mathematical expression that can be used to calculate the number of possible sequences of those nucleotide pairs if only the "usual" bases are present.

16-4 Develop a formula for determining the length in micrometers of a DNA molecule whose number of deoxyribonucleotide pairs is known. (Note: $1 \text{ Å} = 1 \times 10^{-4}$ μm.)

16-5 Phage T2 DNA is estimated to consist of about 200,000 deoxyribonucleotide pairs. What is the length in micrometers of its DNA complement?

16-6 The molecular weight of the DNA of one somatic (i.e., diploid) cell of corn is given as 9×10^{12}. Calculate (a) the number of deoxyribonucleotide pairs, (b) its length in micrometers, and (c) its length in inches. Express answers to (a) and (b) as powers of 10.

16-7 One calculation for the molecular weight of the DNA in human autosome 1 is 0.5 $\times 10^{12}$. (a) Of how many deoxyribonucleotide pairs is this DNA composed? (b) What is its length in micrometers? (c) What is its length in inches? Answer (a) and (b) as powers of 10. Carry calculations to the nearest first decimal place.

16-8 Determine the molecular weight of (a) adenine, (b) thymine, (c) cytosine, (d) guanine.

16-9 What is the molecular weight of each of these deoxyribonucleosides: (a) deoxyadenosine, (b) thymidine, (c) deoxycytidine, (d) deoxyguanosine?

16-10 Calculate the molecular weight of each of the following: (a) deoxyadenylic acid, (b) thymidylic acid, (c) deoxycytidylic acid, (d) deoxyguanylic acid.

16-11 From your answers to the preceding problem, give an average molecular weight for any single deoxyribonucleotide pair. (Consider only the "usual" bases.)

16-12 Based on the answer to the preceding question, and the information given in problem 16-5, what is the molecular weight of phage T2 DNA?

16-13 If the molecular weight of *Escherichia coli* DNA were taken as 2.7×10^9, (a) of how many deoxyribonucleotide pairs would *E. coli* DNA consist, and (b) what would be its length in μm?

16-14 In this chapter the generation time for virulent T2 phage was given as 22 minutes. It appears that there is no replication of DNA for about the first six minutes and that, thereafter, DNA replicates at an exponential rate for five minutes, after which the rate of replication declines to the end of the 22-minute period. In that five-minute period, about 32 phage DNA molecules appear to be formed for each infecting phage particle. At what rate are phage DNA molecules being replicated during this exponential period?

16-15 Based on your answer to the preceding problem and information given in problem 16-5, at what rate are deoxyribonucleotide pairs being synthesized per minute in φT2?

16-16 Assume that you have just determined the adenine content of the DNA of *Bacillus hypotheticus* to be 20 percent. What is the percentage of each of the other bases?

16-17 If you determined the adenine-guanine content of another species to total 25 percent, should you believe it?

16-18 Note the deoxyribonucleotide composition listed in Table 15-1 for φX174. Another report in the literature for this phage gives the following deoxyribonucleotide percentages: A, 27.6; T, 27.8; G, 22.3; C, 22.3. (a) What are the A/T and G/C ratios?

(b) What does your answer to part (a) suggest regarding the structure of the phage DNA whose composition is given in this problem? (c) How would you account for the considerable difference in these ratios from those given in Table 16-1?

16-19 What significance might there be to the fact that *Euglena* chloroplasts contain DNA and that this chloroplast DNA has a markedly different nucleotide composition from nuclear DNA reported for the same organism as given in Table 16-1?

16-20 DNA of a species of yeast, a microscopic fungus, is reported to have a thymidylic acid content of 32.6 percent. The A/T ratio was reported as 0.97. On this basis, what is the percentage composition of deoxyadenylic acid?

16-21 A culture of *Escherichia coli* was grown in a ^{15}N medium until all of the cells' DNA was so labeled. The culture was then transferred to a ^{14}N medium and allowed to grow for exactly two generations, so that all DNA replicated twice in the ^{14}N medium. If the DNA of the latter culture were then subjected to density gradient centrifugation, how much of the DNA should be (a) ^{15}N only, (b) "hybrid," (c) ^{14}N only?

16-22 Differentiate between transformation and transduction.

16-23 Knowing that 25,400 μm = 1 inch, develop a formula for determining the length of a DNA molecule in inches (L_{in}) if (a) the length in μm is known, and (b) if only the number of deoxyribonucleotide pairs is known.

CHAPTER 17

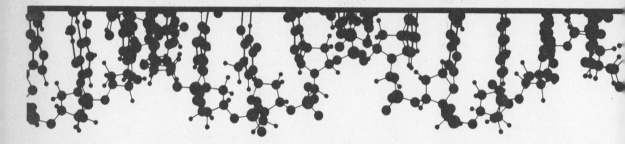

Protein synthesis

With its precise replication, DNA serves to carry genetic information from cell to cell and from generation to generation. The next task will be to determine the way in which this information is translated into phenotype. Virtually all the phenotypic effects examined thus far are results of biochemical reactions that occur in the cell. All of these reactions require enzymes, and enzymes are proteins, either wholly or in part. Other phenotypic effects are due directly to the kinds and amounts of nonenzymatic proteins present, hemoglobin, myoglobin, gamma globulin, insulin, or cytochrome c. Proteins are built up from long-chain linear polymers of amino acid residues (**polypeptide chains**) and are synthesized almost exclusively in the cytoplasm. Because both polypeptide chains and DNA have linear structure, the problem reduces to the question of how DNA, located primarily in nuclear chromosomes, mediates the synthesis of proteins (on ribosomes) in the cytoplasm. In Chapter 3, several cases that appeared to suggest a connection between genes and enzymes were cited. A few additional instances will furnish clear proof of this gene-enzyme relationship.

Genes and enzymes

Phenylalanine metabolism.   The groundwork for a functional relationship between genes and enzymes was laid in 1902, when Bateson reported that a rare human defect, *alkaptonuria*, was inherited as a recessive trait. Then in 1909, the English physician Garrod published a book, *Inborn Errors of Metabolism*, that was far ahead of its time in suggesting a relationship between genes and

with which he dealt at length. This condition, manifested by a darkening of
cartilaginous regions and a proneness to arthritis, results from a failure to break
down alkapton (2,5-dihydrophenylacetic acid, or homogentisic acid). Alkapton
accumulates and is excreted in the urine, which turns black upon exposure to
the air. By the 1920s it was discovered that the blood of alkaptonurics is deficient
in an enzyme, homogentisic acid oxidase, that catalyzes the oxidation of alkapton.

This defect is but one of a group, all related to the body's metabolism of the
essential amino acid phenylalanine. As diagramed in Figure 17-1, human beings
receive their supply of this substance from ingested protein. Once in the body,
phenylalanine may follow any of three paths. It may be (1) incorporated into
cellular proteins, (2) converted to phenylpyruvic acid, or (3) converted to tyrosine, another amino acid.

Persons of genotype *pp* (see Fig. 17-1) fail to produce the enzyme phenylalanine hydroxylase, with the result that phenylalanine accumulates in the blood,
with excesses up to a gram a day being excreted in the urine. Such persons are
said to have *phenylketonuria*, or PKU, which is accompanied by serious mental
and physical retardation. About two-thirds of all PKUs have an I.Q. below 20,
and hence are classed as idiots. However, if symptoms are diagnosed very early
and the infant is put on a diet low in phenylalanine (normal development requires some) for at least the first five years, brain development is more nearly
normal. PKU can be detected at birth; by the end of 1970 forty-three states had
mandatory PKU-screening laws.

Another recessive gene (*t* in Fig. 17-1) blocks conversion of tyrosine to parahydroxyphenylpyruvic acid, probably through failure to produce liver tyrosine
transaminase. This leads to accumulation of tyrosine, excesses of which are
excreted in the urine. This condition is called *tyrosinosis*; no serious symptoms
appear to be associated, although the disorder has been reported for only one
human subject. A somewhat less uncommon, but still rare, disorder, *tyrosinemia*,
is due to inability of persons who are homozygous recessive for another gene (*t'*
in Fig. 17-1) to produce the enzyme parahydroxyphenylpyruvate oxidase. In this
condition large quantities of parahydroxyphenylpyruvic acid, as well as lactic
and acetic acid derivatives, are excreted in the urine. If not treated with low-
tyrosine diets, this condition leads to serious liver problems that culminate in
early death.

Other metabolic blocks associated with faulty utilization of phenylalanine,
and all caused by recessive genes, lead to *albinism* or to *cretinism*. Two kinds of
albinism have been recognized for some time. The majority of albinos are unable
to produce tyrosinase (these are *aa* persons in Fig. 16-1), which catalyzes the
conversion of tyrosine to DOPA: hence they do not produce melanin. Some albino
persons (*a'a'* in Fig. 17-1), however, can produce DOPA but cannot utilize tyrosinase in oxidizing DOPA to DOPA quinone, thus breaking the melanin production process at a later point. Goitrous cretinism is also of several types, each
involving a particular block in the conversion of tyrosine to thyroxine. Cretinism
is accompanied by a considerable degree of physical and mental retardation in
addition to thyroid defects.

In each case described for phenylalanine metabolism defects, a particular
(recessive) gene appears to be associated with nonproduction of a specific enzyme that catalyzes a particular biochemical reaction. Occurrence or nonoc-
currence of the reaction then determines a related phenotypic effect. Heterozygotes
do not display the disorder; one "dose" of the normal gene results in enough
enzyme to permit the reaction to proceed. Because serious inherited metabolic
defects such as PKU are (in the absence of highly uncommon mutation) the
result of marriage between heterozygotes, it is very desirable to be able to detect

[1]See also Harris (1975) and Knudson (1969).

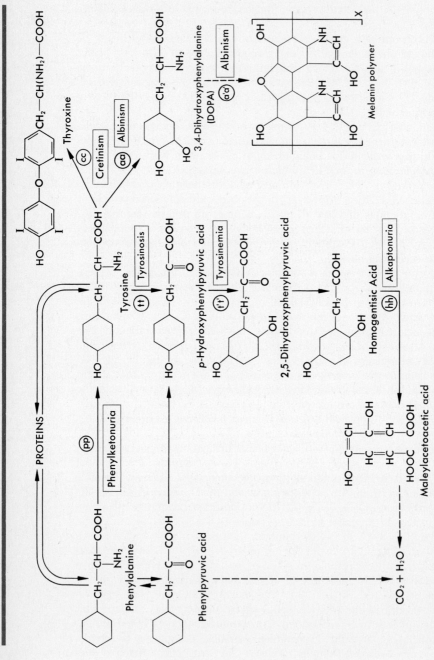

Figure 17-1. Metabolic pathways of the amino acid phenylalanine in humans. Different genotypes cause blocks at various points because of the failure to produce a given enzyme, resulting in such inherited metabolic disorders as phenylketonuria (PKU), tyrosinosis, tyrosinemia, cretinism or albinism.

heterozygosity with certainty in phenotypically normal persons. Considerable progress is being made in this direction, and persons heterozygous for PKU can now be detected by the phenylalanine test. Levels of phenylalanine rise higher and are maintained longer in *Pp* individuals than in homozygotes after administration of a standard dose of this amino acid. Detection of heterozygosity is now possible in more than 100 inherited disorders, although not all tests are completely reliable. Some of the disorders for which carrier tests are available include Tay-Sachs, sickle-cell trait, PKU, galactosemia, cystic fibrosis, glucose-6-phosphate dehydrogenase (G6PD) deficiency, hemophilia A, Huntington's chorea, and Duchenne muscular dystrophy. Depending on the nature of the defect, heterozygosity may be determined even prenatally by amniocentesis. O'Brien et al., for example, reported success in diagnosing the Tay-Sachs disorder by amniocentesis during the sixteenth to twenty-eighth week of pregnancy by testing the sloughed fetal cells present in the amniotic fluid for enzyme level. In this way they were able, in 15 different pregnancies, to detect normal homozygous embryos in seven cases, two heterozygotes, and six with the Tay-Sachs condition. Five of the latter were therapeutically aborted; tests of various organs of these five showed hexosaminidase A to be absent. The remaining one was born and was developing marked symptoms of the Tay-Sachs disorder by the age of nine months.

Neurospora.   Such studies of metabolic defects in humans, important as they are, do not provide a set of experimental conditions for adequate testing of the gene-enzyme hypothesis. In fact, most other studies up to the 1940s had concerned themselves largely with morphological characters, many of which are based on such complex biochemical reactions as to impede analysis, and use genetically inefficient subjects such as humans. The real breakthrough was furnished by Beadle and Tatum in 1941 and in a series of later reports by them and their colleagues. It occurred to Beadle that an ascomycete fungus, the orange-pink bread mold (*Neurospora crassa*), was an ideal organism for this type of investigation. *Neurospora* is easily grown in the laboratory, has a simple life history (Appendix B), in which only the zygote is diploid (hence dominance is not a problem), and it has many easily detected physiological variants. Their aim was to induce metabolic deficiencies by X-irradiation, to identify them by comparing growing ability on minimal and various deficiency media, and so to determine the specific metabolic block. One mutant, for example, was found to be unable to synthesize vitamin B_1 (thiamine). Further tests showed this strain was able to synthesize the pyrimidine half of the thiamine molecule, but not the thiazole portion; the critical chemical difference was one of enzyme-producing ability. So the concept of a one-gene–one-enzyme–one-phenotypic-effect relationship developed.

Protein structure

As indicated earlier, all enzymes are protein, at least in part, and the catalytic property of an enzyme is conferred by its protein makeup, with or without added cofactors. Proteins are large, heavy, generally complex molecules of great variety and biological significance. Their basic structural unit is the amino acid, numbers of which are linked together to form polypeptide chains. Only 20 amino acids are biologically important, although others are necessary in specific instances. Each may be represented by the general structural formula

$$H-\underset{\underset{\textstyle \textcircled{R}}{|}}{\overset{\overset{\textstyle H}{|}}{N}}-\underset{\underset{\textstyle }{}}{\overset{\overset{\textstyle H}{|}}{C}}-\overset{\overset{\textstyle O}{\|}}{C}-OH$$

where Ⓡ represents a side chain. Differences in this side chain determine the specific kind of amino acid. Thus in glycine Ⓡ is merely an H atom:

$$H-N-C-C-OH$$

and in alanine Ⓡ is only a little more complex:

$$H-N-C-C-OH$$

whereas the addition of a benzene ring to the side chain of alanine produces phenylalanine:

$$H-N-C-C-OH$$

and so on. The 20 biologically important amino acids are listed in Appendix C. Amino acids are linked by the **peptide bond** with the loss of the equivalent of a molecule of water. The formation of a dipeptide by condensation of two molecules of alanine may be represented thus:

$$H-N-C-C-(OH + H)-N-C-C-OH \rightarrow$$

$$H-N-C-C-N-C-C-OH + H_2O$$

in which CONH is the peptide bond. A **polypeptide** is thus a series of **amino acid residues,** joined by peptide bonds with an amino end (NH_2) and a carboxyl end (COOH).

Protein structural characteristics, then, are based on the number and kind of amino acid residues and therefore can occur in infinite variety. In these residues, the side chains confer a number of properties. Some are large and complex; others are small and simple. Some carry a positive charge (lysine, arginine), others a negative charge (glutamic acid, aspartic acid), but most carry none. Differences in bulk and electrical charge along a polypeptide, plus an intricate folding of the protein molecule, impart much of the specificity of enzyme-substrate and antigen-antibody relations.

Polypeptide chain synthesis—the components

THE NUCLEIC ACIDS

Because of its infinite variety of possible arrangements of groups of the four nucleotides, DNA is admirably suited for carrying the information needed to

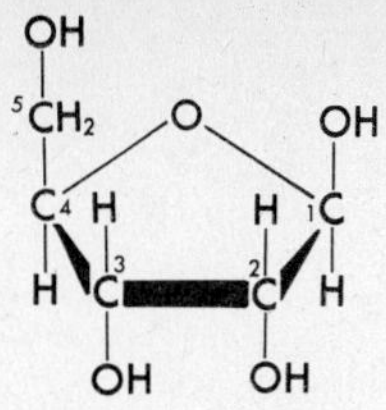

*Figure 17-2. Molecular structure of **ribose,** the sugar of ribo-nucleic acid. Compare the number 2 carbon of ribose with the number 2 carbon of deoxyribose.*

direct the synthesis of an almost unlimited number of different proteins. But DNA, except for small amounts in chloroplasts and mitochondria, is located in the eukaryote nucleus, whereas protein synthesis occurs almost entirely in the cytoplasm. Moreover, DNA is not degraded or "used up" in performing its function, and it differs from the end product for which it is responsible. It may be thought of as directing, through some intermediary substance or substances, protein synthesis out in the cytoplasm. In this sense we may refer to DNA as being *conserved.* The answer to the problem of how chromosomally located, conserved DNA, which carries the genetic information, mediates the synthesis of proteins in the cytoplasm is provided by several kinds of a second nucleic acid, ribonucleic acid (RNA).

RIBONUCLEIC ACID

RNA structure. Ribonucleic acid (RNA) differs from DNA in several important ways. RNA has the following characteristics:

1. Molecules of RNA are single-stranded, linear polymers of mononucleotides, although certain types assume, in part, a secondary double-helix configuration through complementary base pairing.
2. The sugar of the sugar-phosphate strand is **ribose** (Fig. 17-2).
3. Successive **ribonucleotides** are joined by a phosphodiester bond that links the 3′ position of one ribonucleotide to the 5′ position of the next.
4. Nucleotide bases are the purines **adenine** (A) and **guanine** (G), and the pyrimidines **cytosine** (C) and **uracil** (U) (Fig. 17-3); the last replaces the thymine of DNA. However, "unusual" bases occur in certain kinds of RNA; these are chiefly methylated derivatives of A, U, C, and G.

Ribonucleosides and ribonucleotides are named as shown in Table 17-1 (compare with Table 16-2). RNA is synthesized in the nucleus of eukaryotes, using one of the strands of DNA as a template. Therefore the sequence of ribonucleotides should be complementary to one of the DNA strands; that it is, except for the occurrence of uracil instead of thymine, and the modification of some of the usual bases after RNA synthesis. The basic function of RNA is the carrying of the genetic message of DNA out into the cytoplasm, where it is responsible for

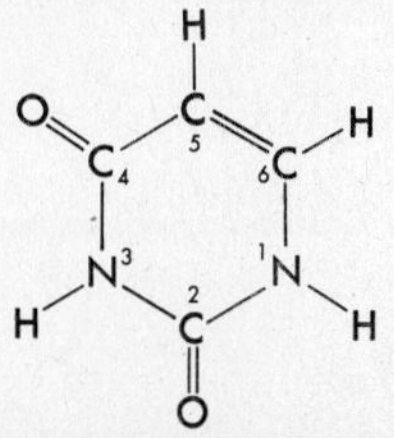

Figure 17-3. Molecular structure of the pyrimidine uracil that replaces thymine in RNA.

building polypeptide chains—the first step in protein synthesis. Three kinds of RNA, which differ in structure and function, occur: **ribosomal, messenger, and transfer.**

Table 17-1. Ribonucleosides and ribonucleotides

Base	Ribonucleoside	Ribonucleotide
Adenine	Adenosine	Adenylic acid
Uracil	Uridine	Uridylic acid
Cytosine	Cytidine	Cytidylic acid
Guanine	Guanosine	Guanylic acid

Ribosomes and ribosomal RNA (rRNA). The site of polypeptide synthesis is the **ribosome,** a small particle that averages approximately 175×225 Angstrom units, composed of protein and **ribosomal RNA.** Ribosomes are asymmetric aggregates of two subunits that can be distinguished by the rates at which they sediment when centrifuged in an appropriate medium. The rate of sedimentation per unit centrifugal field is referred to as the sedimentation coefficient (s); for most proteins s has value of between 1×10^{-13} and 2×10^{-11} second. An s value of 1×10^{-13} is denoted as one **Svedberg unit (S).** The useful point here is that s (and therefore S) for any molecule or small particle such as a ribosome is determined, for any given temperature and solvent, by its shape, mass, and degree of hydration. The larger the S value, the larger the particle, though the relation is not linear. Ribosomes of bacteria have a sedimentation coefficient of 70S, and those of eukaryotes, 80S. Each of these is composed of two unequal subunits, 50S and 30S in bacteria and 60S and 40S in eukaryotes. As suggested by these figures, sedimentation velocities are not additive.

Ribosome structure and composition are becoming better understood, although many details remain to be elucidated. Ribosomal characteristics are best known in *Escherichia coli*. In this organism the 30S subunit is composed of 21 different proteins, each represented by one molecule. The amino acid sequence has been determined for some of these. They range from 74 to 203 amino acid residues; their molecular weights lie in the range of 10,000 to about 65,000. A single 16S molecule of ribosomal RNA (rRNA), which consists of 1,600 ribonucleotides, and whose sequence has been partially determined, is also present.

The 50S ribosomal subunit of *E. coli* consists of 34 proteins; those that have been sequenced are made up of 46 to 178 amino acid residues each, with molecular weights ranging from about 9,000 to 28,000 daltons. The 5S rRNA molecule of some 120 ribonucleotides has been completely sequenced; in *Bacillus subtilis* it is cleaved from a precursor molecule of 179 nucleotides. Sequencing of the 23S rRNA of ~3,200 ribonucleotides, however, is incomplete. Some of these facts are summarized in Table 17-2.

Table 17-2. 70S ribosome composition in Escherichia coli

Ribosomal subunit	Proteins	rRNA		
		Sedimentation coefficient	Molecular weight	Ribonucleotides
30S	21	16S	5.5×10^5	1,600
50S	34	5S	4×10^4	120
		23S	1.1×10^6	~3,200

Through an ingenious technique it has now become possible to locate many of the proteins in the ribosome and to form a good three-dimensional concept of this organelle, at least in prokaryotes. Antibodies have been prepared for each of the 30S and 50S proteins, then mixed with the appropriate subunit, and finally prepared for electron microscopy. Locations of antibodies so visualized correspond to the positions of the several proteins of the ribosome (Fig. 17-4). Ribosomes can be degraded into their components in the laboratory, then reassembled. Reconstruction into a functional ribosome requires addition of the proteins of each subunit in a particular sequence, which indicates the importance of the ribosomal proteins to the structure of those organelles.

Ribosomal proteins are increasingly different in structure as taxonomic affinity decreases. It has been found, for example, that strong immunological cross-reactions occur between ribosomal proteins of taxonomically closely related bacteria. But, as might be expected, there is essentially no structural homology between ribosomal proteins of bacteria and mammals.

We know much less about structural relationships between proteins and subunits of ribosomes in eukaryotes. Considerable differences have been reported which, in the light of the great diversity of higher plants and animals, is hardly surprising. A few data for mammalian rRNA are given in Table 17-3.

Table 17-3. Mammalian rRNA

	rRNA		
Ribosomal subunit	Sedimentation coefficient	Molecular weight	Ribonucleotides
40S	18S	7×10^5	~2,100
60S	5S	4×10^4	120
	28S	1.8×10^6	~5,300

Ribosomes that function in polypeptide synthesis occur in linear groups connected by messenger RNA; such groups of ribosomes are called **polysomes.** In *E. coli* and in other prokaryotes the ribosomes are scattered through the cytoplasm, but in immature erythrocytes (reticulocytes) and in other eukaryotic cells they are situated on the endoplasmic reticulum.

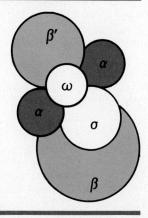

Figure 17-4. *Diagrammatic representation of* Escherichia coli *RNA polymerase.* Sigma (σ) *is required for recognition of the starting point on DNA for transcription.* Psi (ψ) *speeds up transcription of what will become the functional RNA molecule;* rho (ρ) *recognizes the termination site on the DNA template. See text for details.*

RNA of all kinds is *transcribed* from a template strand of DNA, catalyzed by a variety of enzymes collectively referred to as **DNA-dependent RNA polymerases.** As noted earlier, the nucleotide sequence of RNA is complementary to that of the template DNA strand, except that uracil replaces thymine. Why this replacement has occurred is unknown. In transcription one strand of DNA serves as the template for transcription for some RNA molecules, whereas the *other* strand is the template for other RNAs. In brief, precursor RNA is (1) *synthesized* in the process of **transcription,** then (2) *processed* by (a) removal and/or addition of numbers of nucleotides, and (b) modified by methylation of certain of the remaining nucleotides. Secondary structure is attained through some complementary ribonucleotide pairing. In many bacterial cells the DNA sequences that specify rRNA are arranged in long transcriptional sequences in tandem. Some 0.1 to 0.2 percent of the bacterial genome is estimated to be responsible for transcription of rRNA; this would provide as many as 10 copies of the gene(s) for ribosomal RNA. Corresponding numbers for higher organisms are, understandably, greater; for example, 260 in *Drosophila melanogaster*, 1,250 in HeLa cells, and 27,500 in Chinese cabbage!

Ribosomal RNA in *eukaryotes* is synthesized from DNA in the nucleolar organizing region. No primer is needed. The major portion of nuclear RNA is to be found in the nucleolus, which led to early speculation that RNA storage occurred in that body. However, the nucleolar organizing region contains deoxyribonucleotide sequences completely complementary to the ribonucleotide sequence of cytoplasmic rRNA. Based on this evidence, it is now generally accepted that the nucleolar organizing region is the site of rRNA synthesis. Moreover, each nucleolar organizing region includes many identical, repeated DNA sequences responsible for rRNA synthesis.

Ribosomal RNA is nonspecific for the amino acids of polypeptide chains, which are synthesized on polysomes. Three lines of evidence indicate that rRNA is not specific for the amino acid sequences of polypeptide chains: (1) there is no correlation between the ribonucleotide sequence of rRNA and the type of protein synthesized—that is, although different proteins may be produced characteristically in different cells and in different organisms, there is no significant difference in the rRNA of these cells or organisms; (2) although phage infection results in synthesis of new kinds of protein, ribosomes and rRNA in the bacterial cell remain unchanged; (3) rRNA can be shown to be complementary to only very small sections of DNA ($<$ 1 percent), which strongly suggests that rRNA is too limited in size and variety to carry the information needed for many different proteins.

Prokaryotic RNA polymerase.   Prokaryotic RNA polymerase is composed of five core polypeptide chains: two alpha (α) chains, and one each of beta (β), beta prime (β'), and omega (ω). In addition, three accessory chains, sigma (σ), rho (ρ), and psi (ψ), are required for specific functions in transcription (Fig. 17-4). The five-chain core can synthesize RNA in the presence of a DNA template, the four ribonucleotide triphosphates, and available magnesium ions. Sigma is required for recognition by the enzyme of the starting point on the template. In the absence of σ, transcription is random and may start anywhere. As soon as sigma "finds" the starting point (*promoter*), it dissociates from the enzyme molecule and is available for use again. The five-chain rho "recognizes" and binds to a termination site on the DNA template, where transcription ceases, and the RNA molecule diffuses away. Psi causes a speed-up in transcribing the portion that will be the functional RNA molecule. Although the promoter includes nucleotides that are recognized by RNA polymerase, the enzyme binds only loosely to it. Actual transcription begins only after the enzyme has traveled further in the $5' \rightarrow 3'$ direction until the DNA "start transcription" signal is reached. There the enzyme binds more closely and actual transcription gets underway.

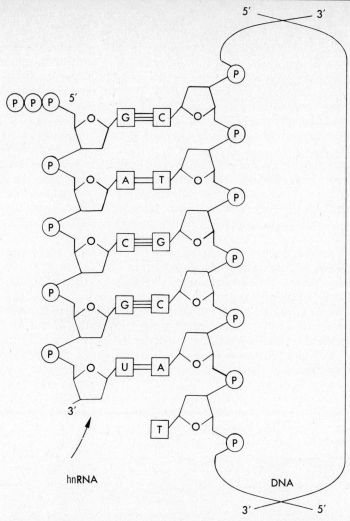

Figure 17-5. *Transcription of* hnRNA *from the template strand of DNA showing that it is antiparallel and complementary to the template DNA strand.* hnRNA *is synthesized in the 5′ → 3′ direction by addition of successive ribonucleotides at the 3′ end of the growing* mRNA *molecule.*

Eukaryotic RNA polymerases. Unlike prokaryotes, eukaryotes produce three different RNA polymerases, each with a distinctive function: *type I* transcribes ribosomal RNA; *type II* catalyzes synthesis of messenger RNA; and *type III* is responsible for transcription of transfer RNA. Because the three kinds of polymerase are distinguished from each other by their degree of inhibition by **α-amanitin** (a toxin derived from a poisonous mushroom of the genus *Amanita*) as well as the kind of necessary ions, it is not difficult to study the role of each separately.

Messenger RNA (mRNA). Specificity for an amino acid sequence is conferred by **messenger RNA** (mRNA), as shown, for example, by the fact that phage mRNA utilizes ribosomes made before infection to bring about synthesis of phage proteins. This ribonucleic acid directs assembly of various amino acids from the intracellular pool at the ribosome surface, where enzymes effect the peptide linkage, and form polypeptide chains.

Transcription of *eukaryotic* mRNA begins with the synthesis of long precursor molecules by RNA polymerase II from the template strand of DNA. This enzyme functions by catalyzing formation of $3' \rightarrow 5'$ phosphodiester bonds of the RNA "backbone" by "reading" the template in the $5' \rightarrow 3'$ direction. The developing mRNA molecule is antiparallel to, and its nucleotides complementary to, those of the DNA template strand (Fig. 17-5). Messenger RNA chain growth is rapid; reports from several laboratories range from 15 to 100 nucleotides per second in vitro.

The immediate product of transcriptions in *eukaryotes*, however, is a molecule of many more ribonucleotides than will make up the ultimate, functional mRNA. This precursor in eukaryotes is generally referred to as **heterogeneous nuclear RNA (hnRNA),** although Watson prefers a somewhat more logical *pre-mRNA*. Molecules of hnRNA are estimated to range from about 500 to 30,000 nucleotides or, by Watson, up to 50,000, nucleotides long. They are very rapidly degraded into much smaller molecules, with the portion toward the 3' end conserved (Fig. 17-6). Watson estimates that only 10 percent of hnRNA eventually reaches the cytoplasm as mRNA. The discarded portion of each "species" of hnRNA contains many unique nucleotide sequences, as well as repeated sequences of several hundred nucleotides that occur randomly both in the 5' terminal region of the precursor and internally. Some of the repeated sequences, in the conserved as well as in the discarded portion, are made up of *palindromes*, inverted repeat sequences in which the same, or nearly the same, series of nucleotides run in opposite directions on the two strands of DNA and which produce hydrogen-

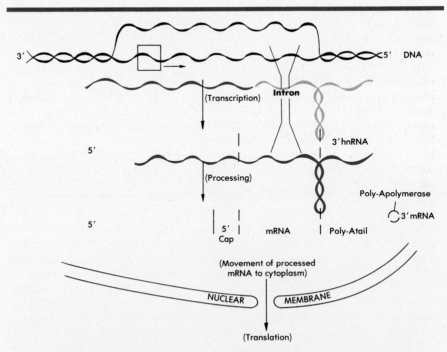

Figure 17-6. *Transcription and processing of heterogeneous nuclear RNA (hnRNA). The square on one DNA strand represents DNA-dependent RNA polymerase. Processing consists of degradation of some of the 5' end of the hnRNA, addition of a "poly-A tail" at the 3' end, and addition of a 5' terminal cap. Following processing the resulting mRNA moves through pores in the nuclear membrane out into the cytoplasm.*

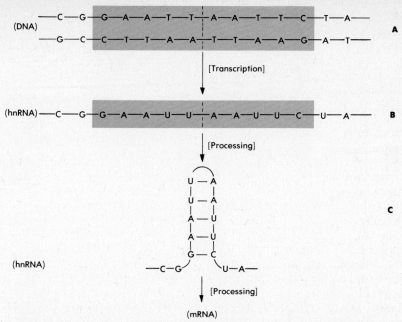

Figure 17-7. *Details of processing of a portion of a hypothetical* hnRNA *molecule showing how palindromes* (boxed in segment) *produce hydrogen-bonded loops in RNA.*

bonded "hairpins" or loops in RNA (Fig. 17-7). These, of course, profoundly affect the three-dimensional shape of the nucleic acid and therefore how and how well it functions. Usefulness of the discarded portions of hnRNA is not understood.

Following its transcription in eukaryotes, hnRNA is *processed* into functional mRNA in three principal steps. Many genes include segments, often very long, which are transcribed as part of hnRNA, but excised by hnRNA processing nucleases and absent from functional mRNA. These intervening segments are called **introns.** Their occurrence requires, of course, that the remaining segments of RNA be spliced together to form functional mRNA molecules. The DNA segments that are transcribed and also present in the final mRNA product are known as **exons.** For example, the β polypeptide chain of human hemoglobin consists of 146 amino acid residues. In the DNA that is responsible for the ultimate synthesis of this polypeptide there are two introns, the larger of which consists of about 900 nucleotide pairs.

Another step involves enzymatic addition of a polyadenylic segment of ~200 nucleotides, the **"poly-A tail,"** to the 3′ end of the molecule. This occurs in the eukaryotic nucleus. A poly-A tail seems to be present on all eukaryotic mRNAs except those responsible for histone synthesis. Similarly, the 3′ terminal poly-A segments have been found in at least some bacterial mRNAs. Poly-A addition appears to be an essential part of post-transcription processing, inasmuch as chemicals that inhibit poly-A synthesis also prevent appearance of mRNA in the cytoplasm. Although the poly-A tail is not required for actual polypeptide synthesis, progressive loss of nucleotides from the poly-A segment ultimately reaches a critical minimum for mRNA stability. Furthermore, length of the poly-A tail differs among different species.

Another step in processing consists of methylation of two or three 5′ terminal nucleotides. To these is added a 5′ terminal "cap" of 7-methylguanosine (7-MG) to the 5′ hydroxyl of the end methylated nucleoside (Fig. 17-8). It appears that

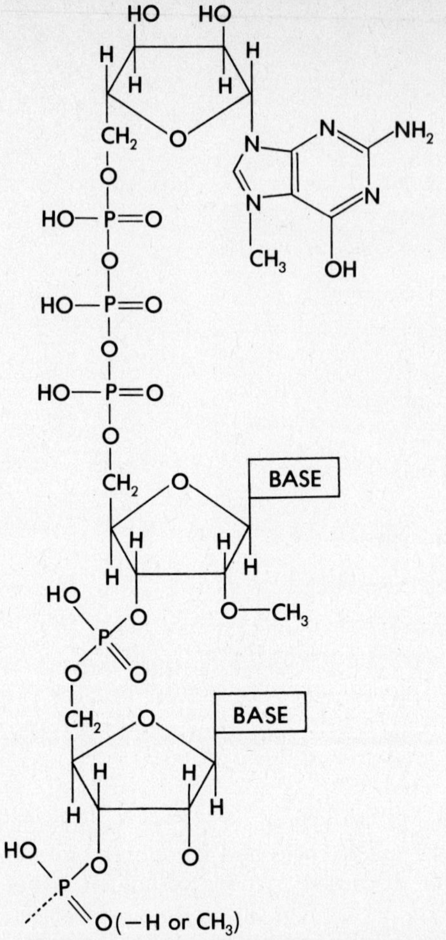

Figure 17-8. *The 5' terminal "cap" of 7-methylguanosine (7-MG), attached by a 5' - 5' triphosphate linkage between the 5' hydroxyl group of 7-MG and the 5' hydroxyl of a methylated nucleoside.*

this cap is required in eukaryotes for formation of a mRNA-ribosome complex and its subsequent operation in polypeptide synthesis. The processing of eukaryotic hnRNA is diagrammed in Figure 17-6.

The number of ribonucleotides and, therefore, the length of any mRNA molecule, depend in large part on the particular polypeptide for whose synthesis that mRNA is responsible. In general, lengths of several hundred to several thousand ribonucleotides have been reported. In bacteria mRNA is produced near, or even at, its functional length. In *E. coli* average mRNA molecular weights fall in the range 300,000 to 500,000. Taking 337 as the mean molecular weight of a ribonucleotide, this would mean an average of 300,000/337 to 500,000/337, or about 900 to 1,500 ribonucleotides per mRNA molecule if the four "usual" bases are equally represented. Of course, the latter assumption may well not be correct in many cases, but this calculation does provide a rough measure of size. Messenger RNA molecules responsible for the α and β polypeptide chains of human hemoglobin consist of more than 400 ribonucleotides each. The matter of size as it relates to specific function is explored in Chapters 18 and 19.

Messenger RNA of *E. coli* is quite short-lived, and functions for only a few minutes. It appears to retain its stability only as long as it is attached to polysomes, with the result that bacterial cells do not become "cluttered" with large amounts of mRNA. For example, the antibiotic actinomycin-D blocks the synthesis of new mRNA in bacteria; studies using actinomycin-treated bacteria in-

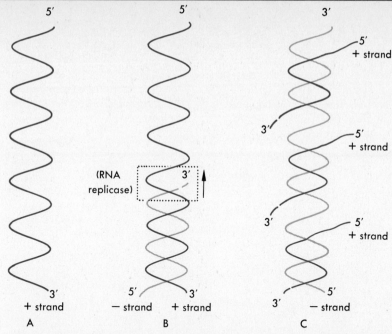

Figure 17-9. *Diagrammatic representation of formation of replicative form and replicative intermediates in the phage QB. Any base pairing is ignored. See text for details. (A) Single stranded plus RNA strand of the phage. (B) Beginning of the formation of the replicative form (RF); RNA replicase has attached at the 3' end of the plus strand and has proceeded a short distance in the 3' → 5' direction along the plus strand to begin synthesis of an anti-parallel, complementary minus strand (replicative form). (C) Synthesis of new plus strands from the minus RNA template, forming a replicative intermediate. Enzyme attachment is not shown.*

dicate that bacterial mRNA is usable only some 10 to 20 times. Therefore if different "species" of mRNA can be selectively produced (see Chap. 20), the cell is able to produce a variety of proteins at different times and under different conditions.

Recent studies on stability of mRNA in eukaryotes indicate somewhat longer functional periods than was earlier suspected. Actinomycin may inhibit initiation of polypeptide synthesis; to obviate this possibility, measurements of decay of uridine-labeled mRNA in HeLa and other eukaryotic cells were made by many investigators. Half-lives of several hours to two or three days were found, a time span quite a bit greater than was determined earlier with actinomycin. The mRNA associated with hemoglobin production in reticulocytes and mature erythrocytes persists and functions for some days after degeneration of the nucleus. In most cases, mRNA is continuously produced, utilized, and degraded.

In those viruses whose genetic material is RNA instead of DNA—for example, tobacco mosaic virus (TMV), Rous sarcoma virus (RSV), polio virus, and phage Qβ—there can be no transcription of mRNA from a DNA template. The RNA viruses as a group behave in a variety of ways; although research in these viruses is being actively pursued, not all details are yet clear. One of the best understood is phage Qβ. The genome of this virus is single-stranded RNA, which has a molecular weight of about 1×10^6. Shortly after infection of susceptible bacterial cells, the viral RNA directs the production of (1) a coat protein, (2) a maturation protein, and (3) an RNA replicase subunit. The latter combines with the three host proteins to form functional *RNA replicase*. This enzyme attaches to the 3' end of the virus RNA molecule (referred to as the + strand), which it transcribes in the $3' \rightarrow 5'$ direction to produce antiparallel polyribonucleotide molecules (the minus strands). Some experimental data suggest that this **replicative form (RF)** of RNA is a double helix through extensive base pairing, but other work has been interpreted as indicating that extensive base pairing does not occur. In any event, the minus strands thus produced then serve as templates for synthesis of several new plus RNA strands (Fig. 17-9). The complex of developing viral plus strands and their complementary RNA templates are called **replicative**

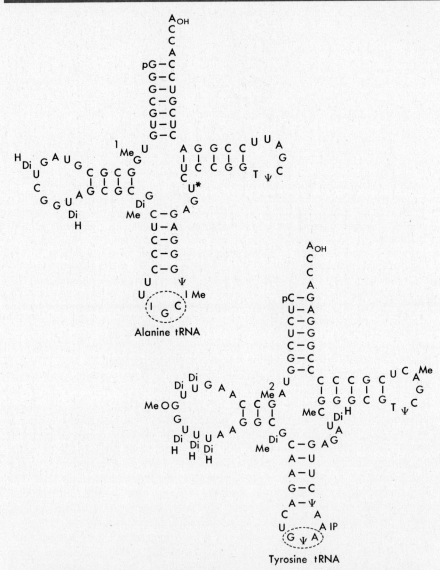

Figure 17-10. "Cloverleaf" model structure for yeast alanine tRNA and tyrosine tRNA. Note that internal pairing is far from complete, and a terminal—A—C—C—A occurs in each molecule. A large number of "unusual" bases, such as ψ, pseudouridine, occurs. The presumed anticodon (see text) is enclosed by the dashed oval. (Courtesy Dr. J. T. Madison, USDA Agricultural Research Service, Ithaca, N.Y. Reprinted, with author's revisions, from Science **153**: 531–534, by permission. Copyright 1966 by the American Association for the Advancement of Science.)

intermediates (RI). In this type of virus, then, the single-stranded RNA genome
(+) serves as a template for transcription of antiparallel, complementary minus
strands, which, in turn, serve as templates for production of numerous new +
strands. Transcription is in the $3' \rightarrow 5'$ direction of the template, and the new
strands are synthesized in their $5' \rightarrow 3'$ direction.

In the *reoviruses*, which are found in the digestive and respiratory tracts of
healthy persons, the genome consists of several separate bits of *double-stranded*
RNA. Each segment of this RNA is transcribed in infected cells into several
"species" of single-stranded mRNA through the action of a viral RNA-dependent
RNA polymerase. The mRNA molecules function directly in synthesis of viral
polypeptides. In this process the double-stranded viral RNA is conserved, just as
is DNA of bacteria and eukaryotes during transcription.

Perhaps the most interesting variation of RNA transcription occurs in a group
of tumor-producing agents, the *leukoviruses*. Rous sarcoma virus particles contain RNA-dependent DNA polymerase, or **reverse transcriptase.** In infected
cells, this enzyme catalyzes the transcription of *DNA* from the viral RNA genome,
which it uses as a template. DNA so synthesized is then transcribed into viral
RNA; moreover, such DNA *may also be incorporated into the host cells' DNA.*
This, of course, alters the genotype of the host cell and is undoubtedly related
to the ability of these viruses to produce tumors.

Transfer RNA (tRNA). The structure of **transfer RNA** is now well understood. This line of research was opened in 1965 when, after seven years of
intensive work, Holley and his colleagues reported the complete nucleotide sequence of alanine tRNA of yeast.[2] Nucleotide sequences are now known for more
than 100 of the different "species" of tRNA. Literature on this molecule has now
become quite voluminous and new papers are appearing frequently.

Transfer RNA has several unique characteristics: (1) It is a relatively small
molecule of 75 to 90 ribonucleotides[3] and is thus much smaller than either mRNA
or any of the rRNAs; (2) a number of "unusual" nucleotides are found in tRNA,
many of them methylated derivatives of the common ones (e.g., 1-methylgua-
nylic acid, 2N-methylguanylic acid, 1-methyladenylic acid, ribothymidylic acid,
5-methylcytosine), whose importance lies in the fact that the presence of methyl
groups prevents formation of complementary base pairs and thus affects the
three-dimensional form of the molecule; (3) hydrogen bonding between usual
bases, and its lack between many of the unusual ones, also affects the shape of
the tRNA molecule. Research from the late 1960s and early 1970s makes it clear
that tRNA molecules may have either three or four lobes, depending on the kind,
number, and sequence of nucleotides. This gives them a so-called cloverleaf
shape. Two dimensional models are shown in Figure 17-10, and a generalized
diagram in Figure 17-11. X-ray diffraction studies are now providing insight into
the three-dimensional structure of tRNA (Fig. 17-12).

There are several common characteristics of the cloverleaf configuration: (1)
the $3'$ end carries a terminal -C-C-A sequence to which an amino acid is cova-
lently attached during polypeptide synthesis; (2) the "stem" or amino acid helix
consists of seven paired bases; (3) the TψC stem is composed of five base pairs;
the last (i.e., nearest the T ψ C loop) is C-G; (4) the anticodon stem includes five
paired bases; (5) the anticodon loop consists of seven bases, the third, fourth,
and fifth of which (from the $3'$ end of the molecule) probably constitute the
anticodon, which permits temporary complementary pairing with three bases
on mRNA; (6) the base on the $3'$ side of the anticodon is a purine: (7) immediately adjacent to the $5'$ side of the anticodon, there occurs uracil and another
pyrimidine; (8) a purine, often dimethylguanylic acid, is located in the "corner"

---

[2]Holley received a Nobel Prize in 1968 for this work.

[3]For example, yeast alanine tRNA, 77 nucleotides; yeast tyrosine tRNA, 78; yeast serine tRNA, 85; *E.
coli* leucine $tRNA_1$, 88.

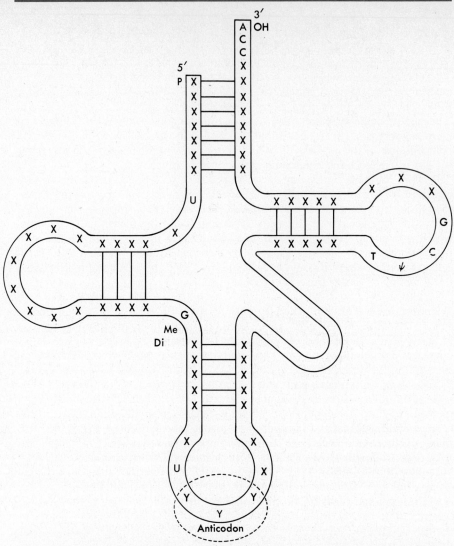

Figure 17-11. *Generalized two-dimensional cloverleaf model of* tRNA, *based on analyses of several yeast* tRNA *molecules by various investigators. The left lobe includes 8 to 12 unpaired bases, the lower lobe 7 (of which the three central ones are believed to be the anticodon), the upper right lobe also 7, and the lower right several unpaired bases that differ in number from one species of* tRNA *to another, or may even be entirely absent. Note the common 3′ terminal -CCA, the TψCG in the bottom of the upper right lobe, the "anticodon-U" (here YYYU) in the lower lobe, the DiMeG between the lower and left lobes, and the U as the first unpaired base on the 5′ strand. As diagramed here, the anticodon is read from right to left (3′ to 5′). The number of paired bases in each lobe shown here appears to be constant for all* tRNA *species so far reported. (A = adenosine, C = cytidine, G = guanosine, T = ribothymidine, U = uridine, ψ = pseudouridine, DiMeG = dimethylguanosine, Y = any base of the anticodon.)*

between the anticodon stem and the DHU stem; and (9) the DHU stem is composed of three or four base pairs (depending on the "species" of tRNA). The extra arm (shown extending toward the lower right in Fig. 17-11) is variable in nucleotide composition, and is lacking entirely in some tRNAs.

Transfer RNA is transcribed from several particular sites on template DNA

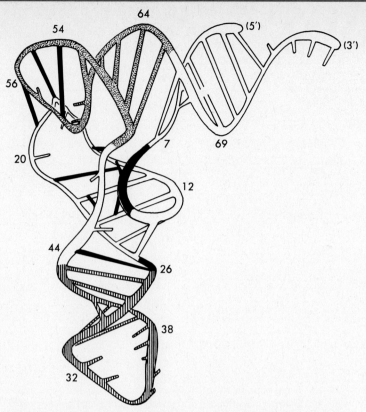

Figure 17-12. Schematic three-dimensional model of yeast phenylalanine tRNA molecule. The ribose-phosphate "backbone" is shown as a continuous cylinder with bars indicating hydrogen bonded base pairs. Unpaired bases are indicated by shorter "branches." The TψC arm is shown in heavy strippling; the anticodon arm is indicated by vertical lines. Black portions represent tertiary interactions. (Redrawn from S. H. Kim, F. L. Suddath, G. J. Quigley, A. McPherson, J. L. Sussman, A. H. J. Wang, N. C. Seeman, and A. Rich, 1974. "Three-Dimensional Tertiary Structure of Yeast Phenylalanine Transfer RNA." *Science,* **185:** 435–440. Copyright 1974 by the American Association for the Advancement of Science. By permission of Dr. S. H. Kim and the American Association for the Advancement of Science.)

and comprises about 15 percent of the RNA present at any one time in an *E. coli* cell.[4]

Transfer RNAs are processed from larger precursors. In prokaryotes, processing includes clipping of nucleotides from the precursor and modification of some internal ones. To simplify, in *E. coli* the precursor for the amino acid tyrosine (pre-tRNA^{tyr}) consists of 128 bases; in processing to tRNA^{tyr} 41 nucleotides are clipped from the 5′ end and two more at the 3′ terminal. In addition to clipping, methylation and inclusion of other "unusual" intercalary bases take place after transcription, as does the addition of the 3′ terminal -C-C-A. The final, functional molecule has 85 bases. In general terms, processing of tRNA precursors involves several major steps, among them: (1) clipping of "extra" nucleotides, (2) methylation and other modifications of many intercalary bases, (3) addition of the 3′ terminal -C-C-A, and (4) assumption of the final three-dimensional shape through

[4]Other approximate quantities are rRNA, 80 percent; mRNA, 5 percent.

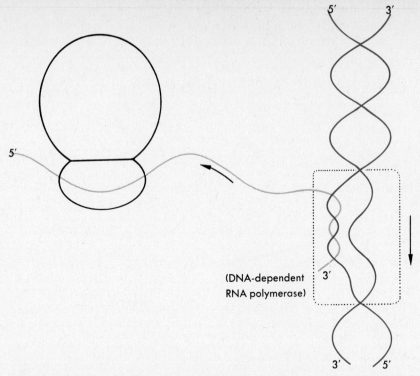

Figure 17-13. *Transcription of* mRNA *from* DNA *template. The enzyme* DNA-*dependent RNA polymerase attaches to* DNA *at a binding site, opening up a short section of the* DNA *double helix by breaking of the hydrogen bonds between deoxyribonucleotide pairs. The enzyme then moves along the* DNA *molecule in the direction shown here by the arrow at the right (i.e., in the 3′ → 5′ direction of the template strand), bringing about transcription of* mRNA, *which is antiparallel and complementary to the* DNA *template strand (except that uracil replaces thymine). New ribonucleotides are added to the* mRNA *molecule at its 3′ end; its 5′ end may already be engaging ribosomes out in the cytoplasm as shown at the left.*

hydrogen bonding. Also in eurkaryotes, tRNAs are processed from larger precursors, many of which contain internal sequences that are clipped out by nucleases. The remaining bases are then joined together, and some of the intercalary bases modified. Processing of eurkaryotic tRNA is often very much more complex than suggested here.

Minimum necessary materials

The stage is now set for an examination of the role of each of these components in the synthesis of polypeptides. Success in such synthesis in in vitro cell-free systems shows that minimal necessary equipment is as follows:

1. Amino acids.
2. Ribosomes containing proteins and rRNA.
3. mRNA.
4. tRNA of several kinds.
5. Enzymes.
    a. Amino-acid-activating system.
    b. Peptide polymerase system.

6. Adenosine triphosphate (ATP) as an energy source.
7. Guanosine triphosphate (GTP) for synthesis of peptide bonds.
8. Soluble protein initiation and transfer factors.
9. Various inorganic cations (e.g., K^+ or NH_4^+, and Mg^{2+}).

Polypeptide chain synthesis—the process

The first critical step in **translation,** that is, the specification by mRNA of the sequence of amino acid residues in a polypeptide chain, involves the transport of amino acids from an intracellular pool to the ribosomes, where they are assembled into polypeptide chains, which are then assembled into proteins elsewhere in the cytoplasm. Some of these amino acids are synthesized in human cells, but others (the *essential* amino acids, Appendix C) cannot be, and must be supplied by the diet. Transfer of amino acids to the ribosome surfaces is accomplished by tRNA and occurs in several stages.

The account that follows is based largely on events in *E. coli.*

Amino acid activation. Each of the 20 kinds of amino acids must be *activated* before it can attach to its tRNA. In activation, an enzyme (aminoacyl-tRNA-synthetase) catalyzes the reaction of a specific amino acid with adenosine triphosphate (ATP) to form aminoacyl-adenosine monophosphate (aminoacyl adenylate) and pyrophosphate:

$$AA + ATP \xrightarrow{\text{amino acyl synthetase}} AA \sim AMP + \text{pyrophosphate}$$

The amino acyl adenylate (AA ~ AMP) is referred to as an **activated amino acid,** which is linked to adenosine by a phosphate ester bond.

As many as 20 different amino acids can be involved; hence each cell must have at least 20 different kinds of aminoacyl synthetases. By its architecture, each kind of enzyme molecule must "recognize" and be able to fit with a particular amino acid. Each of the various "species" of aminoacyl-tRNA-synthetases possesses two binding sites. One of these "recognizes" one, and only one, amino acid of the twenty. The specificity of the binding of the synthetase to its amino acid is dependent, in part at least, on the structure of the latter's side group, though there is not a great size difference between the side groups of, for example, glycine and alanine (see Appendix C). In fact, one estimate places the frequency of wrong insertions as high as 1 in 1,000. However, not all such mistakes are lethal or even disadvantageous.

The activated amino acid, however, is strongly attached to its enzyme and cannot, in this condition, be joined with another to begin formation of a polypeptide chain. This process requires transfer of the activated amino acid to the ribosomes.

Transfer of activated amino acid to tRNA. The same amino acyl synthetase that catalyzed the activation of the amino acid next attaches to a receptor site at the terminal adenine ribonucleotide of a particular tRNA molecule when the two come into contact as a result of a chance collision. That tRNA is then said to be charged:

$$AA \sim AMP + tRNA \longrightarrow AA \sim tRNA + AMP$$

The aminoacyl adenylate attaches to the free 3′-OH of the ribose of the 3′ terminal adenylic acid of tRNA by forming an acyl bond with the α-carboxyl group of the amino acid. Note that the enzymes must be able to bind specifically to both a particular amino acid and also to a particular tRNA molecule. Therefore in addition to the minimum of 20 different kinds of synthetases, each cell must have at least 20 different tRNA species. In fact, there is good experimental evi-

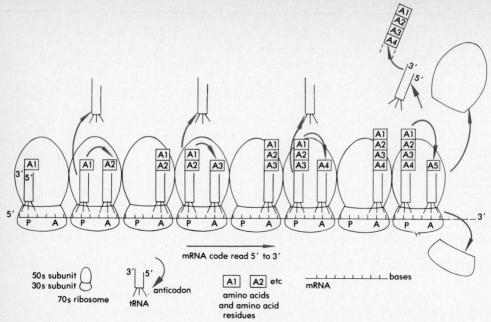

Figure 17-14. *Diagrammatic representation of polypeptide synthesis in* Escherichia coli. *Ribosomes first engage* mRNA *toward its 5′ end at a "start translation" codon, then progress toward the 3′ end. At that point a chain termination codon is encountered and recognized by protein-release factors; the polypeptide is then separated from the ribosome, and the latter dissociates into its component subunits. In this diagram the ribosome shown as having just been separated into its subunits (extreme right) represents the first to have engaged the* mRNA *molecule. See text for detailed outline of the steps involved.*

dence that there are many instances in which two or more different species of tRNA have some degree of specificity for the same amino acid.

Assembly of polypeptides. Following amino acid activation and binding to transfer RNA, the charged tRNA diffuses to the polysome, where actual assembly into polypeptide chains takes place. The following description of polypeptide formation applies to *Escherichia coli* in particular and to bacteria in general, organisms in which the process is best known. It apparently differs only in details from the corresponding series of events in eukaryotes.

Recall that mRNA binds to the smaller (i.e., 30S) subparticles of the bacterial ribosome; it displays there at least two groups of three ribonucleotides each, which constitute sites capable of accepting charged tRNA molecules. (Evidence for *three* ribonucleotides, rather than some other number, is explored in the next chapter.) These sites are designated the **aminoacyl (A) site** and the **peptidyl (P) site,** respectively. At or near the 5′-end of the associated mRNA molecule a chain-initiating group of (three) ribonucleotides must be present. The first ribosome attaches to this 3-base initiating site, often as the mRNA is still being transcribed (Fig. 17-13). In *E. coli,* the first amino acid to be incorporated is N-formylmethionine. This is methionine with a formyl group,

$$H-\overset{\overset{\textstyle O}{\|}}{C}-$$

attached to the amino group:

$$H-\overset{\overset{\displaystyle O}{\|}}{C}-\overset{\overset{\displaystyle H}{|}}{N}-\overset{\overset{\displaystyle H}{|}}{\underset{\underset{\displaystyle CH_3}{\underset{|}{\underset{\displaystyle S}{\underset{|}{\underset{\displaystyle CH_2}{\underset{|}{\underset{\displaystyle CH_2}{|}}}}}}}}{C}}-\overset{\overset{\displaystyle O}{\|}}{C}-OH$$

Methionine is formylated enzymatically after attachment to its tRNA. Not all methionine-tRNA complexes (met-tRNA) can be formylated. There are two kinds of met-tRNA, only one of which permits formylation. The formyl group blocks formation of a peptide bond with the carboxyl group of another amino acid. Depending on the particular polypeptide being synthesized, the N-formylmethionine may or may not later be removed enzymatically before the polypeptide chain is incorporated into a protein, though in most cases it is removed.

Steps in polypeptide synthesis on the polysome in *E. coli* are as follows (see also Fig. 17-14):

1. The 30S subunit of the first ribosome attaches to mRNA at an *initiator* site, coded for by a specific group of three nucleotides.
2. A tRNA, $tRNA_f^{met}$, charged with the first amino acid (N-formylmethionine) binds to the 30S subunit at the P site of three ribonucleotides. Three protein initiation factors (IF) plus guanosine triphosphate (GTP) are required.
3. Immediately upon completion of step 2, the larger 50S subunit attaches to the 30S subunit, and completes assembly of the first ribosome which, of course, still carries the tRNA charged with N-formylmethionine at the P site.
4. The second charged tRNA binds to this first ribosome at the latter's A site, so that both sites are occupied; i.e., the P site with the first tRNA, the A site with the second.
5. A peptide bond is formed enzymatically (by peptidyl synthetase which occurs in the 50S subunit) between the amino group of the second amino acid and the carboxyl group of the first amino acid.
6. A protein elongation factor (EF) is then responsible for transfer of a dipeptide, which consists of N-formylmethionine residue and the second amino acid residue, to the A site.
7. The $tRNA_f^{met}$ is then ejected, to function again later.
8. Translocation, which consists of the movement of the tRNA-dipeptide from the A site to the P site, and movement of the mRNA such that the effect is the apparent movement of the ribosome toward the 3' end of mRNA (by three nucleotides) then takes place.
9. Step 8 brings the previous A site (with its dipeptide) into location at the P site.
10. A third charged tRNA moves into position at the new A site.
11. The process outlined in steps 1-10 is then repeated again and again, with new ribosomes successively engaging the initiation site and moving or being moved along mRNA in the 5' → 3' direction.
12. These steps are continued until the first, and then successive ribosomes, reach a *termination* group of three ribonucleotides, at which point the ribosomes (each bearing a completed polypeptide) are successively ejected from the polysome whose subunits dissociate, and are able to repeat the entire process.

Life of any mRNA molecule is finite, however. It continues to function in the manner described as long as ribosomes attach to its 5'-end, but this terminus is vulnerable to ribonucleases. These are enzymes that hydrolyze RNA, and degrade it into its constituent nucleotides, which may then be used to construct wholly new and different mRNAs. Half-life for much bacterial mRNA appears to be of the order of one to three minutes, but longer in eukaryotes—from a few hours in mouse liver to several or many days in reticulocytes. Relatively little quantitative data are available on the stability of eukaryote mRNA.

Translation occurs with considerable speed. In hemoglobin production, for example, one polypeptide chain of nearly 150 amino acid residues requires about 80 seconds for synthesis. This means addition of one amino acid in a little more than 0.5 second!

Summary. In summary, translation consists of these major steps:

1. Amino acid activation.
2. Charging of tRNA.
3. Polypeptide chain initiation and elongation.
   a. Binding of charged tRNA to the mRNA-ribosome complex by complementarity of the anticodon-codon ribonucleotides.
   b. Synthesis of peptide bonds.
   c. Translocation.
4. Chain termination and release.

In vitro protein synthesis. A superb confirmation of the general outline of protein synthesis was furnished by Von Ehrenstein and Lipmann, who succeeded in synthesizing hemoglobin in a cell-free system. Such a system included

1. Ribosomes in the form of polysomes from rabbit reticulocytes (immature red blood cells in which hemoglobin is almost the only protein synthesized) that included mRNA for rabbit hemoglobin.
2. An energy-yielding triphosphate.
3. tRNA from the bacterium *Escherichia coli*, previously charged with amino acids using *E. coli* activating enzymes.

The most exciting aspect of this work is that Von Ehrenstein and Lipmann were able to synthesize a protein of one species using its mRNA and the tRNA of a vastly different organism. Not only is universality of this general mechanism of protein synthesis strongly suggested, but the fundamental "kinship" of all living organisms through their DNA seems inescapable. It is abundantly clear that DNA is the genetic material of all living organisms (and many viruses, too) and that it is an information vocabulary of four "letters" (adenine, thymine, guanine, and cytosine). Differences among species thus reside in the sequences of the nucleotide "letters" to be formed into certain code "words" (amino acids), which will be combined sequentially into protein "sentences" and phenotypic "paragraphs."

But there is still another problem in translation that must be explored, namely, just how the sequence of nucleotides in mRNA is responsible for the particular sequence of amino acid residues of a given polypeptide and just how synthesis of such chains starts and stops. The nature of the *genetic code* will be examined in the next chapter.

References

Abelson, J., 1979. RNA Processing and the Intervening Sequence Problem. In E. E. Snell, P. D. Boyer, A. Meister, and C. C. Richardson, eds., *Annual Review of Biochemistry*, vol. **48.** Palo Alto, Calif.: Annual Reviews, Inc.

Beadle, G. W., and E. L. Tatum, 1941. Genetic Control of Biochemical Reactions in *Neurospora. Proc. Nat. Acad. Sci. (U.S.)*, **27**: 499–506. Reprinted in J. A. Peters, ed., 1959. *Classic Papers in Genetics*. Englewood Cliffs, N.J.: Prentice-Hall.

Bittar, E. E., ed., 1973. *Cell Biology in Medicine*. New York: Wiley (Interscience Division).

Bosch, L., ed., 1972. *The Mechanism of Protein Synthesis and Its Regulation*. New York: American Elsevier.

Brawerman, G., 1974. Eukaryotic Messenger RNA. In E. E. Snell, P. D. Boyer, A. Meister, and C. C. Richardson, eds. *Annual Review of Biochemistry*, vol. **43**. Palo Alto, Calif.: Annual Reviews, Inc.

Brimacombe, R., R. G. Stöffler, and H. G. Wittman, 1978. Ribosome Structure. In E. E. Snell, P. D. Boyer, A. Meister, C. C. Richardson, eds. *Annual Review of Biochemistry*, vol. **47**. Palo Alto, Calif.: Annual Reviews, Inc.

Busch, H., F. Hirsch, K. K. Gupta, M. Rao, W. Sjohn, and B. C. Wu, 1976. Structural and Functional Studies on the "5′-Cap": A Survey Method for mRNA. In W. E. Cohn and E. Volkin, eds. *Progress in Nucleic Acid Research and Molecular Biology*, vol. **19**. New York: Academic Press.

Chambon, P., 1975. Eukaryotic Nuclear RNA Polymerases. In E. E. Snell, P. D. Boyer, A. Meister, and C. C. Richardson, eds. *Annual Review of Biochemistry*, vol. **43**. Palo Alto, Calif.: Annual Reviews, Inc.

Clark-Walker, G. D., 1973. Translation of Messenger RNA. In P. R. Stewart and D. S. Letham, eds. *The Ribonucleic Acids*. New York: Springer-Verlag.

Furuichi, Y., S. Muthukrishnan, J. Tomasz, and A. J. Shatkin, 1976. The 5′-Terminal Sequence ("Cap") of mRNAs. In W. E. Cohn and E. Volkin, *Progress in Nucleic Acid Research and Molecular Biology*, vol. **19**. New York: Academic Press.

Haenni, A. -L., 1972. Polypeptide Chain Elongation. In L. Bosch, ed. *The Mechanism of Protein Synthesis and Its Regulation*. New York: American Elsevier.

Harris, H., 1975. *The Principles of Human Biochemical Genetics*, 2nd ed. New York: American Elsevier.

Haselkorn, R., and L. B. Rothman-Denes, 1973. Protein Synthesis. In E. E. Snell, ed. *Annual Review of Biochemistry*, vol. **42**. Palo Alto, Calif.: Annual Reviews, Inc.

Howells, A. J., 1973. Messenger RNA. In P. R. Stewart and D. S. Letham, eds. *The Ribonucleic Acids*. New York: Springer-Verlag.

Kaback, M. M., ed., 1977. *Tay-Sachs Disease: Screening and Prevention*. (*Progress in Clinical and Biological Research*, vol. **18**). New York: Alan R. Liss, Inc.

Kameyana, T., ed., 1972. *Selected Papers in Biochemistry*, vol. **5**, *RNA Synthesis*. Baltimore: University Park Press.

Kaziro, Y., ed., 1971. *Selected Papers in Biochemistry*, vol. **7**, *Protein Synthesis*. Baltimore: University Park Press.

Knudson, A. G., Jr., 1969. Inborn Errors of Metabolism. In H. L. Roman, ed. *Annual Review of Genetics*, vol. **3**. Palo Alto, Calif.: Annual Reviews, Inc.

Lucas-Lenard, J., and F. Lipmann, 1971. Protein Biosynthesis. In E. E. Snell, ed. *Annual Review of Biochemistry*, vol. **40**. Palo Alto, Calif.: Annual Reviews, Inc.

Miller, O. L., Jr., and B. A. Hamkalo, 1972. Visualization of Genetic Transcription. In M. Sussman, ed. *Molecular Genetics and Developmental Biology*. Englewood Cliffs, N.J.: Prentice-Hall.

Polya, G. M., 1973. Transcription. In P. R. Stewart and D. S. Letham, eds., 1973. *The Ribonucleic Acids*. New York: Springer-Verlag.

Rich, A., and S. H. Kim, 1978. The Three-dimensional Structure of Transfer RNA. *Scientific American*, **238**, 1: 52–62.

Rich, A., and U. L. RajBhandary, 1976. Transfer RNA: Molecular Structure, Sequence, and Properties. In E. E. Snell, P. D. Boyer, A. Meister, and C. C. Richardson, eds. *Annual Review of Biochemistry*, vol. **45**. Palo Alto, Calif.: Annual Reviews, Inc.

Silvestri, L., 1970. *RNA-Polymerase and Transcription*. New York: Wiley (Interscience Division).

Stewart, P. R., and D. S. Letham, eds., 1973. *The Ribonucleic Acids*. New York: Springer-Verlag.

Problems

17-1 Could PKUs be helped by administration of phenylalanine hydroxylase? (You may have to consult sources outside your textbook to answer this question.)

17-2 If alkaptonurics were given large quantities of parahydroxyphenylpyruvic acid, would they excrete larger quantities of alkapton?

17-3 Would increased intake of maleylacetoacetic acid increase the excretion of alkapton?

17-4 In 1963 P. D. Trevor-Roper reported a family of four normally pigmented children born to a husband and wife, both of whom were albinos. In terms of the gene symbols used in Figure 17-1, give the genotypes of these albino parents and their children.

Use the following information to answer the next four problems.

All microorganisms utilize thiamine (vitamin B_1) in their metabolism. Final steps in its synthesis involve enzymatic synthesis of a thiazole and of a pyrimidine, followed by the enzymatic combination of these two substances into thiamine. Consider the following mutant strains of *Neurospora*; strain 1 requires only simple inorganic raw materials in order to synthesize thiamine; strain 2 grows only if thiamine or thiazole is supplied; strain 3 requires thiamine or pyrimidine; strain 4 grows if thiamine or both thiazole and pyrimidine are supplied. Call the enzyme responsible for synthesis of thiazole from its precursor *enzyme "a,"* the one that catalyzes formation of pyrimidine *enzyme "b"* and the enzyme that catalyzes the combining of thiazole and pyrimidine into thiamine *enzyme "c."*

17-5 Which of the strains described above is prototrophic?

17-6 Which enzyme or enzymes is strain 2 incapable of producing?

17-7 Which enzyme or enzymes cannot be produced by strain 4?

17-8 If we assign gene symbol *a* to any strain incapable of producing enzyme "a," symbol *b* to strains not making enzyme "b," and symbol *c* to those not producing enzyme "c," with + signs denoting the ability to produce a given enzyme, each strain can be represented by three gene symbols, plus signs, and/or letters in the appropriate grouping. Give the genotype, according to this plan, for each of the four strains listed above.

17-9 Beadle and Tatum studied a group of arginine auxotrophs in *Neurospora*. Prototrophs are able to synthesize the required amino acid arginine in a three-step process as follows: precursor → ornithine → citrulline → arginine. Auxotrophic strain 1 grows only on media containing arginine, strain 2 grows provided ornithine, or citrulline, or arginine is supplied, strain 3 grows if either arginine or citrulline is supplied. Call the enzyme that catalyzes the first metabolic step *"a,"* the one that catalyzes the second step *"b,"* and the enzyme that catalyzes the third step *"c."* Which strain is incapable of producing (a) enzyme *a*, (b) enzyme *b*, (c) enzyme *c*?

17-10 A short segment from a long DNA molecule has this sequence of nucleotide pairs:

$$3'\ A\ T\ C\ T\ T\ T\ A\ C\ G\ C\ T\ A\ 5'$$
$$5'\ T\ A\ G\ A\ A\ A\ T\ G\ C\ G\ A\ T\ 3'$$

(a) If the 5', 3' ("lower") DNA strand serves as the template for mRNA synthesis, what will be the sequence of ribonucleotides on the latter?
(b) Give the ribonucleotide at the 5'-end of the mRNA molecule thus transcribed.
(c) What would your answer to part (b) be if the 3', 5' DNA strand served as the template?

17-11 A sedimentation coefficient is calculated as 2×10^{-11} second. Express this value in Svedberg units.

17-12 Assume 101 deoxyribonucleotide pairs are responsible for a particular portion of a certain mRNA molecule. What is the length of that stretch of mRNA in (a) Angstrom units; (b) micrometers?

17-13 Assume a certain RNA molecule consists of only the "usual" ribonucleotides cytidylic, uridylic, adenylic, and guanylic acids, which occur in equal numbers. You determine the molecular weight of that RNA molecule to be about 27,000. (a) Of how many ribonucleotides does it consist? (b) How long would this molecule be in micrometers if it were laid out in a straight line? (c) Identify the kind of RNA molecule.

17-14 Recall that the genome of phage $Q\beta$ is single-stranded RNA that has a molecular weight of 1×10^6. Of about how many ribonucleotides does the genome of this phage consist, if only the four "usual" bases are assumed to occur and in equal number?

17-15 If a portion of the $+$ strand of phage $Q\beta$ has the ribonucleotide sequence $3'$ A U C G G U U A G . . . $5'$, give the ribonucleotide sequence of the $-$ strand.

17-16 A certain codon is determined to be AUG. (a) Of what nucleic acid molecule is this codon a part? (b) What is the corresponding anticodon? (c) Of what nucleic acid molecule is this anticodon a part? (d) What is the deoxyribonucleotide sequence responsible for this codon?

17-17 Differentiate between transcription and translation.

17-18 How do the functions of rRNA, mRNA, and tRNA differ?

17-19 Human hemoglobin polypeptide chain alpha (α) consists of 141 amino acid residues. Would you expect the DNA segment responsible for the ultimate synthesis of this chain to be shorter than, longer than, or about the same length as the functional mRNA molecule for this chain?

CHAPTER 18

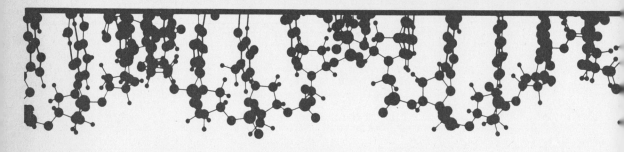

The genetic code

In preceding chapters it has been established that

1. DNA is *the* genetic material in all but some viruses (in which it is RNA).
2. DNA is responsible for phenotypic expression through transcription of mRNA from DNA templates.
3. Specific nucleotide sequences on mRNA interact with complementary base groups of tRNA to translate mRNA base sequences into polypeptides on ribosomal surfaces.

Studies of phenylalanine metabolism, the Tay-Sachs syndrome, nutritionally deficient strains of the fungus *Neurospora*, and a host of others, make it clear that occurrence of a given biochemical reaction depends on the presence of a specific enzyme that, in turn, is due to action of a particular genetic locus. Originally referred to as "one gene-one enzyme," the problem quickly developed into one of determining the precise function of genes in enzyme (and therefore protein) synthesis. Exploration of hemoglobin variants in humans (Chapter 19), for example, soon indicated that genes serve to specify the amino acid sequence of proteins (and therefore their precise structure and function), rather than to act merely as "switches" that determine whether protein will or will not be produced.

Thus there emerges the concept that the classical particulate gene is really a series of deoxyribonucleotides[1] with the relationship *gene → RNA → polypeptide → phenotype*. But this concept raises two fundamental questions: (1) what se-

[1]Or ribonucleotides in the RNA viruses.

quence of how many mRNA nucleotides *codes* for a particular amino acid and for a given polypeptide chain; that is, what is the **genetic code,** and (2) in these terms, then, just how many deoxyribonucleotides equal one gene in eukaryotes? The first of these questions we shall explore in this chapter, the second in Chapter 19.

Problems of the nature of the code

Investigators of the genetic code faced many questions concerning its nature. Those to be examined in this chapter include

1. How many ribonucleotides code for a given amino acid? That is, does a **codon** consist of one, two, three, or more ribonucleotides?
2. Are codons, whatever their nucleotide number, contiguous, or is there some sort of spacer "punctuation" that separates the codons? In other words, is the code **commaless?**
3. If a codon is composed of two or more nucleotides, is the code **overlapping** or **nonoverlapping?** That is, in the mRNA nucleotide sequence beginning, for example, AUCGUA . . ., how many codons are there, AU, CG, UA (nonoverlapping doublet code), or AUC, GUA (triplet, nonoverlapping), or are the codons AU, UC, CG, GU, UA, A− (doublet, overlapping by one base), or AUC, UCG, CGU, GUA, UA−, A− − (triplet, overlapping by two nucleotides), or some other arrangement?
4. What is the **"coding dictionary"?** That is, precisely which codons code for which amino acids?
5. Is a given amino acid coded by more than one codon? That is, is the code **degenerate?**
6. Does one codon code for more than one amino acid? In other words, is the code **ambiguous?**
7. Do the codons of mRNA and the corresponding amino acid residues of polypeptides occur in the same linear order? That is, is the code **colinear?**
8. Are there any codons that serve as "start" or "stop" signals for translation? In other words, are there chain-**initiating** and chain-**terminating** codons?
9. Is the code **universal** for all organisms? Does, in other words, the same codon signal for the same amino acid in phages, bacteria, corn, fruit flies, and human beings?

In succeeding chapters the genetic code as it relates to mutation and recombination will also be considered.

THE BASIC PROBLEM

The basic problem of the genetic code is that there are 20 amino acids that must be coded for by some sequence of four nucleotides in DNA or their complements in mRNA. The mRNA nucleotide or nucleotide sequence that codes for a particular amino acid is called a **codon.** If a codon were to consist of only a single base, the genetic code would have to be quite *ambiguous;* that is, the same codon would have to code for different amino acids under different conditions. In other words, how would four bases code for 20 amino acids? Codons of two bases present the same kind of problem; this provides only $4 \times 4 = 16$ codons for the 20 amino acids. But three-base groups provide 64 codons, *more* than enough for the number of amino acids involved. However, the system will work if the code is *degenerate* in that several codons code for the same amino acid. From the theoretical viewpoint, such a *triplet code* could include both *sense codons* (those that specify particular amino acids) and *nonsense codons* that do not specify any amino acid. Nonsense codons, might, of course, have some other function, such as signaling "start" or "stop" for polypeptide chain synthesis.

Triplet codons.   The first key to the solution was provided in 1955 by Grunberg-Manago and Ochoa, who isolated from bacteria an enzyme that catalyzes polymerization of nucleoside diphosphates into a linear poly-ribonucleotide sequence, chemically and functionally similar to natural messenger RNA. No template is required by this enzyme and the sequence of ribonucleotides in the resulting polymer is random. Hence polyribonucleotides containing one, two, three, or all four of the different bases can be tested in vitro for their protein-synthesizing ability. To be incorporated into a polypeptide chain, some amino acids require a polyribonucleotide that contains only one kind of ribonucleotide (e.g., uridylic acid); others require polyribonucleotides containing two or even three different ones. No amino acid, however, requires all four. This supplies evidence against a four-base code, but does not discriminate among one-, two-, three-, or even five- or six-base codes.

Two other lines of experimental evidence (1) suggested that codons are triplets and (2) permitted the first specific codon assignment. Crick, Barnett, Brenner, and Watts-Tobin provided a significant clue not only to the length of the codon but also to the question of "punctuation" and overlap. To understand the relevance of their experiments one of the important classes of substances that they employed must first be examined.

Frame shifts.   The acridine dyes (proflavin, acridine orange, and acridine yellow, among others) are substances that bind to DNA and, at least in phages, act as mutagens by causing additions or deletions in the nucleotide sequence during DNA replication. An acridine molecule may become intercalated and double the distance between previously adjacent nucleotides. This may either allow later insertion of a new nucleotide during replication or result in the deletion of a base, and therefore alter the sequence. Crick and his colleagues studied a number of acridine-induced mutants in the so-called *rII* region[2] of the DNA of phage T4. In brief, they found that mutant types arose through additions or deletions at any of a large number of sites within the *rII* region, but these may revert to the wild type or a very similar one (pseudo-wild type) through deletions or additions elsewhere in the *rII* region. That is, in many cases, one mutation may be suppressed by another, particularly if the second change is located relatively close to the first. To illustrate, assume the code is triplet, nonoverlapping, and commaless, and that a repeated deoxyribonucleotide sequence is involved:

$$(\text{transcription} \longrightarrow)$$
$$\overline{\text{TAG}}\ \overline{\text{TAG}}\ \overline{\text{TAG}}\ \overline{\text{TAG}}\ \overline{\text{TAG}} \ldots \tag{1}$$

Now if the second T (the fourth base) in this hypothetical series is deleted, sequence (1) becomes

$$\overline{\text{TAG}}\ \overline{\text{AGT}}\ \overline{\text{AGT}}\ \overline{\text{AGT}}\ \overline{\text{AG}-} \ldots \tag{2}$$

A deletion thus alters reading of all following groups; this is termed a *frame shift*. Transcription from the original DNA sequence (1) produces mRNA of repeating AUC codons, and determines a peptide consisting only of the amino acid sequence coded for by AUC; and from (2) the codon sequence AUC UCA UCA UCA UC−. This could well be expected to result in a change of amino acid composition of the peptide chain beyond the deletion, if it is assumed that codons AUC and UCA code for different amino acids. We could represent the amino acid of the wild type (1) as aaX aaX aaX aaX . . ., but after deletion, the mutant type (2) might produce the amino acid sequence aaX aaY aaY aaY

---

[2]The *rII* mutants are described more fully in another context in the next chapter.

... (a *missense* mutant). On the other hand, an a priori possibility might be that UCA codes for *no* amino acid (a *nonsense* mutant).

Now, still carrying this deletion, assume a thymidylic acid (T) to be inserted at a different position in sequence (2), say between the sixth and seventh nucleotides. The DNA sequence then becomes

$$\overline{\text{TAG}} \; \overline{\text{AGT}} \; \overline{\text{TAG}} \; \overline{\text{TAG}} \; \overline{\text{TAG}} \ldots \tag{3}$$

and the wild-type reading is restored with the third triplet; only the second produces a misreading. Insertion of any of the other three nucleotides at the same point produces only a slightly longer faulty segment. Suppose deoxycytidylic acid (C) is inserted instead of thymidylic acid at the same point; the sequence then becomes

$$\overline{\text{TAG}} \; \overline{\text{AGT}} \; \overline{\text{CAG}} \; \overline{\text{TAG}} \; \overline{\text{TAG}} \ldots \tag{4}$$

Wild-type transcription in (4) is restored after two "wrong" triplets instead of one. Thus, an insertion farther down the reading sequence corrects for an earlier deletion, regardless of whether the inserted base is the same as the deleted one or not. A *single frame shift*, therefore, may be expected to result in a protein so altered in amino acid sequence that it is nonfunctional. On the other hand, if the altered reading between two opposing events (e.g., a deletion followed by an insertion) is of small enough magnitude, and if the missense mutation is of such a nature as to alter little or not at all the function of the ultimately produced protein, then the second event suppresses the first. The closer the points at which these opposing alterations occur, the higher the probability that this suppression will take place. In essence, this is precisely what the work of Crick and his group showed.

We may summarize the findings of Crick and his coworkers in this way (D = deletion; I = insertion):

Pseudo-wild type	Mutant
D-I	D
I-D	I
D-D-D	D-D
I-I-I	I-I
D-D-D-D-D-D	D-D-D-D
	I-I-I-I
	D-D-D-D-D
	I-I-I-I-I

A triplet, commaless, nonoverlapping code. Note that several aspects of the genetic code are made clear by these experiments. First, the code is likely to be *triplet*, because a single frame shift results in missense, as do two, four, or five frame shifts, but pairs of opposite kinds of frame shifts restore sense. Likewise, three deletions or three insertions restore sense. Although in a doublet code, for example, one deletion followed by one insertion will also restore sense, such will not always be the case with three (or multiples of three) deletions or insertions. Still other experimental data make it clear that the code is, indeed, triplet.

In addition, this kind of restoration of sense sequences clearly suggests that (1) there is no "punctuation" between the codons, that is, each codon is immediately adjacent to the next with no intervening "spacer" bases, and (2) it is *nonoverlapping*. Another line of evidence for nonoverlap arises in the fact that amino acid residues appear to be arranged in completely random sequence when different polypeptide chains are analyzed; no one amino acid always, or even usually, has the same adjacent neighbors. This could be the case if the code is *nonoverlapping*. If the code did overlap, a given amino acid would always have the same nearest neighbors. An mRNA sequence beginning, for example, A–A–C–C–G–A–G–C–A– . . . consists of three triplets, AAC, CGA, and GCA, which code for asparagine, and arginine, and alanine, respectively (Table 18-1). If the code overlapped by two bases, this sequence of nine bases would consist of the triplets AAC, ACC, CCG, CGA, GAG, AGC, and GCA, coding for asparagine, threonine, proline, arginine, glutamic acid, serine, and alanine. Thus in that particular sequence of ribonucleotides, arginine would always occur between proline and glutamic acid. This is not the case.

Although an overlapping code might seem more efficient in that more "messages" can be coded for by fewer nucleotides, there is also a strong disadvantage to overlap. In the nucleotide sequence previously described, consider the result of a substitution (i.e., a mutation) in the third base (the first deoxycytidylic acid) to deoxyguanylic acid, so that the sequence becomes

A–A–G–C–G–A–G–C–A– . . .

The mRNA codons, if the code were overlapping by two bases, would then become UUC, UCG, CGC, . . ., which codes for phenylalanine, serine, and arginine, respectively, i.e., an entirely different sequence than in the original sequence. Thus, a single base pair substitution in this case would change *three* amino acids (although this does *not* occur) rather than just *one* in a triplet, nonoverlapping code.

Nucleotides functioning in more than one codon in phage G4. Phage G4, like φX174 (Chapter 16), is a single-stranded DNA phage. This single strand is the + ("plus") strand. After infection of *Escherichia coli*, double-stranded replicative forms of DNA are produced in the host cell; one of these strands is identical in nucleotide sequence to the original + strand, the other (the − or "minus" strand) is complementary to it. The latter then serves as a template for transcription of mRNA. Three *overlapping genes* have been described in phage G4. For each gene a given nucleotide functions in only one codon, but in the three overlapping genes the same nucleotide functions in a different reading frame:

+ strand DNA	. . . A A A T G A G G A . . .
− strand DNA	. . . T T T A C T C C T . . .
mRNA	. . . A A A U G A G G A . . .
G4 gene C	|A U G|A G G| . . . methionine-arginine . . .
G4 gene A	|A A A|U G A| . . . lysine-end chain . . .
G4 gene K	|A A U|G A G| . . . asparagine-glutamic acid . . .

Portions of the messenger RNA not included in the codons shown for these three genes serve as parts of codons adjacent to those for which coding is shown. Of course, this is not overlap in the sense described in the preceding section, but it is an interesting "economical" departure from the notion that each ribonucleotide normally serves in only one codon. The "coding dictionary" for all codons is shown in Table 18-1.

THE CODING DICTIONARY

Once the genetic code was established as a *commaless, nonoverlapping, triplet* code, the question of which triplets code for which amino acids was pursued vigorously. Nirenberg and Matthaei pioneered in efforts to crack the code. It is possible to construct short synthetic polyribonucleotides and to test them in cell-free systems for ability to direct incorporation of specific amino acids into polypeptide chains. Using a mixture of amino acids, with a different one radioactively labeled in each run, in a cell-free suspension derived from *Escherichia coli* (tRNA,

Table 18-1. The mRNA code as determined in vitro for E. coli.
(Degeneracies are shown; ambiguities are omitted)

FIRST BASE	SECOND BASE				THIRD BASE
	G	**A**	**C**	**U**	
G	GGG Glycine	GAG Glutamic Acid	GCG Alanine	GUG Valine	**G**
	GGA Glycine	GAA Glutamic Acid	GCA Alanine	GUA Valine	**A**
	GGC Glycine	GAC Aspartic Acid	GCC Alanine	GUC Valine	**C**
	GGU Glycine	GAU Aspartic Acid	GCU Alanine	GUU Valine	**U**
A	AGG Arginine	AAG Lysine	ACG Threonine	AUG Start Chain Methionine	**G**
	AGA Arginine	AAA Lysine	ACA Threonine	AUA Isoleucine	**A**
	AGC Serine	AAC Asparagine	ACC Threonine	AUC Isoleucine	**C**
	AGU Serine	AAU Asparagine	ACU Threonine	AUU Isoleucine	**U**
C	CGG Arginine	CAG Glutamine	CCG Proline	CUG Leucine	**G**
	CGA Arginine	CAA Glutamine	CCA Proline	CUA Leucine	**A**
	CGC Arginine	CAC Histidine	CCC Proline	CUC Leucine	**C**
	CGU Arginine	CAU Histidine	CCU Proline	CUU Leucine	**U**
U	UGG Tryptophan	UAG End Chain *	UCG Serine	UUG Leucine	**G**
	UGA End Chain	UAA End Chain ‡	UCA Serine	UUA Leucine	**A**
	UGC Cysteine	UAC Tyrosine	UCC Serine	UUC Phenylalanine	**C**
	UGU Cysteine	UAU Tyrosine	UCU Serine	UUU Phenylalanine	**U**

*Originally called *amber*.
‡Originally called *ochre*.

Acidic	Aromatic	Basic	Neutral	Contains sulfur

ribosomes, ATP, GTP, necessary enzymes, and, interestingly enough, mRNA from tobacco mosaic virus), these men were able to bring about polypeptide synthesis in vitro. With polyribouridylic acid, for example, they were able to demonstrate the synthesis of a polypeptide consisting only of phenylalanine, even though other amino acids were present. So the mRNA codon for phenylalanine must be UUU.

Testing copolymers of only the ribonucleotides adenylic and cytidylic acids (poly-AC), Nirenberg and Matthaei showed that proline was coded by at least two triplets, one consisting only of cytidylic acid (CCC) and the other of both cytidylic and adenylic acids. Poly-A (AAA) was soon found to code for lysine. These determinations reinforce the concept of a commaless code for, in a code that includes punctuation, U and A would each have to serve both for their respective amino acids as well as to function as commas. From this time on, the notion of a commaless code won general acceptance.

Testing random copolymers provides some additional information on the code. When synthetic mRNA is constructed, it is found that the sequence of nucleotides is random, so that the relative frequency of incorporation of particular nucleotides is mathematically determined. Thus a mixture containing *two parts uracil to one part guanine* is found to produce triplets in these combinations:

$$
\begin{array}{llcll}
\text{UUU} & \frac{2}{3} \times \frac{2}{3} \times \frac{2}{3} = \frac{8}{27} & \qquad & \text{GUU} & \frac{1}{3} \times \frac{2}{3} \times \frac{2}{3} = \frac{4}{27} \\
\text{GGG} & \frac{1}{3} \times \frac{1}{3} \times \frac{1}{3} = \frac{1}{27} & & \text{GGU} & \frac{1}{3} \times \frac{1}{3} \times \frac{2}{3} = \frac{2}{27} \\
\text{UGU} & \frac{2}{3} \times \frac{1}{3} \times \frac{2}{3} = \frac{4}{27} & & \text{UGG} & \frac{2}{3} \times \frac{1}{3} \times \frac{1}{3} = \frac{2}{27} \\
\text{UUG} & \frac{2}{3} \times \frac{2}{3} \times \frac{1}{3} = \frac{4}{27} & & \text{GUG} & \frac{1}{3} \times \frac{2}{3} \times \frac{1}{3} = \frac{2}{27} \\
\end{array}
$$

If each of the possible triplets were to code for a different amino acid, then various ones would be incorporated in the proportions shown. Except for the fact that *each* such three-base group does not code a *different* amino acid, this is essentially what occurs. Use of a synthetic poly-UG, in the relative proportions just given, produces a mixture of polypeptides, of which 8/27 are polyphenylalanine.

In this way it is known that poly-UG that contains 2 U:1 G codes for valine, but the sequence of bases, and therefore the exact codon(s) of those possible, cannot be determined from such random copolymers. Nirenberg and Leder devised a method by which short-chain polyribonucleotides of known sequence could be obtained. They and also Khorana and his group introduced these short-chain polyribonucleotides of known sequence into cell-free systems that included ribosomes and a variety of tRNA molecules charged with their amino acids. As in earlier work, one amino acid in each experimental run was labeled with [14]C. Although the messengers used were too short for protein synthesis, two important considerations emerged: (1) little or no binding of tRNA took place in the presence of dinucleotide messengers but occurred preferentially with trinucleotides, and (2) different sequences of the same three bases stimulated binding of different amino acids. The code was thus shown clearly to be triplet and non-overlapping. Determination of the complete coding dictionary, as shown in Table 18-1, followed quickly. Elucidation of the genetic code has been rightly called "one of the principal triumphs of molecular biology," and a "notable milestone in biology."

DEGENERACY

Although the code is extensively degenerate, as seen in Table 18-1, a certain order to this degeneracy can be discerned. In many instances it is the first two bases that are the characteristic and critical parts of the codon for a given amino acid, whereas the third may be read either as a purine only (e.g., glutamine, CAA and CAG), or as a pyrimidine only (e.g., histidine, CAU and CAC). In other cases, however, the third position is read as any base, for example, AC− for threonine, CC− for proline, and so forth.

In 1966 Crick proposed the **"wobble hypothesis"** to account for redundancy in the 5′ base of the anticodon, which may pair with any of several bases at the 3′ end of the codon. Combinational possibilities are, however, not limitless, but are restricted to those listed in Table 18-2. Because it decreases the fidelity of hydrogen bonding, inosine in the anticodon may pair with uracil, cytosine, or adenine ribonucleotides of the codon. Inosine is derived from adenine by the deamination of the 6-carbon to produce a 6-keto group, as shown in Figure 18-1.

*Table 18-2. **Pairing between codon and anticodon at the third position***

Anticodon	Codon
A	U
C	G
G	U, C
I (inosine)	U, C, A
U	A, G

Based on the work of Crick, 1966.

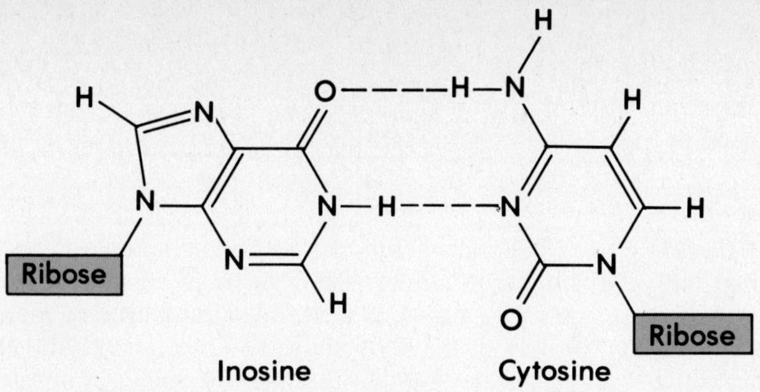

Figure 18-1. *An inosine-cytosine base pair. Inosine also pairs with uracil and adenine by two hydrogen bonds.*

In addition to the "wobble" of the third, or 5'-base, of the anticodon, some amino acids have as many as six codons (e.g., leucine, CU–, UUA, and UUG). Different leucine tRNAs exist for some of the six leucine codons, all of which obey the pairing relationships shown in Table 18-2. For example, pseudouridine (ψ), which occurs as the middle base of the anticodon of yeast tyrosine tRNA, pairs with adenine. Methionine and tryptophan, which have only one codon each (AUG and UGG, respectively), are exceptions to this degeneracy.

AMBIGUITY

Essentially, the code is nonambiguous in vivo under natural conditions. Ambiguity is encountered chiefly in cell-free systems under certain conditions. In such a system prepared from a streptomycin-sensitive strain of *E. coli*, UUU (which ordinarily codes for phenylalanine) may also code for isoleucine, leucine, or serine in the presence of streptomycin. This ambiguity is enhanced at high magnesium-ion concentrations. Poly-U, in a cell-free system from thermophilic bacterial species, has been found to bind leucine at temperatures well below the optimum for growth of living cells of the same species. Changes in pH or the addition of such substances as ethyl alcohol also result in ambiguity. However, ambiguities are not ordinarily encountered in vivo under normal growing conditions for a given species.

COLINEARITY

In a *colinear* code, the sequence of mRNA codons and the corresponding amino acid residues of a polypeptide chain are arranged in the same linear sequence. Now, both DNA and mRNA on the one hand, and polypeptides on the other, are linear, but this, in itself, does not require *colinearity*. That such colinearity does exist, however, is shown by studies of T4 mutants which produce incomplete head protein molecules. These mutants can be shown to map in linear sequence by using recombination techniques. Length of the incomplete head protein molecule made by each mutant is precisely proportional to the map distance involved. Thus the code is determined to be *colinear*. The same conclusions are reached in the work of Yanofsky and his colleagues on the tryptophan synthetase gene system of the colon bacillus. Chapter 19 will examine Yanofsky's studies as well as colinearity in the polypeptide chains of human hemoglobin.

All experimental data clearly indicate that mRNA is synthesized in the $5' \rightarrow 3'$ direction (Chapter 17). The assembly of polypeptide chains is also sequential, from the amino to the carboxyl termini. The question, then, is what sort of signal (if any) directs the initiation and termination of polypeptide synthesis along mRNA?

Chain initiation.   It appears clear that polypeptide chains in *Escherichia coli* are initiated with N-formylmethionine, which, in many instances, is enzymatically removed before assembly into proteins. The same is true of many phages. To illustrate, the coat protein of the RNA phage R17 begins with the N-terminal sequence alanine-serine-asparagine-phenylalanine-threonine. In in vitro systems this sequence is preceded by N-formylmethionine; in such systems it is evident that the enzyme necessary to remove the formylated methionine is absent.

Only one codon, AUG, exists for methionine (Table 18-1) so the question naturally arises of how chain-initiating N-formylmethionine and internally located methionine are distinguished in the biosynthesis of polypeptides. The answer in *E. coli*, at least, lies in the occurrence of two different tRNAs for methionine. One of these, symbolized as $tRNA_f^{met}$ or as fmet-tRNA, is formylated and serves only for initiation of polypeptide synthesis. The other, methionyl-tRNA ($tRNA_m^{met}$ or met-tRNA), does not serve as a substrate for the formylating enzyme and inserts its methionine only into intercalary positions; thus it functions in elongation rather than in initiation. The anticodon for both kinds of tRNA is $3'UAC5'$; therefore the same AUG codon codes for both formylated and nonformylated methionine. Fmet-tRNA and met-tRNA do, however, differ in many of their internal nucleotides (Fig. 18-2). It is reported by some investigators that fmet-tRNA is required for the initiation of all bacterial proteins, although the formyl group, and in some cases the entire methionine residue, is enzymatically removed from the polypeptide chain at some point prior to its incorporation into protein. There is some evidence that methionine, but not N-formylmethionine, functions in initiation in mammalian cells. A special met-tRNA is responsible for insertion of methionine into the N-terminal position, although it is removed in some cases, e.g., rabbit globin, before the polypeptide chain is incorporated into hemoglobin. Initiation is more than a simple case of an initiation codon alone; at least three protein initiation factors (IF) are involved as well.

AUG determines the reading frames of mRNA. The synthetic ribonucleotide AUGGUUUUUUU . . . is translated only as N-formylmethionine-valine-phenylalanine-phenylalanine . . .; that is, the reading is AUG–GUU–UUU–UUU . . . and GGU (glycine) is not read as such. Translation, once initiated, continues until a "stop" codon (see next section) is encountered. It would thus appear logical that a chain-initiating AUG should occur adjacent to, or at least fairly close to, a preceding chain-terminating codon. Otherwise, AUG would be read as an internal site by $tRNA_m^{met}$.

The RNA of phage R17 consists of about 3,300 ribonucleotides, of which some 3,000 code for three proteins (phage coat, release enzyme, and a maturation protein involved in assembling coat and RNA into a mature virus. The beginning sections of each of the three genes have been isolated with their initiator regions. Each initiator region begins with AUG but is not immediately preceded by any of the chain terminators that are described in the next section. Thus introns intervene between the three structural genes of R17; initiation and termination codons may therefore be rather widely spaced.

Chain termination.   Three codons, UAA, UAG, and UGA do not code for any amino acids and hence are termed *nonsense codons*. Before their base sequences were determined, UAA was known as "ochre," UAG as "amber," and UGA as "opal"—terms that are still employed for these codons as a matter of conven-

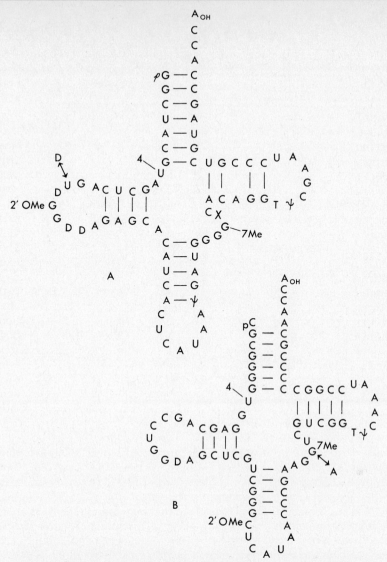

Figure 18-2. *Ribonucleotide sequences of (A) tRNA$_m$, and (B) tRNA$_f$. Unusual ribonucleotides are symbolized as follows: D, dihydrouridylic acid; ψ, pseudouridylic acid; T, thymidylic acid; 7MeG, 7-methylguanylic acid; 2 MeC, 2 O-methylcytidylic acid; 2 OMeG, 2 O-methylguanylic acid; 4U, 4-thiouridylic acid. (Based on work of S. K. Dube, K. A. Marcker, B. F. C. Clark, and S. Cory, 1968. "Nucleotide Sequence of N-Formyl-Methionyl-Transfer RNA." Nature **218**: 232–233, and S. Cory, K. A. Marcker, S. K. Dube, and B. F. C. Clark, 1968. "Primary Structure of a Methionine Transfer RNA from Escherichia coli." Nature **220**: 1039–1040.)*

ience. Evidence from both in vitro and in vivo experiments demonstrates that all three nonsense codons are involved in both chain termination and release (the latter process also requires release factors).

UNIVERSALITY

Significantly, the current evidence is that the code is universal for all living organisms and for viruses; the same triplets code similarly in a wide variety of

Table 18-3. Comparison of mRNA codons for twenty amino acids in different organisms*

Amino acid	Escherichia coli (a bacterium)	Rat liver	Wheat embryo (flowering plant)	Chlamydomonas (a green alga)
Alanine	GCU GCA GCC GCG			
Arginine	CGU CGA CGC CGG AGA AGG			
Asparagine	AAU AAC			
Aspartic acid	GAU GAC		GAU	
Cysteine	UGU UGC AGU AGC			
Glutamine	CAA CAG			
Glutamic acid	GAA GAG		GAU(?)	
Glycine	GGU GGA GGC GGG	GGU	GGU	
Histidine	CAU CAC			
Isoleucine	AAU AUC AUA	AUU	AUU	AUU
Leucine	CUU CUA CUC CUG UUA UUG		CUU CUC	
Lysine	AAA AAG	AAA	AAU(?)	
Methionine	AUG		AUG	
Phenylalanine	UUU UUC	UUU	UUU	UUU
Proline	CCU CCA CCC CCG		CCU	
Serine	UCU UCA UCC UCG AGU AGC	UCU	UCC	UCU
Threonine	ACU ACA ACC ACG			
Tryptophan	UGG	UGG	UGG	
Tyrosine	UAU UAC	UAU	UAU	UAU
Valine	GUU GUA GUC GUG	GUU	GUU GUG	GUU

*After Groves and Kempner (1967).

organisms (Table 18-3). Compare codons for isoleucine, as an illustration, for the four very different species listed.

DNA from different species is expected to exhibit degrees of similarity in nucleotide sequences in proportion to the closeness of evolutionary relationship of those species. Zoologists have long agreed that humans and monkeys are more closely related than either are to fish or bacteria, for example. Significant chemical proof was demonstrated in 1964 by Hoyer, McCarthy, and Bolton.

Their method was simple in concept but delicate in operation. DNA to be tested was first made to undergo strand separation by heating (denaturation), then cooled quickly to prevent recombining, and immobilized in agar. To this were added from another species short strands of denatured DNA that had previously been made radioactive by incorporation of carbon-14 or phosphorus-32. Pairing of homologous nucleotides from the two species occurred during several hours of incubation. The "hybrid" DNA was then recovered and assayed for radioactivity; this resulted in a measure of base-pair homology between the two species. The results are summarized in Table 18-4.

The significance of these data lies in the similarity of probably fairly long sequences between human and monkey and between mouse and rat, and the evident dissimilarity of either with such taxonomically distant species as salmon or the colon bacillus. Hoyer and his colleagues concluded

This observation raises the question of whether there exists among the various animals a particular class of nucleotide sequences which have been retained during the diversification of the vertebrate forms. . . . It is clear . . . that there exist homologies among polynucleotide sequences in the DNAs of such diverse forms as fish and man. These sequences represent genes that have been conserved with relatively little change throughout the long history of vertebrate evolution. Al-

Table 18-4. *Percentage recombining of* [14]*C-labeled human DNA and* [32]*P-labeled mouse DNA with unlabeled DNA of other species**

	Percentage labeled DNA bound	
Unlabeled DNA source	[14]C human	[32]P mouse
Human	18*	5
Rhesus monkey	14	8
Mouse	6	22*
Cattle	5	4
Rat	4	14
Guinea pig	4	3
Hamster	4	12
Rabbit	4	3
Salmon	1.5	1.5
Colon bacillus	0.4	0.4

*Human-human and mouse-mouse results serve as a base of comparison.
Data from Hoyer, McCarthy, and Bolton (1961).

though we have no means yet of relating such genes to particular phenotypic expressions, it is conceivable that they are the determinants of the fundamental conservative characteristics of the vertebrate form.

These latter would include such fundamental traits as skeletal structure and hemoglobin production. Human and mouse are mammals with many traits in common; but within the primate group, human and monkey are phenotypically similar with rather long segments of like genetic material. Recall also that, in Chapter 4, it was shown that humans and the great apes (chimpanzee, gorilla, and orangutan) have many karyotypic homologies.

Certainly of equal significance is the fact that in vitro protein synthesis, which uses such disparate ingredients as rabbit reticulocyte mRNA and *E. coli* tRNA (plus other necessary components), is successful. The result is production of normal rabbit hemoglobin, which shows that bacterial tRNA can "recognize" mRNA and polysomes from a taxonomically very different organism. It is true that taxonomically divergent organisms do differ in the degree to which a given codon responds to a particular species of tRNA. For example, AAG codes readily for lysine in vertebrates, but does so more weakly in *E. coli*. But this is only a difference in *degree*, not in kind.

Furthermore, such proteins as cytochrome c show uniformity in several sequences of amino acid residues in very different groups as mammals, fishes, yeasts, and bacteria, even though their evolutionary divergence must have taken place many hundreds of millions of years back in geologic time.

The question of *evolution* of the genetic code is, as might be surmised, replete with problems. Its very universality complicates the question, as does the paradoxical situation wherein operation of the code requires precise functioning of many enzymes that, apparently could not be produced without the translation mechanism for which they are required.

In the next chapter the concept of the ultimate structure of the gene will be examined through the process of mutation at the molecular level.

Summary of code characteristics

In summary, the evidence shows that the genetic code *is triplet, commaless, nonoverlapping, degenerate, essentially nonambiguous* under natural conditions, *colinear,* and *universal.* In addition, polypeptide chain *initiation* is signaled by certain codons (notably AUG) that bind tRNAs which carry blocked amino acids, although specific protein initiation factors are also required. Chain *termination* is governed by three nonsense codons (UAA, UAG, and UGA); chain *release* requires protein release factors as well.

References

Adams, J. M., and M. R. Capecchi, 1965. N-formylmethionine-sRNA as the Initiator of Protein Synthesis. *Proc. Nat. Acad. Sci. (U.S.),* **55:** 147–155.

Beaudet, A. L., and C. T. Caskey, 1972. Polypeptide Chain Termination. In. L. Bosch, ed. *The Mechanism of Protein Synthesis and Its Regulations.* New York: American Elsevier.

Brenner, S., A. O. W. Stretton, and S. Kaplan, 1965. Genetic Code: The "Nonsense" Triplets for Chain Termination and Their Suppression. *Nature,* **206:** 994–998.

Crick, F. H. C., 1963. On the Genetic Code. *Science,* **139:** 461–464.

Crick, F. H. C., 1966. Codon-Anticodon Pairing: The Wobble Hypothesis. *Jour. Molec. Biol.,* **19:** 548–555.

Crick, F. H. C., L. Barnett, S. Brenner, and R. J. Watts-Tobin, 1961. General Nature of the Genetic Code for Proteins. *Nature,* **192:** 1227–1232.

Frisch, L., ed., 1967. The Genetic Code. *Cold Spring Harbor Symposia Quant. Biol.,* **31** (1966). Cold Spring Harbor Laboratory of Quantitative Biology, Cold Spring Harbor, Long Island, New York.

Garen, A., 1968. Sense and Nonsense in the Genetic Code. *Science,* **160:** 149–159.

Grunberg-Manago, M., and S. Ochoa, 1955. Enzymatic Synthesis and Breakdown of Polynucleotides: Polynucleotide Phosphorylase, *Jour. Amer. Chem. Soc.,* **77:** 3165–3166.

Hosman, D., D. Gillespie, and H. F. Lodish, 1972. Removal of Formyl-methionine Residue from Nascent Bacteriophage f2 Protein. *Jour. Molec. Biol.,* **65:** 163–166.

Hoyer, B. H., B. J. McCarthy, and E. T. Bolton, 1964. A Molecular Approach in the Systematics of Higher Organisms. *Science,* **144:** 959–967.

Leder, P., and M. W. Nirenberg, 1964. RNA Codewords and Protein Synthesis III. On the Nucleotide Sequence of a Cysteine and a Leucine RNA Codeword. *Proc. Nat. Acad. Sci. (U.S.),* **52:** 1521–1529.

Nirenberg, M., and P. Leder, 1964. RNA Codewords and Protein Synthesis. *Science,* **145:** 1319–1407.

Nirenberg, M. W., and J. H. Matthaei, 1961. The Dependence of Cell-Free Protein Synthesis in *E. coli* upon Naturally Occurring or Synthetic Polyribonucleotides. *Proc. Nat. Acad. Sci. (U.S.),* **47:** 1588–1602.

Revel, M., 1972. Polypeptide Chain Initiation: the Role of Ribosomal Protein Factors and Ribosomal Subunits. In L. Bosch, ed. *The Mechanism of Protein Synthesis and its Regulation.* New York: American Elsevier.

Sarabhai, A. S., A. O. W. Stretton, and S. Brenner, 1964. Co-linearity of the Gene with the Polypeptide Chain. *Nature,* **201:** 13–17.

Shaw, D. C., J. E. Walker, F. D. Northrop, B. G. Barrell, G. N. Godson, and J. C. Fidders, 1978. Gene *K,* a New Overlapping Gene in Bacteriophage G4. *Nature,* **272:** 510–515.

Watson, J. D., 1976. *Molecular Biology of the Gene,* 3rd ed. Menlo Park, Calif.: W. A. Benjamin, Inc.

Whitfield, H., 1972. Suppression of Nonsense, Frameshift, and Missense Mutation. In L. Bosch, ed. *The Mechanism of Protein Synthesis and its Regulation.* New York: American Elsevier.

Woese, C. R., 1967. *The Genetic Code—the Molecular Basis for Genetic Expression.* New York: Harper & Row.

Woese, C. R., 1970. The Problem of Evolving a Genetic Code. *BioScience,* **20:** 471–485.

Yanofsky, C., B. C. Carlton, J. R. Guest, D. R. Helinski, and U. Henning, 1964. On the Colinearity of Gene Structure and Protein Structure. *Proc. Nat. Acad. Sci. (U.S.),* **51:** 266–272.

Zinder, N. D., D. L. Englehardt, and R. E. Webster, 1966. Punctuation in the Genetic Code. *Cold Spring Harbor Symposia Quant. Biol.,* **31:** 251–256.

18-1   Assume a length of template DNA with the deoxyribonucleotide sequence 3′ T A C C G G A A T T G C 5′. (a) If the code is triplet, nonoverlapping, and commaless, of which amino acid residues (in sequence) will the polypeptide chain for which this stretch of DNA is responsible consist? (b) If the code is triplet, overlapping by two bases, and commaless, how would you answer the preceding question? (c) Of what significance is the TAC triplet in the DNA template?

18-2   For the DNA template of the preceding problem assume the second C to be deleted. What now is the sequence of amino acid residues coded for if the code is assumed to be triplet, nonoverlapping, and commaless?

18-3   In the DNA length of problem 18-1 assume the second C to be deleted and a T to be inserted after the GG sequence so that the DNA strand now reads 3′ T A C G G T A A T T G C 5′. (a) How does the amino acid sequence now coded for compare with your answer to part (a) of problem 18-1? (Assume the code to be triplet, nonoverlapping, and commaless.) (b) Does your answer to the preceding part of this problem illustrate a sense, a missense, a nonsense mutation, or none of these?

18-4   Synthetic mRNA is constructed from a mixture of ribonucleotides supplied to a cell-free system in this relative proportion: 3 uracil:2 guanine:1 adenine. What fraction of the resulting triplets would be (a) UGA; (b) UUU?

18-5   The human hemoglobin molecule includes four polypeptide chains, two α chains of 141 amino acid residues each, and two β chains of 146 amino acid residues each. Neglecting chain-initiating and chain-terminating codons, as well as introns, (a) of how many ribonucleotides does the mRNA molecule responsible for the α chain consist? (b) What is the length in micrometers of that molecule? (c) Is your answer to part (b) likely to be too high, too low, or about right?

18-6   Assume an alanine tRNA charged with labeled alanine is isolated, and the amino acid is chemically treated to change it to labeled glycine. The treated amino acid–enzyme-tRNA complex is then introduced into a cell-free peptide synthesizing system. At which of two mRNA triplets, say GCU or GGU, would this tRNA now become bound? Why?

18-7   If single-base changes occur in DNA (and therefore in mRNA), which amino acid, tryptophan or arginine, is most likely to be replaced by another in protein synthesis?

18-8   Wittmann-Liebold and Wittman studied a number of mutants in the coat protein of the RNA-containing tobacco mosaic virus. Two of their mutants were

Mutant	Amino acid position	Replacement
A-14	129	isoleucine → threonine
Ni-1055	21	isoleucine → methionine

By reference to Table 18-1, explain what has happened in each of these cases.

18-9   Yanofsky et al. studied a large number of mutants for the tryptophan synthetase A polypeptide chain of *Escherichia coli*. This polypeptide chain consists of 267 amino acid residues. In the wild-type enzyme, a part of the amino acid sequence is: –tyrosine–leucine–threonine–glycine–glycine–glycine–glycine–glycine–serine–. In their mutant A446, cysteine replaces tyrosine; in mutant A187, the third glycine is replaced by valine. By reference to the genetic code, suggest a mechanism for each of these amino acid replacements.

18-10 How many different mRNA nucleotide codon combinations can exist for the internal pentapeptide threonine–proline–tryptophan–leucine–isoleucine?

18-11 From Table 18-1, note that (a) UUU and UUC code for phenylalanine; on the other hand, (b) So and Davie report that, with relatively high concentrations of ethyl alcohol, the incorporation of leucine, and isoleucine to a lesser degree, is sharply increased, whereas incorporation of phenylalanine is decreased, for these same codons. Which of the described situations, (a) or (b), represents ambiguity, and which degeneracy?

18-12 Assume a series of different one-base changes in the codon GGA, which produces these several new codons: (a) UGA, (b) GAA, (c) GGC, (d) CGA. Which of these represent(s) degeneracy, which missense, and which nonsense?

18-13 Assume the average molecular weight of an amino acid residue to be 125 daltons, and a eukaryotic polypeptide of 50,000 daltons. Assume further the mRNA transcript specifying that polypeptide to include one initiation and one termination codon. (a) Of how many ribonucleotides does that messenger RNA molecule consist? (b) Do introns need to be taken into account in answering part (a)? Why?

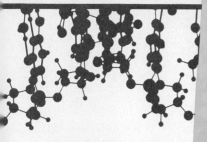

Molecul ... ture
of the g ...

In the previous discussion of the genetic code, an important question was raised: in terms of the operation of the genetic code, just how much of the nucleotide sequence comprises a gene? Even by raising this question, scientists have come a long way from the early, understandably vague concept of the gene as a "bead on a string," separated from adjacent genes by nongenetic material (although it is now known that many genes *are* separated by at least untranslated sequences and many also contain introns). To begin to solve this problem at the molecular level, the process of mutation must first be examined.

Mutations are sudden changes in genotype that involve **qualitative or quantitative changes in the genetic material itself.** Under this concept, recombination is excluded, for that process merely redistributes existing genetic material among different individuals; usually it makes no *change* in it. Geneticists often distinguish between two kinds of mutation, chromosomal and "point." So-called chromosomal mutations, or, preferably, chromosomal aberrations, may result in a change in the amount or position of genetic material. These have already been described in Chapters 13 and 14. "Point" mutations, however, are *changes within the DNA molecule*, and the term *mutation* is now generally employed in this more restricted sense. The term will be used here in this narrower meaning.

Of course, mutations may occur in any living cell that contains genetic material, either a somatic or a reproductive cell, and at any stage in its life cycle. Somatic mutations are perpetuated only in cells descended from the one in which the mutation originally took place. If the mutant trait is clearly detectable, a patch or sector of cells all of which have this new characteristic, will result. For example, Fig. 19-1 shows the result of a somatic mutation from "peppermint" to solid color that occurred in a zinnia in the author's garden. Whether or not somatic mutations produce an immediate effect depends on the nature of the change, the dominance relations of the original and mutant genes, and the stage of development of the individual or part at which the mutation occurs. Unless mutation takes place in reproductive cells or in tissues that will give rise to reproductive cells, and is maintained therein, the mutation will not be passed on to succeeding generations. Even so, whether it is detectable in a later generation depends on many factors, such as dominance, the environment, mating patterns, and so forth.

The discussion of protein synthesis and coding suggested that a gene may be identifiable as a *functional unit*, which has a definite cellular locus and consists of many nucleotides. But recall that a single nucleotide change will, in many instances, code for a different amino acid. Thus, base changes of a certain kind and amount (an addition, a deletion, or a replacement) might be expected to cause one or several amino acid substitutions. A quite different protein or even a nonfunctional one might result and therefore produce a lethal change or at least a different phenotype. Therefore, a gene may also be thought of as a *mutational unit*, which can be as little as a single deoxyribonucleotide pair. A functional gene might thus consist of very many mutable sites. Also, the genetic *recombinational unit* may be identifiable at the molecular level and therefore constitute a third view of the gene. So, many recombinational sites may exist within a gene, and perhaps even involve intra-codon recombination. It is therefore expected that the gene, when considered from the functional, mutational, and recombinational points of view, may show somewhat different characteristics.

Mutagenic agents

Although their mode of action is not always fully understood, a wide variety of agents have been implicated in the production of mutations. These mutagens may be broadly grouped into three classes, i.e., *radiation*, *chemicals*, and *temperature shock*. Some examples of the first two groups are the following:

1. Radiation.
   A. Ionizing.
      1. X-rays (wavelength 10^{-8} to 10^{-9} cm).
      2. γ rays (wavelength 10^{-9} to 10^{-10} cm).
      3. Cosmic rays (wavelength about 10^{-11} to 10^{-14} cm).
   B. Nonionizing (e.g., ultraviolet, wavelength about 10^{-4} to 10^{-6} cm).
2. Chemicals.
   A. Reactants and base analogs.
      1. Reactants with purines and pyrimidines.
         a. Nitrous acid.
         b. Formaldehyde.
      2. Base analogs.
         a. 5-bromouracil (a pyrimidine analog).
         b. 2-amino purine (a purine analog).
   B. Acridine dyes (e.g., acridine orange, acridine yellow, proflavin).
   C. Aklylating agents (e.g., mustard gases).
   D. Others.
      1. Carcinogens (e.g., methyl cholanthrene).
      2. Acids (e.g., phenols).

Not all of these are of equal effectiveness. Some are more effective in bacteria, for example, than in mammals, and some are more effective at certain loci than

Figure 19-1. *Mutant flower heads* (inflorescences) *of* Zinnia. *The somatic mutation was from "peppermint"* (splotched color) *to solid color.*

at others in the same organism. Other agents, not listed here, produce chromosomal aberrations (e.g., colchicine). It is not yet possible to induce specific mutations, although the effect of laser microirradiation in damaging small, preselected portions of the DNA of specific chromosomes is being explored.

RADIATION

Ionizing radiation. The spectrum of electromagnetic radiation is a broad one; it extends from the long radio waves, which may have wavelengths as great as several kilometers, down through cosmic rays, which may be as short as 10^{-14} cm. As wavelength becomes progressively shorter, the energy that the radiation particles contain becomes progressively greater, with the result that they penetrate cells and tissues. In this process particles of sufficient energy content may collide with one of the orbiting electrons of an atom, and knock it out of orbit. In this way a previously neutral atom becomes positively charged (an *ion*), because there is then a greater positive charge in the atomic nucleus than there are negative charges in the orbiting electrons. Ionized atoms, and the molecules in which they occur, are chemically much more reactive than neutral ones. In addition, the electrons lost from an atom in this process move off at high speed and cause other atoms to become ionized. Each electron lost from an atom is ultimately gained by another, which causes it to become a negatively charged ion. The result is a train of *ion pairs* along the path of the high-energy particles. Mutagenic effects result from the chemical reactions undergone by ions as their charges are neutralized.

One outcome of the ionization process is breakage of the sugar-phosphate strand(s) of DNA. This, of course, can lead to some of the chromosomal structural aberrations (such as deletions). But should strand breakage occur at two or more closely spaced points, one or more nucleotides or nucleotide pairs, may be lost, and as a result, the reading frame and the transcribed mRNA are altered. The ultimate outcome, of course, can then be a protein or an enzyme so altered as to be somewhat reduced in capacity to function, or even inoperative. The latter event may well be lethal, particularly in homozygotes. If this occurs in a reproductive cell, a phenotype, new for that particular line of descent, is created; in short, a *mutation* occurs.

There is compelling evidence also that some mutagenic effects of irradiation operate indirectly. Irradiated proteins or amino acids within a cell may themselves act as mutagenic agents. Seeding previously irradiated culture media also results in an increase in mutation rate with bacteria and *Neurospora*. In this situation it appears that hydrogen peroxide (H_2O_2) and other peroxides are produced; these are the actual mutagens.

Radiation applied to tissues is commonly measured in *roentgens* (r), named after the discoverer of X-rays, Wilhelm Roentgen. (It is worth noting that the unit of measurement is not capitalized, even though it represents the discoverer's name.) One roentgen (1r) produces about two ion pairs per cubic micrometer (μm^3), or about 1.6×10^{12} ion pairs per cubic centimeter (cm^3), of tissue. Radiation absorbed dose is expressed in *rads*; one rad is the amount of radiation that liberates 100 ergs of energy in a gram of matter. The rad is slightly larger than the roentgen, inasmuch as a gram of tissue exposed to 1r of γ rays absorbs about 93 ergs.

In many experimental organisms mutation rate is proportional to radiation dose, whether received as a "one-time" massive exposure or gradually over a period of time. Figure 19-2 shows this relationship for *Drosophila*. However, insect are more resistant to the effects of radiation than many organisms. For example, for given amounts of radiation applied to mouse spermatogonia, chronic, low-intensity exposure produces a significantly smaller amount of detectable mutations than the same total dose applied acutely. Exposure of mouse spermatogonia to 90r per minute produces about four times the number of mutations as does the same amount of radiation administered at the rate of 90r per week. It is believed that enzymatically controlled repair processes can take place here

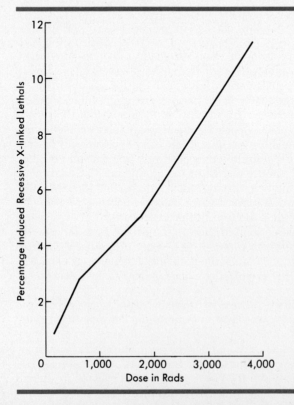

Figure 19-2. Percentage of recessive X-linked lethal mutations in Drosophila *sperm exposed to different doses of ionizing radiation. There is approximately a 1 percent increase in (point) mutation frequency for each 380 rad increase in dosage. (Based on I. H. Herskowitz, H. J. Muller, and J. S. Laughlin, 1959. "The Mutability of 18MEV Electrons Applied to* Drosophilia *Spermatozoa."* Genetics, **44:** 321–327.)

between successive low-intensity irradiations, whereas at the higher rate of exposure, these processes either cannot cope with the larger number of points of simultaneous damage or are themselves impaired.

All persons receive some radiation during their lifetimes from cosmic rays and from radioactive materials on the earth's surface.[1] In addition, medical X-ray examinations add to the accumulation of ionizing radiation, although only that received by the gonads contributes to the *genetic load* of future generations. The average person in developed countries probably receives less than 50r of radiation to the gonads during a 30-year reproductive period as a result of medical examinations. For example, a single X-ray examination of the teeth is reported to provide about 0.0008r to the gonads; a chest X-ray about 0.0006r in males, but some 0.002r in females; an abdominal X-ray 0.13r in males, and 0.25r in females; and a fluoroscopic examination of the pelvic region about 4 to 6r.

A number of calculations for the amount of radiation required to double the rate of mutation have been accumulated for mice and for *Drosophila*. Applying such figures to human beings requires several assumptions, the bases for which are not yet clear. Nevertheless, many authorities presently use the estimate of about 50r as the **doubling dose** for humans.[2] Although the amount of radiation received by the gonads for a given type of examination will vary among different institutions, it is helpful for each person to keep a record of the number and kind of X-ray and fluoroscopic examinations received.

Fallout of radioactive isotopes following above-ground nuclear explosions adds a small amount of gonadal exposure. In the 1960s, after testing of nuclear weapons in the air was incorrectly assumed to have ceased, it was estimated that such sources would contribute only about 0.1r to the gonads of each person during a reproductive lifetime. But the fission products of this kind of explosion remain in the upper atmosphere for long periods, settle slowly to earth and provide a continuing source of radioactive contamination. By the beginning of the twenty-first century, it has been estimated, some 60 percent of these fission products will still be present in the atmosphere. Among these are carbon-14, strontium-90, and cesium-137, which contaminate foodstuffs and water, and therefore serve as a source of at least somatic mutations in persons now living. So although the total amount of irradiation received per person from testing of military nuclear devices may appear small, experiments already conducted will serve as a source of danger for many years to come and to persons as yet unborn, an unsavory heritage for our children.

Nonionizing radiation. Ultraviolet light is a fairly effective mutagenic agent, although much less so than the ionizing radiations. Its effects on the pyrimidines cytosine and thymine may produce (1) photoproducts that cause local strand separation in DNA, or (2) dimers that may prevent strand separation and replication or interfere with normal base-pairing.

Thus ultraviolet may weaken the double bond between the fourth and fifth carbon of cytosine (Fig. 19-3), and allow water to be added at these points. The resulting photoproduct is unable to form hydrogen bonds properly with guanine, and this leads to strand separation. Ultraviolet radiation of about 2,800 Å likewise weakens the 4–5 C double bond, and permits two thymines to link as a dimer (Fig. 19-4). Such dimers may link thymine *between* strands, which interferes with replication, or connect adjacent thymines of the same strand, which disrupts normal T-A pairing.

Photoproducts and dimers, induced by ultraviolet irradiation, may be "repaired" (in *Escherichia coli* and phage) by (1) photoreactivation or (2) dark

[1]Such unusual factors as the famous radioactive monazite sands of small beach areas in the state of Kerala in extreme southwestern India, which the author has visited, increase this background radiation more than 10 times in that area.

[2]But see Chapter 22, pages 428–429.

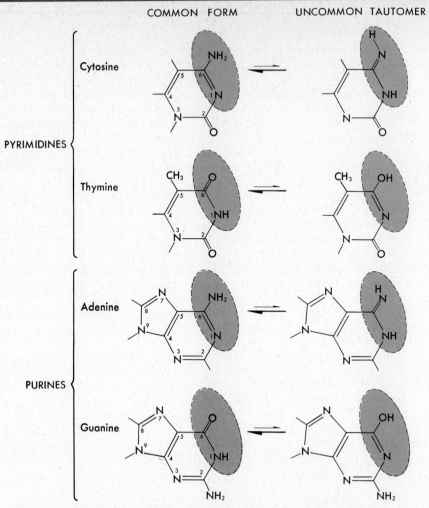

Figure 19-3. (A) Comparison of common forms of DNA bases with their rare tautomers. Tautomers, which may occur in several forms, differ from each other by rearrangements of protons and electrons. This is symbolized in the structural formula by changes in positions of H atoms and double bonds within the dashed ovals. Only the tautomer thought to be important in each case as a mutagen is shown here; its occurrence is quite rare.

reactivation. In photoreactivation an enzyme that has been bound to DNA during ultraviolet irradiation is activated by intense visible light supplied following ultraviolet exposure. The enzyme then catalyzes the cleavage of pyrimidine dimers, and restores their normal structure. Dark reactivation involves instead the enzymatic removal of dimers from the DNA molecule and synthesis of normal replacement segments for those excised, at least in certain genetic strains of bacteria.

CHEMICALS AND SUBNUCLEOTIDE CHANGES

Deamination. It is well known that various chemicals produce mutations; these are especially well documented in bacteria, yeasts, and phages. Nitrous acid (HNO_2) is one such *mutagenic* substance. It brings about changes in DNA

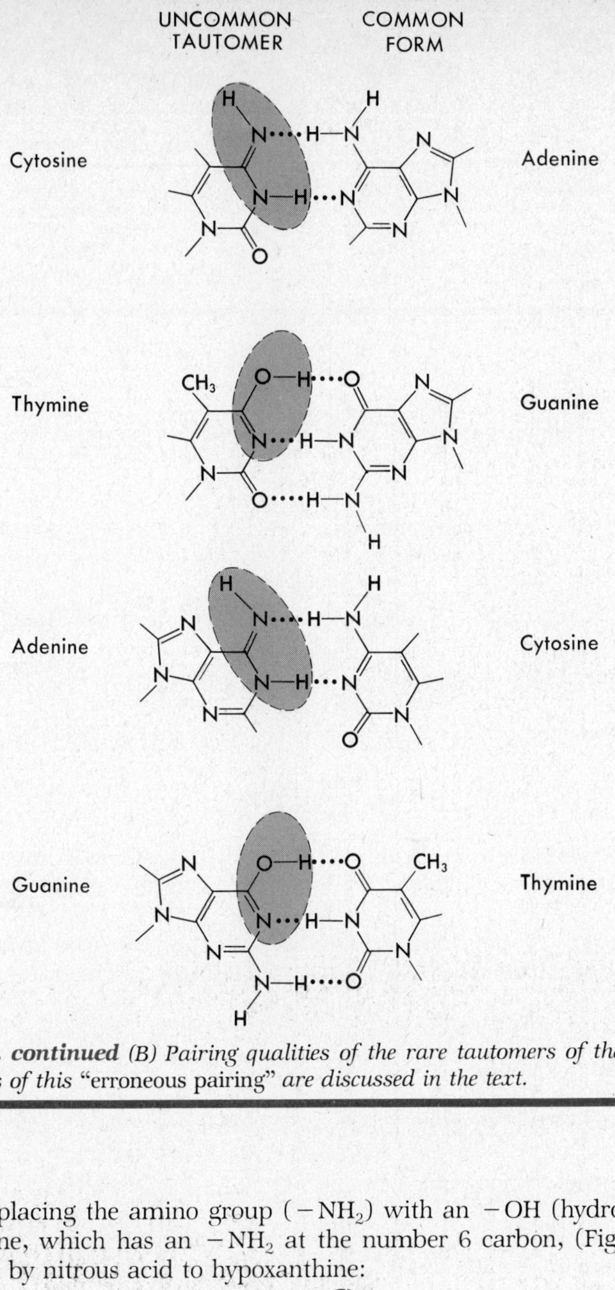

	UNCOMMON TAUTOMER	COMMON FORM	

Figure 19-3. continued (B) Pairing qualities of the rare tautomers of the four bases. Consequences of this "erroneous pairing" are discussed in the text.

bases by replacing the amino group ($-NH_2$) with an $-OH$ (hydroxyl) group. Thus adenine, which has an $-NH_2$ at the number 6 carbon, (Fig. 19-3A), is deaminated by nitrous acid to hypoxanthine:

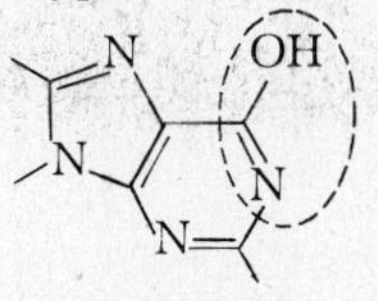

By a tautomeric shift, a more common (keto) tautomer is formed,

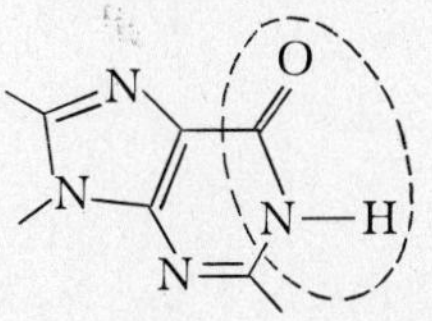

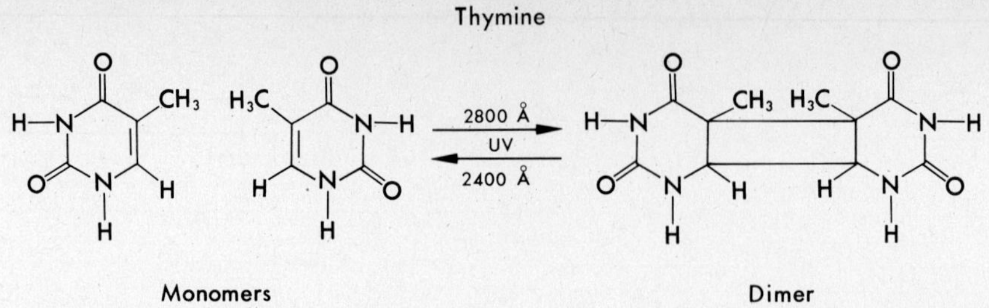

Figure 19-4. *The monomer ⇌ dimer conversion in thymine under ultraviolet light.*

which pairs with cytosine. Thus, an A:T pair can be converted to a G:C pair:

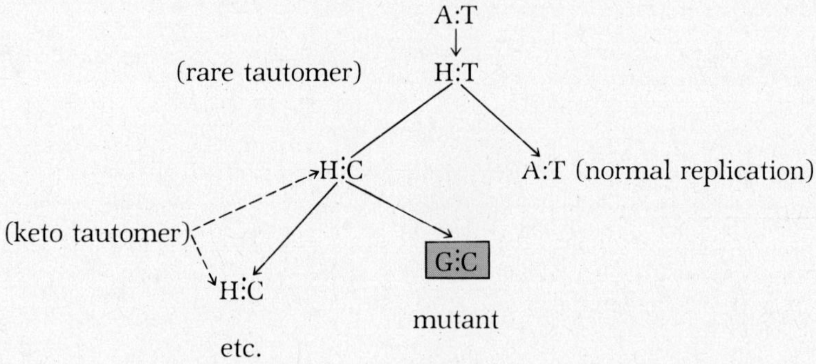

Similarly, deamination converts cytosine to uracil, which pairs with adenine (thus CG becomes TA), and guanine to xanthine, which pairs by *two* H bonds with cytosine.

Base analogs. Certain substances have molecular structures so similar to the usual bases that, if they are available, such **analogs** may be incorporated into a replicating DNA strand. One example will suffice to indicate the process and consequence. For instance, 5-bromouracil in its usual (keto) form

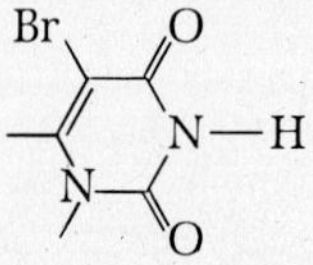

will substitute for thymine, which it closely resembles structurally. Thus an AT pair becomes and remains ABu (Fig. 19-5). There is some in vitro evidence to indicate that Bu immediately adajcent to an adenine in one of the DNA strands causes the latter to pair with guanine. But in its rarer (enol) state

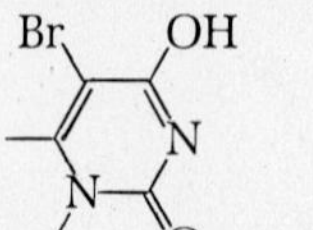

5-Bu behaves similarly to the tautomer of thymine (Fig. 19-3) and pairs with guanine. This converts AT to GC, as shown in Figure 19-5. Studies show that 5-Bu increases the mutation rate by a factor of 10^4 in bacteria.

Nitrous acid and base analogs like 5-Bu can produce transitions as well as cause reversion of transition mutants in phages and bacteria to their original

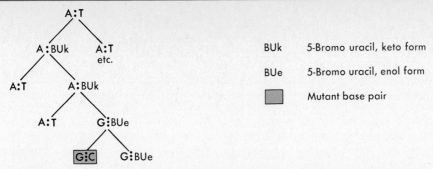

BUk — 5-Bromo uracil, keto form

BUe — 5-Bromo uracil, enol form

▢ — Mutant base pair

Figure 19-5. Substitution of the common keto tautomer of 5-bromouracil for thymine; its subsequent tautomerization to the rarer enol form, if it occurs, converts an A:T pair to a G:C pair. Moreover, the presence of 5-BU itself in the DNA sequence may cause imperfections in RNA synthesis.

state, regardless of what the initial mutagen may have been. Some *rII* mutants of phage T4 are transitions and can be reversed in this fashion.

Tautomerization. The purines and pyrimidines of DNA and RNA may exist in several alternate forms, or **tautomers.** Tautomerism occurs through rearrangements of electrons and protons in the molecule. Uncommon tautomers of adenine, cytosine, guanine, and thymine, shown in Figure 19-3(A), differ from the common form in the position at which one H atom is attached. As a result, some single bonds become double bonds, and vice versa.

Transitions. The significance of these tautomeric shifts lies in the changed pairing qualities they impart. The normal tautomer of adenine pairs with thymine in DNA; the rare (imino) form pairs with the normal tautomer of cytosine, as depicted in Fig. 19-3(B). The rare tautomer is unstable and usually reverts to its common form by the next replication. If tautomerism occurs in an already incorporated adenine, the result is a conversion of an AT pair to GC (Fig. 19-6A). On the other hand, if tautomerism occurs in an adenine about to be in-

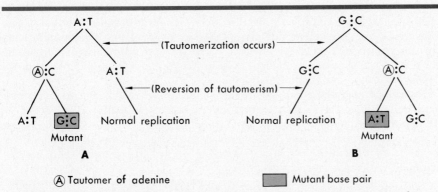

Ⓐ Tautomer of adenine   ▢ Mutant base pair

Figure 19-6. Conversion of an A:T pair to a G:C pair by tautomerization of an already incorporated adenine during the first replication of DNA. Reversion of the tautomerism in the second replication leads to the production of one mutant G:C strand. (B) Conversion of a G:C pair to an A:T pair by incorporation of a tautomer of adenine in the first replication of DNA. Assuming reversion of the tautomerism in the second replication, a mutant A:T pair is produced in place of the normal G:C pair. In both figures, the tautomer of adenine is indicated by the circled A; the mutant base pair is boxed.

corporated, the result is a conversion of a GC pair to AT (Fig. 19-6B). Such substitution of one purine for another, or of one pyrimidine for another, is termed a **transition.** Transitions may come about in a number of other ways, as described in the following sections.

Acridine dyes: whole nucleotide changes. The acridine dyes are effective as mutagens. As described in Chapter 18, they act by permitting base additions and/or deletions. If the intercalation of an acridine and consequent stretching of the DNA molecule occurs at the time of crossing-over between DNA molecules, the result can be unequal crossing-over. As a result, one strand may have one more nucleotide than its complement, and thus produce one daughter helix with one base pair more than the other. In general, the acridines are mutagenic in bacteria only at the time of recombination.

Mutation rate in human beings

If one considers the most reasonable estimates of the numerical values involved, some idea of the number of mutations that occur per human generation may be calculated. If one assumes a mean mutation rate per locus of 1×10^{-6} and accepts McKusick's estimate of at least 100,000 genes per diploid human cell, then the probability that any individual will produce a mutation during his or her reproductive lifetime is $1 \times 10^{-6} \times 1 \times 10^5 = 0.1$. Or it can be stated that one person in 10 may be expected to produce a mutation during reproductive life. For a United States population of some 226.5 million (U.S. Census Bureau), which is composed of two generations, those in the reproductive age bracket and those not, an average of $(2.265 \times 10^8) \times 1 \times 10^{-1}/2 = 1.1325 \times 10^7$, or more than 11.3 million mutations occur each generation. Because many of these involve nonsense or missense in the code, deletions in DNA, and the like, most of these >11 million mutations per generation may be expected to be deleterious or lethal. Fortunately, however, most mutant genes are recessive, which at least prevents their expression in any but those homozygous for the mutant gene. Each of us, it has been estimated, possesses at least 5 to 10 disadvantageous, deleterious, or lethal genes; because of the relatively large total number of genes that each of us is calculated to have, the probability of two unrelated persons having the *same* such (recessive) gene(s) is quite low. In a randomly mating population, the likelihood of homozygous recessives is thereby still further lessened when we are considering low-frequency disadvantageous or lethal genes. However, even rare deleterious genes become very real to a couple who have them in common. To a couple, when both persons are normal but heterozygous for the Tay-Sachs gene or for PKU, for example, the problem suddenly becomes much more than a statistical abstraction.

Not all genes mutate at a rate even as high as 1×10^{-6} per 25- or 30-year reproductive life. The estimated mutation rate for calf thymus histone 4 is 6×10^{-2} per 100 amino acid residues *per 100 million years*. Because the calf histone 4 molecule consists of only 102 amino acid residues, this reliable estimate is valuable for the entire polypeptide chain and shows an extreme level of conservation of molecular genetic structure.

Genetic polymorphism

The simultaneous occurrence of two or more discontinuously different genotypes in the same interbreeding population is referred to as **genetic polymorphism.** The Rh+ and Rh− genotypes and the major histocompatibility complex (MHC) with its human leukocyte antigens (HLA), illustrate this principle. The latter is presently the most complex genetic system known in humans. These and some others to be described provide evidence for the foregoing mechanisms of change in nucleic acids and indicate that (1) mutation can, and often does, involve a single base change, and (2) at the mutational level a gene may be as little as one nucleotide. Two illustrations in human beings, glucose-6-phosphate dehydrogenase and hemoglobin polypeptide variants (hemoglobinopathies), and one in bacteria, which involves the tryptophan synthetase enzyme system will be examined.

GENETICS OF GLUCOSE-6-PHOSPHATE DEHYDROGENASE (G6PD)

The enzyme glucose-6-phosphate dehydrogenase is involved in glucose metabolism. Production of this enzyme is determined by an X-linked locus that is closely linked to loci for hemophilia and one of the color-blindnesses. Its importance at this point centers in the extreme polymorphism of this locus. McKusick (1978) lists 199 variants of G6PD deficiency. Activity of so-called normal G6PD, which is determined by gene Gd^B, is set arbitrarily at 100 percent; activity of variant forms ranges from 0 to 400 percent of that of the normal enzyme. When present, the enzyme occurs in a variety of cells and tissues—red cells, white cells, platelets, spleen, liver, and skin, among others. It also occurs in saliva. G6PD is required for stability of glutathione, a tripeptide capable of alternate oxidation and reduction, which fills an important role in cellular oxidations. Inhalation of the pollen of the broad bean (*Vicia faba*) or ingestion of its seed (which is used as food in some areas of the world) or of such drugs as sulfanilamide, sulfapyridine, aspirin, or primaquine (an antimalarial drug) bring on sudden hemolytic anemia in persons of some, but not all, G6PD deficiency genotypes. In the presence of such agents as these, glutathione levels drop and the red cells are destroyed. Persons so affected recover when the triggering substance is removed. It should be stressed, however, that many variants of G6PD deficiency can be detected only in laboratory tests (e.g., electrophoretic mobility) and these variant individuals manifest no overt symptoms or discomfort (Table 19-1).

Table 19-1. Some variants of G6PD deficiency

Variant	Gene symbol	Red cell activity (percent of normal)	Incidence
Normal (B)	Gd^B	100	Common in all populations
A+	Gd^{A+}	80–100	Common in blacks
A−	Gd^{A-}	8–20	Common in blacks
Mediterranean	$Gd^{Mediterranean}$	0–7	Common in southern Europe
Athens	Gd^{Athens}	25	Common in Greece
Hektoen	$Gd^{Hektoen}$	400	Rare, primarily in U.S. whites
*Ohio	Gd^{Ohio}	2–16	Rare, primarily in persons of Italian descent
*Chicago	$Gd^{Chicago}$	9–26	Rare, primarily in persons of western European descent
*Worcester	$Gd^{Worcester}$	0	Probably rare

*Associated with hemolytic anemia.

The important point for consideration here is that the many variants of G6PD deficiency are the result of a large number of different amino acid substitutions in the enzyme. These have been determined thus far, however, for only a few forms, among them A+ and Hektoen. The A+ substitution has been found to be a replacement of asparagine by aspartic acid. The major problem in this work has been securing enough of the enzyme for the necessary analyses.

GENETICS OF HEMOGLOBIN

Structure of the hemoglobin molecule. That mutation may involve as little as a single deoxyribonucleotide pair has been clearly shown by the pioneering work of Ingram and the work of Hunt and Ingram on the chemical differences between normal and variant hemoglobins. Human hemoglobin is a protein with a molecular weight of about 67,000. The globin consists of four polypeptide chains, two alpha chains and two beta chains, each with iron-containing heme groups. As pointed out in Chapter 18, the α chain includes 141 amino acid residues, the β 146. In one molecule there are thus $(2 \times 141) + (2 \times 146)$, or 574 amino acid residues. Nineteen of the 20 biologically important amino acids are included, and their exact sequence in both α and β chains has been determined.

Tryptic digestion. Before biochemists had analyzed the complete sequence of amino acids in the α and β chains of hemoglobin, Ingram was able to report on chemical differences between hemoglobin of normal persons (hemoglobin A, or Hb-A) and that of individuals suffering sickle-cell anemia (hemoglobin S, or Hb-S). Because the molecule was too large and complex for total analysis at that time, Ingram digested it with trypsin. This enzyme breaks the peptide bonds between the carboxyl group of either arginine or lysine and the amino group of the next amino acid. Because there are about 60 of these amino acids in the hemoglobin molecule, approximately 30 shorter polypeptide segments (in duplicate) are produced in this way. Each of these was then analyzed by Ingram for amino acid content.

"Fingerprinting" peptides. Ingram placed small samples of the trypsin-digested hemoglobin (A or S) on one edge of a large square of filter paper, then subjected the peptide mixture to an electrical field in the process called electrophoresis. Under these conditions, differently charged portions will migrate characteristically. Next the filter paper with its as yet invisible, spread-out "peptide spots" was dried, turned 90° and placed with one edge in a solvent (normal butyl alcohol, acetic acid, and water). Because of differences in solubility of the peptides in this solvent, additional migration and spreading ensued. The paper was finally sprayed with ninhydrin, which produces a blue color in reaction with amino acids. The resulting chromatogram was called a "fingerprint" by Ingram, an apt description because differences in charge and amino acid content can thereby be picked out (Fig. 19-7) readily and characteristically.

When Ingram fingerprinted the peptides produced from hemoglobin A and S by tryptic digestion, he discovered all "peptide spots," except one which he called "peptide 4," of each to be identical in their locations on filter paper. This means that the long α and β chains of hemoglobins A and S are identical *except for one peptide* of the 30 kinds. This one carries a positive charge in Hb-S and no charge in Hb-A, a difference that by 1960 Hunt and Ingram were able to relate directly to a *single amino acid change* in the sequence of 146 composing the β chain (see Table 19-2).

Analyses of several hemoglobins. With the development of the electrophoretic-chromatographic technique for peptide analysis, studies of many different hemoglobins progressed rapidly. In addition, the complete amino acid sequence for both polypeptide chains is now known. The amino acid sequence for the β chain of hemoglobin A begins with these eight, starting with the NH_2 end: valine, histidine, leucine, threonine, proline, glutamic acid, glutamic acid, and lysine. The electrophoretic behavior of the peptides consisting of these eight amino acids is explainable by their electrical charge differences. Glutamic acid carries a negative charge, valine and glycine none, and lysine a positive charge.

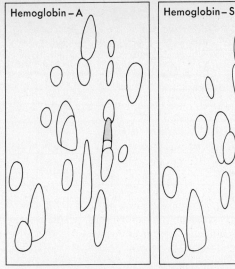

*Figure 19-7. Drawings of "fingerprint" chromatogram of the
hemoglobins A (normal) and S (sickle cell). Each enclosed
area represents the final location of a specific peptide; peptide
4 of Ingram is indicated by shading.*

Many other types of hemoglobinopathies have been similarly analyzed. Each
differs from Hb-A in the change of at least one amino acid; some of these are
in the α chain, others in the β. An extensive list of specific amino acid substitutions is given in McKusick (1978).

Transversions. Hemoglobin S differs from hemoglobin A (Table 19-2) by the
substitution of valine for glutamic acid as the sixth amino acid in the β chain.
Now the codons for glutamic acid (Table 18-1) are GAA and GAG, where valine
is coded for by GUA, GUG, GUC, and GUU. If, say, a GAA (glutamic acid) codon
undergoes a replacement to become GUA, valine will, of course, be coded for
(Fig. 19-8). Such a replacement in mRNA will occur if the DNA trinucleotide
CTT suffers a change to CTA; that is, the purine adenine replaces the pyrimidine
thymine. The substitution of a purine for a pyrimidine (or vice versa) is called
a **transversion.**
   So the far-reaching and important phenotypic difference between hemoglobin
A and that of persons with either sickle-cell trait or sickle-cell anemia is the

*Table 19-2. Comparison of the first eight amino acids of the β chain
of four hemoglobins*

Hb-A	Hb-S	Hb-C	Hb-G
Valine	Valine	Valine	Valine
Histidine	Histidine	Histidine	Histidine
Leucine	Leucine	Leucine	Leucine
Threonine	Threonine	Threonine	Threonine
Proline	Proline	Proline	Proline
Glutamic acid	VALINE	LYSINE	Glutamic acid
Glutamic acid	Glutamic acid	Glutamic acid	GLYCINE
Lysine	Lysine	Lysine	Lysine

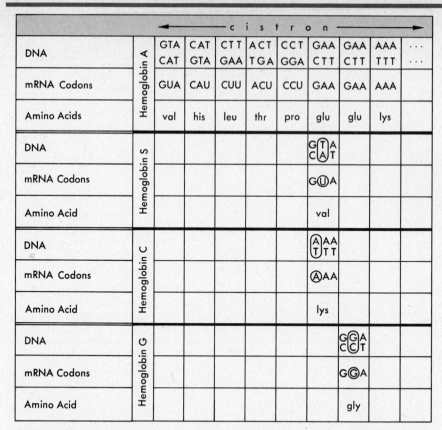

Figure 19-8. *Suggested derivation of three mutant β-chain hemoglobins from hemoglobin A by single nucleotide changes. The codon for particular amino acids is arbitrarily chosen when several may code for the same amino acid. Altered nucleotides are enclosed by solid lines.*

result of transversion, the change of one nucleotide pair in DNA from AT to TA. A *single nucleotide* in mRNA thus may spell the difference between life and death.

Transitions versus transversions. By definition, transitions and transversions represent different kinds of base substitutions. A comparison of the two may be represented thus:

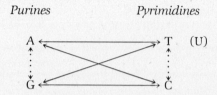

The dotted arrows represent transitions, the solid ones transversions.

Inheritance of hemoglobin variants. As described in Chapter 2, production of hemoglobin S is caused by a single gene that is codominant with the gene for hemoglobin A. It is now clear that genes for hemoglobins A, S, C, and G, all of which affect the β chain, are multiple alleles, and that the α chain is specified by a different series of multiple alleles. The α and β loci are not linked and produce changes in the alpha chain, therefore, do not themselves cause changes in the beta chain, and vice versa.

THE TRYPTOPHAN SYNTHETASE SYSTEM

In a sense, Ingram's work on hemoglobin really set the stage for the important work of Yanofsky on the tryptophan synthetase enzyme system of *Escherichia coli*. In this bacterium the complete enzyme molecule is a complex of two different subunits, A and B. The A component is a single polypeptide chain of 267 amino acid residues whose exact sequence is now known; the B component is a dimer of two identical polypeptide chains. The enzyme catalyzes three of the steps in tryptophan synthesis (Fig. 19-9). Both the A and B subunits are required for reaction (AB), whereas only the A component is necessary for indole production (reaction (A)), and the B component alone is needed for reaction (B). Both the A and B components are required, then, for synthesis of tryptophan.

Some mutants that are incapable of forming functional tryptophan synthetase do, however, produce a protein that, although enzymatically inactive, does give an immunological reaction with tryptophan synthetase immune serum prepared from rabbit blood. This protein has been named *cross-reacting material* (CRM); both *CRM*⁺ (producing cross-reacting material) and *CRM*⁻ (not producing cross-reacting material) mutants have been detected. It is believed that the difference between *CRM*⁺ and *CRM*⁻ mutants is that the latter produce either small frag-

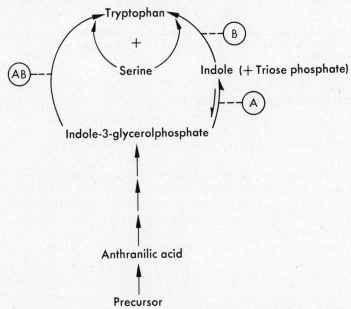

Figure 19-9. *Pathways of tryptophan synthesis involving tryptophan synthetase. The enzyme consists of two subunits, A and B, both of which are required for the reaction indole-3-glycerol-phosphate + serine → tryptophan (reaction* (AB) *). The A subunit is required for production of indole and triose phosphate (reaction* (A)*), and the B subunit for the reaction indole + serine → trypotophan (reaction* (B)*). Thus, A-deficient mutants require indole or tryptophan for growth, and B-mutants must be supplied with tryptophan.*

ments of the enzyme molecule or grossly abnormal molecules as the result of nonsense and missense mutations.

Mutants defective in the A polypeptide occur, as do others defective in the B component. Complementation tests make it clear that two different *functional genes* are involved, one for each of the components (A and B).

Yanofsky and his colleagues have determined the amino acid substitutions for a number of A mutants (Table 19-3). These not only clearly confirm the co-linearity of the genetic code, but also show that so-called point mutations may involve a change in as little as a single ribonucleotide (and therefore in a single deoxyribonucleotide pair). These changes, then, constitute the basis of genetic polymorphism.

Table 19-3. Some A mutants in the tryptophan synthetase system of Escherichia coli

Mutant	Amino acid position*	Wild type amino acid residue	Mutant amino acid residue
A3	48	Glutamic acid	Valine
A33	48	Glutamic acid	Methionine
A446	174	Tyrosine	Cysteine
A487	176	Leucine	Arginine
A223	182	Threonine	Isoleucine
A23	210	Glycine	Arginine
A46	210	Glycine	Glutamic acid
A187	212	Glycine	Valine
A78	233	Glycine	Cysteine
A58	233	Glycine	Aspartic acid
A169	234	Serine	Leucine

*Amino acid residues are numbered consecutively from 1 (amino end) to 267 (carboxyl end). Based on work of Yanofsky et al.

Fine structure of the gene

The older concept of the gene as a single, indivisible entity can, as predicted in Chapter 8, no longer be valid. On the one hand, there is the series of nucleotides that specifies the sequence of amino acid residues of a polypeptide chain (such as those of the A and B chains of the tryptophan synthetase enzyme, or the α and β chains of hemoglobin). But a change in as little as one nucleotide of the polypeptide-specifying gene may change (mutate) and produce a variant of the wild-type chain that differs in one amino acid residue. So the functional gene is not the same as the mutational gene, but appears to consist of many mutable sites. The gene must also be considered from the standpoint of the nature of the sites at which recombination may occur. The functional gene therefore appears to be composed of many mutational as well as recombinational subunits.

THE CISTRON

Phage T4 is a virus that infects the colon bacillus *Escherichia coli*. Plaques, or clear areas on agar plates of the host, indicate areas of infection and lysis. Those formed by the wild-type phage (r^+) are small and have rough or fuzzy edges. Many phage mutations that alter plaque morphology are known; the *r* (rapid lysis) mutants are easily recognizable because they produce larger plaques with sharp edges. The *r* mutants have been shown to involve three major map regions in the phage DNA and are designated *rI*, *rII*, and *rIII*. These loci were originally so numbered because it was thought that they belonged to three different linkage groups. Later investigations demonstrated that the T4 "chromosome" is a circular DNA molecule of close to 200,000 base pairs; the three rapid lysis loci occupy fairly widely separated positions on this unit DNA molecule (Fig. 19-10). Distances between them are great enough that early studies *appeared* to show independent assortment. The *rII* mutants cannot grow in strains of *E. coli* K12 that carry an integrated phage lambda (λ) "chromosome." This resistant strain is designated *E. coli* K12 (λ). The basis of resistance of this strain of the bacterium remains unknown. The *rII* mutants lyse *E. coli* B, but *rI* and *rIII* mutants lyse both strains.

The *rII* mutants fall into two groups, *rIIA* and *rIIB*. Each of these groups has been found to have an internal linear arrangement and to be immediately adjacent, but not overlapping, on the phage DNA molecule (Fig. 19-10). The *rIIA*

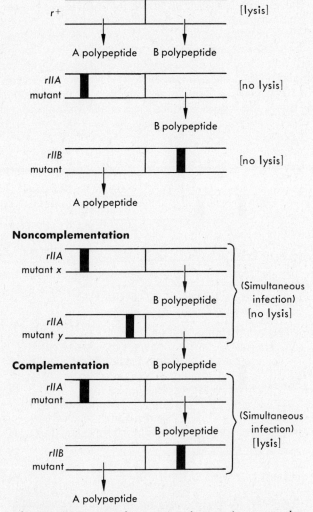

Figure 19-10. Complementation (see text) occurs when Escherichia coli *is infected simultaneously with an rIIA and an rIIB mutant phage; complementation does not occur if two different A mutants, for example, simultaneously infect the host. For reference, r⁺, rIIA, and rIIB mutants are diagramed also. Diagrams are not to any scale.*

region consists of about 2,000 deoxyribonucleotide pairs, or around 1/100 of the T4 genome. The A region transcribes a mRNA that translates an *A* polypeptide; the shorter *B* region is similarly responsible for a *B* polypeptide. Both are necessary for lysis of susceptible cells. The r^+ (wild-type) phage produces both *A* and *B* polypeptides; *A* mutants produce normal *B* polypeptide but not *A*, and vice versa. So infection with identical *rIIA* mutants, or by identical *rIIB* mutants alone fails to result in lysis. Nor does lysis ensue in cases of infection by mutants, each of which contains a different *rIIA* (or *rIIB*) mutation; none of these latter phages can produce *both A* and *B* polypeptides (Fig. 19-11). On the other hand, infection by two different phage strains, one an *rIIA* mutant and the other an *rIIB* mutant, does result in lysis (Fig. 19-11). That is, regions A and B are nonallelic and functionally different in that each is responsible for a different product, but they exhibit **complementation** in that the wild-type phenotype is expressed because each mutant strain makes up, as it were, for the defect in the other. Benzer found that with infection by two phage strains—one the wild-type (r^+) and the other doubly mutant in either the A or the B region, that is, with the mutations in the *cis* position—lysis occurred. But if the A (or B) mutations were in the *trans* configuration, lysis did *not* follow (Fig. 19-12). Thus it was seen that (1) mutations in one functional gene (A or B) are complementary only to mutations in the other region, and (2) complementation is detectable by the *cis-trans test*. Each functional region responsible for the production of a given polypeptide chain and defined by the *cis-trans* test was named a **cistron** by Benzer. The cistron therefore may be thought of as the gene at the functional level with three times as many deoxyribonucleotide pairs as the amino acid residues in the polypeptide coded for, if start and stop signals as well as any introns are ne-

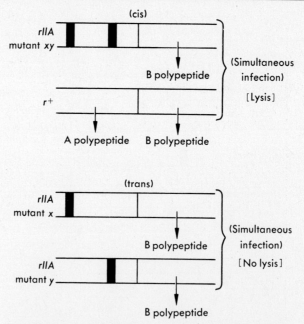

Figure 19-11. *Simultaneous infection of* Escherichia coli *by* rIIA *(or* rIIB*) double mutants and wild-type* r^+ *phages* (top diagram), *where the cistron defects of the mutant are in the* cis *position (i.e., in the same DNA molecule), results in lysis. When the mutations are in the* trans *position (lower diagram), i.e., in different DNA molecules, no lysis takes place.*

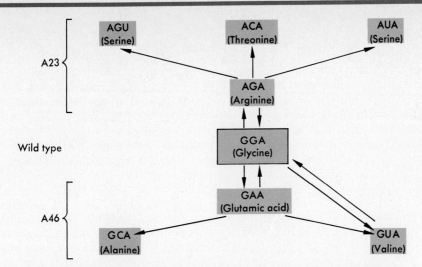

*Figure 19-12. Derivation of A23 and A46 tryptophan synthetase mutants in
E. coli by single nucleotide substitutions. Based on the work of Yanofsky.*

glected. Thus the length of a cistron (either in number of deoxyribonucleotides
or in units of length such as the micrometer) is defined by the number of amino
acid residues of the polypeptide chain for which it is responsible, plus initiating,
terminating, and any untranscribed nucleotides.

However, the globin genes, and many others of mammals, are not so arranged.
One of the beta globin chains of the mouse is not coded for by an uninterrupted
series of 146 codons, but rather is interrupted between the codons for amino
acid residues 104 and 105 by an intron as large as the gene itself. The human δ
and β globin chains are separated by about 7,000 deoxyribonucleotide pairs.
Nienhuis gives the sequence as follows: (1) codon sequence for the N-terminal
part of δ globin, (2) an intervening sequence, (3) codon sequence for the C-
terminal part of the δ globin, (4) 7,000 intervening nucleotide pairs, (5) codon
sequence for the N-terminal portion of β globin, (6) an intervening sequence,
and (7) codon sequence for the C-terminal portion of β globin. The human
proinsulin molecule contains a sequence of 35 amino acid residues that are
enzymatically removed, then discarded, in the conversion of proinsulin to the 30
and 21 amino acid residue chains of insulin (see Chapter 22). In the proinsulin
case, at least, the intervening sequences are transcribed *and* translated, but the
translational product is subsequently discarded. Significance of these intervening
sequences is unknown, but they do clearly demonstrate the cistrons (genes) in
at least some cases are discontinuous.

THE MUTON

Study of the genetic code (Table 18-1) and the tryptophan synthetase mutants
listed in Table 19-3 makes it clear that a change in a single deoxyribonucleotide
pair may result in a *missense* codon in transcribed mRNA, (e.g., AGC → AGA)
or *nonsense* (e.g., UGC → UGA). So a cistron may be expected to consist of many
mutable sites, or **mutons,** as they have been termed by Benzer. *A muton is thus
the smallest length of DNA capable of mutational change.*

As previously noted, many mutant hemoglobins differ from normal hemoglobin A by single amino acid substitutions and these may be accounted for in
many instances by single nucleotide changes (both transitions and transversions).
By the same kind of reasoning and based on clear experimental evidence, Yan-

ofsky has shown how each of the many tryptophan synthetase mutants may occur by nucleotide substitutions. Two ultraviolet-induced mutants, A23 and A46, have been especially revealing. In one of the A23 mutants, glycine is replaced by arginine, whereas in one of the A46 mutants, the *same* glycine (at position 210, as indicated in Table 19-3) is replaced by glutamic acid. The codon changes responsible may be designated, for example, as GGA (glycine, wildtype), AGA (argenine, A23 mutant), and GAA (glutamic acid, A46 mutant), as diagramed in Figure 19-13. These examples clearly indicate that the muton often is as little as one nucleotide, and that a cistron includes many mutons. Nucleotide changes in these mutable sites may, of course, produce sense, missense, or nonsense reading changes.

THE RECON

Tryptophan synthetase system. Recall that the discussion of pseudodoalleles in Chapter 8 suggested that crossing-over may occur within a functional gene or, to use Benzer's terminology, a **cistron.** The A23 and A46 tryptophan

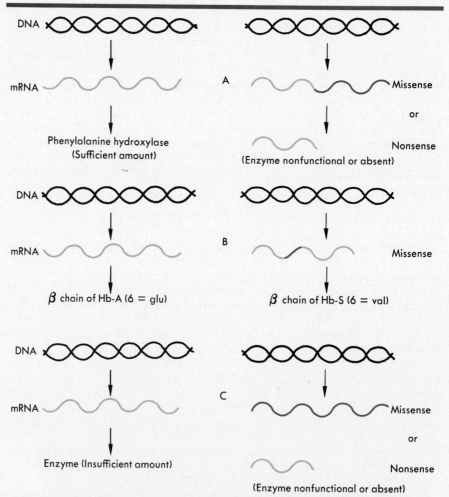

Figure 19-13. *Diagrammatic representation of (A) complete dominance, (B) codominance and (C) incomplete dominance in terms of the nature of DNA and its responsibility for polypeptide translation via transcribed mRNA. See text for details.*

synthetase mutants of *Escherichia coli* confirm this suggestion. These two mutants, which involve the same glycine of the A protein, have been shown by recombinational studies *not* to map at precisely the same point, but to be 0.001 map unit apart. In fact, Yanofsky comes to some interesting conclusions in comparing genetic map length with distances in the polypeptide chain. He finds the map length for 187 amino acid residues (positions 48 through 234) of the A protein in the *E. coli* tryptophan synthetase system to be about three map units, or $3/187 = 0.016$ map unit per amino acid residue, and thus also for a group of three adjacent nucleotides. Distances between adjacent nucleotides are thus of the order of 0.001 to 0.005 map unit.

It is therefore evident that, just as mutation may occur *within* a triplet codon, so may recombination. Furthermore, Yanofsky's studies of both mutation and recombination (by means of transduction) clearly show that in each protein with a different amino acid at a given position, the amino acids on either side remain unchanged. To quote Yanofsky, "this is perhaps the most convincing evidence available that excludes overlapping codes." Thus we are provided with still another subdivisional concept of the cistron, namely the **recon.** *A recon is the smallest unit of DNA capable of recombination* or (in bacteria) of being integrated by transformation or transduction.

The rII region again. Confirmation of this concept of the recon is provided also by Benzer's work on the *rII* region of phage T4. If two different A mutations or two different B mutations are located far enough apart (and this distance need not be very large) in their respective cistrons, *recombination* may occur between them. For example, infection of B-type hosts with two *different* A mutants results in lysis (*rII* mutants can lyse B cells) and release of progeny phage particles. These latter are then used to infect K12 hosts that do not have the integrated λ DNA. Any lysis must, therefore, result from recombination between two different A mutations to produce wild-type phage. The same effect is noted in the B cistron. Careful study has made it possible to detect recombinants with frequencies as low as 0.00001 percent (1×10^{-7}). Some of the A × A and B × B "crosses" fail, however, to result in recombination; these have been found to be deletions.

From recombination experiments involving both "point mutations" and deletions, Benzer has found over 3,000 mutants, which have more than 300 sites identifiable by recombination in the entire *rII* region. Subsequent work has extended these findings to the point where at least 500 recombinational sites have been identified in the *rIIA* region alone. From a variety of evidence, the T4 DNA molecule is calculated to have a molecular weight of about 130,000,000. With 650 as the average molecular weight of a deoxyribonucleotide *pair*, the number of nucleotide pairs in the entire T4 genome is 130,000,000/650, or about 200,000 as previously noted. Estimates of the number of base pairs in the A cistron range up to 2,500, and generally between 1,000 and 1,500 for the B cistron. With an estimate of 2,000 nucleotide pairs for the *rIIA* region, the unit of recombination (the *recon*) in the A cistron can include no more than an average of 2,000/500, or four nucleotide pairs. If one considers the degeneracy of the code and the probable fact that not all recombinational instances have been found, it appears that the recon, like the muton, is very small and must consist of as little as one nucleotide pair. Mutons and recons are, therefore, structurally indistinguishable.

Dominance and recessiveness

As a result of the consideration of DNA, the genetic code, polypeptide synthesis, and the molecular nature of the gene, it is now possible to understand how dominance, incomplete dominance, and codominance operate in terms of the structure and function of the genetic material.

Phenylketonuria will serve as an illustration of complete dominance. Recall that PKUs, which are homozygous recessives, are unable to synthesize the enzyme phenylalanine hydroxylase, with the result that phenylalanine accumulates. Heterozygotes are normal, although they have a defective cistron in one set of their chromosomes. Apparently the wild-type cistron in the other set of chromosomes is able to mediate synthesis of a sufficient amount of enzyme to convert phenylalanine to tyrosine. The defective cistron may be assumed to contain a gross missense, or even a nonsense mutation (Fig. 19-13A) so that its product is completely nonfunctional.

Genes for hemoglobins A and S are, as noted earlier, codominant. These two hemoglobins differ by a single amino acid substitution in the sixth position of the beta chain (page 379). In this case, clearly a missense mutation is responsible for producing the aberrant hemoglobin, whereas the other cistron in heterozygotes gives rise to the normally functioning hemoglobin A (Fig. 19-13B).

In Chapter 2 an example of incomplete dominance in the red, pink, and white phenotypes in the flowering plant snapdragon was cited. Here it is quite possible that one cistron, which can be designated a^1, produces the appropriate enzyme needed for synthesis of the red pigment, but, by itself, in an amount insufficient to give the petals the full red color. Thus, if the defective cistron is designated a^2, a^1a^1 plants produce sufficient functional enzyme to give enough pigment to result in red petals, the heterozygotes (a^1a^2) produce only enough to give a paler color, and a^2a^2 plants fail to produce any functional enzyme at all so that the petals remain unpigmented (Fig. 19-14C).

Because defects may occur at so many different intracistronic sites, it is not surprising that forward mutations generally have a higher incidence than do back mutations, that is why $u > v$ in many cases (Chapter 15).

Conclusions

The last four chapters have given the following important conclusions regarding the gene:

1. The genetic material is DNA in all organisms except a few viruses, where it is RNA.
2. A gene is a specific linear sequence of nucleotides and may be considered at more than one level of organization.
   a. The functional unit, the *cistron*, is responsible for specifying a particular polypeptide chain and consists of three adjacent deoxyribonucleotide pairs for each amino acid residue in the chain, plus a minimum of one chain-initiating triplet and one or more chain-terminating triplets.
   b. Eukaryotic cistrons often consist of transcribed and translated sequences (exons), interrupted by untranscribed and/or untranslated sequences (introns).
   c. Each cistron is divisible into
      1. *Mutons*, the smallest number of nucleotides independently capable of producing a mutant phenotype. In many cases this is a single nucleotide or nucleotide pair.
      2. *Recons*, the smallest number of nucleotides capable of recombination. This likewise is as small as one nucleotide. Recons and mutons are structurally identical.
3. Specification of the amino acid sequence of a polypeptide chain by a cistron is made possible through the latter's particular deoxyribonucleotide sequence.
   a. Mediation of polypeptide synthesis operates through an intermediate, messenger RNA, which is ordinarily a single-stranded, base-for-base,

complementary transcription of the nucleotide sequence on one DNA strand, plus a 5' "cap" and a poly-A "tail."

b. Messenger RNA, in conjunction with polysomes, tRNA, energy sources, and a battery of enzymes, is directly responsible through its sequence of ribonucleotides (the genetic code) for polypeptide biosynthesis at ribosomal surfaces, chiefly in the cytoplasm (exceptions will be taken up in Chapter 20). Proteins are then synthesized from the polypeptide chains thus formed.

c. The genetic code is
  1. Triplet.
  2. "Commaless."
  3. Nonoverlapping.
  4. Degenerate.
  5. Essentially nonambiguous.
  6. Colinear.
  7. Universal.

References

Benzer, S., 1955. Fine Structure of a Genetic Region in Bacteriophage. *Proc. Nat. Acad. Sci. (U.S.)*, **41:** 344–354.

Benzer, S., 1961. On the Topography of the Genetic Fine Structure. *Proc. Nat. Acad. Sci. (U.S.)*, **47:** 403–415.

Benzer, S., 1962. The Fine Structure of the Gene. *Sci. Amer.*, **206**(1): 70–84.

Benzer, S., and E. Freese, 1958. Induction of Specific Mutations with 5-Bromouracil. *Proc. Nat. Acad. Sci. (U.S.)*, **44:** 112–119. Reprinted in G. S. Stent, ed., 1965, 2nd ed. *Papers on Bacterial Viruses*. Boston: Little Brown.

Hunt, J. A., and V. M. Ingram, 1958. Allelomorphism and the Chemical Differences of the Human Haemoglobins A, S, C. *Nature*, **181:** 1062.

Hunt, J. A., and V. M. Ingram, 1960. Abnormal Haemoglobins, IV. The Chemical Difference Between Normal Human Haemoglobin and Haemoglobin C. *Biochim. Biophys. Acta*, **42:** 409–421. Reprinted in J. H. Taylor, ed., 1965. *Selected Papers on Molecular Genetics*. New York: Academic Press.

Ingram, V. M., 1956. A Specific Chemical Difference Between the Globins of Normal and Sickle-Cell Anaemia Haemoglobin. *Nature*, **178:** 792–794.

Ingram, V. M., 1957. Gene Mutations in Human Haemoglobin: the Chemical Difference Between Normal and Sickle-Cell Haemoglobin. *Nature*, **180:** 326–328. Reprinted in S. H. Boyer, ed., 1963. *Papers on Human Genetics*. Englewood Cliffs, N.J.: Prentice-Hall.

McKusick, V. A., 1978. *Mendelian Inheritance in Man*, 5th ed. Baltimore: Johns Hopkins Press.

Muller, H. J., 1927. Artificial Transmutation of the Gene. *Science*, **66:** 84–87.

Nienhuis, A. W., 1978. Mapping the Human Genome. *New England Jour. Med.*, **299:** 195–196.

Radding, C. M., 1973. Molecular Mechanisms in Genetic Recombination. In H. L. Roman, ed. *Annual Review of Genetics*, vol. 7. Palo Alto, Calif.: Annual Reviews, Inc.

Russell, W. L., L. B. Russell, and E. M. Kelly, 1958. Radiation Dose Rate and Mutation Frequence, *Science*, **128:** 1546–1550.

Stadler, D. R., 1973. The Mechanism of Intragenic Recombination. In H. L. Roman, ed. *Annual Review of Genetics*, vol. 7. Palo Alto, Calif.: Annual Reviews, Inc.

Stamatoyannopoulos, G., 1972. The Molecular Basis of Hemoglobin Disease. In H. L. Roman, ed., *Annual Review of Genetics*, vol. 6. Palo Alto, Calif.: Annual Reviews, Inc.

Yanofsky, C., 1963. Amino Acid Replacements Associated with Mutation and Recombination in the A Gene and Their Relationship to in Vitro Coding Data. *Cold Spring Harbor Symposia Quant. Biol.*, **28:** 581–588.

Yanofsky, C., 1967. Structural Relationships Between Gene and Protein. In H. L. Roman, ed. *Annual Review of Genetics*, vol. 1. Palo Alto, Calif.: Annual Reviews, Inc.

Yanofsky, C., G. R. Drapeau, J. R. Guest, and B. C. Carlton, 1967. The Complete Amino Acid Sequence of the Tryptophan Synthetase A Protein (α Subunit) and its Colinear

Relationship with the Genetic Map of the A Gene. *Proc. Nat. Acad. Sci. (U.S.),* **57:** 296–298.

Yoshida, A., 1967. A Single Amino Acid Substitution (Asparagine to Aspartic Acid) Between Normal (B+) and the Common Negro Variant (A+) of Human Glucose-6-Phosphate Dehydrogenase. *Proc. Nat. Acad. Sci. (U.S.),* **57:** 835–840.

Yoshida, A., G. Stamatoyannopoulos, and A. G. Motulsky, 1967. Negro Variant of Glucose-6-Phosphate Dehydrogenase Deficiency (A−) in Man. *Science,* **155:** 97–99.

Problems

19-1   From this chapter and any other sources available to you, explain why most mutations are deleterious.

19-2   From this chapter and any other sources available to you, explain why most mutations are recessive.

19-3   From this chapter and any other sources available to you, evaluate the short-term and long-term effects of mass irradiation of the human population.

19-4   Look again at Figure 19-2. Would you say that there is a threshold dose of irradiation below which no mutation is induced?

19-5   X-linked recessive mutations are more easily studied in appropriate organisms than are autosomal ones. Why?

19-6   Which of the following would be likely to suffer the greatest, and which the least, genetic damage from radiation exposure: (a) a monoploid, (b) a diploid, (c) a polyploid?

19-7   Three species of oats (*Avena*) are listed in Table 4-1. Irradiating many samples of these three species with the same amount of X-irradiation resulted in the following rates of induced detectable mutations (given as $\bar{x} \pm s$ per sample):

Species	Irradiated	Control
A. brevis	4.1 ± 1.1	0.2 ± 0.03
A. barbata	2.4 ± 0.6	0.05 ± 0.01
A. sativa	0.0	0.0

Give a reason for this kind of result.

19-8   Suppose samples of the species of wheat (*Triticum*) listed in Table 4-1 were each given the same dose of X-irradiation. Assuming that some radiation-induced mutation occurred in at least some of those species, which species would you expect to show the highest frequency of mutation?

19-9   In an imaginary flowering plant, petal color may be either red, white, or blue. White is due to the presence of a colorless precursor from which red and blue pigments may be synthesized in two consecutive, enzyme-controlled processes. The cross of two blue-flowered plants yields an F_1 ratio of nine blue, three red, and four white. (a) How many pairs of genes are involved? (b) Suggest the correct sequence of enzyme-controlled steps and products. (c) Starting with the first letter of the alphabet and using as many more in sequence as needed, give the genotype of (1) the P individuals; (2) the F_1 blue plants. (d) Give genotypes for (1) red and (2) white F_1 plants. (e) According to the system you have worked out, which pair of genes controls the first step of the process you have worked out in part (b) of this problem?

19-10   By deamination a $\begin{smallmatrix} 5' \ T\,T\,T\ 3' \\ 3' \ A\,A\,A\ 5' \end{smallmatrix}$ segment of a DNA molecule is changed to a $\begin{smallmatrix} 5' \ T\,T\,G\ 3' \\ 3' \ A\,A\,C\ 5' \end{smallmatrix}$ segment. What change in amino acid incorporation does this produce, assuming the 3′, 5′ ("lower") strand serves as the template for transcription?

19-11 Look again at Figure 19-5. Does the enol form of 5-bromouracil cause a transition or a transversion?

19-12 Table 19-3 shows two tryptophan synthetase mutants, A78 and A58, that affect amino acid residue 233. In A78 wild-type glycine at this position is replaced by cysteine; in A58 glycine is replaced by aspartic acid. Among the mRNA codons for glycine is GGC; one of the two for cysteine is UGC; and one of the two for aspartic acid is GAC. Which of the mutant strains, A78 or A58, involves a transition and which a transversion in the DNA template?

19-13 Tryptophan synthetase mutant A446 has cysteine at amino acid position 174 instead of wild-type tyrosine. This could come about by a change in the DNA template strand from ATG to ACG. Suppose strain A446 were irradiated and the same DNA template triplet were converted from ACG to ACT. (a) What kind of code reading error does this produce? (b) Would you expect a functional polypeptide to be formed or not? (c) If it is assumed that a polypeptide chain of some length is formed as a result of the change described here, of how many amino acid residues would it consist?

19-14 Suppose a certain cistron is found to consist of 1,500 deoxyribonucleotides in sequence. (a) What is the maximum number of mutons of which this cistron could consist? (b) Is this number likely to be too high, too low, or about right for an actual organism? Why?

19-15 For how many codons is the tryptophan synthetase A cistron responsible, if initiating, terminating, and possible introns are neglected?

19-16 If initiating, terminating, and possible introns are neglected, how many nucleotides are there in the tryptophan synthetase A cistron?

19-17 One calculation for the molecular weight of the DNA in a single (*monoploid*) set of human chromosomes is 1.625×10^{12}. (a) If 650 is assumed as the average molecular weight of a pair of deoxyribonucleotides, how many nucleotide pairs comprise the DNA of one *diploid* somatic nucleus? (b) What, then, is the maximum number of mutons in each of your somatic cells? (c) What is the total length in micrometers (microns) of the DNA in the nucleus of one of your *diploid* somatic cells? (d) At these values, for how many mRNA codons is the DNA of *one* of your (*monoploid*) sets of chromosomes responsible? (e) If you assume an average of 300 amino acid residues per polypeptide chain, and neglect introns and start-stop signals, how many cistrons do you have in *one* of your sets of 23 chromosomes?

19-18 Explain why one mutation in a given cistron may be lethal, whereas another in the same cistron may produce no adverse phenotypic effect and be detectable only by laboratory tests.

19-19 How might a multiple-allele series arise at a given locus during the evolution of a particular species?

19-20 Why are not chemical and radiational mutagens used more widely in developing improved varieties of crop plants and domestic animals?

19-21 Distinguish among these three concepts: cistron, muton, and recon.

19-22 How do complementation and recombination differ?

19-23 A group of mutants shows a series of one-base substitution mutations at a site which, in the wild-type, is occupied by serine: serine → asparagine → threonine → isoleucine. Sequencing of wild-type mRNA shows the serine codon to be AGU. (a) Give the sequence of mutant codons; start with the wild-type codon. (b) The change from which codon(s) to another constitutes a transversion?

CHAPTER 20

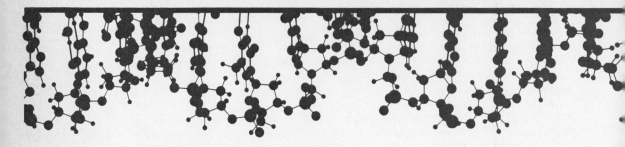

Regulation of gene action

Multicellular organisms normally develop by an orderly process of differentiation from a single cell, the zygote, to an adult form of many different kinds of cells, tissues, and organs. But because of the behavior of chromosomes in mitosis, all the cells of a complex, multicellular soma may be presumed to have the same genotype. In short, between zygote and adult stages, cells of the organism *differentiate*, both physiologically and physically, yet the genome of all cells should be identical under normal conditions. The problem may be quite simply stated: *how do genetically identical cells become functionally different?*

Up to at least a certain point in development, the cells of a multicellular organism are *totipotent*; that is, they are capable of forming complete bodies if isolated after development has begun. For example, in carrot, excised, mature, differentiated cells from phloem (the food conducting vascular tissue) form an irregular mass of large, thin-walled, undifferentiated cells (called a callus) on a solid nutrient medium. Transfer of the callus to a liquid medium, which is kept gently shaken, separates the cells. Upon subsequent transfer to new cultures, these callus cells develop into embryolike structures, or *embryoids*. Liquid media induce root formation by the embryoid; later transfer to a solid medium results in stem and flower production, in short, a fully differentiated, mature plant. A small section of the body of the flatworm *Planaria* or the mature leaf of a plant like the African violet (*Saintpaulia ionantha*) regenerates the entire organism under proper environmental conditions, though this is by no means true for all animals and plants. Many experiments with very young embryos of sea urchins and frogs indicate that even after the zygote has divided to produce two to eight cells, separation of cells or forcing them to divide in only one plane is followed by normal differentiation and embryo formation. Separation of the cells of two-celled rabbit embryos and then reimplantation into another female similarly produces normal embryos.

A closely related procedure is now used to produce vastly larger numbers of calves from prize cows than would be possible by the usual biological procedures. Such cows are first hormonally induced to superovulate, then inseminated (either naturally or artificially), with the result that most of the ova are fertilized. Within a week pregnancy can be determined and the very small embryos flushed out by buffered saline solution. These are then implanted singly in dams of undistinguished ancestry. Whereas a cow can be mated but once a year, and generally produces one calf each time, this procedure assures large numbers of offspring. In 1980, for example, some 20,000 calves were born to unmated heifers. A year earlier a prize Simmental cow produced 89 calves!

In the tobacco plant, investigators have reported in vitro development of microspores into monoploid plants that reach flowering stage, although seeds are not produced. (Why?) An exciting development involves protoplast fusion, which uses protoplasts (naked or wall-less cells) from plant species that cannot be crossed. Walls are first destroyed enzymatically, then, with polyethylene glycol added to the culture medium, some one to ten percent of the protoplasts fuse. The resulting cells are next stimulated to reform a common wall, then (by still another culture medium) induced to differentiate into new plants. In this way, for example, potato plants with the resistance to late blight from a tomato genome have been developed in the laboratory. Such techniques will soon be widely available to growers.

The key to orderly structural and functional differentiation of multicellular organisms lies in the fact that their cells do not always produce the same proteins all the time. Even unicellular forms such as bacteria display a temporal shift in protein synthesis. Apparently some mechanism exists in the cell to turn genes "on" or "off" at different times and/or in different environments. In fact, it is now clear from considerable experimental evidence that differentiation is the consequence of orderly, temporal activation and repression of genetic material. Gene regulatory mechanisms may operate at either the transcriptional level or at some posttranscriptional stage. As might be expected, regulatory systems are dissimilar in eukaryotes and prokaryotes. This *regulation of gene action* is examined in this chapter.

Evidence of regulation of gene action

CHROMOSOME PUFFS

Giant chromosomes.   In dipteran flies, the chromosomes of larval tissues such as the salivary glands regularly exhibit unusual behavior that has shed some light on gene regulation. Although the cells of these tissues do not divide, their chromosomes replicate repeatedly while permanently synapsed in homologous pairs, and produce polytene or many-stranded giant chromosomes. If these chromosomes are examined at several stages of the individual's development, specific areas (sets of bands) are seen to enlarge into prominent "puffs" or Balbiani rings (Fig. 20-1) from electron micrographs. The puff is described as a loosening of the tightly coiled DNA into long, looped structures not unlike the lampbrush chromosomes described in the next section. These puffs appear and disappear in a given tissue at certain chromosomal locations as development proceeds; those at particular locations are correlated with specific developmental stages of the insect. The pattern of puffing varies in a regular and characteristic way with the tissue and its stage of maturation.

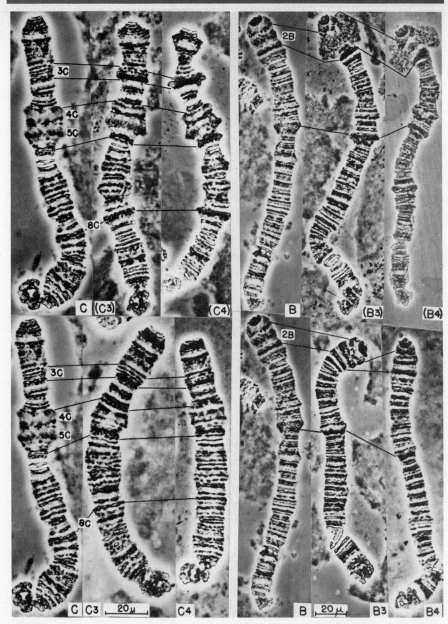

Figure 20-1. *Chromosome puffs in polytene chromosomes of* Rhynchosciara, *at 4C and 5C. Top row, chromosomes C and B from controls (no ligation) and their appearance after 3 (C3 and B3) and 4 (C4 and B4) days. Note especially the development of puff 2B (upper right photograph). Bottom row, chromosomes B and C before ligation (C and B); C3 and B3, chromosomes C and B 3 days after ligation; C4 and B4, the same chromosomes 4 days after ligation. Note particularly the failure of puff 2B to develop in the experimental series* [Courtesy Drs. J. M. Amabis and D. Cabral, Universidade de São Paulo, São Paulo, Brazil. Reprinted from *Science,* **169:** 694 (14 August 1970) by permission. Copyright 1970 by the American Association for the Advancement of Science.]

By means of differential staining techniques, biochemical tests, and use of radioactive isotopes, it has been demonstrated that each puff is an active site of transcription. If one recalls that genes (i.e., cistrons) are associated with particular bands, this temporal puffing clearly indicates changes in gene activity over time. Tests show that the mRNA synthesized in one puff is characteristic and differs from that produced by other puffs.

Injection of very small amounts of the hormone ecdysone, which is produced by the prothoracic glands and which induces molting, causes formation of the same puffs that occur normally prior to molting in untreated larvae. The prothoracic glands are activated by the flow of brain hormone from neurosecretory cells. This has been demonstrated in an ingenious experiment reported by Amabis and Cabral. Tying off the anterior part of the larva of the dipteran *Rhynchosciara* just behind the brain, they found, resulted both in failure of normal puffing and a decrease in size of puffs already initiated at the time of ligation (Fig. 20-1). In other experiments, injection of the antibiotic actinomycin D (which inhibits mRNA synthesis) prevents puff formation for several hours, even when ecdysone is used simultaneously. Radioactive uridine, injected into larvae, accumulates only in the puffs and nucleoli, but fails to do so if it is preceded by injection of actinomycin D.

All these results clearly point to the puffs as sites of RNA synthesis, according to a pattern closely associated with the development of the individual. Evidently, then, genes (cistrons) undergo reversible changes in activity (principally mRNA synthesis) that are related to the developmental stage of the organism.

Lampbrush chromosomes. This "turning on and off" of genes is evident also in the large chromosomes of amphibian oocytes (Fig. 20-2). Here the DNA strands of the long axes form lateral loops. Ribonucleic acid is produced in the loops of these chromosomes in cyclic fashion, very much as in the puffs of dipteran chromosomes. It seems likely that cyclic mRNA production is a general phenomenon.

Transcriptional control and chromosomal proteins in eukaryotes. As was noted in Chapter 4, DNA of eukaryotic cells is complexed with low molecular weight, basic proteins, the histones.[1] These positively charged molecules owe

---

[1]Except in sperm nuclei, where protamines are involved.

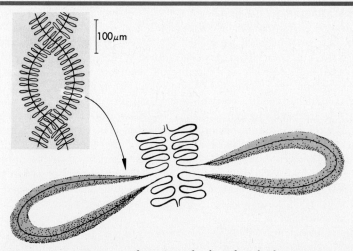

100 µm

Figure 20-2. *Diagram of portion of a lampbrush chromosome as seen in an amphibian oöcyte.*

their charge to high levels of arginine and lysine (see Table 18-1), and complex with the negatively charged PO_4 groups of the DNA molecule. Their presence increases the diameter of DNA (20 Å) to the 35 Å diameter of nucleoprotein. Histones enhance coiling of DNA, which inhibits transcription. But only five kinds of histone occur in a given cell; their lack of structural variety suggests a general, rather than gene-specific, regulatory function. On the other hand, the more acidic nonhistone proteins[2] occur in great variety. They differ markedly in composition among different eukaryotic tissues and species, and thus suggest a more specific role in gene regulation than would be possible for the histones. Nonhistone proteins may combine with histones to bring about the decomplexing of the latter with DNA, which permits transcription. It is possible that the σ subunit of RNA polymerase (Chapter 17), which is responsible for initiation of transcription, is a phosphorylated nonhistone protein. Which cistrons are actively transcribing at any given time, then, appears to be closely tied to the histone and nonhistone protein that complexes with DNA. The former class of proteins probably serves as general regulators, the latter as more specific regulators of particular genes. Indirect evidence for this role of nonhistone proteins is extensive; they (1) possess considerable structural diversity, (2) exhibit tissue and nucleotide sequence binding specificity, (3) occur in much higher levels in transcriptionally active cells, whereas histone levels remain essentially constant, and (4) are closely associated with initiation and continuation of transcription both in vivo and in vitro.

The degree of histone-DNA complexing appears, in at least some organisms, to vary with the body part and with time. In developing pea embryos, for example, the cotyledons (food storage organs of the embryo) develop rather rapidly in embrogeny, then cease to grow further, even in germination. Tests indicate that almost all the DNA of mature cotyledons is histone-complexed, whereas the terminal meristems, active growth regions that, at germination, produce all the root and shoot of the seedling, have more noncomplexed DNA.

This suggests an inverse relationship between histone complexing and protein synthesis. Bonner and his colleagues have presented excellent confirmation of this hypothesis. Pea cotyledons produce a protein, seed reserve globulin, that is not produced in vegetative tissues. DNA from cotyledons can be made to produce the mRNA necessary for in vitro synthesis of this protein. DNA from vegetative parts of the plant does not yield mRNA for globulin unless the histone is chemically removed! It has also been shown that maize embryos begin growth and development into seedlings only when most of the DNA is not histone complexed. Histone removal may therefore be involved in the physiological mechanism that initiates germination of the grain.

The question of regulation of eukaryotic genes is still not completely solved. Although there is some evidence for the occurrence of operons (see next section) in some of the fungi, they have not been found in higher eukaryotes. Chromatin in these organisms consists of DNA literally covered with proteins. Although histones have long been suspected of playing a part in gene regulation, a promising approach lies in the discovery of unusual conformations of DNA "upstream" from actively transcribing genes. Coiling of transcriptionally active regions of DNA is more relaxed than nontranscribing regions. The more open coils of transcribing DNA are sensitive to **DNase I,** an enzyme that attacks and digests active euchromatin. Weintraub and his colleagues have found that active globin genes in chicken red blood cells are ten times more sensitive to this enzyme than are inactive globin genes. The DNase I-sensitive regions in these erythrocytes is very long (some 100,000 bases), whereas the globin genes themselves are only 1,000 to 2,000 bases in length.

"Hot spots" for action of DNase I have been found; these sequences are only 100 to 200 nucleotides long, but are a hundred times more sensitive to DNase I

[2]Nonhistone proteins are rich in glutamic and aspartic acids (Table 18-1).

than other stretches of chromatin. Such hypersensitive zones may occur "before," within, as well as "after" genes. Those located ahead of genes are detectable whenever a gene is transcriptionally active, and have been found in a variety of eukaryotes—rat, chicken, yeasts, and *Drosophila*. Interestingly enough, these hypersensitive regions upstream appear *before* transcription begins. These "hot spots" can be cut by the enzyme **S1 nuclease,** which cuts single-stranded DNA but does not ordinarily cut chromatin. This suggests looser coiling in the hypersensitive regions, and has been verified in a number of species. Because of their increased accessibility to enzymes, it is also likely that these regions are more accessible to protein initiation factors. Deletions outside the hypersensitive regions still permit accurate and efficient transcription. It is quite possible that gene expression depends on relaxing of the coils of the nucleosomal unit. Although other factors are undoubtedly involved in eukaryotic gene regulation, it is clear that chromatin structure is an important part of the story.

Heterochromatin. **Heterochromatin,** as was noted in Chapter 4, is tightly coiled nucleoprotein and is therefore less available for transcription. Blocks of genes in heterochromatic regions (e.g., in the centromeric region, in the inactivated X chromosome, or in the Y chromosome) may be repressed either permanently or at certain stages of development. Noncentromeric heterochromatin differs chemically from euchromatin only in its nonhistone protein composition. The DNA of the centromeric region includes highly repetitive nucleotide sequences (satellite DNA); other reiterated sequences occur between cistrons. Generally, they are too short to function as cistrons themselves; in the mouse some 200 to 300 nucleotide pairs comprise most of the repeated segments. The functions of centromeric repetitive sequences is not understood. In the African clawed toad (*Xenopus laevis*) different sequences of about 300 nucleotides each are repeated from 20 to more than 30,000 times, and alternate with nonrepetitive segments of around 800 nucleotide pairs each. The interspersed reiterated sequences may play a part in regulation of transcription, but their operation has not been determined.

Enzyme regulatory mechanisms

Inasmuch as the ultimate product of a cistron is a polypeptide chain, which is frequently incorporated into an enzyme, a simple system of control would operate at the level of enzyme regulation. Evidence accumulated in recent years indicates that there are, in fact, two basic mechanisms of enzyme regulation. One controls the *activity* of enzymes whose production is not altered; the other regulates the *synthesis* of enzymes. The first of these is generally termed **end-product inhibition;** the second involves the concept of the **operon.** The operon is a group of adjacent cistrons whose function is regulated by other sites, called the operator and the promoter. Operons are common in bacteria; they will be examined shortly.

END-PRODUCT INHIBITION

In studies on isoleucine synthesis in *Escherichia coli*, it was demonstrated that addition of isoleucine (the end product of a five-step conversion of threonine, Figure 20-3) to a culture of the bacterium resulted in immediate blocking of the threonine → isoleucine pathway. In the presence of added isoleucine, the cells preferentially use this *exogenous* end product and cease their own isoleucine synthesis. Moreover, it has been shown that *production* of each of the five enzymes is *not* interfered with, but action of the enzyme responsible for the deamination of threonine to α-ketobutyrate (Fig. 20-3) is *inhibited* by the end product, isoleucine.

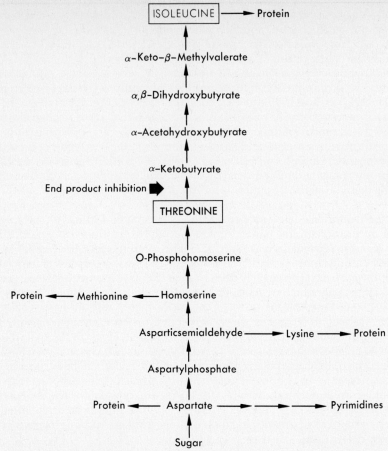

Figure 20-3. *Some steps in isoleucine synthesis. In end-product inhibition in* E. coli, *enzyme inhibition occurs in the deamination of threonine to α-keto-butyrate at the point indicated.*

Interestingly, this inhibition results from a binding of end product to the enzyme so that the inhibitor appears to compete with the substrate for a site on the enzyme molecule. But it is known that this competition is not for the same site; the enzyme apparently has two specific recognition sites, one for its substrate and another for the inhibiting end product. However, attachment at the inhibitor site affects the substrate site. The term *allosteric interaction* is applied to such changes in enzyme activity produced by binding of a second substance at a different and nonoverlapping site.

THE OPERON

The lac operon in Escherichia coli. Among the many strains of the colon bacillus, two in particular have shed considerable light on another system of gene regulation. Two enzymes are required for lactose metabolism in this organism: β-*galactoside permease*, responsible for transport of lactose into the cell and its concentration there, and β-*galactosidase*, which catalyzes the hydrolysis of lactose to galactose and glucose. The wild-type *E. coli* produces these enzymes only in the presence of lactose; this is an *inductive* strain. One might say the wild-type cells are *induced* to synthesize the enzymes that function in lactose metabolism only when the enzymes' substrate is present. When the source of carbon available to *E. coli* does not include lactose, β-galactoside permease and β-galactosidase are present in very small amounts—about 10 molecules per cell. When and if lactose is supplied, the rate of synthesis of these enzymes increases as much as 1,000 times.

Other strains of the colon bacillus, derived by mutation from the inductive wild type, produce these enzymes continuously, whether lactose is present or not. These are *constitutive* strains. After some initial uncertainty over whether enzyme synthesis or enzyme activity was affected by lactose, it has become clear that the presence of lactose does indeed *induce* synthesis of enzymes in the inductive strain. Lactose, the substrate, is acting here as an enzyme inducer. Enzymes produced only when a substrate is present are designated as *inducible enzymes.* Although inducers are quite specific, structural analogs may often function in the same way.

The *lac* segment of the *E. coli* chromosome is known to include three cistrons, *z* for β-galactosidase, *y* for β-galactoside permease, and *a* for thiogalactoside transacetylase.[3] The three cistrons are closely linked and regulated coordinately. Jacob and Monod and their associates developed a model of a genetic system that regulates lactose metabolism, to which they gave the name **operon.** With minor refinements, this model explains all observations in this bacterium. Several other similar regulatory systems have subsequently been described for this and other bacteria. Operons are now known to be widespread in bacteria.

Three cistrons, *z*, *y*, and *a*, are each responsible for one enzyme (Fig. 20-4); they occupy adjoining positions on the DNA molecule and map in that sequence. The genetic information of these three cistrons is transcribed into a single polycistronic mRNA molecule that is subsequently translated into the polypeptide chains of the three enzymes. RNA polymerase attaches to the promoter site (*p*), then transcription proceeds in the $5' \rightarrow 3'$ direction, beginning with the operator site, *o*, which lies to the "right" or "downstream" of *p*, that is, in the direction of the three structural genes (Fig. 20-4) and exercises a control over transcription. Mapping experiments show the sequence of these genetic elements of DNA to be

p o z y a

The three cistrons, together with the promoter and operator sites, comprise the lac *operon.* **An operon, then, consists of a system of cistrons, operator, and promoter sites.** The basis for inductive and constitutive production of the three enzymes of the *lac* operon lies in still another site, the regulator (*i*), and the interaction of its protein product with lactose and with the operator site. The regulator site is near, if not directly adjacent to, the promoter. The complete map sequence involved, then, may be represented as

i p o z y a

and is composed of about 4,700 nucleotides. On the linkage map of *E. coli* (Fig. 6-7) this sequence maps at approximately 7.8 minutes.

The regulator site produces a repressor protein of 360 amino acid residues and a molecular weight of about 150,000. In the inductive strain of the bacterium, the repressor protein binds to the wild-type operator (o^+), and thereby halts progression of RNA polymerase so that transcription of the three cistrons is prevented. But lactose, if present, combines with the repressor, inactivates it, with the result that the three cistrons are transcribed. Lactose, the substrate for β-galactosidase and β-galactoside permease, thus acts as an *inducer,* or effector, so that the operator is unrepressed (Fig. 20-4). Transcription is initiated in the operator site, which is transcribed as part of the overall mRNA product. *Trans-*

[3]The function of thiogalactoside transacetylase in vivo is unknown.

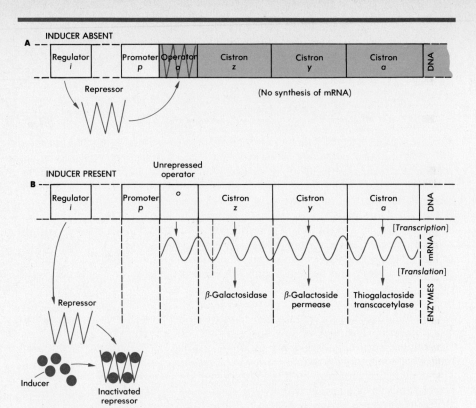

Figure 20-4. Diagrammatic representation of the lac *operon of* Escherichia coli. *Transcription is initiated at the operator site* (o). *The enzyme* DNA-*dependent* RNA *polymerase attaches to the promoter site* (p) *in the presence of cyclic* AMP *receptor which must bind first to the promoter at or near the 5' end of the latter region. Transcription proceeds in the 5' → 3' direction beginning as shown within the operator. If the repressor protein is present in the absence of lactose, it will bind to the operator, preventing transcription of the three cistrons. If lactose is present, it combines with the repressor to inactivate, allowing transcription to proceed. The regulator* (i) *may or may not be directly adjacent to the promoter. For further details, see text.*

lation, however, begins in the z cistron with the *eleventh* nucleotide rather than with the first (Fig. 20-5).

Constitutive strains of *E. coli* are characterized by either a defective regulator (i^-) or a mutant operator that is unable to bind the repressor. For example, i^+ p^+ o^+ z^+ y^+ a^+ is an inductive strain, i^+ produces a repressor that binds to o^+, and prevents transcription of z^+, y^+, and a^+ (Fig. 20-4). Presence of lactose, of course, inactivates the repressor so that o^+ permits transcription. On the other hand, i^- p^+ o^+ z^+ y^+ a^+ is constitutive because a defective repressor, which will not bind to o^+, is produced, so transcription of the three cistrons is continuous. An i^+ p^+ o^c z^+ y^+ a^+ (an *operator-constitutive* strain) is also constitutive because the repressor binding site, rather than the repressor, is defective.

Interestingly enough, studies of merozygotes (partial diploids) produced through conjugation show i^+ to be dominant to i^-. In a cell of genotype i^+ p^+ o^+ z^+ $\ldots/i^-$ p^+ o^+ z^- $\ldots$ one might expect normal β-galactosidase to be produced inductively and a modified form of the enzyme (inactive or only weakly active) constitutively. Not so; both types of β-galactosidase are produced inductively. The repressor is able to diffuse from the i^+ p^+ o^+ z^+ $\ldots$ "chromosome" to the other and repress z^-, even though the latter is linked to i^-. The regulator gene

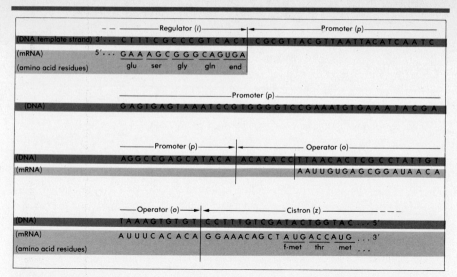

Figure 20-5. *Nucleotide sequences of the template* DNA *strand and the transcribed* RNA *molecule for a portion of the regulator* (i), *all of the promoter* (p) *and operator* (o), *and the beginning of the* z *cistron of the* lac *operon. Note the translation begins only at the eleventh ribonucleotide of the* z *cistron. Amino acids are abbreviated as follows:* f-met, N-*formylmethionine;* gln, *glutamine;* glu, *glutamic acid;* gly, *glycine;* met, *methionine;* ser, *serine;* thr *theronine. (Read all nucleotide sequences from left to right.)*

thus functions in either the *cis* or *trans* position. On the other hand, the operator (*o*) functions only in the *cis* position; for example, the partial diploid i^+ p^+ o^c z^+ .../i^- p^+ o^+ z^- ... produces normal β-galactosidase constitutively, but no modified enzyme. This is to be expected, inasmuch as the operator is the place where transcription begins.

This aspect of the *lac* operon illustrates *negative control*, the basis for which is production of a *repressor*. In this case, the operator is repressed, and the cistrons turned off, unless the inducer (lactose) is present to inactivate the repressor.

Another group of mutants involves the repressor gene, *i*. These mutations in *i* act on the *z-y-a* structural genes in either the *cis* or *trans* configuration in merozygotes. The i^s mutant produces a *superrepressor* that prevents production of the enzyme products of all three cistrons. From study of partial diploids, i^s is seen to be dominant to i^+ so the dominance relationship among these three genotypes is $i^s > i^+ > i^-$.

Because the direction of transcription is from promoter to operator to cistrons z, y, and a, in that order, *polar mutations* affect all reading frames downstream. A defect in z affects all three cistrons, one in y affects only genes y and a, and one in a affects only a.

Behavior of several genotypes is summarized in Table 20-1.

"Turning on" of the lactose operon also requires molecules of catabolite receptor protein (CRP) bound to molecules of cyclic adenosine monophosphate (cAMP). CRP is the product of gene *crp*; cAMP is made from ATP (adenosine triphosphate) with the aid of the enzyme adenylate cyclase which is produced by gene *cya* (Fig. 20-6). Both genes are to the left of (upstream from) the promoter (Fig. 20-5). If glucose is *not* present, the cAMP-CRP complex is produced and binds to a site just to the left of (upstream from) the lactose promoter (Fig. 20-5). This *enhances* RNA polymerase binding and consequent transcription of the lactose cistrons. *Presence* of glucose interferes with production of cAMP with the result that the cAMP-CRP complex cannot be formed. This halts transcription

Table 20-1. *Production of beta-galactosidase in several lac genotypes in E. coli**

Genotype	Constitutive Normal	Constitutive Modified	Inductive Normal	Inductive Modified	Basis
$i^+ o^+ z^+ \dots$	−	−	+	−	Wild-type
$i^- o^+ z^+ \dots$	+	−	−	−	Defective regulator
$i^+ o^c z^+ \dots$	+	−	−	−	Operator constitutive
$i^s o^+ z^+ \dots$	−	−	−	−	Superrepressor
$i^+ o^+ z^+ \dots / i^- o^+ z^+ \dots$			+	+	Wild-type/defective regulator
$i^+ o^+ z^- \dots / i^- o^+ z^+ \dots$	−	−	+	−	i^+ dominant to i^-
$i^+ o^+ z^+ \dots / i^s o^+ z^+ \dots$	−	−	−	−	i^s dominant to i^+

*A plus sign indicates beta-galactosidase production; a minus sign indicates lack of production of the enzyme.

of the lactose cistrons. Defective mutants of both *crp* and *cya* have been produced; either *crp*⁻ or *cya*⁻ individuals produce such low levels of their respective products that their effect is minimal or absent. Because the *z*, *y*, and *a* cistrons are turned *on* in the presence of the cAMP-CRP complex, this represents *positive* control.

OTHER OPERONS

The his operon.   The *lac* operon involves an active repressor; however, not all operons function in this manner. The *his* operon, in *Salmonella typhimurium* (closely related to *E. coli*) is responsible for the synthesis of the amino acid histidine. It includes some 13,000 deoxyribonucleotide pairs. Synthesis of histidine in this species occurs in ten enzymatically controlled steps, which involve nine enzymes (one functions in two different steps) produced by as many cistrons. The regulator is responsible for an *inactive* repressor protein which, in the *absence* of histidine, is unable to repress the operator. The result, then, is that RNA polymerase proceeds through the nine cistrons, transcribes a polycistronic messenger and leads to histidine synthesis. If exogenous histidine is available, it combines with the repressor, and changes the latter to an active form. This active form represses the operator, and blocks transcription as well as histidine synthesis. Although the operation of the *his* operon differs somewhat from that of the *lac* operon, it is, nevertheless, a *negative* control because a *repressor* protein is involved.

The ara operon.   Regulation of the three cistrons responsible for the production of the three enzymes that catalyzes the three-step breakdown of the pentose sugar arabinose to xylulose-5-phosphate (another pentose sugar, involved in cellular oxidations) in *E. coli*, represents an interesting variation in operon operation. The segment of the DNA molecule involved (*E. coli* K-12 map position is about 1.3 minutes; see Fig. 6-7) consists of three structural genes, each of which is responsible for one of the three enzymes, an initiator site, an operator, and a regulator. The regulator produces an allosteric protein that serves both as an activator and a repressor of the three cistrons. In the absence of arabinose, this protein serves as an active repressor, binds to the operator and blocks transcription. Arabinose will bind to the repressor and/or the repressor-operator complex, and thus produce a conformational change in the regulator protein, which con-

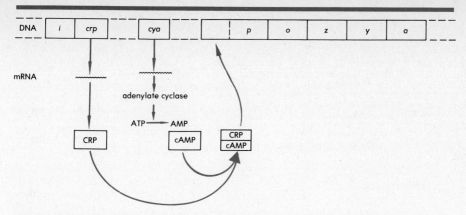

Figure 20-6. *The* cAMP-CRP *positive control of the* lac *operon in* Escherichia coli. *See text for details.*

verts it to an activator. The latter then interacts with the promoter, to permit transcription of the three cistrons. So, in the *absence* of the inducer (here arabinose, the metabolite), the protein of the regulator site functions as a *repressor* by binding to the operator and preventing transcription; this represents, of course, *negative* control. The same protein of the regulator is converted to an *activator* in the presence of the inducer by binding to it, and allowing transcription to take place; this represents *positive* control.

Present evidence indicates that gene regulation by operons is widespread in bacteria. Many have been described, and some 100 to 200 have been estimated for *Escherichia coli* alone. For most of the known operons in bacteria the operation has been determined. Other operons have been reported for *Neurospora*, the orange-pink bread mold so useful in early biochemical genetics work, for the mold *Aspergillus*, and for the yeast *Saccharomyces*. However, in these instances, an operator site appears to be lacking, and operons are not at all characteristic of eukaryotes.

Summary of gene regulation

Genetic material has turned out to be fairly complicated in its regulation as well as in its action and structure. Yet it is a relatively simple system for the production of an endless variety of phenotypes. Gene action appears to be related to certain proteins and/or to other genes or even to certain metabolites. So to our summary of gene function and fine structure (Chapter 19) we need to add the following points regarding the regulation of gene action:

1. The activity of many cistrons is regulated so that their polypeptide product is formed only under certain chemical conditions and/or at certain developmental stages.
2. Although the function of histones and nonhistone proteins is not yet fully known, it has been suggested on credible grounds that they regulate gene activity in those cells in which they occur by affecting DNA coiling and RNA transcription.
3. In other organisms, notably bacteria, there is clear evidence for a regulator-operon complex in which the regulator produces a protein that
    a. Represses an operator (active repressor), with the result that transcription does not occur, but can be inactivated by an inducer, which

allows transcription to take place (negative control, as in the *lac* operon), or

b. The operon may be under positive control, as in the effect of cAMP-CRP in the *lac* operon, or

c. The repressor fails to repress the operator (inactive repressor) unless activated by combining with the inducer (negative control, as in the *his* operon), or

d. Represses the operator site in the absence of the inducer (negative control), or serves as an activator after combining with the inducer (positive control), as in the *ara* operon.

References

Amabis, J. M., and D. Cabral, 1970. RNA and DNA Puffs in Polytene Chromosomes of *Rhynchosciara:* Inhibition by Extirpation of Prothorax. *Science*, **169:** 692–694.

Ames, B. N., and P. E. Hartman, 1963. The Histidine Operon. *Cold Spring Harbor Symposia Quant. Biol.*, **28:** 349–356. Reprinted in E. A. Adelberg, ed., 1966. *Papers on Bacterial Genetics.* Boston: Little, Brown.

Beckwith, J. R., 1967. Regulation of the Lac Operon. *Science*, **156:** 596–604.

Beckwith, J. R., and D. Zipser, eds., 1970. *The Lactose Operon.* Cold Spring Harbor, N.Y.: Cold Spring Harbor Laboratory.

Beckwith, J., and P. Rossow, 1974. Analysis of Genetic Regulatory Mechanisms. In H. L. Roman, ed. *Annual Review of Genetics*, vol. **8.** Palo Alto, Calif.: Annual Reviews, Inc.

Blair, J. G., 1982. Test-Tube Gardens. *Science*, **82**(1): 71–76.

Dickson, R. C., J. Abelson, W. M. Barnes, and W. S. Reznikoff, 1975. Genetic Regulation: The Lac Control Region. *Science*, **187:** 27–35.

Englesberg, E., and G. Wilcox, 1974. Regulation: Positive Control. In H. L. Roman, ed., *Annual Review of Genetics*, vol. **8.** Palo Alto, Calif.: Annual Reviews, Inc.

Epstein, W., and J. R. Beckwith, 1968. Regulation of Gene Expression. *Ann. Rev. Biochem.*, **37:** 411–436. Palo Alto, Calif.: Annual Reviews, Inc.

Jacob, F., and J. Monod, 1961. Genetic Regulatory Mechanisms in the Synthesis of Proteins. *Jour. Molec. Biol.*, **3:** 318–356.

Jacob, F., D. Perrin, C. Sanchez, and J. Monod, 1960. The Operon: A Group of Genes Whose Expression Is Coordinated by an Operator. *Compt. Rend. Acad. Sci.*, **250:** 1727–1729.

Lewin, B., 1974. *Gene Expression—1: Bacterial Genomes.* New York: John Wiley & Sons.

Maclean, N., 1976. *Control of Gene Expression.* New York: Academic Press.

Nitsch, J. P., and C. Nitsch, 1969. Haploid Plants from Pollen Grains. *Science*, **163:** 85–87.

O'Malley, B. W., H. C. Towle, and R. J. Schwartz, 1977. Regulation of Gene Expression in Eucaryotes. In H. L. Roman, ed., *Annual Review of Genetics*, vol. **11.** Palo Alto, Calif.: Annual Reviews, Inc.

Pardee, A. B., F. Jacob, and J. Monod, 1959. The Genetic Control and Cytoplasmic Expression of "Inducibility" in the Synthesis of β-galactosidase by *E. coli. Jour. Molec. Biol.*, **1;** 165–178.

Randal, J., 1981. Breeding the Perfect Cow. *Science*, **81, 2**(9): 86–92.

Reznikoff, W. S., 1972. The Operon Revisited. In H. L. Roman, ed. *Annual Review of Genetics*, vol. **6.** Palo Alto, Calif.: Annual Reviews, Inc.

Wang, T. Y., N. C. Kostraba, and R. S. Newman, 1976. Selective Transcription of DNA Mediated by Nonhistone Proteins. In W. E. Cohn and E. Volkin, eds., *Progress in Nucleic Acid Research and Molecular Biology.* New York: Academic Press.

Wilcox, G., K. J. Clemetson, P. Cleary, and E. Englesberg, 1974. Interaction of the Regulatory Gene Product with the Operator Site in the L-Arabinose Operon of *Escherichia coli. Jour. Molec. Biol.*, **85:** 589–602.

Problems

20-1 In what way or ways are operator and regulator sites similar? Dissimilar?

20-2 Differentiate between repressors and inducers.

20-3 Distinguish between positive and negative control.

20-4 Synthesis of mRNA by the *lac* operon of *Escherichia coli* increases with addition of the inducer. Is this evidence for action of the inducer at the transcriptional or at the translational level?

20-5 Will production of *normal* beta-galactosidase be constitutive, inductive, or absent for each of the following genotypes:
(a) $i^+ p^+ o^+ z^+ y^+ a^+$, (b) $i^- p^+ o^+ z^+ y^+ a^+$, (c) $i^+ p^+ o^c z^+ y^+ a^+$, (d) $i^+ p^+ o^+ z^- y^+ a^+$, (e) $i^+ p^+ o^+ z^- y^+ a^+ / i^+ p^+ o^c z^+ y^+ a^+$?

20-6 Will each of the three enzymes coded for by the *lac* operon be produced constitutively or inductively, or not produced in each of the following merozygote genotypes:
(a) $i^s p^+ o^+ z^+ y^+ a^+ / i^+ p^+ o^+ z^+ y^+ a^+$; (b) $i^+ p^+ o^c z^+ y^- a^+ / i^+ p^+ o^+ z^- y^+ a^-$?

20-7 Assume an *Escherichia coli* culture in which all the individuals are of genotype $crp^+ cya^+ \ldots i^+ p^+ o^+ z^+ y^+ a^+$; will the mRNA for the three *lac* enzymes be transcribed or not? (a) Glucose and lactose both present; (b) glucose and lactose both absent; (c) glucose absent, lactose present.

20-8 Assume that the fourth transcribed group of three deoxyribonucleotide pairs in the wild type z cistron of the *lac* operon of *Escherichia coli* is $\dfrac{5' \ldots C\,A\,A \ldots 3'}{3' \ldots G\,T\,T \ldots 5'}$. Through tautomerization this group is changed to $\dfrac{5' \ldots T\,A\,A \ldots 3'}{3' \ldots A\,T\,T \ldots 5'}$. The 3', 5' strand is the template for transcription. (a) Will β-galactosidase be produced following this mutation? (b) Will β-galactoside permease and/or thiogalactoside transacetylase be produced after this mutation has occurred?

20-9 An experimental strain of *Escherichia coli* is developed in which the 35 deoxyribonucleotide pairs of the operator are deleted. No other changes occur. Will production of β-galactosidase, β-galactoside permease, and/or thiogalactoside transacetylase be inductive, constitutive, or lacking?

20-10 Would you expect a nonsense mutation in one of the structural genes of the *his* operon to affect transcription or translation?

20-11 The tryptophan (*trp*) operon of *Escherichia coli* is responsible for coordinate production of the enzymes involved in the synthesis of the amino acid tryptophan. The operon includes five structural genes, an operator, and a promoter. A regulator site produces a repressor that is activated by tryptophan. (a) Is this an illustration of positive or of negative control? (b) In the absence of tryptophan in the cell, are the enzymes of the *trp* operon produced or not?

CHAPTER 21

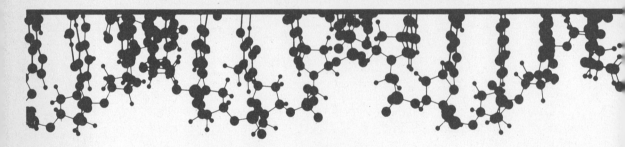

Cytoplasmic genetic systems

The existence of genes as segments of nucleic acid molecules, located in chromosomes and controlling phenotypes in known and predictable fashion, has been amply demonstrated on sound, observable, verifiable bases. But the firm establishment of such a chromosomal mechanism of inheritance does not necessarily preclude a role by other cell parts. From time to time observations have been reported that suggest a cytoplasmic role in genetics. To evaluate these reports, a clear understanding of what cytoplasmic inheritance does or should mean is needed.

SUGGESTED CRITERIA FOR CYTOPLASMIC GENES

Before a particular genetic phenomenon is clearly acceptable as having a cytoplasmic basis, the following criteria should be applied:

1. *Is there evidence of genes physically and/or physiologically independent of any nuclear genetic material?*
2. *What is the chemical nature of the suspected cytoplasmic gene?*
   a. *Is it DNA? RNA? Something else?*
   b. *Is it operationally genetic, that is, is it a conserved structure whose activities govern the trait, or is it the trait itself that is being called a cytoplasmic gene?*
   c. *Is it a normal and usual part of the cell's cytoplasm (if we consider that "normal" may not be easy to define)?*

3. *What are its mutational properties and methods?*
4. *Do reciprocal crosses show clear evidence of a cytoplasmic role as opposed to sex (or other) chromosomes?*
5. *Does its pattern of inheritance bear no relation to chromosomal segregation?*
6. *Is it unequivocally not a part of any chromosomal linkage group?*
7. *Are there alternative phenotypic expressions and are they stable through the life cycle?*
8. *Is transmission of the trait unaffected by nuclear transplants?*

Although this list is neither exhaustive nor universally accepted, cytoplasmic inheritance will be understood here to be based on cytoplasmically located, independent, self-replicating nucleic acids, which differ from chromosomal genes only in their location within the cell and often have their own unique nucleotide sequences. Not many of the purported examples of cytoplasmic inheritance fit this concept, and only a few have been unequivocally established. Some of them will be examined in the following sections.

CHLOROPLASTS

Plastids are cytoplasmic organelles in the cells of many different kinds of plants. Those that contain the green photosynthetic pigments, the chlorophylls, are called chloroplastids or simply *chloroplasts*. Plastids arise by differentiation from simpler structures, the proplastids. These are able to divide and so increase in number in a manner only in part under the control of nuclear DNA. Electron microscopy reveals chloroplasts as fairly complex structures (Fig. 21-1), and they do contain rather significant amounts of DNA. Green and Burton were able to release highly supercoiled plastid DNA by osmotic shock in the large unicellular green alga *Acetabularia* (Fig. 21-2). The largest intact segment sufficiently untwisted to permit measurement they reported to be 419 μm in length, which corresponds to about 1.23×10^6 base pairs. In general, molecular weights of (double-stranded) chloroplast DNA (**cpDNA**) in a wide variety of higher plant species[1] run from 85×10^6 to 97×10^6. This corresponds to a range of about 1.3×10^5 to 1.5×10^5 deoxyribonucleotide pairs. The small unicellular green alga *Chlamydomonas* contains about 4×10^9 daltons of cpDNA in its single chloroplast, and thus corresponds to 6.15×10^6 base pairs. Chloroplast DNA of *Euglena*, which has a molecular weight of 9.2×10^7, consists of 1.4×10^5 nucleotide pairs. By comparison, the double-stranded nucleoid of *Escherichia coli* consists of about 4.3×10^6 base pairs (Chapter 16). A large number of investigators have provided definitive evidence that cpDNA is a closed, circular, double-stranded molecule.

It is now clear that cpDNA differs, often greatly, in nucleotide composition from nuclear DNA of the same species. Ultraviolet microbeam irradiation of the cytoplasm of the familiar one-celled *Euglena* results in plastid destruction, whereas nuclear irradiation does not. This organism's chloroplasts contain a unique DNA component of low guanine-cytosine content that is absent in plastidless mutants. Plastids also contain a DNA polymerase and are able to replicate their own DNA. This process has been shown to be semiconservative in *Chlamydomonas* and to take place at a time quite different from that for nuclear DNA replication.

Replication of cpDNA in a number of higher plant species has been found not only to be semiconservative, but to be accomplished by the rolling circle model, as was described in Chapter 16 for *E. coli*. Such facts, in addition to the circular nature of chloroplast DNA and its general size range, have led to the building of a considerable case for the thought that cpDNA (and mitochondrial DNA) may well have originated as symbiotic prokaryotes.

---

[1]E.g., peas (*Pisum*), tobacco (*Nicotiana*), corn (*Zea*), spinach (*Spinacea*), and wheat (*Triticum*).

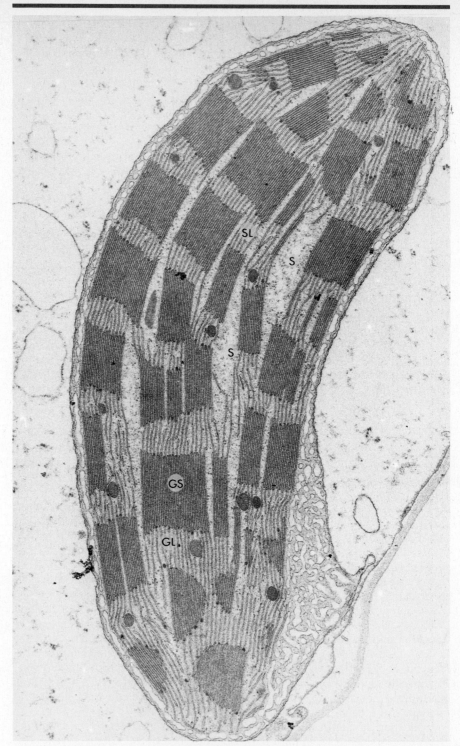

Figure 21-1. *Electron micrograph of a thin section of a chloroplast from corn leaf. The internal membrane system contains the chlorphylls and is the actual site of photosynthesis; DNA is contained within the stroma (S). GS, grana stacks; GL, grana lamellae; SL, stroma lamellae; S, stroma containing DNA and ribosomes.* × *80,000.* (Photo by Dr. L. K. Shumway, Program in Genetics, Washington State University. From *Cell Ultrastructure* by William Jensen and Roderic Park, copyright 1967 by Wadsworth Publishing Co., Inc., Belmont, California. Reproduced by permission of publisher.)

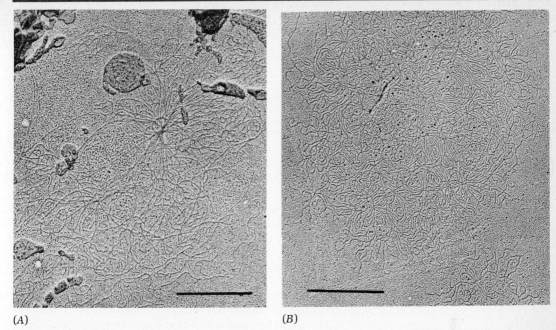

(A)   (B)

Figure 21-2. *Electron micrographs of chloroplast DNA from the green alga* Acetabularia mediterranea. *(A) Lysed chloroplast showing chloroplast DNA fibrils around a "center." (B) Large array of DNA released from a chloroplast. The bar represents 1 μm.* [Courtesy Dr. Beverly R. Green, University of British Columbia. Reprinted by permission of authors and publisher from Green, B. R., and H. Burton, "*Acetabularia* Choroplast DNA: Electron Microscope Visualization," *Science* (22 May 1970), **168:** 981–982. Copyright 1970 by The American Association for the Advancement of Science.]

Chloroplasts also contain RNA and ribosomes. The latter, at least in higher plants, are smaller in mass than those in the cytoplasm of the same cell. They resemble much more closely bacterial ribosomes in size, as demonstrated by determinations of their sedimentation coefficients. They are also inactivated by chloramphenicol (as are bacterial ribosomes), which does not affect cytoplasmic ribosomes. Chloroplasts, then, contain their own machinery for protein synthesis, and it is clear that their DNA does play a part in incorporating amino acids into the characteristic plastid proteins. So here are self-replicating structures, which contain their own unique DNA and an array of protein-synthesizing equipment, although, at the same time, under some control by nuclear DNA as to their existence. However, on the basis of their unique cpDNA, their complement of polypeptide-synthesizing apparatus, and their production of proteins, chloroplasts meet the criteria listed at the beginning of this chapter for cytoplasmic inheritance.

But chloroplasts themselves also constitute particular phenotypic traits, e.g., green color and ability to photosynthesize, which are quite complex. Actually, there is a semantic problem here; that is, nuclear-controlled *presence or absence* of plastids per se and their ability to play a part in the synthesis of their own proteins must be carefully distinguished. From the former viewpoint, plastids are just another aspect of phenotype, i.e., no different from those that have been considered previously; from the latter point of view, however, *they do demonstrate extranuclear control by normally occurring, DNA-containing organelles,* regardless of their ultimate origin and evolution.

Chiang has summed up the case for chloroplast DNA as an example of true cytoplasmic inheritance; "Following the demonstration that chloroplast DNA replication is semiconservative and is temporally separated from the nuclear

DNA replication in the vegetative cell of *Chlamydomonas*, it has become evident that organelle DNA molecules themselves may be responsible for cytoplasmic inheritance, since the physical preservation, and thus informational continuity, of the organelle DNA in eukaryotes is maintained in much the same manner as that of the chromosomal (nuclear) DNA through the mitotic divisions and/or vegetative multiplication." One can question only Chiang's understated conclusion that ". . . organelle DNA molecules themselves *may* be responsible for cytoplasmic inheritance . . ." (emphasis added).

On the other hand, plastid *inheritance* in the four o'clock plant (*Mirabilis jalapa*, Fig. 21-3) is a classic illustration of *purported* cytoplasmic inheritance. This plant may exist in three forms: normal green, variegated (patches of green and of nongreen tissue), and white (no chlorophyll). Now and then a plant with branches of two or three of these types occurs. Plastids in the white areas have no chlorophyll. Offspring of crosses are phenotypically like that of the pistillate (egg-producing) parent except where the pistillate parent is variegated, in which case the offspring are of all three types, in irregular ratio. Egg cells from green plants carry normal green plastids; those from white plants contain only white plastids; but eggs produced by variegated parents may have both plastid types. Pollen (which produces the sperm and rarely contains any plastids) has no effect on progeny phenotype.

Note that all that is really involved here is the kind of plastid and proplastid present in the egg cytoplasm. If that plastid is defective with regard to chlorophyll synthesis, the F_1 plant will be nongreen; if not, it will be green. Variegated parents produce eggs which contain both normal and defective plastids. These eggs give rise to plants some of whose cells fortuitously receive a majority of green plastids, and some that receive larger numbers of white plastids, which results in variegated plants. Recall that division of the cytoplasm following mitosis is not a quantitatively exact process. Yet on the other hand, the situation

Figure 21-3. *The four o'clock plant,* Mirabilis jalapa, *green-leaved variety.* (Photo courtesy Burpee Seeds.)

here could be rationally explained on the basis of an extranuclear determinant that might or might not be located in the chloroplast. In fact, in theory at least, it could be located anywhere in the cytoplasm; but, like some others to be described shortly, no site has ever been demonstrated. Most observational evidence, however, suggests that the chloroplast itself is *not* the actual genetic determiner in this case.

The evening primrose (*Oenothera* spp.) provides an interesting variant of plastid inheritance in which nuclear DNA plays an important part. Many species of *Oenothera* are characterized by their own unique combinations of reciprocal translocations, which result in the production of only two types of functional gametes, i.e., those that contain only an individual's maternal chromosomes or only its paternal chromosomes. Therefore, chromosomes in these species are transmitted as unitary groups.

Both *O. muricata* and *O. hookeri* are normal green plants. Crosses that utilize *O. muricata* as the pistillate parent produce normal green F_1 individuals. The reciprocal cross produces a nongreen F_1, all of which die. Thus *muricata* plastids can develop in the presence of a *muricata-hookeri* nucleus, whereas *hookeri* plastids cannot. Again, *these* latter two illustrations do not appear to provide examples of an *independent* gene system in the cytoplasm.

Mitochondria. All living cells except bacteria, blue-green algae, and mature erythrocytes contain *mitochondria*—small, self-replicating organelles of considerable internal structural complexity, which are centers of aerobic respiration. A few are shown in sectional view in Fig. 4-2. Although the shape of a given mitochondrion is not constant, mitochondria often appear as elongated, slender rods of rather variable dimensions, and average some 0.5 μm in diameter and 3 to 5 μm in length. Extremes of 0.2 to 7 μm in width and 0.3 to 40 μm in length have been reported. Their number per cell varies considerably, from one in the unicellular green alga *Micrasterias* to as many as about a half million in the giant amoeba *Chaos chaos*.

Like chloroplasts, mitochondria contain their own unique DNA (**mtDNA**), whose base composition is known to differ from that of both nuclear and plastid DNA. Mitochondrial DNA is replicated independently of nuclear DNA through the action of a DNA polymerase whose chromatographic properties differ from those of the polymerase of the nucleus. In many multicellular animal tissues (e.g., rat liver and mouse fibroblast cells) mitochondrial DNA is present as a double-stranded, circular structure, strongly reminiscent of bacterial DNA, with a molecular weight of 9 to 10×10^6. This would indicate roughly 14,000 base pairs and a length of nearly 5 μm. Circular mitochondrial DNA has also been reported for humans. Circular DNA molecules several times larger than 5 μm from mouse fibroplast cells (Fig. 21-4) have been described in the literature.

Many of the mitochondrial proteins are synthesized within that organelle itself, and use mtDNA as a template for unique mitochondrial RNA; however, some other proteins, notably cytochrome c, are made elsewhere in the cell. Among the tRNAs identified in mitochondria is N-formylmethionyl tRNA (tRNA$_f^{met}$), known otherwise only from bacterial cells (Chapter 18). Ribosomes, and *even a cell-free protein synthesizing system*, have been found in yeast. So, as in the case of chloroplasts, this is an independent, DNA-containing organelle that is able to synthesize at least some of its own proteins.

Studies of respiratory-deficient strains of baker's yeast (*Saccharomyces cerevisiae*) indicate clearly the genetic role of mitochondria. Yeasts are unicellular ascomycete fungi. In the life cycle of some species (Appendix B), diploid and monoploid adults alternate; the former reproduces by meiospores called ascospores, the latter by isogametes. Respiratory-deficient strains are able to respire only anaerobically and, on agar, produce characteristically small colonies known as "petites." Both nuclear and cytoplasmic controls affect this trait.

In one type of petite (*segregational petite*), respiratory deficiencies have been

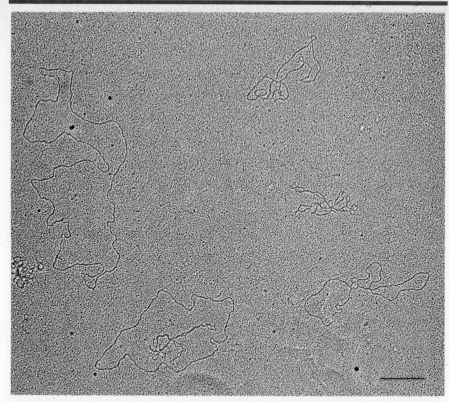

Figure 21-4. *Electron micrograph of DNA isolated from mitochondria of mouse fibro-blast cells. Note the circularity of these molecules. A highly supercoiled molecule lies inside an open monomer* (Lower left), *and three loosely twisted molecules are located at* the right. *The bar represents 1* μm. (Courtesy Dr. Margit M. K. Nass, University of Pennsylvania. Reprinted by permission of author and publisher from Nass, M. M. K., "Mitochondrial DNA: Advances, Problems, and Goals," *Science*, **165:** 25–35, 4 July 1970. Copyright 1970 by The American Association for the Advancementof Science.)

clearly shown to be caused by a recessive nuclear gene. Crosses between monoploid petite and normal result in all normal diploid F_1 progeny, but ascospores produced by the latter segregate 1:1 for petite and normal. On the other hand, another strain (*neutral petite*) has defective mitochondrial DNA. In these the petite character fails to segregate and F_1 and F_2 of the cross petite × normal are all normal; the cytoplasm that contains normal mitochondria is incorporated into F_1 zygotes and vegetative cells and distributed to ascospores and monoploid vegetative cells of the next generation. In a third type of petite (*suppressive petite*), crosses with normals produce a highly variable fraction of petites in the progeny. It is thought that suppressives have rapidly replicating, abnormal mitochondrial DNA. So petites may possess defective mitochondria because of either mutant nuclear DNA or mutant mitochondrial DNA, or both. Respiratory enzymes (except for cytochrome c) are here under a double genetic control.

Mutation of mtDNA probably involves a relative increase of adenine-thymine pairs, and it has been suggested that crossing-over may occur between molecules of cytoplasmic DNA. The action of various mutagens in producing the suppressive petite mutants results from incomplete mtDNA molecules produced because of premature detachment of DNA polymerase in replication. The shorter, defective DNA molecules are replicated more rapidly, and therefore cause the suppressive trait to appear pseudodominant.

Here, as in the case of plastid DNA, there is a degree of cytoplasmic genetic control by DNA of unique and characteristic base sequence not, however, always and wholly independent of nuclear DNA. Whether or not mitochondria (and chloroplasts) have originated as independent, invading organisms that have now become normally occurring organelles in almost all living cells through progressively higher degrees of symbiosis by loss of some of their functions to the nucleus, as has been suggested, there are in these two examples evidence of cytoplasmic genetic systems.

EXTRANUCLEAR INHERITANCE WITHOUT KNOWN CYTOPLASMIC STRUCTURES

A number of rather perplexing cases of apparent cytoplasmic inheritance that cannot at present be related to any identifiable structures or determinants have been reported from time to time. Examination of one of these will suffice to indicate the scope and nature of the problem.

Uniparental inheritance in Chlamydomonas.   The ubiquitous green alga *Chlamydomonas reinhardi* has a relatively simple life cycle (Appendix B). The motile vegetative cells are monoploid (haploid); the zygote is the only diploid cell. In germination the zygote undergoes meiosis to form four monoploid zoospores that quickly mature into vegetative adults. Sexual reproduction in most species represents morphological isogamy but physiological anisogamy in that the gametes are, as noted in Chapter 11, differentiated into $+$ and $-$ mating strains that are physically indistinguishable from each other. Mating type depends on nuclear genes, the locus for which has been determined to belong to linkage group 6 (there are 16 linkage groups in *C. reinhardi*; see Table 4-1). Chemical attraction between gametes of opposite mating strain appears to be exerted through the flagella.

Two types of reaction to the antibiotic streptomycin occur in *C. reinhardi*. Vegetative cells of genotype *sr-1* are resistant to 100 μg per milliliter of streptomycin; those of genotype *ss* (sensitives) are killed at this concentration. This pair of genes shows regular Mendelian segregation. Another genotype, *sr-2*, confers resistance to 500 μg per milliliter of streptomycin; sensitives are again designated as *ss*. Reaction to this higher concentration of the drug is *not* transmitted in Mendelian fashion, but streptomycin response of the progeny is that of the $+$ parent, even though the progeny segregate normally for mating type, as seen in this series of crosses:

(a)   P $sr\text{-}2^+ \times ss^-$
       F_1 $\frac{1}{2}sr\text{-}2^+ + \frac{1}{2}sr\text{-}2^-$
(b) F_1 $sr\text{-}2^+ \times P\ ss^- \rightarrow \frac{1}{2}sr\text{-}2^+ + \frac{1}{2}sr\text{-}2^-$
       F_1 $sr\text{-}2^- \times ss^+ \rightarrow \frac{1}{2}ss^+ + \frac{1}{2}ss^-$

Notice that in each case, mating strain segregates in normal Mendelian fashion, whereas streptomycin response of the progeny follows that of the $+$ parent.

PLASMIDS

Genetic elements composed of circular, closed DNA molecules occur in bacteria and have been most clearly characterized in *Escherichia coli*. These are now generally and collectively termed **plasmids,** although the term *episome* was originally introduced by Jacob and his colleagues in 1960 to designate genetic material such as the F (fertility, or sex) factor, which can exist either integrated into the bacterial "chromosome" or separately and autonomously in the cytoplasm. The term *plasmid* was reserved for those bits of genetic material that exist *only* extrachromosomally and cannot be integrated into the nucleoid. Episomes were said to be integrable, whereas plasmids were not. However, the two terms are now often used synonymously. For simplicity's sake, all bacterial extrachromosomal genetic material, whether integrable or not, will be designated *plasmids*. There are three important classes of plasmids: **F** (fertility) factors, **R** (resistance) factors, and **col** (colicigenic) factors.

The F factor. Attention has already been directed to the F factor and its important function in bacterial conjugation. Under the strict semantic separation, the F factor is an episome, for it is integrable into the nucleoid. When so integrated, the donor bacterial cell is designated a high-frequency recombinant (*Hfr*); when it exists as a separate cytoplasmic organelle, the donor cell is designated F^+. The latter behaves differently from the *Hfr* cell in conjugation. The fertility factor is composed of DNA; in *E. coli* it is composed of $\sim 1 \times 10^5$ base pairs. It determines by its presence that the cell in which it is located will serve as a donor in conjugation, and by its location (i.e., whether integrated or not), whether that cell will be *Hfr* or F^+. (Cells without the F factor are receptors in conjugation.) Therefore, the fertility factor does fulfill the criteria at the opening of this chapter for a true cytoplasmic genetic determinant.

R factors. R factors are plasmids carrying genes for resistance to one or more chemicals—for present purposes these include many drugs used to combat infection (e.g., chloramphenicol, neomycin, penicillin, streptomycin, the sulfonamides, and tetracyclines, among others). Resistance to one or more (usually several) drugs is infectious, in the sense that blocks of DNA-carrying resistance markers are often transferred during conjugation.

R factors are composed of two parts, the *resistance transfer factor* (RTF), responsible for transfer, and various *drug resistance cistrons*, which comprise the r factor. The latter includes all drug resistance factors that may be present, except for tetracycline resistance, which appears to be in the RTF portion. Lewin and others have described the discovery of bacterial drug resistance during treatment of cases of bacillary dysentery (caused by species of the bacterial genus *Shigella*) in Japan in the late 1950s. Four drugs in common use for treatment of this disease are chloramphenicol, streptomycin, sulfonamides, and tetracycline. Strains of the causal organism, isolated from human cases of the disease, were observed to acquire resistance to one or, more commonly, to all four drugs as a result of two events. Japanese investigators, in the late 1950s and early 1960s, established that *Shigella* strains with multiple drug resistance arose in the human gut by conjugation with resistant strains of *E. coli*. Drug resistance can be transferred only by cell-to-cell contact but may be passed both from *E. coli* to *Shigella* and from the latter to *E. coli*. Intergeneric transfer of this kind is widespread in bacteria, especially in the members of the family Enterobacteriaceae, to which both *E. coli* and *Shigella* belong. At least a low level of resistant *E. coli* cells are often present in the intestinal tract, and treatment of dysentery with any of the four drugs kills off all the susceptible *E. coli* and *Shigella*. The remaining resistants multiply and "trade" RTFs by conjugation in the gut.

The R factors are plasmids; they range in size from an apparent low of $\sim 1.5 \times 10^4$ to a high of $\sim 1 \times 10^5$ deoxyribonucleotide pairs. They are most often present in one to three copies per cell, although somewhat higher numbers have been reported for a few R factors.

Colicinogenic factors. The colicinogenic (*col*) factors are plasmids that carry markers for production of **colicins,** which are highly specific protein antibodies produced by some strains of intestinal bacteria. Other strains are susceptible if their walls bear specific colicin receptor sites, and resistant if the receptors are absent. Colicinogenic cells are resistant to the colicins that they themselves synthesize. There are many, many kinds of colicins, each of which has its own killing action. One of the most interesting, colicin K, inhibits DNA replication, RNA transcription, and polypeptide synthesis. Many, but not all of these plasmids are transferable in conjugation. They are closed, double-stranded, circular DNA molecules ranging roughly between 6,000 and 21,000 base pairs in size.

Detection of plasmids.   The occurrence of plasmids in a given cell line can be determined in several ways. First, and most convincingly, they can be photographed by use of an electron microscope. Second, some are "infectious," that is, they can be transferred in conjugation. This is notably true in the case of drug resistance. Third, because they replicate at their own speed, independently of the chromosome, they may be lost from a cell line in a random pattern. Fourth, they can be genetically mapped, and their loci show no linkage to any markers on the nucleoid.

MATERNAL EFFECTS

Shell coiling in the snail Limnaea.   The direction of coiling of the shell in such snails as *Limnaea* illustrates the influence of nuclear genes acting through effects produced in the cytoplasm.

The shells of snails coil either to the right (dextral) or to the left (sinistral), as seen in Figure 21-5. A shell held so that the opening through which the snail's body protrudes is on the right and facing the observer is termed dextral; if the opening is on the left, coiling is sinistral. Direction of coiling is determined by a pair of nuclear genes; dextral (+) is dominant to sinistral (*s*). But expression of the trait depends on the maternal genotype. Because snails may be self-fertilized, the following cross is possible:

$$
\begin{array}{ccc}
\text{P} & + + \, (\female) \times & ss \, (\male) \\
& \text{right} & \text{left} \\
\text{F}_1 & +s & \\
& \text{right} & \\
& \text{(self-fertilized)} & \\
\text{F}_2 & \tfrac{1}{4} + + \; : \; \tfrac{2}{4} + s \; : \; \tfrac{1}{4} ss & \\
& \text{right} \quad\;\; \text{right} \quad\; \text{right!} &
\end{array}
$$

When each of the F$_2$ genotypes is again self-fertilized, progeny of the + + and + *s* animals are dextral, but those of the *ss* individuals are sinistral.

Additional investigation has disclosed that direction of coiling depends on the orientation of the spindle in the first mitosis of the zygote. Spindle orientation, in turn, is controlled by the genotype of the oocyte from which the egg develops and appears to be built into the egg before meiosis or syngamy occurs. The exact basis of this rather unusual control is unknown.

Water fleas and flour moths.   A closely similar situation occurs in at least two very different invertebrates, the water flea (*Gammarus*) and the flour moth (*Ephestia*). Pigment production in eyes of young individuals depends on a pair of nuclear genes, *A* and *a*. The dominant gene directs production of **kynurenine,** a diffusible substance that is involved in pigment synthesis. The cross *Aa* (♀) × *aa* (♂), for example, produces progeny all of which have dark eyes while young. Upon reaching the adult stage, half the offspring (those of genotype *aa*)

Figure 21-5. Dextral and sinistral coiling of the shell in the snail, Limnea.

Dextral   **Sinistral**

become light-eyed. The explanation, of course, is that kynurenine diffuses from the *Aa* mother into all the young, enabling them to manufacture pigment regardless of their genotype. The *aa* progeny, however, have no means of continuing the supply of kynurenine, with the result that their eyes eventually become light. This is obviously a cytoplasmic effect, but wholly without a cytoplasmic genetic mechanism.

INFECTIVE PARTICLES

Carbon dioxide sensitivity in Drosophila.   A certain strain of *Drosophila melanogaster* shows a high degree of sensitivity to carbon dioxide. Whereas the wild type can be exposed for long periods to pure carbon dioxide without permanent damage, the sensitive strain quickly becomes uncoordinated in even brief exposures to low concentrations. This trait is transmitted primarily, but not exclusively, through the maternal parent. Tests have disclosed that sensitivity is dependent on an infective, viruslike particle, called *sigma*, in the cytoplasm. It is normally transmitted via the eggs' larger amount of cytoplasm but occasionally through the sperms as well. Sensitivity may even be induced by injection of a cell-free extract from sensitives.

Sigma contains DNA and is mutable, but it is clearly an infective, "foreign" particle. Multiplication is independent of any nuclear gene, but the mechanism of its sensitizing action is unknown.

Female production in Drosophila.   In another case in *Drosophila*, almost all male offspring die soon after zygote formation. This trait is transmitted by the female and is independent of nuclear genes. Studies have disclosed the trait to be dependent on a spirochete (one of the classes of bacteria) in the hemolymph of the female.

Killer trait in Paramecium.   Some races ("killers") of the common ciliate *Paramecium aurelia* produce a substance called paramecin that is lethal to other individuals ("sensitives"). Paramecin is water-soluble, diffusable, and depends for its production on particles called *kappa* in the cytoplasm. Kappa contains DNA and RNA and is mutable, but its presence is dependent on the nuclear gene *K*. Animals of nuclear genotype *kk* are unable to harbor kappa.

K− individuals do not possess kappa unless and until it is introduced through a cytoplasmic bridge during conjugation (Fig. 21-6). Nonkiller *K*− animals may also be derived from killers by decreasing the number of kappa particles. This may be accomplished either by starving the culture or subjecting it to low temperatures or, on the other hand, by causing killers to multiply more rapidly than kappa.

Kappa particles can be seen with the microscope. Electron microscopy reveals them to have minute amounts of cytoplasm and to be bounded by a membrane. They can be transferred to other ciliates by feeding. Far from being the illustration of the cytoplasmic gene it was first surmised to be, kappa must be regarded as an infectious organism that has attained a high degree of symbiosis with its host.

Similarly, the mate-killer trait in *Paramecium* is imparted by a *mu* particle that, in turn, exists only in those cells whose micronucleus contains at least one dominant of either of two pairs of chromosomal genes. *Mu* particles, too, appear to be endosymbiotes.

Milk factor in mice.   The females of certain lines of mice are highly susceptible to mammary cancer. Results of reciprocal crosses between these and animals of low-cancer-incidence strain depend on the characteristic of the female parent. Allowing young mice of a low-incidence strain to be nursed by suscep-

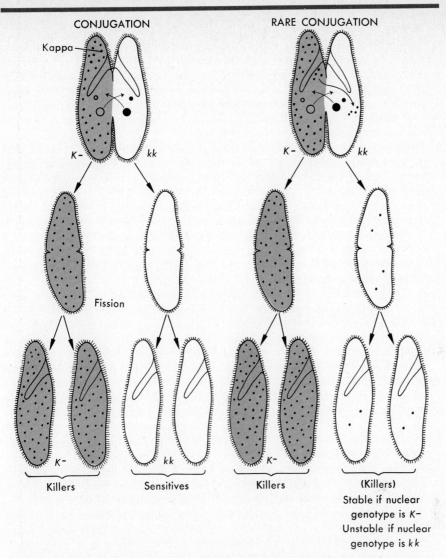

CONJUGATION   RARE CONJUGATION

Kappa

$K-$ kk $K-$ kk

Fission

$K-$ kk $K-$

Killers   Sensitives   Killers   (Killers)

Stable if nuclear
genotype is $K-$
Unstable if nuclear
genotype is kk

Figure 21-6. Conjugation in Paramecium *and the killer trait. Progeny of sensitives
are killers only in rare situations where conjugation persists for a longer period so
that kappa-containing cytoplasm is introduced into the conjugating sensitive. Kappa
particles, however, are maintained only in the presence of a* $K-$ *nuclear genotype.*

tible foster mothers produces a high rate of cancer in these low-incidence young.
Apparently this is a case of an infective agent that is transmitted in the milk.
This so-called milk factor has many of the characteristics of a virus and has been
discovered to be transmissible also by saliva and semen. Its presence in body
fluids, moreover, is again dependent on certain nuclear genes.

    Situations of the kind involving the milk factor, which often show a maternal
inheritance pattern, are certainly not examples of cytoplasmic inheritance as we
have defined it. Rather than being components of the individual's genetic mechanism and results of independent, conserved nucleic acids located in the cytoplasm, they are, in fact, simply acquired infective agents.

Concluding view

In this chapter the problem of the existence of cytoplasmic genetic mechanisms has been examined through some of the many purported illustrations from the literature, viewed against the backdrop of a working concept of what such mechanisms ought to entail. If the broad notion of transmission of heritable traits themselves via the cytoplasm were to be accepted, then from the examples cited here it would have to be acknowledged that *cytoplasmic inheritance* is of widespread occurrence.

But application of more rigorous criteria to a heterogeneous collection of cases eliminates many, *although not all*, of these as instances of cytoplasmic inheritance. It is clear the chloroplasts and mitochondria do possess their own DNA genetic mechanisms that operate in the same way as nuclear genetic material in eukaryotes. In addition, the bacterial plasmids, F factor, R factor and the *col* factors also illustrate true cytoplasmic inheritance. In all of these instances the extrachromosomal genetic material is typical DNA but with a base sequence unlike that of the nucleus or nucleoid. Although it is in the form of closed, circular molecules of comparatively limited length, it has all the other characteristics of chromosomal DNA.

On the other hand, such purely maternal effects as shell coiling in the snail or color in the young water flea can be dismissed rationally as the lingering effects of a parental genotype operating on Mendelian bases. They certainly give no indication that a genetic system even remotely resembling DNA, exists in the cytoplasm.

When maternal effects, infective particles and the like are considered, it is necessary to distinguish carefully between fortuitous parasites and "normal" cell components. At one end of the spectrum is the obvious, identifiable, dispensable parasite (such as spirochetes) on through the viroid milk factor and sigma to asynchronously multiplying kappa to nonintegrable DNA molecules (e.g., RTF and *col*) and the integrable F factor at the other. The very fact that such a series can be arranged invites speculation regarding the possible evolution of such cytoplasmic genetic mechanisms as the F factor, a speculation that is still without any experimental proof and is likely to remain unproven for some time to come. Has there been a gradual evolution from virulent parasite through at first dispensable then indispensable symbionts to a cytoplasmic genetic system that may even integrate into the chromosomal mechanism? Or has the episomic state derived from originally purely chromosomal genes? If there is a relationship, then why have not episome/plasmid systems been discovered outside the bacteria? Have virulent and temperate phages had anything to do with the development of extrachromosomal genetic systems? Where in this whole problem do transforming and transducing DNAs fit? The answers to these questions are still being sought but will surely one day be in hand.

Scientists have begun to make striking practical use of plasmids, which will be described in the next chapter.

References

Bernardi, G., 1976. The Mitochondrial Genome of Yeast; Organization and Recombination. In T. Bücher, W. Neupert, W. Sebald, and S. Werner, eds., *Genetics and Biogenesis of Chloroplasts and Mitochondria*. New York: North-Holland Publishing Co.

Bücher, T., W. Neupert, W. Sebald, and S. Werner, 1976. *Genetics and Biogenesis of Chloroplasts and Mitochondria*. New York: North-Holland Publishing Co.

Chiang, K. S., 1976. On the Search for a Molecular Mechanism of Cytoplasmic Inheritance: Past Controversy, Present Progress and Future Outlook. In T. Bücher, W. Neupert, W. Sebald, and S. Werner, eds., *Genetics and Biogenesis of Chloroplasts and Mitochondria*. New York: North-Holland Publishing Co.

Ephrussi, B., 1953. *Nucleo-Cytoplasmic Relations in Micro-Organisms.* Oxford: Oxford University Press.

Ephrussi, B., H. de Margerie-Hottinguer, and H. Roman, 1955. Suppressiveness: A New Factor in the Genetic Determinism of the Synthesis of Respiratory Enzymes in Yeast. *Proc. Nat. Acad. Sci. (U.S.),* **41:** 1065–1070.

Falkow, S., E. M. Johnson, and L. S. Baron, 1967. *Bacterial Conjugation and Extrachromosomal Elements.* In H. L. Roman, ed. *Annual Review of Genetics,* vol. **1.** Palo Alto: Annual Reviews, Inc.

Green, B. R., and H. Burton, 1970. *Acetabularia* Chloroplast DNA: Electron Microscopic Visualization. *Science,* **168:** 981–982.

Green, B. R., and M. P. Gordon, 1966. Replication of Chloroplast DNA of Tobacco. *Science,* **152:** 1071–1074.

Lewin, B., 1977. *Gene Expression—3: Plasmids and Phages.* New York: John Wiley & Sons, Inc.

Nanney, D. L., 1958. Epigenetic Control Systems. *Proc. Nat. Acad. Sci. (U.S.),* **44:** 712–717.

Nass, M. M. K., 1969. Mitochondrial DNA: Advances, Problems, and Goals. *Science,* **165:** 25–35.

Preer, J. R., Jr., 1971. *Extrachromosomal Inheritance.* In H. L. Roman, ed. *Annual Review of Genetics,* vol. **5.** Palo Alto, Calif.: Annual Reviews, Inc.

Sager, R., 1976. The Circular Diploid Model of Chloroplast DNA in *Chlamydomonas.* In T. Bücher, W. Neupert, W. Sebald, and S. Werner, eds., *Genetics and Biogenesis of Chloroplasts and Mitochondria.* New York: North-Holland Publishing Co.

Sager, R., and Z. Ramanis, 1963. The Particulate Nature of Nonchromosomal Genes in *Chlamydomonas. Proc. Nat. Acad. Sci. (U.S.),* **50:** 260–268.

Problems

21-1 You have just discovered a new trait in *Drosophila* and find that reciprocal crosses give different results. How would you determine whether this trait was sex-linked, a purely maternal effect, or the result of an extranuclear genetic system?

21-2 What phenotype would be exhibited by each of the following genotypes in the snail: + s, s +, ss, + +?

21-3 What kind(s) of progeny with respect to eye color result from these crosses in the flour moth (*Ephestia*): (a) light ♀ × homozygous dark ♂, (b) homozygous dark ♀ × light ♂?

21-4 A four-o'clock plant with three kinds of branches (green, variegated, and "white") is used in a breeding experiment. What kinds of progeny are to be expected from each of these crosses: (a) green ♀ × white ♂, (b) white ♀ × green ♂, (c) variegated ♀ × green ♂?

21-5 Green and Burton (1970) were able to secure intact segments of *Acetabularia* cpDNA as long as 419 μm. As noted in this chapter, such DNA would consist of about 1.23×10^6 deoxyribonucleotide pairs. (a) If it is assumed that all this cpDNA is transcribed, for how many codons could it be responsible? (b) If this entire amount of cpDNA is transcribed and the resulting codons are all translated, how many polypeptide chains could be produced if each consists of 400 amino acid residues?

21-6 Although a neutral petite in yeast has defective mitochondrial DNA, it may have a nuclear gene for normal mitochondrial function. As indicated in the text, a monoploid segregational petite carries a recessive nuclear gene for defective mitochondria, but it may possess normal mitochondrial DNA. If such a neutral is crossed with segregational petite of the type described here, what is the phenotype of (a) the diploid F_1, (b) the monoploid generation that develops from ascospores produced by these diploid cells?

21-7 Employing different substrains of *Escherichia coli* strain K-12 as *Hfr* conjugants produces different results (see linkage map, Fig. 6-7):

	Chromosomal genes transferred in conjugation	
K-12 Substrain	First	Last
C	lys + met	gal
H	pil	pyr-B

For each substrain, give (a) the location of F and (b) the second chromosomal gene that would be transferred.

21-8 For the following crosses in the four o'clock plant give the progeny phenotypes:

Pistillate parent (♀)	Staminate parent (♂)
(a) White	Green
(b) Green	White
(c) Green	Variegated
(d) Variegated	Green
(e) Green	Variegated

21-9 In isogamous sexual reproduction in the unicellular green alga *Chlamydomonas* the chloroplast of the minus mating strain is lost, whereas that of the plus parent is retained, becomes the chloroplast of the zygote and, by division, the chloroplasts of the four cells derived by meiosis from the zygote. Isogametes have about equal amounts of cytoplasm and are combined in the zygote. The sm4 strain of this organism *requires* the antibiotic streptomycin for survival. This trait is passed to progeny only through the plus mating strain. Where is the *sm4* gene most likely located?

21-10 Nitrosoguanidine is a potent mutagen. Its application to a culture of *Chlamydomonas* just prior to nuclear division causes mutation in a variety of nuclear genes, but its application to a culture tube of cells of the plus mating strain just prior to their being allowed to mix with and engage in sexual reproduction with untreated minus cells results in many sm4 mutants. On the other hand, application of nitrosoguanidine to a tube of only the minus strain just before sexual reproduction fails to produce any sm4 mutants. What effect do these facts have on your answer to 21-9?

CHAPTER 22

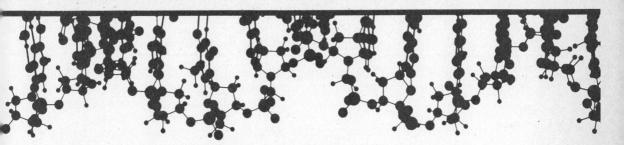

The new genetics, recombinant DNA, and the future

The exciting progress of the science of genetics following the rediscovery of Mendel's principles at the turn of the century has at no point been more apparent than in recent investigations at the molecular level of the nature and action of genetic material. Yet these studies, fruitful as they have been, have lifted the veil but slightly on many still unresolved problems. Both their solution and their application to the welfare of humanity constitute some of the fascinating, and to a few, frightening challenges confronting not only this and the next generation of the world's scientists, but you and your children as well. In this concluding chapter, this prologue to the future will be examined.

BACKGROUND

Human beings have made almost exponential progress during their lifetime on this planet in developing better varieties of plants and lower animals and, even more important, in understanding the mechanisms of genetics and genetic change. For example, in 1974 Fletcher pointed out that the doubling time for scientific knowledge in general is 10 years, but 5 years for the life sciences and 2 *years for genetics*. If anything, the doubling rate of knowledge in the science of genetics has decreased to little more than *one year. At these rates, our genetic knowledge*

422

CHAPTER 22
THE NEW
GENETICS,
RECOMBINANT
DNA, AND THE
FUTURE

will increase during your college career by a factor somewhere BETWEEN 4 AND 16 TIMES! Although much remains to be learned, of course, a good deal is already known about the molecular processes involved in recombination, transduction, mutation, transcription, translation, DNA replication, and, most importantly, recombinant DNA. With the exception of recombinant DNA, a considerable degree of chance is involved in the application of these processes, and the outcomes are uncertain.

Progress in medical science has compensated, to some extent, for deficiencies resulting from deleterious genes (diabetes is a case in point), but the lethal or disabling consequences of defective genotypes and karyotypes is still fairly considerable. Roberts and his colleagues, in a study of the causes of death of 1,041 children over a seven-year period in the hospitals of Newcastle in Great Britain, found gene and chromosome defects to be responsible for 42 percent of the deaths in their sample. Single gene defects accounted for 8.5 percent of these childhood deaths; chromosome aberrations, 2.5 percent; and those probably resulting from complex genetic causes, 31 percent. It is estimated that some type of chromosomal abnormality is present in almost 0.5 percent of live-born infants and to occur in nearly one-fourth of all spontaneously aborted fetuses. Polyploidy or aneuploidy has been found in almost two out of three such fetuses. Chapter 11 pointed out that sex chromosome anomalies occur with frequencies up to 2 per 1,000 live births, and that such aberrations as aneuploidy (trisomy in particular) and even polyploidy are surprisingly frequent (Chapters 11 and 13). Moreover, Chapter 19 indicated that each person carries several deleterious or lethal genes, and roughly 11.3 million mutations occur in each human generation in the U.S., almost none of which are neutral or advantageous.

From the personal as well as the humanitarian viewpoint, it is desirable to minimize these mutations and these chromosomal aberrations and to negate the effect of defective genes. Some would go so far as to raise the question of whether "desirable" genotypes can be made up to order; i.e., *can human beings control their own evolution?* In fact, this query is being raised with increasing frequency; several symposia have addressed themselves solely to this question, and a large number of papers and books, some of which are listed at the end of this chapter, are appearing on the same topic. Two major recent developments with lower organisms make it appear that the question of our capability to shape in some degree our own evolution must soon be answered in the affirmative. These developments are (1) the construction, in vitro, of biologically active, recombinant ("hybrid") DNA molecules, and (2) the production of exact genotypic copies of individuals, or cloning. True, our efforts in these directions have been successful *principally* in lower organisms. For example, cloning has been accomplished in carrots, frogs, and mice. In frogs and mice the monoploid nucleus of an unfertilized egg was removed, and a (diploid) somatic nucleus from the same species inserted in its place. Until recently, **recombinant DNA,** whereby a segment of one species' DNA is inserted into that of another and annealed to it, involved transfer between such disparate species as *Escherichia coli* and *Drosophila melanogaster.* Then, in mid-1978, Harvard's Gilbert and his colleagues announced the successful insertion of rat DNA coding for rat proinsulin into the *E. coli* plasmid that codes for the penicillin-destroying enzyme penicillinase. Introduction of the modified plasmid into other *E. coli* cells was followed by synthesis of about 100 proinsulin polypeptide chains per cell. These then excreted the penicillinase + proinsulin into the culture medium. In this work, the rat proinsulin DNA was "copied" from the mRNA of an insulin-producing rat tumor. A year earlier, a small California pharmaceutical laboratory had succeeded in synthesizing somatostatin (a brain hormone) DNA and inserting it into the DNA of bacterial cells, which then began producing the hormone. Then in September of 1978 came the exciting announcement by the same California laboratory that a synthetic DNA sequence for *human* proinsulin had been inserted into *E. coli* cells, which then began manufacturing human proin-

423

*THE NEW
GENETICS,
RECOMBINANT
DNA, AND THE
FUTURE*

sulin. Conversion of the single proinsulin polypeptide chain to the two chain insulin molecule (Fig. 22-1) was next successfully undertaken. The process by which all this was done was based on work by molecular biologist Arthur D. Riggs of the City of Hope Medical Center (Duarte, California) and the University of California (San Francisco).

Only the time needed for testing and approval by the federal Food and Drug Administration appears to stand between this experimental procedure and the availability of almost unlimited quantities of relatively inexpensive human insulin for diabetes.[1] Not only are supply and price important, but allergic reactions of some people to insulin from cattle and pigs, which must presently be used, will be obviated. (Although the insulin molecule from these species differs from human insulin by some nine amino acid residues, it is effective in human use.) Then in mid-1979, Baxter, Martial, Hallewell, and Goodman, of the University of California at San Francisco, succeeded in producing a strain of *Escherichia coli* bearing a recombinant plasmid that carried a cistron for pituitary human growth hormone. The cistron had been obtained from pituitary tumors. At present, pituitary glands from 50 cadavers are required to secure enough of the hormone to treat one child for pituitary dwarfism for a year.

Transformation by gene transfer has begun to yield important clues in understanding—and perhaps preventing—at least some kinds of cancer. Briefly, cultured cancer cells from human and animal tumors have been found to carry *transforming genes* that convert normal cells to cancerous ones.

Earlier editions of this text cited the prediction that ". . . most geneticists and biochemists believe extension of the (recombinant DNA) technique to mammals, including man, will soon be feasible." That prediction has now been at least partially fulfilled, and in an even shorter time than was generally accepted a few years ago.

Should recombinant DNA techniques be extended directly to human beings (e.g., inserting a proinsulin cistron directly into the DNA of living persons), we shall, indeed, be able to control our own evolution. Such procedures as those involving proinsulin could, then, be extended, at least theoretically, to almost any heritable human trait.[2] As might be imagined, with the press of population and competition for the world's resources becoming ever more acute, and with the rising hope of eradicating genetic disorders, the question of control of human evolution is becoming increasingly important to leaders from the natural sciences, the social sciences, philosophy, and religion. It is a problem that will affect you, and one to which, as an educated, intelligent citizen, you must be prepared to give earnest consideration.

There are two broad facets to the question of control of our own evolution. Simply stated, they are, "Can we?" and "Should we?" The second, which raises profound moral, legal, and ethical considerations, lies somewhat beyond the immediate scope of an introductory course in genetics. But concern with the first should help us develop a sense of values based on knowledge, as well as help initiate the sober reflection for which there may still be time.

RATIONALE OF HUMAN EVOLUTION CONTROL

The bank of human genetic material is undergoing a slow but inexorable decline in quality due to several reasons that may at first be rather surprising. Several dysgenic influences are contributing in different degrees to this qualitative dilution, which must be counteracted by conscious efforts at quality control. Disagreement lies not so much with the need for genetic improvement as with methods of achieving it.

---

[1]FDA approval was granted late in 1982.

[2]This does *not* imply that such techniques, if and when perfected, will be uniformly feasible across the entire spectrum of the human genome.

424

CHAPTER 22
THE NEW
GENETICS,
RECOMBINANT
DNA, AND THE
FUTURE

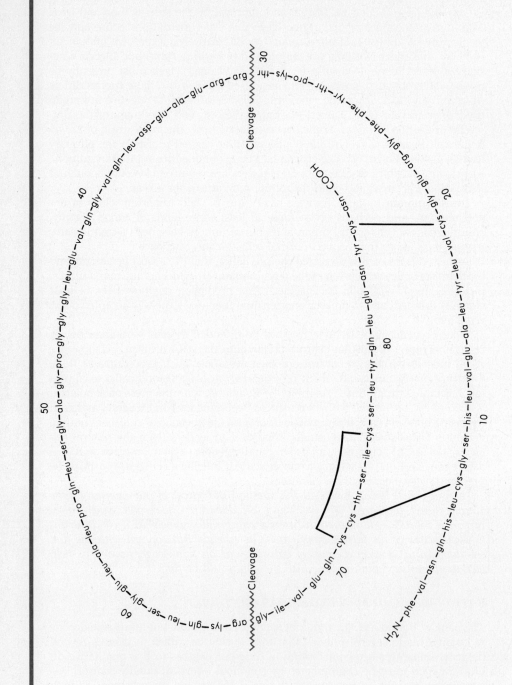

425

THE NEW
GENETICS,
RECOMBINANT
DNA, AND THE
FUTURE

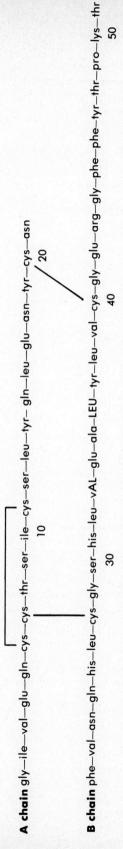

A chain gly—ile—val—glu—gln—cys—cys—thr—ser—ile—cys—ser—leu—tyr—gln—leu—glu—asn—tyr—cys—asn

10   20

B chain phe—val—asn—gln—his—leu—cys—gly—ser—his—leu—vAL—glu—ala—LEU—tyr—leu—val—cys—gly—glu—arg—gly—phe—phe—tyr—thr—pro—lys—thr

30   40   50

Figure 22-1. *(A) The sequence of 86 amino acid residues of the human proinsulin polypeptide, which is the product of a single cistron. In formation of the two shorter peptide chains of insulin, the proinsulin molecule is enzymatically cleaved between its 30th and 31st amino acid residues and also between its 65th and 66th, as shown. The intercalary segments (amino acid residues 31 to 65, inclusive) forms an inactive peptide and is discarded. The remaining portions are joined to form the insulin molecule. (B) Amino acid residue sequence of the human insulin molecule; lines connecting cysteine residues are disulfide bridges.*

The fact that here, as in other eukaryotes, a segment of the polypeptide sequence produced by a single cistron is "clipped out," with the remainder rejoined, raises some interesting evolutionary questions about the relationship between the longer (proinsulin) chain and the shorter (insulin) chains (Doolittle, 1978).

Abbreviations for amino acids follow the standard system (see Appendix C).

426

CHAPTER 22
THE NEW
GENETICS,
RECOMBINANT
DNA, AND THE
FUTURE

In spite of a recent decrease in the rate of population growth in some western countries, the general rate of increase cannot be ignored (Table 22-1).

The overall world average in 1981 was +1.70, up from +1.64 in 1975 (Table 22-2). But when one considers the time required for doubling of populations at these rates (Fig. 22-2), the pressing nature of the problem is obvious. The theme for Brown's *The Twenty-ninth Day* is a French riddle: a lily pond contains a single leaf. Each day the number of leaves doubles. . . . If the pond is completely full on the thirtieth day, when is it half full? The answer is *on the twenty-ninth day.* Brown goes on to point out that "the global lily pond in which four billion of us live may already be half full."

Table 22-1. *Annual rate of population change for selected countries*

Annual percentage change	Country
4.10	Jordan
3.75	Mexico
3.51	Ecuador
3.45	Venezuela
3.24	Israel
3.14	Chile
3.05	Costa Rica
2.78	Brazil
2.75	India
1.75	Australia
1.30	People's Republic of China
1.26	Japan
0.94	U.S.S.R.
0.90	U.S.A.
0.86	France
0.70	Norway
0.26	Sweden
0.06	United Kingdom
−0.04	Austria
−0.24	Federal Republic of Germany
−0.35	German Democratic Republic

The resulting competition for available resources is even now strikingly evident in many countries. Moreover, the doubling time for the world's population is decreasing at an alarming rate (Fig. 22-3). *Worldwatch Paper 8, World Population Trends: Signs of Hope, Signs of Stress* pointed out that the United Nations projections show "world population increasing from the current four billion to some 10 to 16 billion before eventually leveling off." We may well be in the "twenty-ninth day." But population growth is perhaps more a function of decreased death rate than of increased birth rate. In India, for example, the birth rate has shown a gradual decline for much of this century, yet the population continues to grow. People are simply living longer. Life expectancy did not exceed 25 to 30 years well into the Middle Ages; since then it has risen sharply, especially in the "have" nations, to around 70. A great deal of this increase is the result of medical and health advances. But every successful technique that lengthens the life span of persons with inherited defects increases the likelihood that such individuals will reproduce and pass on their defective genes to the genetic load of future generations. As Fleming put it, "conventional medicine is now seen by the biological revolutionaries as one of the greatest threats to the human race."

427

*THE NEW
GENETICS,
RECOMBINANT
DNA, AND THE
FUTURE*

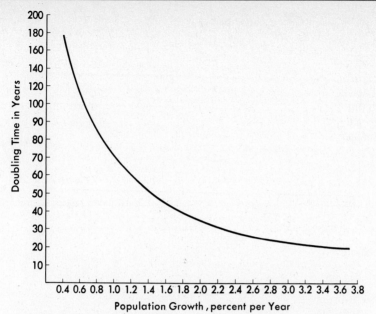

Figure 22-2. *Curve showing approximate doubling time in years plotted against rate of population growth in percent per year. Compare with figures for actual annual growth for selected countries in Table 22-1.*

Selection and survival of the genetically less fit does, indeed, slowly but relentlessly increase the frequencies of these genes and the probability that they will occur in their descendents.

Table 22-2. World population growth by geographic regions

Region	Natural percentage increase
Middle East	2.72
Africa	2.71
Latin America	2.65
Southeast Asia	2.33
South Asia	2.13
East Asia	1.18
Eastern Europe	0.86
North America	0.60
Western Europe	0.32
World	1.70

Data adapted from figures published by Population Reference Bureau, Inc., Washington, D.C.

Moreover, the nearly continuous warfare in which "civilized" humanity has engaged throughout virtually its entire history, now with sophisticated weaponry of terrible efficiency from which no one is safe any longer, tends to siphon off the physically and often the intellectually fit. In addition, the danger of accident or misuse of atomic energy is but a hand's motion away from wreaking tre-

428

CHAPTER 22
THE NEW
GENETICS,
RECOMBINANT
DNA, AND THE
FUTURE

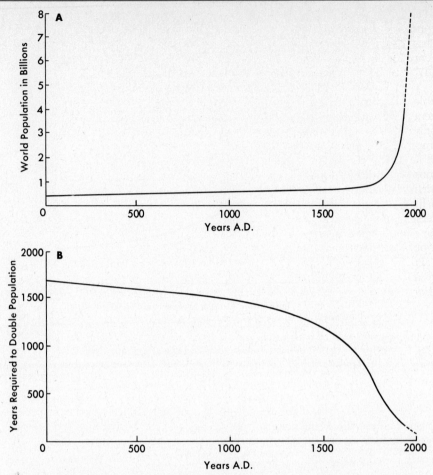

Figure 22-3. *(A) The rise of world population from the time of Christ projected to the year 2000, based on present rate of increase. (B) The diminishing time required to double the world's population from the beginning of the Christian era to the year 2000.*

mendous genetic damage on a large share of the world's inhabitants. Not only does nuclear weaponry constitute a Damocles' sword to the world, but disposal of radioactive wastes from the nuclear power industry, as well as problems of operation of nuclear power plants, are ever present hazards. Most people are aware of the Love Canal problem caused by buried radioactive wastes and other chemicals, and also the Three Mile Island accident (as well as others of presumably lesser magnitude). And who can forget the occasions on which faulty "early" warning devices signaled the approach of nuclear warheads from the U.S.S.R.? In these latter cases the U.S. was minutes away from a decision to retaliate. The narrow margin by which the world escaped a war that would, indeed, end all wars, and perhaps the human race as well, is frightening. However, it should be noted that the doubling dose of radiation for mutation in human beings may be higher than the figure presently accepted. In a 1981 preliminary report on continuing studies of children born to survivors of the atomic bombings of Hiroshima and Nagasaki, the average doubling dose for humans is calculated at 156 rems. A **rem** (**r**oentgen **e**quivalent **m**an) is that dosage of ionizing radiation that will produce a biological effect approximately equal to that produced by one roentgen of X- or gamma-ray irradiation. This is about three times *higher* than the estimate in use in the early 1980s, based largely on experiments with mice. The average of 156 rems is simply the mean of the doubling dose for three components developed from this study: untoward pregnancy outcomes ("mutations, nonfamilial pediatric syndromes, and/or an unusual physiognamy;" see Schull et al. in the references), 69 rems; death during infancy and childhood, 147 rems; and for sex chromosome aneuploidy, 252 rems.

429

THE NEW
GENETICS,
RECOMBINANT
DNA, AND THE
FUTURE

As previously noted, a significant percentage of all births carries some detectable, genetically based defect. Many of the bearers of these defect-producing genes die, some very early, others a little later, and a great many can look ahead only to a life of greater or lesser misery, if their deficiencies are not such as to impair their mental processes. An additional and apparently rather large number of fetuses, homozygous for recessive lethal genes or carrying some serious chromosomal aberration, abort spontaneously. Quite aside from considerations of birth control and death control, those who *are* born ought to have a chance for a life *free from this type of defect*. Some view the problem as the right to life versus the *right to what kind of life*. Moreover, is it fair to yet unborn generations for persons carrying genes for genetic disorders to serve as additional sources of input for deleterious genes?

But the fallacy of a single, absolute solution to the variety of problems imposed by an at-risk pregnancy is best seen by comparing the question of whether to abort, say, a Tay-Sachs fetus with that in the following exchange:*

> ONE DOCTOR TO ANOTHER: "About the terminating of pregnancy, I want your opinion. The father was syphilitic. The mother was tuberculous. Of the four children born, the first was blind, the second died, the third was deaf and dumb, the fourth was also tuberculous. What would you have done?"
> SECOND DOCTOR: "I would have ended the pregnancy."
> FIRST DOCTOR: "Then you would have murdered Beethoven."

POSSIBLE SOLUTIONS

Three general approaches, not all presently of equal feasibility or usefulness, and certainly not all equally accepted, are at least theoretically available for gene quality control:

1. *Overriding the expression* of defect-producing genes without trying to control their frequencies.
2. *Directed recombination;* that is, selecting those genes that will be permitted to be passed on to future generations.
3. *Genetic engineering;* that is, modifying or even replacing the defective genes themselves or, as a spin-off of the future, cloning individuals to order.

Overriding gene expression.   Well over 1,500 hereditary disorders affect human beings, and many more remain to be discovered. In some of these it is possible to supply affected persons with the necessary substance that they are unable to produce, compensating for, or overriding, one or another expression of their genotypes. For example, diabetes mellitus is an inherited disorder that affects several million persons in this country. Although the genetics of this condition is complex and not well understood (autosomal recessive, autosomal dominant, and polygenic bases have been proposed), the physiological basis is failure to produce insulin, as noted earlier. Its precursor, proinsulin, is formed only in the beta cells of the islets of Langerhans in the pancreas. Proinsulin is the product of a single cistron of 258 nucleotides (plus start and stop sequences, as well as a sizable intron). It is manufactured in the usual biological fashion, in which the appropriate sequence of deoxyribonucleotides is transcribed into

*From *Life or Death: Ethics and Options*, ed. D. H. Labby (Seattle: University of Washington Press, 1970).

430

CHAPTER 22
THE NEW
GENETICS,
RECOMBINANT
DNA, AND THE
FUTURE

mRNA which, together with the necessary amino acids, ribosomes, tRNA, and so forth, is translated into the proinsulin polypeptide of 86 amino acid residues. The latter is cleaved and folded to make up the two peptide chains of insulin (Fig. 22-1B).

All somatic cells contain all the genetic information of the individual, yet only a very restricted group of cells is able to manufacture proinsulin. In all other cells of the body the proinsulin cistron is permanently repressed. Knowledge of the mechanism of gene repression in higher organisms is progressing, and a future approach in treating diabetes mellitus may lie in our learning how to derepress the proinsulin cistron, either in the cells where it is normally produced or in other cells. Once such activation has been accomplished, there remains the problem of conversion of proinsulin to insulin and its release from the cells in which its synthesis is effected. By comparison, this is not very difficult. On the other hand, should the genetic basis of even some cases be nonsense or missense in the codons affecting proinsulin synthesis or the conversion of proinsulin to insulin, or production of an insulin inhibitor, rather than repression, an entirely different dimension arises.

Presently, diabetes mellitus can be largely controlled through injection of the insulin that diabetics fail to synthesize. This, however, permits more affected persons to be kept alive and has no effect in upgrading the human gene pool. Rather, it increases the number of persons who can and do serve as additional sources of deleterious genes for succeeding generations. In terms of the human *species*, as opposed to the *individual*, this control measure is actually counter-productive, as is the case with so many medical advances. So, with reference to humankind, other avenues must be explored; that is, we need to distinguish genetic intervention, such as employment of recombinant DNA in one way or another for purposes of individual therapy, from measures designed to alter the course of human evolution. Therein lies the moral and ethical dilemma.

Directed recombination. In some ways, directed recombination, or gene selection, is mechanically the easiest, though not necessarily the most satisfactory, solution to the problem of quality control in the human gene pool. It requires no sophisticated equipment, no techniques that have not been available, even practiced in other species, for years. Basically, it involves only the selection of parental stocks, those whose genes are to be transmitted to future generations. The oldest technique would merely restrict childbearing to those best fitted to perpetuate the species. But this raises significant problems. What are the standards? Who determines those standards? Is our knowledge of human genetics sufficiently advanced to permit standards to be set up? For what purposes would those standards actually be devised and applied? Are human societal values sophisticated enough to replace selfish interests by selfless goals?

Heterozygotes ("carriers") of defect-producing genes can now be identified by relatively simple tests in such varied inherited metabolic disorders as sickle-cell trait, cystic fibrosis, phenylketonuria, and galactosemia, and the list is growing. Could reproduction by such persons be limited or prevented? The problem of how that control might be accomplished becomes a very personal one when it is recognized that each person carries at least one, and probably several, such lethal, semilethal, or disabling genes. Should control be exercised through advice and counseling? On a distressingly small scale, this is being done now, but only for a small, generally upper stratum of the human population. Should control be by governmental restraint, by legislation, with penalties for breaking the law—or even by compulsory sterilization? Once again, the question of who should make these judgments is an exceedingly difficult one and may well have no acceptable answer. But, viewed coldly and impersonally, this approach does offer the advantage of being possible now in many instances, for it requires little more genetic knowledge than is presently available.

431

THE NEW
GENETICS,
RECOMBINANT
DNA, AND THE
FUTURE

If the concept of selective parenthood seems repugnant or impractical or dangerous, the additional problem of whether to allow fetuses that express defective genes or that have been damaged by such drugs as thalidomide, alcohol, or LSD, or by radiation exposure to be carried to term, must also be considered. To permit them to be born is regarded by some as the deepest kind of cruelty, both to the child and to its family. By others, *not* to do so is viewed as murder. Public attitudes toward abortion have changed to the point that some question whether its prohibition under professional medical care is even a proper concern of the state, particularly when it is determined that the fetus carries a gross genetic defect. Accordingly, laws have been markedly liberalized since a 1973 Supreme Court decision declared it unconstitutional for any state to forbid abortion in the first trimester of pregnancy. This decision negated statutes in 46 states. By the early 1980s, however, the whole issue was again being raised.

On the other hand, still another avenue of directed recombination is, even now, open to us. Before long it may become much more practical as our knowledge of the human genotype and linkage groups increases. Rather than select certain *persons* who shall or shall not become parents, or which *embryos* will be allowed to mature, we may select the *reproductive cells* themselves. To a degree this is practiced now. Artificial insemination, so long applied with rather successful results in animal breeding, is currently used to a much more limited extent in humans. But all too often, no account is taken of the genetic constitution of the prospective mother and little more of the sperm donor. Ordinarily the goal is merely the *fact* of child production in families in which it would not otherwise be possible and, incidentally, to produce children who could be mistaken for those of the nonbiological father.

The late Herman Muller in 1965 proposed setting up sperm and egg banks, because sex cells can be kept alive and functional for considerable periods by freezing. Muller proposed such banks, with the most desirable male and female subjects serving as donors. By whatever standards that might be set up, the best sources could continue to serve for a considerable period after their deaths and, with suitable precautions, even after a nuclear war. The question of determining genetic desirability here is, of course, no easier than in other systems of restrictive parenthood. But Muller expressed hope in these words: "With the coming of a better understanding of genetics and evolution, the individual's fixation on the attempted perpetuation of just *his* particular genes will be bound to fade. It will be superseded by a more rational view . . . he will condemn as childish conceit the notion that there is any reason for his unessential peculiarities, idiosyncrasies, and foibles to be expressed generation after generation." In this context, recall the private establishment of a sperm bank that utilizes Nobel laureates as donors. Children, apparently normal, have already been born to women who have been artificially inseminated with sperm from this bank. As expected, some controversy has arisen, but the procedure continues to be employed.

Use of sperm, frozen in liquid nitrogen at $-321°F$, was first reported to have resulted in successful pregnancies in 1953; since that time, by Fletcher's estimates, several hundred children have been conceived through use of sperm frozen for as long as two-and-a-half years. Incidence of abnormalities in children so conceived appears not to differ significantly from that of comparable samples of children produced in the customary biological fashion. Of course, use of stored, frozen sperm from selected donors still entails an element of chance in recombination of disadvantageous genes, though the overall probability might be lower in a system of random mating if donors can be selected with a skill and wisdom that we may not yet possess.

Implantation of selected eggs, fertilized or unfertilized, is already a fact in both laboratory animals and in human beings. Edwards, Steptoe, and Purdy, as long ago as 1970, were able to bring about fertilization of human ova in vitro and succeeded in carrying the resulting zygote to at least a 16-cell embryo stage.

432

CHAPTER 22
THE NEW
GENETICS,
RECOMBINANT
DNA, AND THE
FUTURE

In 1971 Shettles of Columbia University was able to implant a human embryo at the blastocyst stage. Then, in 1978, Edwards and Steptoe disclosed that a baby girl, conceived in vitro (using egg and sperm from the two parents-to-be), and implanted at a very early stage of development, was born—the world's first known human to be conceived in vitro. Happily, this child appeared normal in every respect. A second such pregnancy, followed by birth of a normal child, occurred in the United States in 1979. Since then the procedure has become almost routine and a number of children, all apparently normal, resulting from in vitro fertilization and implantation in the uterus of the woman who supplied the egg, had been born by the early 1980s.

The prospects are virtually limitless. A system of volunteer "host" mothers who would bear other people's children for a fee has been suggested for instances where it may be physically unwise for a woman to undertake the risks and rigors of pregnancy, or even in case she simply does not wish to interrupt a career. In fact, surrogate mothers have already been hired to bear children for women who should not or cannot bear their own, and have borne normal children. The process is quite simple, although it meets with less than complete acceptance. The surrogate is impregnated either naturally or artificially by the man whose wife cannot bear a child. She then gives up the child at its birth. That is the theory. In 1981 such a surrogate mother changed her mind and refused to part with the baby. A suit for breach of contract ensued, but the surrogate won custody of the child.

Some writers, only partly with tongue in cheek, have compared human germinal choice to selection of a packet of desirable flower seeds in the supermarket. The container would carry a brief statement of the most probable traits, from intelligence to physical perfection and even sex[3]; selection could be made as simply as deciding on the kind of plants to grow. From the psychological standpoint, there is considerable doubt that we are ready for this, and our genetic knowledge of the human species is not yet up to our understanding of zinnias and cattle. When it is—and some scientists have estimated that this will be before the end of this century—perhaps rather than implant a fertilized egg in a human female, egg and sperm could be selected on a pedigree basis. In vitro fertilization would follow, with the embryo even raised to term in a glass womb.

In fact, although some scientists have expressed doubt, Petrucci in Italy is said to have raised human embryos in this way for as long as two months, and then deliberately terminated the experiment because the embryo was grossly deformed. The moral question "Is it murder?" is reported to have deeply disturbed him as a Catholic. Others have worked and are working along these lines and someday, somewhere, someone is going to be successful in producing such a person. Think of the cognate problems: Who is this person? Who is or are the parents? What are his or her legal rights? Will we mass-produce a new race of slaves? One day these questions, and more, are going to have to be answered.

The possibility of vegetative multiplication, or cloning, is also interesting from the point of speculation. It has long been possible to maintain cultures of human cells. Will it eventually be possible to induce differentiation and thereby create unlimited numbers of custom-made, identical individuals to certain specific genetic designs? Or will it be useful to have identical clonants to serve as organ donors to each other, at least until effective immunosuppressives are perfected? Sinsheimer is quoted by Fletcher in these words: "[Cloning would] permit the preservation and perpetuation of the finest genotypes that arise in our species— just as the invention of writing has enabled us to preserve the fruits of their life work." Differentiation in cell cultures, culminating in the formation of complete organisms, has already been accomplished with lower forms of life, as noted earlier, and some scientists feel it is more a question of *when* than *if*. Nobel laureate Lederberg is quoted by Toffler (*Future Shock*) as putting "the time scale

---

[3]However, predetermination of sex does not appear imminent.

Though that may become a technical possibility, experimenters may simply elect not to do it with human beings.

433

THE NEW
GENETICS,
RECOMBINANT
DNA, AND THE
FUTURE

Another avenue to such asexual reproduction lies in removal of the monoploid nucleus from a human egg and replacing it with a diploid nucleus from the female who supplied the egg, or from some other individual. Theoretically, human eggs so manipulated could then either be implanted or even raised to term in vitro. Individuals produced in this way would be genotypically identical to the one serving as the nuclear donor and, of course, of the same sex. Such immortality of the individual is thought to be remote for human beings. As late as 1980 informed opinion held that cloning of a mammal by such nuclear transfer was at least five years in the future. But in early 1981—less than a year later—it was reported that body cells from a very early embryo could serve successfully as nuclear donors for this kind of cloning in mice. As is so often the case, predictions in rapidly moving fields are far too conservative. Incidentally, the reported successful production in 1978 of a clone of an aging man with no direct descendants was not taken seriously by geneticists and medical scientists, and in 1982 both author and publisher admitted that the report was a hoax.

Genetic engineering.   Certainly in some genetic disorders, and very possibly in many, the problem is a cistron whose nucleotide sequence specifies a "wrong" series of amino acids. Even a single base-pair change—for example, a switch from an adenine-thymine pair to one of guanine-cytosine, or even from an adenine-thymine to a thymine-adenine pair, or the deletion or insertion of one nucleotide pair—can, as seen in Chapter 19, code for an incorrect amino acid at a given position in the polypeptide chain. The result could be the formation of an aberrant or a nonfunctional protein. This is the sort of change, you will recall, that spells the difference between normal hemoglobin A and the hemoglobin S of sickle-cell disease. Could such defective cistrons be replaced by genetic surgery? In brief, the answer today for human beings is "No, not yet." But it can be done in bacteria and in cell cultures, though it is not yet possible to control *completely* which genes are replaced. Present progress is rapid, and a high level of control is well within the lifetime of today's college student.

As noted in earlier chapters, transduction involves transfer of DNA from one cell to another through the mediation of a phage. In the infection of a bacterium by a virulent phage, the vast majority of the new virus particles released upon lysis of the host contain DNA identical to that of the original infecting phage. But a very few contain a segment of bacterial DNA from the host cell, which replaces a corresponding bit of viral DNA. On the other hand, recall that the temperate phages do not regularly lyse and destroy the bacterial cell. Infection of a cell by temperate phage *lambda,* for example, results in incorporation of some of the viral DNA into the bacterial cell's chromosome. The bacterial cell survives and multiplies; some of its descendents contain DNA that includes sequences of nucleotides received from the phage.

The question here is whether this same kind of transfer of genetic material could be effected in the cells of higher organisms. There is evidence to suggest an affirmative answer. Aaronson and Todaro reported in 1969 that DNA isolated from simian virus 40 can become established in human fibroblast cells in vitro and appears to express itself in protein synthesis in such cells. SV40 is a small virus that produces tumors in appropriate animal hosts. Human cells in which SV40 DNA has been incorporated undergo certain characteristic changes, including loss of sensitivity to contact inhibition of cell division and the production of SV40-specific mRNA. Such human cells also produce a new protein, the so-called T-antigen, which persists in clonal cells. Aaronson and Todaro conclude their report with these words: "There is considerable evidence [to show] that SV40 DNA can become a permanent part of the host cell genome. Most the SV40 DNA in transformed cells is associated with the chromosomes. . . . However, if

434

CHAPTER 22
THE NEW
GENETICS,
RECOMBINANT
DNA, AND THE
FUTURE

the viral DNA's ability to integrate into the human cell genome can be separated from its [tumor-producing property], it may then be possible to use 'integrating' viral DNA to insert specific information into human cells." The important point here is the incorporation and persistence of viral DNA in the human gene complement and its subsequent manifestation in specific new protein synthesis.

Three additional discoveries of tremendous significance (two of which have already been referred to in Chapters 16 and 20) complete the groundwork, although their application to human genetic surgery must still hurdle some problems. But, by comparison, these by no means appear insurmountable. Although it is, of course, risky to speculate, timetables for discoveries and their utilization are frequently much too conservative. To address ourselves to the question of human genetic modification—rather than hope to come upon a convenient transducing virus carrying, for example, a cistron for proinsulin or phenylalanine hydroxylase—it might be more practical to make it to order.

In 1967 Kornberg and his associates were able to synthesize the single-stranded DNA ($\sim$ 5,400 nucleotides) of phage X-174. This synthetic DNA was found to be fully infective for the usual host, *E. coli*, and the virus particles resulting from that infection were completely indistinguishable from wild-type phage. Two years later some 4,700 nucleotides of the operator, promoter, and z cistron of the *lac* operon of *E. coli* were isolated. Since then the ability to manufacture cistrons to order by "copying" them from their mRNA transcripts has been developed and expanded.

Finally, it is reported that infection of cultured mouse kidney cells by polyoma virus results in production not only of normal polyoma virus but also of pseudoviruses. These consist of *fragments of host cell DNA* contained within normal polyoma virus protein coats.

To summarize, prospects are promising that it may soon be possible to isolate and synthesize cistrons coding for almost *any* trait, of packaging these in protein coats, and of finding that they will express themselves in the recipient cells by the usual biological method of protein synthesis. Once the original isolation and synthesis have been accomplished, they will be done for all time; following established methods, they can be copied accurately forever. If the question of "whether" seems answerable in the affirmative, what about the question of "when"? Fleming, while acknowledging that "gene manipulation and substitution in human beings . . . is the remotest prospect of all" in the biological revolution of which he writes, does suggest its feasibility "maybe by the year 2000." Quantum leaps in that direction have already been made, and it seems likely that any delay will be for ethical, rather than technical reasons.

If a few small cistrons can be synthesized, next may well come the possibility of replicating the entire human DNA complement; i.e., literally making human beings to order with almost any set of predetermined characteristics. If we combine this kind of genetic engineering with the glass womb, we will, indeed then control our own evolution, by shaping it to suit whatever the goal, good or evil. And it *will* be possible in the foreseeable future.

THE ETHICS OF THE "NEW GENETICS"

There are, of course, many puzzling issues in genetic engineering, whether it involves recombinant DNA, fertilization in vitro and implantation, or nuclear transfer. Like so many advances, the "new genetics" is viewed with alarm by some, but by others as potentially of great benefit to the human species. It should be only a short time until human interferon, insulin, and pituitary growth hormone (and other substances), produced by vats of microorganisms (e.g., *Escherichia coli*), will have been tested for safe use in human beings and available on the pharmaceutical market. These techniques are also being investigated in numerous laboratories for faster, more controllable and specific improvements in farm animals and crop plants. Many new commercial laboratories, funded by

433

. . . anywhere from zero to fifteen years from now," that is, by the mid-1980s. Though that may become a technical possibility, experimenters may simply elect not to do it with human beings.

both private and public investment, have been formed to pursue these objectives. Some of these goals have already been realized, others are still in the experimental stage and give promise of soon being commercially feasible, but others (e.g., development of nitrogen-fixing corn, wheat, and other crop plants) appear to most investigators to be farther in the future. There are few concerns here by the public, although scientists are alarmed about the potential accelerated loss of valuable genotypes from the world crop gene pool.

435

THE NEW
GENETICS,
RECOMBINANT
DNA, AND THE
FUTURE

On the other hand, there is legitimate concern over some of the ethical issues raised by continued application to *Homo sapiens,* but a great lack of understanding by lay persons. Some heat, but little light, is produced by such magazine cover scare headlines as "What's Science Doing To the Human Race? A Shocking Report on Genetic Experimentation" (Parents magazine, May, 1981). As pointed out earlier, medical science is already using a number of techniques in humans, e.g., sperm storage, artificial insemination, superovulation and in vitro fertilization, and prenatal screening by amniocentesis. Future possibilities are sex selection by sperm typing (X versus Y), parthenogenesis (either by nuclear transfer or egg fusion), and obviation of genetic disorders by recombinant DNA techniques.

A number of measures are presently in experimental use for introduction of exogenous DNA. One rather inefficient method introduces chromosomes into recipient cells, generally by cell fusion, with a success rate of only about one in a million cells. Transformation utilizing crystallized DNA has been effective, but only for one in 100,000 to 1,000,000 cells. Limited success has been achieved in injecting DNA directly into the nucleus of the target cell. Although as much as 80 percent of cells so treated experience transient transformation, only 1 percent or less of them express the new genetic information stably.

Our increasing control over genotype and phenotype and for altering those properties in directed fashion is, for some, an exciting challenge, for others a source of vague uneasiness, and for still others an alarming threat. Of course, this growing capability can be viewed as usurpation of the responsibility that has been historically entrusted to forces other than *Homo sapiens* for maintaining our genetic integrity. On the other hand, it can also be seen in a positive light as a means of improving our lot, both now and in the future.

Religious arguments have been made on both sides of this question. Pope John Paul II has held genetic engineering to be in opposition to natural law. On the other hand, a prominent Catholic philosopher, Robert Francoeur, has written:

"Man has played God in the past, creating a whole new artificial world for his comfort and enjoyment. Obviously we have not always displayed the necessary wisdom and foresight in that creation; so it seems to me a waste of time for scientists, ethicists, and laymen alike to beat their breasts today, continually pleading the question of whether or not we have the wisdom to play God with human nature and our future."

The Office of Technology Assessment of the 97th Congress of the United States, in a report *Impacts of Applied Genetics* makes this significant statement:

> "Genetics thus poses social dilemmas that most other technologies based in the physical sciences do not. Issues such as sex selection, the abortion of a genetically defective fetus, and in vitro fertilization raise conflicts between individual rights and social responsibility, and they challenge the religious or moral beliefs of many."

By such techniques and procedures as have been outlined, then, almost any faulty or "undesirable" gene could perhaps be changed, first in the somatic cells and next in the reproductive cells; for there are two facets to the task here: Modification of the genetic material of somatic cells can, of course, have no effect on future generations; to exercise quality control over human evolution, such modification would also have to be made in the sex cells, either directly by

436

CHAPTER 22
THE NEW
GENETICS,
RECOMBINANT
DNA, AND THE
FUTURE

genetic engineering or by the more inefficient and problem-laden re-combination route.

We are still left, however, with several grave questions: (1) To what extent will diminution of negative selection pressures (e.g., keeping alive those whose genotypes are lethal today) "pollute" the human gene pool? (2) How is "desirability" in the genome to be defined; that is, desirability from whose viewpoint? (3) What presently unforeseen dangers lie ahead in genetic engineering experimentation, even with microorganisms? (4) Should research in genetic engineering therefore be halted—either permanently or until our ability to use this new knowledge has matured? (5) What is the responsibility of the scientist in helping the policy makers and the public sort out and understand the complex issues involved? (6) How can ethical and legal questions be answered? Scientists themselves are devoting more and more thought to these and similar questions, and, as might be surmised, agreement is less than complete.

There is consensus that medical progress, in keeping alive to reproductive age persons whose genotypes presently doom them to death or incapacitation before reproducing, will have only a small quantitative effect on gene frequencies. These deleterious genes have, for the most part, very low frequencies in the large human gene pool. The number of generations required to double the number of heterozygotes, for example, depends on both the present frequency of the gene under consideration and the number of births. Dr. Arthur Steinberg has arrived at the figures shown in Table 22-3, where q_0 represents the present frequency of the (recessive) deleterious gene and n the number of children per family. If we assume a human generation of 30 years, and only two children per family, the doubling time for heterozygote frequency ranges from 600 years if $q_0 = 0.1$ to nearly half a million years if $q_0 = 0.0001$. Even without future medical progress to alleviate lethal genetic disorders (which is certainly not to be expected), the effect on the species is far from great. It is only when one focuses on a particular family that concerns become more immediate.

Table 22-3. Number of generations required to double the frequency of heterozygotes

n	q_0			
	0.1	0.01	0.001	0.0001
2	20	156	1,517	15,127
4	14	110	1,070	10,675

Data of Steinberg.

Definitions of "desirability" are simply not possible for at least two reasons. First, human history has been one of using discoveries for evil as well as for good; atomic fission is a recent illustration, and many others doubtless occur to you. Second, future applications of seemingly harmless or even beneficial discoveries cannot always be anticipated. Signer cites a case in point. The Ph.D research of Arthur Galston disclosed that 2,3,5-triiodobenzoic acid increases the number of flowers and fruits in soybean, which results in greater yield per acre—an important discovery in an age of food shortages. However, government chemical warfare research teams found that higher concentrations of this substance produced defoliation. This discovery was followed by development of defoliants widely used in Vietnam. One of these, "Agent Orange," which is a mixture of 2,4-dichlorophenoxyacetic acid and 2,4,5-trichlorophenoxyacetic acid, accounted for 58 percent of the spraying in Viet Nam, where it was applied at a rate of 20 to 30 pounds per acre—10 times the amount used in the U.S. It produces malformations in fetuses of rats and mice. Unusual increases in stillbirths and in birth defects (e.g., cleft palate and spina bifida—a defect in the walls of the spinal canal resulting in tumor formation) were reported in Tay Ninh Provincial Hospital and in Saigon in 1971 after heavy and widespread spraying by the United States.

RECOMBINANT DNA AND THE FUTURE

Construction of recombinant DNA is simple in concept but delicate with regard to technique. We owe our ability to create recombinant DNA to the discovery of *restriction endonucleases*, enzymes that break internal bonds of DNA to make double-stranded breaks only within short sequences that show twofold symmetry around a given point.[4] These enzymes are highly specific, recognizing only certain sequences. For example, the restriction endonuclease Eco RI (derived from *Escherichia coli KY13*) recognizes only this sequence:

$$5' \ldots G A A T T C \ldots 3'$$
$$3' \ldots C T T A A G \ldots 5'$$

Here the axis of symmetry is between the third and fourth nucleotides from the 5' end. Note that both strands "read" alike, but in opposite directions (i.e., in their 5' to 3' directions). Eco RI breaks this sequence between deoxyguanylic and deoxyadenylic acids (Fig. 22-4). Such **staggered breaks** create "sticky ends" because of the strong affinity of unpaired bases for their complementary nucleotides. Into this gap with its sticky ends a segment of DNA, either synthesized in the laboratory or derived from another species' natural DNA by other restriction endonucleases, can be inserted. The whole is then annealed by DNA ligase. One such recognition site for Eco RI occurs in the genome of SV40 virus, and five each in adenovirus 2 and phage lambda. Restriction endonuclease Hind II, derived from the bacterium *Hemophilus influenzae Rd*, recognizes this sequence:

[4]Drs. Daniel Nathans and Hamilton Smith of Johns Hopkins University, together with Dr. Werner Arber of the University of Basel, shared the 1978 Nobel Prize in medicine for their pioneering accomplishments in this work.

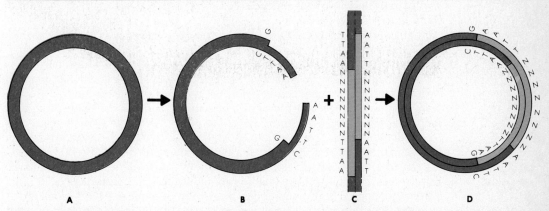

Figure 22-4. *Diagrammatic representation of the formation of recombinant DNA from a bacterial plasmid (pSC101). (A) Intact double-stranded plasmid. (B) Plasmid with staggered gap produced by restriction endonuclease Eco RI. (C) Segment of exogenous DNA from which a portion (dark stippling) is excised by a restriction endonuclease. (D) Recombinant DNA plasmid. The inserted segment may be of almost any length, but must pair complementarily with the plasmid's "sticky ends." A, deoxyadenylic acid; T, thymidylic acid; C, deoxycytidylic acid; G, deoxyguanylic acid; N, any deoxyribonucleotide of a complementarily matched pair.*

438

CHAPTER 22
THE NEW
GENETICS,
RECOMBINANT
DNA, AND THE
FUTURE

$$5' \ldots G\ T\ Py\ Pu\ A\ C \ldots 3'$$
$$3' \ldots C\ A\ Pu\ Py\ T\ G \ldots 5'$$

cleaving DNA at the axis of symmetry, which is between the pyrimidine and purine nucleotides in both strands. Py can be either pyrimidine, and Pu either purine. Five such sequences occur in virus SV40, 34 in phage λ, and 720 in adenovirus 2.

An important plasmid in the early development of recombinant DNA techniques is pSC101 (p for plasmid, SC for Stanley Cohen, in whose laboratory it was isolated). This plasmid consists of only a few cistrons in its approximately 9,000 base pairs and one recognition site for Eco RI. Action of the enzyme opens up the circular plasmid into a linear molecule with sticky ends; into this gap is inserted the desired deoxyribonucleotide sequence. Free ends of the recombinant molecule are next joined through action of DNA ligase. Under suitable culture conditions, cells of *E. coli* are made permeable to the "hybrid" plasmids, whose cistrons, native and inserted, then carry out their respective normal functions in their new cellular home.

By 1974, when techniques for producing recombinant DNA from *E. coli* and humans were rapidly developing, the implied question "Is recombinant DNA prologue or epilogue for the human race?" arose in the minds of many of the scientists working in this field. Accordingly, in 1974, the National Research Council's Committee on Recombinant DNA Molecules proposed voluntary deferral of certain kinds of experiments dealing with recombinant DNA pending assessment of the dangers that might be involved. Proscribed work included "construction of new, autonomously replicating bacterial plasmids that might result in introduction of genetic determinants for antibiotic resistance or bacterial toxin formation into bacterial strains that do not at present carry such determinants, e.g., *E. coli*; or construction of new bacterial plasmids containing combinations of resistance to clinically useful antibodies" and linkage of DNA from tumor-producing viruses to autonomously replicating DNA or to other viral DNA. Members of this group were deeply concerned that our ability to construct biologically active recombinant DNA, using *E. coli*, for example, to multiply the number of the recombinant molecules, might permit "new DNA elements introduced into *E. coli* to become widely disseminated among human, bacterial, plant, or animal populations with unpredictable results," according to Dr. Paul Berg, one of the original proponents of a cautious approach. Dangers cited by this group were felt to be very real, for in this kind of research we appeared to be moving rapidly toward results whose consequences could not be predicted and, at that time, could be but dimly perceived.

At a conference in Pacific Grove, California, late in February 1975, more than 100 of the world's leading biological scientists voted to replace the voluntary ban with a set of stringent safety precautions under which all but the most hazardous experimentation might be resumed. These strict controls were eventually adopted, with few substantive changes, as *guidelines* by the National Institutes of Health. They include four levels of physical containment and three of biological. Both the moratorium and the recommendations of the Pacific Grove conference represented an unusual, if not unique, effort to impose voluntary safeguards before, rather than after, the occurrence of a hazardous event.

Even in 1974, however, not all scientists saw a totally dark outlook. Davis pointed out then that genetic engineering in humans and bacteria present quite different biological and moral issues, that ability to clone any but a small group of lower organisms at that time was far behind ability to develop recombinant DNA molecules, and that success in molecular genetics has been confined to single-gene traits as opposed to polygenic characters (which are probably numerous in human beings). He took the position that "since we cannot predict when a particular kind of manipulation may become feasible, and since moral standards and social needs change with time, it would be presumptuous for us to try to guide future generations by our present wisdom." He did agree that the scientist has an important responsibility to aid in public education, but expressed the fear that public anxiety over unlikely developments "could lead to pruning of valuable major limbs on the tree of knowledge, rather than branches with dangerous fruit."

439

THE NEW
GENETICS,
RECOMBINANT
DNA, AND THE
FUTURE

Accordingly, the NIH guidelines have been progessively liberalized, although a 1981 proposal by the National Institutes of Health Recombinant DNA Advisory Committee to make the guidelines wholly voluntary was abandoned in early 1982. (Committee proposals were published in the *Federal Register* of December 7, 1981.)

Many hold the view that it is the compelling duty of the scientist to continue ceaselessly to push back the boundaries of the unknown and to seek out the truth, wherever the search might lead. Under this philosophy, which certainly has some logic, no scientist should ever withhold new facts or techniques because of fear of what their use might hold for humanity. Applications of laboratory findings, it is argued, are made by society, by politicians, and by the business community, for example, rather than by scientists themselves. Wilson (1970), a scientist-novelist, put it this way:

> Because of the scientist's inability to look over the walls of history and foresee what subsequent generations will do with the fruits of his discovery, society today blames the scientist for what it wrenched from his hand and turned into engines of evil. . . . The scientist is bewildered to find himself considered the villain. . . . The very scientists who are being considered bogeymen are the ones who must still be called upon to use their ingenuity to help undo the damage which society has done to itself.

But others have pointed out that, although scientists discovered the structure of LSD and how to bring about atomic fission, they are not without responsibility in publicizing the consequences of the use or misuse of new discoveries. Both LSD and atomic fission can be and are put to both "good" and "bad" uses. Similarly, the internal combustion engine has given us many benefits, but automobiles propelled by them still kill tens of thousands of people annually and pollute the air to some extent.

The guidelines, suggested by responsible scientists, and adopted by the National Institutes of Health, represented conscientious efforts by informed specialists to set up a series of safeguards. But if guidelines become *controls* or laws under which certain avenues of investigation are *forbidden*, freedom of inquiry is in peril of being stifled. *Responsible* inquiry must, in a free and truly democratic society, *never* be prevented or even discouraged. That this is no idle specter is reflected in two newspaper headlines of the mid-1970s: "Congress Takes Look at Cloning" and "Gene Research Fight Expected in Congress." Congressional committee hearings have been held on recombinant DNA research, as the latter headline implies. Transcripts of those hearings show clearly how legal specialists, woefully ignorant in even basic life science (in which genetics represents to them a vast *terra incognita*), have considered writing laws on subjects about which they know nothing and whose implications for humanity they seem unable to grasp. Humanity may be in danger of seeing freedom of inquiry legally narrowed to the point of probable grave detriment of future generations, as Davis's viewpoint so well puts it.

Fortunately, many of the scientists who were properly cautious in 1974 have now reached the conclusion that the dangers envisaged are much fewer and less severe. As one of many illustrations of this change of view, a short 1978 paper by Pritchard bears the title *Recombinant DNA Is Safe*.

440

CHAPTER 22
THE NEW
GENETICS,
RECOMBINANT
DNA, AND THE
FUTURE

Conclusions

Although the successful application of genetic engineering still lies largely in the future, the future quickly becomes the present, and it is not at all unlikely that it will one day soon be possible in human beings. As Friedmann (1979) has pointed out:

> The scientific dreams and expectations of one generation become the techniques that later generations take for granted. It has always been so, and it should be no surprise that startling new techniques continue to appear, making possible work that was previously unimaginable. . . . Already these techniques have shed light on the mechanisms of gene control in prokaryotic and eukaryotic systems, and they promise to be used at increasing rates for the study of gene expression, for potential treatment of human disease, for agriculture, and all other fields of genetics.

The new genetics—recombinant DNA—prologue or epilogue? With our past record of inept handling of new technologies, the outlook may not be good, but with wisdom and sober, logical regard to the ethical issues involved, the new genetics offers the human race an exciting prologue. For you and your children to be a part of that stimulating future, however, you must become informed on the *facts* and weigh the ethical problems with care and logic. If you adopt an informed, unprejudiced position, you will be a witness, possibly even a participant, in perhaps humanity's greatest achievement of all time.

References

Aaronson, S. A., and G. J. Todaro, 1969. Human Diploid Cell Transformation by DNA Extracted from the Tumor Virus SV40. *Science,* **166:** 390–391.

Augenstein, L., 1969. *Come, Let Us Play God.* New York: Harper & Row.

Bresler, J. B., 1973. *Genetics and Society.* Reading, Mass.: Addison-Wesley.

Brown, L. R., 1976. *World Population Trends: Signs of Hope, Signs of Stress.* Worldwatch Paper 8.

Brown, L. R., 1978. *The Twenty-Ninth Day.* New York: W. W. Norton & Co.

Burns, G. W., 1970. Tomorrow—or the Day After. *Ohio Jour. Sci.,* **70:** 193–198.

Congress of the United States, Office of Technology Assessment, 1981. *Impacts of Applied Genetics.* Washington, U.S. Government Printing Office.

Fleming, D., 1969. On Living in a Biological Revolution. *The Atlantic Monthly,* **223:** 64–70.

Fletcher, J., 1974. *The Ethics of Genetic Control.* Garden City, N.Y.: Anchor Press/Doubleday.

Friedmann, T., 1979. Rapid Nucleotide Sequencing of DNA. *Amer. Jour. Hum. Genet.,* **31:** 19–28.

German, J., 1970. Studying Human Chromosomes Today. *Amer. Scientist.* **58:** 182–201.

Goodfield, J., 1977. *Playing God: Genetic Engineering and the Manipulation of Life.* New York: Random House.

Goulian, M. A., A. Kornberg, and R. L. Sinsheimer, 1967. Enzymatic Synthesis of DNA, XXIV. Synthesis of Infectious φX-174 DNA. *Proc. Nat. Acad. Sci. (U.S.),* **58(6):** 2321–2328.

Grobman, A. B., ed., 1970. *Social Implications of Biological Education.* Princeton, N.J.: The Darwin Press.

Halacy, D. S., Jr., 1974. *Genetic Revolution: Shaping Life for Tomorrow.* New York: Harper & Row.

Hamilton, M., ed., 1972. *The New Genetics and the Future of Man.* Grand Rapids, Mich.: Eerdmans.

Hilton, B., D. Callahan, M. Harris, P. Condliffe, and B. Berkley, 1973. *Ethical Issues in Human Genetics.* New York: Plenum Press.

Huisingh, D., 1973. Should Man Control His Genetic Future? In J. B. Bresler, ed. *Genetics and Society.* Reading, Mass.: Addison-Wesley.

Hutton, R., 1978. *Bio-Revolution. DNA and the Ethics of Man-Made Life.* New York: New American Library.

Jaroff, L., ed., 1972. Man into Superman; the Promise and Peril of the New Genetics. *Time*, April 19, 1971, 33–52.

Karp, L. E., 1976. *Genetic Engineering: Threat or Promise?* Chicago: Nelson-Hall.

Keenberg, M., ed., 1981. *Genetic Engineering*. Proceedings, 1981 Battelle Conference on Genetic Engineering, vol. **IV.** Seattle: Battelle Seminars and Study Program.

Lewin, B., 1977. *Gene Expression, volume 3, Plasmids and Phages*. New York: John Wiley & Sons.

Lipkin, M., Jr., and P. T. Rowley, 1974. *Genetic Responsibility*. New York: Plenum Press.

Luria, S. E., 1973. *Life—The Unfinished Experiment*. New York: Scribner.

Marx, J. L., 1982. Gene Transfer Yields Cancer Clues. *Science*, **215**(4535): 955–957.

Milunsky, A., and G. J. Annas, eds., 1976. *Genetics and the Law*. New York: Plenum Press.

Milunsky, A., and G. J. Annas, eds., 1980. *Genetics and the Law*, vol. **2.** New York: Plenum Press.

Neumann, M., 1978. *The Tricentennial People: Human Applications of the New Genetics*. Ames: Iowa State University Press.

Pritchard, R. H., 1978. Recombinant DNA Is Safe. *Nature*, **273:** 696.

Ramsey, P., 1973. The Moral and Religious Implications of Genetic Control. In A. S. Baer, ed. *Heredity and Society: Readings in Social Genetics*. New York: Macmillan.

Reilly, P., 1977. *Genetics, Law, and Social Policy*. Cambridge: Harvard University Press.

Richards, J., 1978. *Recombinant DNA. Science, Ethics and Politics*. New York: Academic Press.

Rogers, M., 1977. *Biohazard*. New York: Alfred A. Knopf.

Rosenfeld, A., 1965. Will Man Direct His Own Evolution? *Life Magazine*, **59(14),** Oct. 1, 1965.

Rosenfeld, A., 1969. *The Second Genesis*, Englewood Cliffs, N.J.: Prentice-Hall.

Schull, W. J., M. Otake, and J. V. Neel, 1981. Genetic Effects of the Atomic Bombs: A Reappraisal. *Science*, **213:** 1220–1227.

Setlow, J. K., and A. Hollaender, eds., 1979 and 1980. *Genetic Engineering Principles and Methods*, vols. 1 and 2. New York: Plenum Press.

Sonneborn, T. M., 1973. Ethical Issues Arising from the Possible Uses of Genetic Knowledge. In B. Hilton, D. Callahan, M. Harris, P. Condliffe, and B. Berkley, eds. *Ethical Issues in Human Genetics*. New York: Plenum Press.

Toffler, A., 1970. *Future Shock*. New York: Random House.

U.S. House of Representatives, Ninety-Third Congress, Report for the Subcommittee on Science, Research, and Development, 1974. *Genetic Engineering: Evolution of a Technological Issue*. Washington, U.S. Government Printing Office.

Wade, N., 1977. *The Ultimate Experiment, Man-Made Evolution*. New York: Walker and Company.

Wallace, B., 1970. Genetics and Genetic Manipulation. In A. B. Grobman, ed. *Social Implications of Biological Education*. Princeton, N.J.: The Darwin Press.

Watson, J. D., 1968. *The Double Helix*. New York: Athenum.

Watson, J. D., 1971. Moving Toward the Clonal Man—Is This What We Want? *Congressional Record, Senate*, April 29, 1971: 12751–12752.

Wilson, M., 1970. On Being a Scientist. *The Atlantic Monthly*, **226:** 101–106.

Questions for reflection

(No claim is made for originality in the following questions—many writers have raised similar ones, implicitly or explicitly. Moreover, few of them can be given absolute answers at our present stage of understanding. Most of them are opinion questions to which, it is hoped, your genetics course will have contributed some knowledge as well as a sense of priorities and values. Therefore, no answers will be found in Appendix A. However, all these questions are important to you personally.)

22-1   In a recent case of child abuse, the child was so severely beaten by the parents that it suffered permanent brain damage and was crippled for life. Do you recommend taking children away from such parents?

22-2   A couple has a child who develops the Tay-Sachs syndrome; tests confirm that both husband and wife are heterozygous. Does the probability in this case indicate they should "try again"?

442

CHAPTER 22
THE NEW
GENETICS,
RECOMBINANT
DNA, AND THE
FUTURE

22-3   The couple in the preceding question does "try again"; amniocentesis discloses that the second child will also have the Tay-Sachs syndrome. Should they abort?

22-4   How would you answer the two preceding questions if one member of the couple is yourself?

22-5   Suppose a child of yours is born with the Tay-Sachs syndrome, but needs a respirator to sustain life during the period shortly after birth. Would you want the respirator used?

22-6   You are one of approximately three in 100 who is heterozygous for cystic fibrosis. In this condition, which is due to a recessive autosomal gene, affected persons are unable to digest food properly and are highly subject to infection; their ligaments do not form properly, and the lungs fill with fluid that has to be removed (sometimes painfully) almost daily. Death often occurs in the teens. (a) Would you marry? (b) If so, would you elect to have children of your own? (c) If you did, and amniocentesis disclosed the fetus to be homozygous recessive (which, of course, discloses your spouse also to be heterozygous), would you opt for an abortion?

22-7   You are heterozygous for several lethal genes. Would you like to subject yourself to genetic engineering by transducing viruses in order to change your genotype with respect to those genes? Explain the bases for your choice.

22-8   Some people are unable to synthesize arginase and consequently have high blood levels of the amino acid arginine. As a result, they suffer spastic paraplegia, epileptic seizures, and severe mental retardation. This condition, called arginemia, is caused by a recessive autosomal gene. Infection with the Shope virus, which carries a cistron for arginase synthesis, has resulted in elevated levels of the enzyme in both normal and affected persons. The virus produces skin cancer in rabbits, though its carcinogenic effect on humans has not been established. If you had a child born with arginemia, would you want it treated with the Shope virus?

22-9   Should parents in general have freedom to choose whether or not to have children of their own (a) if both are heterozygous for several different lethal or disabling genes, (b) if both are heterozygous for the same lethal or disabling gene?

22-10 If you knew you and your spouse were both heterozygous for the same recessive lethal gene (which, when homozygous, causes intense physical suffering, then death between the ages of five and 10), (a) would you want the freedom to make your own choice as to whether or not to have children? (b) Would you want the decision to be made for you, say by a government commission?

22-11 A woman is carrying fraternal (dizygotic) twins; amniocentesis discloses one of the fetuses to be homozygous for a lethal gene that will kill the child some time between the ages of five and 10 after intense physical suffering. Assume the other fetus to be normal. Assuming that an abortion cannot be selective in aborting only the defective fetus, what option would you accept if (a) the woman were unknown to you, (b) the woman were your sister, (c) the woman were yourself or your spouse?

22-12 Would you like to be able to choose the sex of your offspring?

22-13 If it were possible to choose the sex of one's offspring with near certainty, can you foresee any possible disadvantages for the human species?

22-14 Would you like to be able to choose the intelligence range of your children within, say, about 10 I.Q. points?

22-15 Would you like to have one or more clonants? Give the bases for your answer.

22-16 Do you feel that we should try to alter human genotypes (a) now or (b) in the future? Give the rationale for your viewpoint.

22-17 What is recombinant DNA and how is it made?

22-18 What potential benefits and/or dangers for the human race do you see in recombinant DNA? Evaluate those benefits and/or dangers critically.

22-19 Do you feel that research on recombinant DNA should be (a) carried out without interruption but with all possible precaution against contamination of the human gene pool, (b) temporarily and voluntarily halted until risks and advantages can be evaluated, (c) permanently and voluntarily halted, or (d) permanently halted by statutory fiat? Explain your view in the light of your present genetic knowledge and your assessment of the likelihood of future progress and its potential uses.

22-20 Assuming that this planet can no longer sustain its population, that agriculture-related remedies have failed, and that time has run out, do you recommend (a) mass starvation, (b) starvation of selected populations, or (c) the elimination of "substandard" or noncontributing members of the human race? Do you have any other options to suggest?

APPENDIX A

Answers to problems

Chapter 1

1-1   Bacteria display many detectable variations, especially physiological ones; they have several means of genetic recombination (taken up in later chapters); they have a short life cycle (as short as twenty minutes or even less); they produce large numbers of progeny in a short time (measured in hours); they are easily cultured in the laboratory.

1-2   Obscuring of recessive traits by dominant genes does not occur.

1-3   (a) They are multiples of 7. What does this suggest?
(b) Einkorn; why?

1-4   About as well as bacteria, except for a longer life cycle (one or two generations per season in outdoor culture) and a smaller number of progeny, although the number is still large enough to permit analysis of inherited traits with little difficulty.

1-5   We cannot mate members of the human race as we can laboratory animals and plants; generation time in humans is some 25 to 30 years (pedigree analysis can offset this to some degree); numbers of progeny per mating is small; humans are hardly convenient to culture (!).

1-6   It has been disproven by many experiments.

1-9   Both. "Sun red" corn kernels turn red only if exposed to sunlight; they have the genetic capability to do so. Other varieties of corn that lack this genetic capability do not turn red even if exposed to sunlight.

1-10   It would appear that both genetic susceptibility to the polio virus and exposure to it are required. (The existence of a genetic basis to polio susceptibility is well established.)

1-11 True reproduction is purely asexual in the blue-green algae (cell division, fragmentation, asexual reproductive cells); they lack efficient means of recombination.

1-12 More difficult, inasmuch as symptoms do not develop ordinarily until middle life, whereas ear lobe morphology can be seen at birth.

1-13 Both. With some exceptions (taken up later in this book), tall parents are more likely to have tall children; smaller parents, shorter children. Yet it is well established that environmental factors such as diet also play a part.

1-14 No, because the child's genotype is not changed.

1-15 No, because insulin injection does not change the individual's genotype.

Chapter 2

2-2 *ww.*

2-3 (a) 1. *aa*; 2. *Aa*; 3. *Aa*; 4. *aa*; 5. *A–*.
(b) 1:1.

2-4 Incomplete dominance.

2-5 (a) 25%.
(b) 50%.
(c) *Probably zero; evidence suggests* woman is homozygous normal.

2-6 (a) Incompletely dominant as relates to chloride excretion.
(b) Recessive lethal.

2-7 (a) None.
(b) Yes; this couple's children have a 50% probability of being heterozygotes. If one of such heterozygotes marries another, *their* children (grandchildren of the original couple) have a 25% chance of having the disease.

2-8 (a) Purple is heterozygous; blue is homozygous.
(b) Purple.

2-9 Hornless is the dominant character. Hornless animals producing horned offspring are heterozygotes; any that do so should not be bred. Because he needs to get rid of a recessive, and cattle usually produce but one offspring per year, the problem is not going to be solved quickly. On the other hand, red animals are homozygous, so roans and whites can be excluded from the breeding program.

2-10 (a) *hh.*   (e) *h.*
(b) *HH.*   (f) *Hh.*
(c) *H.*   (g) *Hhh.*
(d) *h.*   (h) *hh.*

2-11 Curly is the heterozygous expression of a recessive lethal; 341:162 is a close approximation of a 2:1 ratio.

2-12 *Ff; Ff; ff; Ff; F—; ff; ff; Ff.*

2-13 Testcross.

2-14 *Aa; Aa; aa.*

2-15　On the basis of her daughter, III-4, who has to be aa, inheriting one a from each parent.

2-16　Rh positive.

2-17　(a) Both completely recessive.
　　　(b) Both are heterozygous.

2-18　Codominance.

2-19　Incomplete dominance, but with the gene for thalassemia major recessive with regard to lethality.

2-20　(a) $\frac{1}{2}$. (b) $\frac{1}{4}$.

2-21　No; the tranfused blood does not affect the recipient's genotype.

2-22　(a) Recessive.
　　　(b) It is maintained in and transmitted by heterozygotes, some of whose children are PKUs. Mutation is another factor, and probably balances the occasional loss of recessive genes in PKUs.

2-23　(a) $\frac{2}{3}$ (*not* $\frac{1}{2}$, because his normal phenotype eliminates the possibility that he might be homozygous recessive for PKU).
　　　(b) Probably none; it appears highly likely that the woman is homozygous dominant for normal.
　　　(c) That any heterozygous child of theirs (which could occur if the husband is heterozygous) has one chance in four of having a PKU child if he or she marries another heterozygote. This eventuality is more likely if such a heterozygous child marries a relative, such as a cousin.

2-24　(a) $(\frac{2}{3})^2 = \frac{4}{9}$. (b) $(\frac{1}{9})$. (c) $\frac{1}{4}$.

2-25　(a) Completely dominant.
　　　(b) 1:1.
　　　(c) Testcross.

2-26　Genes for antigens A and B are codominant, and both are dominant to the gene that results in production of neither antigen.

Chapter 3

3-1　6 (i.e., $AA \times AA$; $AA \times Aa$; $AA \times aa$; $Aa \times Aa$; $Aa \times aa$; $aa \times aa$).

3-2　(a) $\frac{3}{16}$.　　　　(c) $\frac{2}{16}$.
　　　(b) $\frac{3}{16}$.　　　　(d) $\frac{1}{16}$.

3-3　(a) 8.
　　　(b) 2^{12}.

3-4　(a) 1:1.
　　　(b) 1:1:1:1:1:1:1:1.

3-5　16.

3-6　(a) 16.
　　　(b) 81.

3-7　256.

3-8　(a) 4.　　　　(c) 16.
　　　(b) 8.　　　　(d) 2^n.

3-9　$(\frac{1}{4})^n$.

3-10　(a) $\frac{1}{64}$.
　　　(b) $\frac{1}{16}$.

3-11 (a) 24.

(b) $\frac{1}{256}$.

(c) $\frac{1}{128}$.

3-12 Letting Y represent red and y yellow, the parental genotypes are Yyh^1h^2 $\times Yyh^1h^2$ (red, scattered hairs).

3-13 (a) Cream.

(b) $\frac{2}{16}$.

3-14 3:6:3:1:2:1.

3-15 (a) Two pairs, both incompletely dominant.

(b) Broad red, narrow red, broad white, narrow white.

3-16 9 red:3 pink:4 white.

3-17 9 normal:7 deaf.

3-18 (a) 9:7. (c) $AaBb$.

(b) $A - B -$. (d) $AAbb \times aaBB$.

3-19 (a) $aabb$.

(b) $AaBb$.

(c) Anything *except aabb*.

3-20 9 black:3 brown:4 white.

3-21 12 white:3 black:1 brown.

3-22 12 white:3 brown:1 black.

3-23 9 brown:3 black:4 white.

3-24 (a) 2.

(b) $AaBb$.

(c) Purple, $A - B -$; red, $A - bb$; white, $aa - -$, if one assumes gene A is responsible for the enzyme converting colorless precursor to cyanidin, and B for the enzyme converting cyanidin to dephinidin.

3-25 9 both enzymes:3 enzyme number 1 only:3 enzyme number 2 only:1 neither enzyme.

3-26 $aaBB \times AAbb$; $aaBb \times Aabb$.

3-27 Four: red long, red round, white long, and white round.

3-28 (a) 3:6:3:1:2:1.

(b) 1:2:1.

3-29 (a) Let A represent a color inhibitor gene, a the gene for color, B yellow, b green. Then $A - - -$ is white, $aaB -$ yellow, and $aabb$ green.

(b) P: $AAbb$ (white) $\times aaBB$ (yellow)

F_1: $AaBb$

F_2: 9 $A - B - $ $\left.\right\}$ white

 3 $A - bb$

 3 $aaB -$ yellow

 1 $aabb$ green.

3-30 (a) Disk $C - D -$, sphere $C - dd$ and $ccD -$, elongate $ccdd$.

(b) P: $CCdd \times ccDD$

F_1: $CcDd$

F_2: 9 $C - D -$ disk

 3 $C - dd$ $\left.\right\}$ sphere

 3 $ccD -$

 1 $ccdd$ elongate.

3-31 (a) 8.
     (b) 1.
     (c) 24.

3-32 (a) 9.
     (b) $\frac{108}{256}$.

3-33 (a) red $R - S -$; sandy $rrS -$ and $R - ss$; white $rrss$.
     (b) case 1 $RRSS \times RRSS$.
         case 2 $RrSS \times RRSS$, or $RRSs \times RRSs$, or $RrSs \times RrSS$, or $RrSs \times$
         $RRSs$.
         case 3 $RRSS \times rrss$.
         case 4 $rrSS \times RRss$.
         case 5 $rrSs \times Rrss$.

3-34
	Genotypic:	*Phenotypic:*
(a)	1:2:1:2:4:2:1:2:1	9:3:3:1
(b)	1:2:1:2:4:2:1:2:1	3:6:3:1:2:1
(c)	1:2:1:2:4:2:1:2:1	1:2:1:2:4:2:1:2:1
(d)	1:2:1:2:4:2	3:1
(e)	1:2:1:2:4:2	1:2:1
(f)	1:2:2:4	all alike.

3-35 3 red:6 purple:3 blue:4 white.

Chapter 4

4-1 (a) 48.   (e) 48.
     (b) 24.   (f) 24.
     (c) 48.   (g) 12.
     (d) None.   (h) 12.

4-2 (a) 20.   (d) None.
     (b) 40.   (e) 40.
     (c) 40.

4-3 (a) 20.   (e) 10.
     (b) 10.   (f) 10.
     (c) 30.   (g) 20.
     (d) 10.

4-4 (a) 40.
     (b) 40.
     (c) 40.

4-5 160

4-6 (a) 80.
     (b) 160.

4-7 (a) $(\frac{1}{2})^{30}$.
     (b) $(\frac{1}{4})^{30}$.

4-8 (a) 33.
     (b) Irregularities of pairing at synapsis lead to defective gametes with more
         or less than one complete set of chromosomes.

4-9 All Aa.

4-10 Two A and two a.

4-11 Prophase longest, next telophase, next metaphase, with anaphase shortest;
      why?

4-12 $(\frac{1}{2})^7$.

4-13 (a) 1. (d) 4.
  (b) 2. (e) 2.
  (c) 2. (f) 32.

4-14 (a) 2.
  (b) *AB* and *ab*.

4-15 (a) 4. (c) 0.1 each.
  (b) *AB, ab, Ab, aB*. (d) 0.4 each.

4-16 (a) Yes.
  (b) Crossing-over between the two pairs of genes.

4-17 (a) 0.000,005 m. (c) 5,000 nm.
  (b) 0.005 mm. (d) 50,000 Å.

4-18 Increases variability by recombining characters from two parents.

4-19 May give rise to deleterious gene combinations in progeny or disrupt gene combinations (supergenes) that confer selective advantage on the organism.

4-20 Chromosomes usually occur in pairs in diploid organisms. So, regardless of the number of *pairs* (odd or even), the total number of chromosomes (= number of pairs × 2) will be an even number except in a few cases like the male grasshopper.

Chapter 5

5-1 (a) $(\frac{1}{2})^3$, or $\frac{1}{8}$.
  (b) $\frac{3}{8}$.

5-2 $\frac{1}{2}$.

5-3 $\frac{1}{2}$.

5-4 (a) $\frac{1}{6}$.
  (b) $\frac{1}{36}$.
  (c) $\frac{1}{6}$.

5-5 (a) $\frac{3}{4}$.
  (b) $\frac{1}{4}$.

5-6 (a) $\frac{4}{16}$. (c) $\frac{9}{16}$.
  (b) $\frac{1}{16}$. (d) $\frac{3}{16}$.

5-7 (a) $28a^6b^2$.
  (b) $\frac{28}{256}$.
  (c) $\frac{252}{65,536}$, or about 1 in 260.

5-8 18.

5-9 (a) $\frac{1}{8}$.
  (b) $\frac{1}{4}$.
  (c) 6.

5-10 $\frac{270}{32,768}$, or about 1 in 121.

5-11 (a) $\frac{243}{32,768}$, or about 1 in 135.
  (b) No, because other phenotypes, not included here, are possible.

5-12 $\frac{24}{81}$, or roughly 3 in 10.

5-13 (a) 0.02.
     (b) 0.0392.

5-14 Complete correspondence; i.e., the number observed equals the number calculated.

5-15 $P = 1$.

5-16 (a) 1.
     (b) Yes, for 3:1.
     (c) A recessive lethal.

5-17 P lies between 0.5 and 0.3 ($\chi^2 = 0.993$).

5-18 (a) P is between 0.2 and 0.05 ($\chi^2 = 2.0$).
     (b) P is between 0.8 and 0.7 ($\chi^2 = 0.127$).
     (c) No.
     (d) Either accept results as a better reflection of a 9:7 expectancy, or obtain a larger sample.

5-19 (a) P <0.01 ($\chi^2 = 20.0$).
     (b) P lies between 0.30 and 0.20 ($\chi^2 = 1.27$).
     (c) Yes; significant for a 1:1 expectancy.
     (d) The larger the sample the greater the usefulness of the chi-square test.

5-20 (a) $\chi^2 = 0.015$; P = 0.80–0.95; not significant.
     (b) $\chi^2 = 0.451$; P = 0.50–0.70; not significant.
     (c) $\chi^2 = 0.563$; P = 0.30–0.50; not significant.
     (d) $\chi^2 = 0.618$; P = 0.80–0.95; not significant.

Chapter 6

6-1 (a) *R Ro* and *r ro* each 0.4375; *R ro* and *r Ro* each 0.0625.
     (b) 0.1914.
     (c) 0.6912.

6-2 (a) El_1d/el_1D.
     (b) *Trans.*

6-3 (a) 0.485.
     (b) 0.015.

6-4 (a) *wo dil o aw*.
     (b) Double (or any even number of) crossovers are not detected in genes as far apart as *wo* and *aw* are here. The more accurate *wo aw* distance is 22 map units, i.e., 9 + 6 + 7.

6-5 (a) *jvl fl e*.
     (b) Double crossovers are missed in a dihybrid cross involving only *jvl* and *e*.

6-6 It could be either 19 map units to the "left" of *jvl* or 19 to the "right" of *jvl*. If it is the latter (see answer to 6-5), it is also to the "right" of *e*.

6-7 To the "right" of *e*.

6-8 (a) Yes.
     (b) 0.77 percent.

6-9 4.

6-10 12.

6-11 (a) 12.
     (b) 12.
     (c) 23.
     (d) 24.

6-12   Because genes for *some* of the traits he followed are on different chromosome pairs; linked traits he worked with (seed color/flower color and pod contour, that is, wrinkled versus smooth/plant height) are so far apart on their chromosomes that genes for them appear to segregate randomly. In one other case Mendel did not make the cross that would have shown linkage.

6-13   (a) Yes.
       (b) No.

6-14   (a) 4.
       (b) 2.

6-15   (a) 8 (2 noncrossover, 4 single crossover, 2 double crossover.)
       (b) 2.

6-16   Yes. Normal beaked plants are doubly homozygous recessive; in this cross, with unlinked genes, about 6 percent ($= \frac{1}{16}$) of the progeny should have this phenotype. The number actually observed is about 3.68 times greater than is to be expected with unlinked genes.

6-17   (a) Data indicate *cis* linkage (*Cu Bk/cu bk*) in each parent and that the frequency of *cu bk* gametes in each was 0.48.
       (b) Crossover gametes were then produced with a frequency of 0.02 **each,** so the genes are 4 map units apart.

6-18   0.4 *Pl Py*; 0.4 *pl py*; 0.1 *Pl py*; 0.1 *pl Py*.

6-19   16 percent.

6-20   9 percent.

6-21   34 percent.

6-22   (a) 88.2 percent.
       (b) 0.2 percent.

6-23   (a) *h fz eg*; *h-fz*, 14 map units; *fz-eg*, 6 map units.
       (b) 0.238.

6-24   0.5.

6-25   (a) *d + +* and *+ m p*.
       (b) *p d m* (or, of course, *m d p*).
       (c) *p-d*, 4.5 map units; *d-m*, 4.5 map units.
       (d) Yes; coincidence = 0.5.

6-26   (a) *b + +* and *+ cn vg*.
       (b) *b cn +* and *+ + vg*.
       (c) *Trans.*
       (d) *cn.*
       (e) *cn − b*, 9 cM; *vg − cn*, 9.5 cM.
       (f) Yes.
       (g) 0.35.

6-27   (a) 0.85.   (c) 0.10.
       (b) 0.05.   (d) None.

6-28

	(a)	(b)
+ + +	0.4275	0.42625
$pg_{12}gl_{15}bk_2$	0.4275	0.42625
+ $gl_{15}bk_2$	0.0225	0.02375
pg_{12} + +	0.0225	0.02375
+ + bk_2	0.0475	0.04875
$pg_{12}gl_{15}$ +	0.0475	0.04875
+ gl_{15} +	0.0025	0.00125
pg_{12} + bk_2	0.0025	0.00125

Chapter 7

7-1 No.

7-2 $c^{ch}c \times c^h c$.

7-3 (a) 3.
   (b) Superdouble, double, single.
   (c) One parent heterozygous for superdouble and single, the other heterozygous for double and single.

7-4 Alexandra, normal, Blue Moon, Primrose Queen.

7-5 4.

7-6 1.

7-7 3 (heterozygous Alexandra-Normal × the same, heterozygous Alexandra-Normal × heterozygous Alexandra-Blue Moon, and heterozygous Alexandra-Normal × heterozygous Alexandra-Primrose Queen).

7-8 (a) 0.0126.
   (b) 0.0042.

7-9 15.

7-10 0.5.

7-11 MN, secretor.

7-12 20.

7-13 210.

7-14 8.

7-15 A_1B.

7-16 Yes, if the woman is $I^{A_1}i$ and the man $I^B i$.

7-17 Yes, he is eliminated on the basis of the MNSs test (only). Neither the woman nor the alleged father could have contributed Ns to the child.

7-18 (a) No.
   (b) The HLA test.

7-19 Monozygotic twin; dizygotic twin, sibling or parent; uncle; first cousin.

7-20 A1B8/A3B7.

7-21 3.

Chapter 8

8-1 (a) Lozenge.
   (b) Wild.

8-2 The *cd* pair because of the *cis-trans* effect.

8-3 I-1, *Dd*.
   I-2, *dd*.
   II-1, *Dd*.
   II-2, *Dd*.
   II-3, *Dd*.
   II-4, *dd*.

8-4 II-1 must be Rh+ (*Dd*) because the child evidently sensitized the mother, with the result that II-2 and II-3 were erythroblastotic. Had II-1 been *dd*, II-2 would not have been erythroblastotic.

8-5 No cases of erythroblastosis fetalis would have occurred in succeeding children.

8-6 A − ♀ × O + ♂.

8-7 16.

8-8 48.

8-9 162.

8-10 Yes.

8-11 Yes.

8-12 Yes.

8-13 Yes.

8-14 No. The claimant's M-N, S-s type would not be possible with the purported parentage.

8-15 Children numbers four, five, and six cannot be those of the husband.

8-16 $I^B i \, Dd$.

Chapter 9

9-1 Intelligence, height, skin color, eye color.

9-2 Any parental genotypes that can produce at least some progeny genotypes with a greater number of contributing alleles than they themselves have are possible, e.g., *AaBbCcDd* × *AaBbCcDd*, *AaBbccdd* × *aabbCcDd*, etc.

9-3

(a) $\frac{1}{4,096}$.

(b) 13.

(c) $\frac{924}{4,096}$.

9-4 8 polygenes (4 pairs); 3 inches per contributing allele.

9-5 10 polygenes (5 pairs); 4.8 inches per contributing allele.

9-6 $\frac{1}{4,096}$.

9-7 *AaBbCcDd* × *AaBbCcDd*.

9-8 Any in which each parent is homozygous effective for two of the four pairs, e.g., *AABBccdd* × *aabbCCDD*, etc.

9-9 (a) *AABBCCDD*.
(b) *aabbccdd*.
(c) Only green.

9-10 (a) $\frac{1}{256}$.
(b) $\frac{28}{256}$.
(c) $\frac{56}{256}$.

9-11 (a) Green.
      (b) $\frac{70}{256}$.

9-12 Transgressive variation.

9-13 769.

9-14 8.

9-15 (a) 25 and 5 cm.
      (b) 15 cm.
      (c) 1 (25):4 (20):6 (15):4 (10):1 (5).

9-16 (a) 15 and 5 cm.
      (b) 15 cm.
      (c) 9 (15):6 (10):1 (5).

9-17 $(a + b)^{12}$.

Chapter 10

10-1 23.0 (actually, 23.04).

10-2 20.04.

10-3 (a) 4.477, or approximately 4.5.
      (b) That 0.6826 of the sample should lie in the range 23 $\pm$ 4.5, and so
          forth. See text and Appendix D.

10-4 0.6826, or about $\frac{2}{3}$.

10-5 (a) 0.895.
      (b) That there is a 0.6826 probability that the population mean falls in
          the range 23 $\pm$ 0.895, and so forth.

10-6 0.9544.

10-7 1.2

10-8 0.6826, or about two chances in three.

10-9 No; S_d = only $(\bar{x}_1 - \bar{x}_2)$, and to be significant S_d must exceed
      $2(\bar{x}_1 - \bar{x}_2)$.

10-10 8.

Chapter 11

11-1 (a) Metafemale.   (d) Intersex.
      (b) Metamale.   (e) Female (tetraploid).
      (c) Metafemale.   (f) Metamale.

11-2 (a) Female.   (d) Male.
      (b) Male.   (e) Male.
      (c) Female.

11-3 9 normal monoecious:3 pistillate (ears terminal and lateral):3 staminate:1
      pistillate (ears terminal only); i.e., 9 monoecious:3 staminate:4 pistillate.

11-4 $\frac{1}{4}$.

11-5 3 male:1 female.

11-6 (a) *bW* (or simply *b*).   (c) *BW* (or *B*).
(b) *B* − (*BB* or *Bb*).   (d) *bb*.

11-7 (a) $\frac{1}{4}$ each of the following: barred ♀, nonbarred ♀, barred ♂, and non-barred ♂.
(b) All ♂s barred, ♀s 1 barred:1 nonbarred.

11-8 $\frac{1}{3}$ male, $\frac{2}{3}$ female.

11-9 0.1.

11-10 45 percent (i.e., 20 percent XX + 25 percent XXY).

11-11 (a) $\frac{1}{4}$.
(b) $\frac{1}{4}$.
(c) $\frac{1}{2}$, assuming the metafemales do not survive to maturity.

11-12 (a) White.
(b) Red.

11-13 Fluorescence pattern, especially the bright longer arm, is the best identification criterion; also, the longer arms are close together, and satellites are absent. Its length relative to members of the F and G groups of autosomes is more variable.

11-14 From the theoretical standpoint several alternative explanations are possible. Among the most likely are
(a) Nondisjunction of Y in the second meiotic division in spermatogenesis;
(b) Nondisjunction in the first meiotic division in spermatogenesis, giving rise to an XY sperm, which fertilizes an X egg;
(c) First or second division nondisjunction in either spermatogenesis or oogenesis, producing either O sperm or egg, which then fuses with an X gamete from the other sex. The XXY condition could also arise through nondisjunction in the first cleavage division of a normal XY zygote, whereby one daughter cell receives XXY and the other (nonviable) OY.

11-15 Most likely by lagging of an X chromosome in early mitoses of an XX zygote.

11-16 The X chromosome carries a large number of genetic loci, most or all of which appear to be necessary for normal development.

11-17 Humans. Single genes (*Asparagus*) are subject to mutation that could result in a sex imbalance in small populations or a lethal condition, but in humans there appear to be many genes governing sex on the X and Y chromosomes. Furthermore, the occasional "male × male" crosses that occur in *Asparagus* increase the likelihood of homozygosity of deleterious genes in the progeny. Also, a 1:1 sex ratio in Asparagus occurs only in populations where "males" are heterozygous.

11-18 (a) XX.   (c) XY.
(b) XY.   (d) XX.

11-19 Because of irregularities in meiosis.

11-20 Maternal.

11-21 Mother, *Xg Xg*; father, *Xg^a* (Y); daughter, *Xg* O.

11-22 (a) Mother.
(b) Father.

11-23 Father, *Xg^a* (Y); mother, *Xg Xg*; son, *Xg Xg* (Y).

Chapter 12

12-1 1 bent-tail female:1 normal male.

12-2 50 percent probability that any girls will be heterozygous (slight nystagmus) and 50 percent chance that any boys will have severe nystagmus.

12-3 Sex-linked recessive.

12-4 1 barred rose male:1 nonbarred rose female.

12-5 $\frac{6}{16}$; equally divided between male and female.

12-6 1 barred:1 nonbarred; 3 rose:1 single.

12-7 No chance that any children of the boy will develop the disease as long as he marries a $+$ $+$ girl; each girl has a probability of 0.5 of being heterozygous and therefore transmitting the trait to half her sons.

12-8 Sex-linked recessive lethal.

12-9 2 female:1 male, all normal.

12-10 $\frac{1}{2}$.

12-11 $\frac{3}{64}$.

12-12 (a) 0.   (c) $\frac{1}{4}$.
        (b) $\frac{1}{8}$.   (d) $\frac{1}{2}$.

12-13 (a) Early bald.
        (b) Nonbald.
        (c) 9 early bald:3 late bald:4 nonbald.
        (d) 12 nonbald:3 late bald:1 early bald.

12-14 Female.

12-15 Male.

12-16 Sex-limited.

12-17 All males yellow; females 3 white:1 yellow.

12-18 Sex-limited.

12-19 (a) All short.
        (b) All males short, all females long.
        (c) Males: 3 short:1 long; females: 3 long:1 short.
        (d) All males short; females 1 short:1 long.

12-20 Sex-linked recessive may appear in either sex, but much more frequently in males; it is often transmitted from father to half the grandsons via a female. Holandric genes appear only in the male sex and are transmitted directly from father to all his sons.

12-21 Traits determined by holandric genes appear only in the heterogametic sex; these normally cannot be heterozygous.

12-22 A sex-linked dominant will be passed from a heterozygous affected mother to children of either sex with a probability of 0.5 or, from a homozygous affected mother, to all her children of either sex. It will be passed from an affected father to all his daughters, but to none of his sons. A holandric gene is passed from an affected father to all his sons and to none of his daughters.

12-23 All boys would have hairy ears; none of the girls would.

12-24 *HhZW* × *hhZZ*, or *hhZW* × *HhZZ*.

12-25 I-1, II-3, II-9, III-2, III-4, III-5, III-8.

12-26 $\frac{1}{2}$.

12-27 It would be possible, but very unlikely. It *appears* that all of the persons listed were free of the gene for hemophilia; if this is so, then Elizabeth II is homozygous normal. Under those circumstances, only a mutation, such as apparently occurred with Queen Victoria, could produce the defect in any children of Elizabeth.

12-28 (a) $\frac{3}{16}$.
(b) $\frac{1}{16}$.
(c) All males.

12-29 (a) *Nn* × *NY*.
(b) $\frac{1}{4}$.
(c) All staminate.

12-30 (a) Hemophilia A.
(b) Sons only.
(c) Yes, all for B, $\frac{1}{2}$ for A and B.
(d) $\frac{1}{4}$.

Chapter 13

13-1 *DDDD*.

13-2 *DD*.

13-3 *DDd*.

13-4 *DDdd*.

13-5 1 *DD*:4 *Dd*:1 *dd*.

13-6 (a) Autotetraploidy.
(b) $\frac{1}{36}$, or $(\frac{1}{6})^2$.

13-7 $\frac{1}{1,296}$, or $(\frac{1}{6})^4$.

13-8 $\frac{1}{46,656}$, or $(\frac{1}{6})^6$.

13-9 $(\frac{1}{6})^{2n}$.

13-10 Reduces it.

13-11 Those having 28 represent diploids; 56, tetraploids; 70, pentaploids; 84, hexaploids.

13-12 A different series of chromosomal aberrations in each species, so that normal pairing is impossible. Translocations are probably the most frequent of these aberrations.

13-13 Yes, by creating an allotetraploid hybrid.

13-14 Euploidy.

13-15 13.

13-16 12 (dodecaploid).

13-17 (a) Male.
(b) Female.

13-18 69,XXY.

13-19 (a) 92,XXYY.
(b) No.

13-20 (a) 1*P*:2*p*:   (d) All *P*.
2*Pp*:1*pp*.
(b) 1*P*:2*p*.   (e) 2*P*:1*p*.
(c) 1*P*:1*PP*.

13-21 (a) 15 purple:3 white (= 5 purple:1 white).
(b) 11 purple:1 white.
(c) 12 purple:6 white (= 2 purple:1 white).

13-22 (a) 3 normal:1 eyeless.
(b) 11 normal:1 eyeless.
(c) 35 normal:1 eyeless.

13-23 Autosomal monosomy probably constitutes a lethal genic imbalance.

13-24 Trisomy for other chromosomes appears to produce a lethal imbalance.

Chapter 14

14-1

14-2

(a)

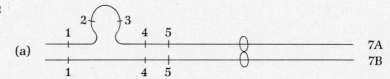

(b)

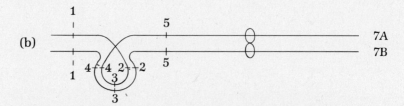

(c)

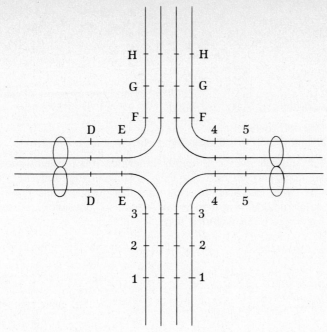

(d)

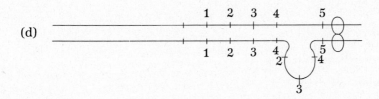

14-3 (a)

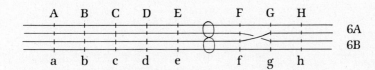

(b)

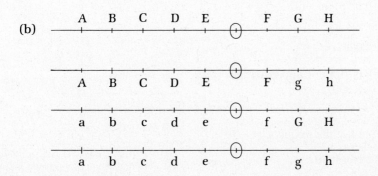

14-4 A paracentric inversion does not include the centromere, whereas a pericentric one does include the centromere.

14-5 (a) Deletion.
(b) Inversion.
(c) Duplication.

14-6 *AAA/AA* between heterozygous ultrabar and homozygous ultrabar; *AAAA/AAA* below homozygous ultrabar.

14-7 (a) Deletion (deficiency).
(b) Deletion loop in salivary gland chromosomes.

14-8 See text and references. Although the evidence is presently contradictory, neither drug can yet be completely absolved. Moreover, the time factor must be taken into account. Note that this question and its answer refer only to cytological and genetic consequences; physiological, mental, and other effects are not included here.

14-9 Farther from the centromere, which has an interfering effect on crossing-over.

Chapter 15

15-1 (a) *MN.*
(b) *NN.*

15-2 *M*, 0.6; *N*, 0.4.

15-3 *M*, 0.546; *N*, 0.454.

15-4 *M*, 0.19; *N*, 0.81.

15-5 *T*, 0.6; *t*, 0.4.

15-6 (a) 48.
(b) 36.

15-7 About 1.4 percent or one in 71 persons.

15-8 One in 500.

15-9 I^A, 0.3; I^B, 0.1; *i*, 0.6.

15-10 $I^A I^A$, 0.04; $I^A i$, 0.28; $I^B I^B$, 0.01; $I^B i$, 0.14; $I^A I^B$, 0.04; *ii*, 0.49.

15-11 Deviation is highly significant (chi-square = 35.69).

15-12 A, 39.36 percent; B, 8.76 percent; AB, 2.88 percent; O, 49.0 percent.

15-13 (a) 5.76 percent.
(b) 8.4 percent.

15-14 0.00004, or $\frac{1}{25,000}$.

15-15 (a) 0.0198, or approximately 0.02.
(b) 0.0001.

15-16 0.35.

15-17 Decrease the frequency of Hb^S.

15-18 Deterioration.

15-19 0.25.

15-20 (a) 0.045.

(b) the 198th generation.

15-21 0.22.

15-22 0.267.

15-23 0.8.

15-24 0.83.

15-25 0.04.

15-26 (a) 0.02.

(b) 2×10^{-3}.

15-27 (a) 1,500 years.

(b) Probably not, because of mutation from dominant normal to recessive lethal.

15-28 (a) p^2.

(b) p^4.

(c) $2pq$.

(d) $4p^2q^2$.

(e) $4pq^3$.

(f) $p^4 + 4p^3q + 6p^2q^2 + 4pq^3 + q^4$.

15-29 (a) $p^2 + 2pq$.

(b) q^2.

15-30 (a) $p^2q^2 + 2pq^3 + q^4$.

(b) $0.0625 + 0.125 + 0.0625 = 0.25$.

15-31 $\frac{2}{2,048} = 0.0009766$, or approximately 0.00098.

15-32 0.4995117, or approximately 0.4995.

15-33 (a) 0.006.

(b) 0.988.

(c) 0.012.

(d) 0.002.

(e) 1.6×10^{-11}; i.e., u^2, or $(4 \times 10^{-6})^2$.

15-34 2×10^{-3}.

15-35 3.317×10^{-3}, or slightly more than 0.003.

15-36 (a) 0.775.

(b) 0.447.

(c) 12.

Chapter 16

16-1 Dominance, epistasis, sex linkage.

16-2 $3'$ T T G C A T G A C G $5'$.

16-3 4^n.

16-4 $L_{\mu m} = 3.4 \times 10^{-4}P$, where $L_{\mu m}$ = length in micrometers and P = the number of pairs of nucleotides.

16-5 68 μm.

16-6 (a) 1.38×10^{10}.
(b) 4.7×10^6.
(c) 185.

16-7 (a) 7.7×10^8.
(b) 2.6×10^5.
(c) 10.2.

16-8 (a) 135.   (c) 111.
(b) 126.   (d) 151.

16-9 (a) 251.   (c) 227.
(b) 242.   (d) 267.

16-10 (a) 329.   (c) 305.
(b) 320.   (d) 345.

16-11 649.5, but use **650** for easier calculations.

16-12 About 1.3×10^8.

16-13 (a) 4.15×10^6.
(b) 1,411.

16-14 One per minute.

16-15 2×10^5.

16-16 Thymine 20 percent, cytosine and guanine 30 percent each.

16-17 No. Why?

16-18 (a) A/T $= 0.99$; G/C $= 1.00$.
(b) That it is double stranded.
(c) This is the replicative form.

16-19 This suggests that not all gene function is nuclear in nature. The matter is explored in Chapter 21.

16-20 31.6 percent.

16-21 (a) None.
(b) Half.
(c) Half.

16-22 Both processes may result in recombination, but transformation involves naked DNA from one cell becoming incorporated into another's DNA, whereas in transduction a virus serves as the vector transferring DNA from one cell to another.

16-23 (a) $L_{in} = \dfrac{L_{\mu m}}{2.54 \times 10^4}$

(b) $L_{in} = \dfrac{3.4 \times 10^{-4}\, P}{2.54 \times 10^4}$ or $L_{in} = 1.3386 \times 10^{-8}\, P$.

Chapter 17

17-1 No. Why?

17-2 Yes.

17-3 No.

17-4 Parents, $aaA'A' \times AAa'a'$; children all $AaA'a'$.

17-5 Strain 1.

17-6 Enzyme "a."

17-7 Enzymes "a" and "b."

17-8 1, + + +; 2, a + +; 3, + b +; 4, ab +.

17-9 (a) Strain 2.
(b) Strain 3.
(c) Strain 1.

17-10 (a) A U C U U U A C G C U A.
(b) Adenylic acid (A).
(c) Uridylic acid (U).

17-11 200.

17-12 (a) 340.
(b) 0.034.

17-13 (a) 80.
(b) 0.027.
(c) tRNA.

17-14 About 2,967.

17-15 5′UAGCCAAUC . . . 3′.

17-16 (a) mRNA.   (c) tRNA.
(b) UAC.   (d) TAC.

17-17 Transcription: formation of mRNA from a DNA template. Translation: formation of a particular polypeptide chain consisting of specific amino acid residues in a specific sequence as determined by the sequence of mRNA codons.

17-18 rRNA is required for correct structure and functioning of the ribosomes; mRNA essentially carries the genetic "message" encoded in DNA out into the cytoplasm; tRNA transfers activated amino acids from the cytoplasmic pool to the polysome "assembly line" for incorporation into polypeptide chains (whose sequence is determined by the codon sequence of mRNA).

17-19 Longer. Why?

Chapter 18

18-1 (a) Methionine, alanine, leucine, threonine.
(b) Methionine, tryptophan, glycine, alanine, proline, leucine, leucine, end chain.
(c) DNA trinucleotide TAC transcribes into AUG of mRNA, which serves as a chain-initiating codon.

18-2 Methionine, proline, end chain.

18-3 (a) It differs by one amino acid residue; the second is now proline instead of alanine. Sense is restored starting with the third amino acid residue.
(b) Missense (in the second amino acid residue).

18-4 (a) $\frac{6}{216}$, or $\frac{1}{36}$.
(b) $\frac{27}{216}$, or $\frac{1}{8}$.

18-5 (a) 423.
(b) 0.14 μm.
(c) Too low. Why?

18-6 At GCU, an mRNA codon for alanine. tRNA-mRNA pairing is determined by anticodon-codon complementarity, not by the amino acid.

18-7 Tryptophan; it has only one codon, whereas arginine has six.

18-8 A-14 mutant has undergone a base change in the second base of the isoleucine codon (AUU, AUC, or AUA) to one coding for threonine (ACU, ACC, or ACA); Ni-1055 mutant has undergone a change in the third base of AUU, AUC, or AUA (coding for isoleucine) to AUG (coding for methionine).

18-9 In A446, UAU or UAC has been changed to UGU or UGC (second base); in A187, GG— has been changed at the second position to GU—. These represent base changes at positions 2 and 17, respectively, of the stretch of mRNA involved.

18-10 288.

18-11 (a) Represents degeneracy.
(b) Represents ambiguity.

18-12 (a) Nonsense.   (c) Degeneracy.
(b) Missense.   (d) Missense.

18-13 (a) 402.
(b) No, because they are clipped out in processing.

Chapter 19

19-1 For the environment of a given species, those mutations that have either a positive or a neutral selection value should be expected to increase in frequency, although deleterious recessive mutations may be expected to persist at a low frequency in heterozygotes. In addition, many (but not all) mutations result in proteins of lowered functional capability; such mutant individuals are usually at a disadvantage in survival and reproduction.

19-2 See Figure 19-13 and accompanying text.

19-3 Short-term effects include radiation sickness, surface and deep tissue burning, loss of hair, and so on. Long-term effects include an increased incidence of leukemia and a variety of mutations.

19-4 No.

19-5 Recessive mutations are detected more readily in the hemizygous males.

19-6 Monoploid greatest, polyploid least. Why?

19-7 *A. brevis* is a diploid, *A. barbata* is a tetraploid, and *A. sativa* is a hexaploid. Because most mutations are recessive, frequency of *detectable* mutations may be expected to be inversely related to ploidy.

19-8 *T. monococcum.* Why?

19-9 (a) 2.
(b) White→red→blue.
(c) (1) *AaBb.*
    (2) $A - B -$.
(d) (1) $A - bb.$
    (2) $aa - -$.
(e) The $A - a$ pair.

19-10 Phenylalanine is coded before deamination; leucine will be coded after deamination.

19-11 A transition.

19-12 A58: transition; A78: transversion.

19-13 (a) Nonsense.
(b) No.
(c) 173.

19-14 (a) 1,500.
(b) Too high.

19-15 267.

19-16 801.

19-17 (a) 5×10^9.

(b) 5×10^9.

(c) 1.7×10^6.

(d) 8.33×10^8.

(e) 2.78×10^6.

19-18 Depending on the amino acid position involved, some missense mutations result in substitutions that do not materially affect the functioning of the resulting protein but, for example, they may result in peptides with different electrophoretic mobilities because of charge differences.

19-19 By accumulation of a series of missense mutations affecting different amino acid positions in a given polypeptide; those not lethal or disabling might be expected to be perpetuated and passed on to later generations. This is the basis of genetic polymorphism.

19-20 Because of the present impossibility of developing specific mutations, one simply takes what one gets in induced mutations. Techniques for directed mutation are not yet adequate outside the experimental laboratory.

19-21 *Cistron:* a segment of DNA specifying one polypeptide chain. *Muton:* the smallest segment of DNA that can be changed and thereby bring about a mutation; can be as small as one deoxyribonucleotide pair. *Recon:* the smallest segment of DNA that is capable of recombination; can be as small as one deoxyribonucleotide pair.

19-22 *Complementation:* the ability of linearly adjacent segments of DNA to supplement each other in phenotypic effect. *Recombination:* a new association of genes in a recombinant individual, arising from (1) independent assortment of unlinked genes, (2) crossing-over between linked genes, or (3) intracistronic crossing-over.

19-23 Thr—trp—leu—val—cys (or reverse sequence).

19-24 (a) AGU → AAU → ACU → AUU.
(b) AAU → ACU.

Chapter 20

20-1 They are similar, in that each is composed of a given segment of deoxyribonucleotides and each participates in regulatory control over cistrons. On the other hand, although the operator is transcribed in large part, it is not translated. The regulator is responsible for production of a regulatory protein (either a repressor or an activator); no such product has been demonstrated for the operator.

20-2 The repressor is a protein, the product of a regular site, which inhibits the action of an operator site so that cistrons of the operon are "turned off." An effector is any substance, often the substrate of the enzyme(s) for which the cistron(s) of the operon are responsible; it binds to the repressor protein, which thereby suffers a change in shape so that it can no longer bind to the operator.

20-3 Both involve control by a regulatory protein that binds to an operator. In positive control the regulatory protein serves as an activator, permitting translation of the cistrons of the operon by binding to the operator. In negative control the regulatory protein represses the operator, either in the absence of an effector (*lac* operon) or in the presence of an effector (*his* operon).

20-4 Transcriptional level.

20-5 (a) Inductive. (d) Absent.
(b) Constitutive. (e) Constitutive.
(c) Constitutive.

20-6 (a) Absent.
(b) Beta-galactosidase constitutive, beta galactoside permease inductive, thiogalactoside transacetylase constitutive.

20-7 (a) Not transcribed.
(b) Not transcribed.
(c) Transcribed.

20-8 (a) No.
(b) Neither. Why?

20-9 All three are constitutive. Why?

20-10 Translation. Why?

20-11 (a) Negative.
(b) Produced.

Chapter 21

21-1 Sex-linked recessive traits show a characteristic inheritance sequence from affected father to "carrier" daughter to about half her sons; sex-linked dominants are transmitted by an affected mother (× normal father) to about half her sons and half her daughters or by an affected father (× normal mother) to all his daughters and none of his sons. Purely maternal effects are transmitted from mother to all her progeny but do not persist in certain nuclear genotypes. Extranuclear genetic systems would ordinarily operate through the maternal line. If the trait is repeatedly transmitted through backcrosses of F_1 individuals with maternal parent but not with paternal parent, an extranuclear genetic system may be involved. It should be identified and located, and such guiding criteria as those listed at the outset of this chapter applied.

21-2 Any genotype will have the phenotype determined by the maternal genotype (a maternal effect).

21-3 (a) Young all light-eyed, adults all dark-eyed.
(b) Both young and adult dark-eyed.

21-4 (a) All green.
(b) All "white."
(c) All three types in irregular ratio.

21-5 (a) 4.1×10^5.
(b) 1,025.

21-6 (a) All normal.
(b) 1:1 petite:normal.

21-7 (a) In substrain C, between *lys* + *met* and *gal;* in substrain H, between *pil* and *pyr B*.
(b) In C, *muc;* in H, *thr*.

21-8 (a) White.
(b) Green.
(c) Green.
(d) Green, variegated, or white.
(e) Green.

21-9 In the chloroplast DNA. Why?

21-10 Confirms the chloroplast location of the gene.

APPENDIX B

Selected life cycles

1. Bacteria

Bacteria reproduce most frequently by asexual cell division, but at least certain genera, notably *Escherichia* and closely related genera, may also engage in a type of reproduction called conjugation. Conjugants are of two general types, F^- ("female" or, better, **recipient** cells) and either F^+ or *Hfr* ("male" or, better, **donor** cells). Donor cells possess the fertility factor F; in F^+ cells F is an independent cytoplasmic plasmid, whereas in *Hfr* cells F is integrated into the cell's large DNA molecule (its "chromosome"). Conjugation between an *Hfr* (**high frequency recombinant**) and an F^- cell includes the following steps:

a. Chance contact between *Hfr* and F^- cells in pairs.

b. Formation of a conjugation tube connecting the two conjugants.

c. One strand of the double-stranded DNA "chromosome" of the *Hfr* cell is cut enzymatically at a point within F. DNA replication occurs by the rolling circle method and a single strand with its $5'$ end enters the F^- cell through the conjugation bridge.

d. The integrated F is at the trailing end ($3'$) of the strand being transferred.

e. Cell-to-cell contact is usually broken by external forces before most of the $5'$, $3'$ strand is transferred from donor to recipient.

f. Synapsis between donated segment and the homologous section of the F^- "chromosome" occurs.

g. Donor strand segment is integrated into the recipient "chromosome"; the displaced single-stranded segment from the recipient cell is enzymatically degraded after its excision from the F^- "chromosome."

h. Replication of DNA strand complementary to the now integrated donor strand occurs. Recipient cell is now a *recombinant* if the donated DNA segment bore alleles of those of the recipient's original "chromosome" (e.g., *donor: leu$^+$ pan$^+$ met D$^-$ pro A$^-$*; and *recipient: leu$^-$ pan$^-$ met D$^+$ pro A$^+$* and so forth).

i. The *F* factor sometimes carries a variable number of chromosomal cistrons in its own circular form. This type of fertility factor is designated *F'*. The recipient cell in conjugation with an *F'* cell can and usually does become diploid (a **merozygote**) for those chromosomal genes that are included within the fertility factor. Conjugation involving *F'* donor cells is called **sexduction.**

In the case of *F$^+$* × *F$^-$* conjugation almost always (>99.99 percent of the cases) only the plasmid *F* is transferred to the recipient, which thereby becomes an *F$^+$* cell. The *F$^+$* donor remains *F$^+$* because plasmid *F* replicates just prior to the actual conjugation process. On the other hand, in *Hfr* × *F$^-$* conjugation, the *Hfr* donor remains *Hfr* except in very rare instances when all the donor strand (i.e., including *F*) is transferred.

2. Neurospora

Like most fungi, *Neurospora* produces large numbers of asexual spores (here called conidia) but also reproduces sexually if + and − mating strains come into contact. In *Neurospora*, as in many fungi, pairing of nuclei of sex cells does not result in immediate syngamy. The nuclei that ultimately fuse are daughter nuclei of the original pairing nuclei. The essential steps are diagramed in Figure B-1 and include:

a. Contact between filaments (hyphae) of + and − strains.

b. Pairing (not fusion) of + and − nuclei.

c. Development of the "fruiting body" (ascocarp), called a perithecium, which consists of n^+, n^-, and dikaryon (n^+/n^-) hyphae.

d. Development of large numbers of elongate, saclike sporangia called asci (sing., ascus) in the perithecium.

e. Fusion in the young asci of a + and a − nucleus that were derived through several mitoses from the original pairing nuclei, thus forming a diploid zygote.

f. Meiosis of zygote soon after formation, in the developing ascus, to form four meiospores.

g. Mitosis of the four meiospores to form eight (monoploid) spores called ascospores. Four of these will give rise to + mating strain plants, the other four to − strain plants.

h. Release of ascospores and their germination to form new adults.

3. Saccharomyces (yeast)

Yeasts are one-celled ascomycete fungi. Multiplication is by "budding," in which the nucleus divides by mitosis. In the full life cycle, however, morphologically identical diploid and monoploid generations alternate, as shown in Figure B-2. The life history consists of the following steps:

a. Diploid adults multiply by "budding" but, under certain environmental conditions, undergo meiosis to form four monoploid meiospores within the old cell wall.

b. Maturation of the four meiospores to become four ascospores (baker's

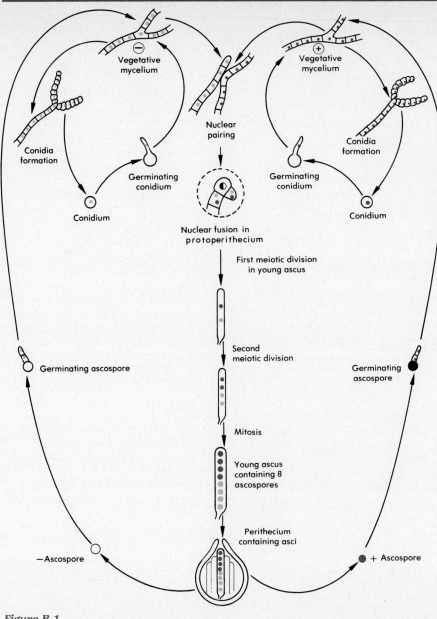

Figure B-1

yeast) or, in some other species, mitosis of each of the four meiospores to form eight ascospores.

c. Liberation of ascospores from ascus (old vegetative cell wall).

d. Germination of ascospores to form monoploid adult cells. Half the ascospores from any one ascus give rise to + mating strain adults and half to − mating strain.

e. Multiplication of monoploid + and − adults by "budding."

f. Contact between + and − cells.

g. Formation of intercellular cytoplasmic bridge.

h. Fusion of + and − nuclei (each monoploid cell in the pair furnishes a single gamete).

i. Formation of diploid vegetative cell, often from the cytoplasmic bridge.

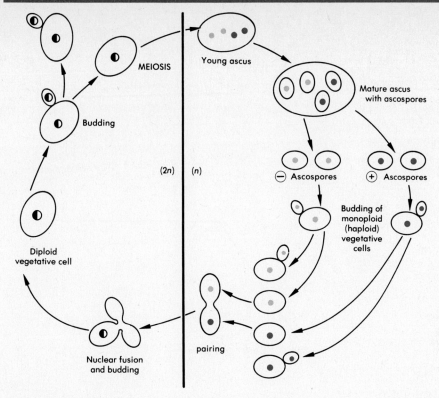

Figure B-2

4. Chlamydomonas

The small, unicellular, motile green alga *Chlamydomonas* reproduces freely by cell division (mitosis and cytokinesis). The vegetative cells are monoploid (haploid). In sexual reproduction, the vegetative cell functions as a gametangium; its protoplast divides mitotically to produce 4, 8, 16, or 32 gametes. These sex cells are morphologically similar to the vegetative cells but smaller in size. In many species the gametes are identical in appearance, hence may be referred to as isogametes. In other species varying degrees of morphological differentiation of gametes occur. Chemical differences among gametes, and the cells producing them, occur and mating strains are designated as $+$ and $-$. Gametes of opposite mating strain come into contact at their flagellar ends; the protoplasts fuse to form a four-flagellate zygote. The zygote soon loses its flagella, develops a wall, and becomes dormant. Germination of the zygote begins with meiosis of its diploid nucleus and ends with liberation of biflagellate zoospores from the old zygote wall. In some species only four zoospores are thus formed, but in others meiosis is followed by one or more mitoses so that 8, 16, or more zoospores are produced. Zoospores resemble the vegetative cells into which they will develop; and of the number produced from a single zygote, half are of each mating strain. The process is diagramed in Figure B-3.

5. Sphaerocarpos (liverwort)

Vegetative plants are small, thin, and lobed; these are monoploid (haploid), unisexual gametophytes (gamete-producing plants). Males produce motile sperms in antheridia; females develop one egg in each of several archegonia. Syngamy

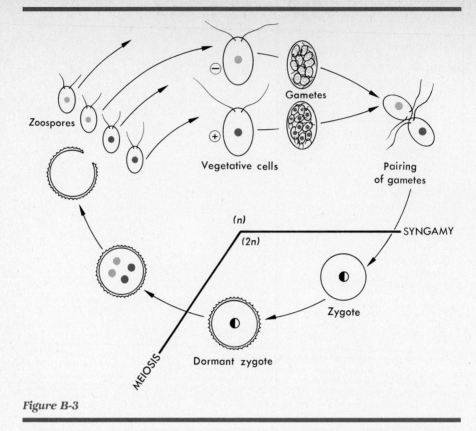

Figure B-3

occurs when liquid water (from rain or dew) is present, and thus allows sperms to swim to the archegonia. The resulting zygote develops into a small, multi-cellular sporophyte (spore-bearing plant), which remains permanently attached to the parent gametophyte. It ultimately protrudes from the remains of the archegonium and produces internally a large number of diploid sporocytes that undergo meiosis to produce four meiospores each. Two of each of these will produce male gametophytes, and two female. The life cycle is diagrammed in Figure B-4.

6. Flowering plants (class Angiospermae)

The plant we recognize by name in the angiosperms is the diploid sporophyte. The monoploid (haploid) gametophyte is microscopic and contained almost entirely within various floral structures. A (usually) triploid food storage tissue, the endosperm, also occurs and is cytologically unique to the angiosperms. The life cycle of a representative angiosperm is diagrammed in Figure B-5 and consists of the following structures and steps:

a. Production of flowers by the sporophyte.
b. Meiosis of microsporocytes (pollen mother cells) in anthers of stamens to form four functional, uninucleate microspores each.
c. Development of young male gametophyte by mitosis of the microspore nucleus within the microspore wall, inside the anther. This two nucleate structure (tube nucleus and generative nucleus) is sometimes called a pollen grain.

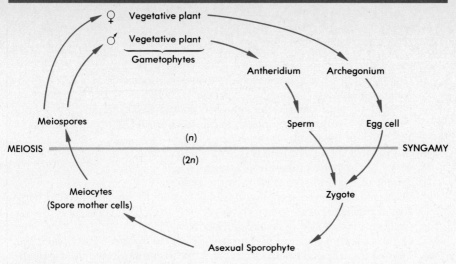

Figure B-4

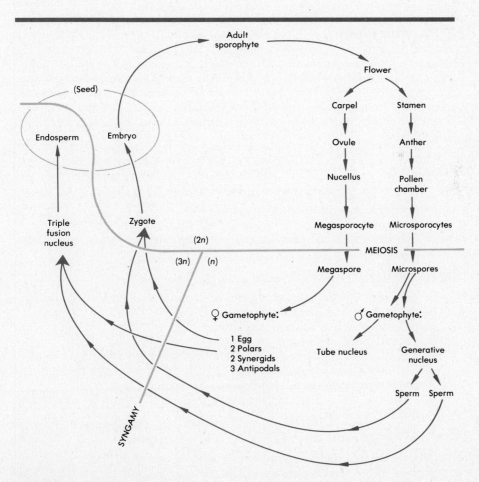

Figure B-5

d. Transfer of pollen grains to stigmas of the pistils, where each grain produces a tubular outgrowth, the pollen tube, which grows down through structures of the pistil to the ovule, which it enters via the micropyle.

e. Development in each ovule of a megasporocyte, which then undergoes meiosis to form four megaspores, three of which degenerate.

f. Development of the female gametophyte by mitosis from the one functional megaspore. In the classical case the mature female gametophyte consists of eight nuclei (one egg, two polars, two synergids, three antipodals) in a common cytoplasm within each ovule, and contained within the ovary of the pistil.

g. Mitosis of generative nucleus to form two sperms in the pollen tube.

h. Entry of the pollen tube into the embryo sac.

i. Fusion of egg and sperm to form the zygote.

j. Fusion of the second sperm with the two polars to form the triploid triple fusion nucleus.

k. Degeneration of synergids and antipodals.

l. Development of multicellular embryo from the zygote.

m. Development of endosperm from the triple fusion nucleus. In some plants this tissue is absorbed by the cotyledons of the embryo during the latter's development.

n. Development of seed coat, primarily from the integuments of the ovule.

o. Development of a fruit from the ovary.

7. Paramecium

Paramecia are elongate ciliates of the phylum Protozoa. Each animal contains a large macronucleus, which exerts phenotypic control for that individual, and two micronuclei, which function in the sexual process. The macronucleus is polyploid, the micronucleus diploid in the vegetative animal. Paramecia increase in number only by fission; other processes, important in genetics, also occur and are as follows:

FISSION

a. Mitosis of micronuclei.

b. Constriction of macronucleus to form two.

c. Movement of one macronucleus and one micronucleus to each end of the animal, which then constricts in the middle to form two new individuals.

CONJUGATION (FIG. B-6)

a. Pairing of two animals (conjugants) and formation of intercellular bridge.

b. Disintegration of macronucleus.

c. Meiosis of each of the two micronuclei.

d. Disintegration of seven of the eight products of meiosis.

e. Mitosis of the remaining monoploid nucleus to form two.

f. One of the two monoploid nuclei of each conjugant passes through the connecting bridge to the other animal (reciprocal transfer).

g. Fusion of the two monoploid nuclei in each conjugant, to restore the diploid condition.

h. Two mitoses of the fertilization nuclei, which results in four diploid nuclei per conjugant.

i. Separation of the conjugants, which are now genetically alike.

j. Two of the four nuclei in each ex-conjugant become macronuclei, two become micronuclei.

k. Distribution of two macronuclei to each daughter cell at next fission.

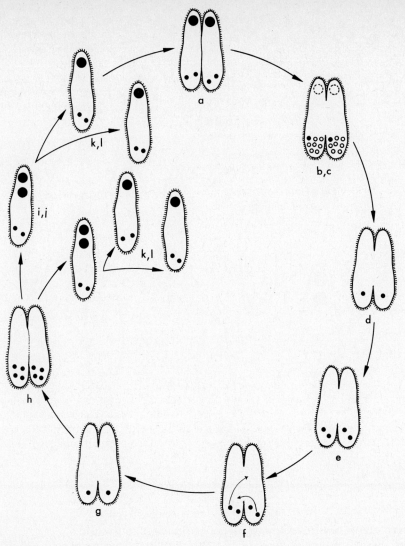

Figure B-6

l. Mitosis of the two micronuclei at next fission, two being distributed to each daughter cell.

m. Conjugation is usually a short-term event with little cytoplasmic transfer. Under certain conditions the intercellular connection may persist for a longer period, and allow exchange of considerable cytoplasm.

AUTOGAMY (FIG. B-7)

This is a type of internal self-fertilization that resembles somewhat the events of conjugation but involves only a single animal. It results in homozygosity of the individual undergoing the process.

a. Macronucleus behaves as in conjugation.

b. Meiosis of the two micronuclei and disintegration of seven of the eight resulting nuclei as in conjugation.

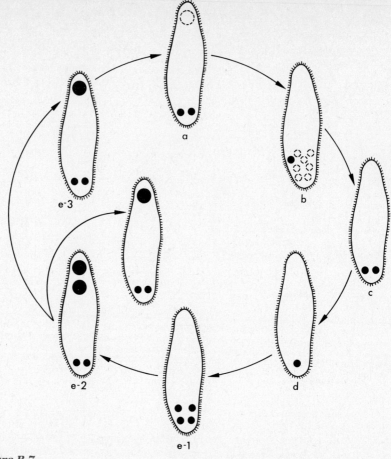

Figure B-7

c. Mitosis of the remaining monoploid nucleus to form two.
d. Fusion of the two monoploid nuclei resulting from step (c).
e. Restoration of micronuclei and the macronucleus as for conjugation.

8. Mammals

In mammals, the somatic cells (except for some, such as liver cells, which may be polyploid) are diploid; meiosis immediately precedes the formation of gametes. The diploid condition is restored at syngamy (Fig. B-8). The following steps are involved in the male:

a. Development of diploid primary spermatocytes.
b. Meiosis. The two monoploid products of the first meiotic division are called secondary spermatocytes. The four cells resulting from the second meiotic division are called spermatids.
c. Maturation of spermatids into sperms.

In the female:

a. Development of diploid primary oocytes.

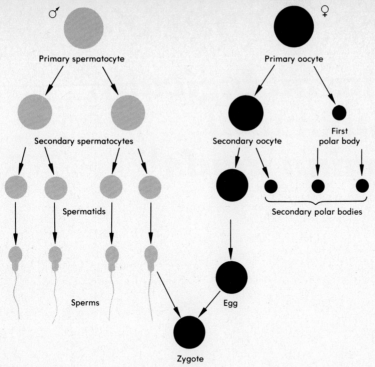

Figure B-8

b. Meiosis. The first division produces two unequal cells, a smaller first polar body and a larger secondary oocyte. The second division produces two second polar bodies from the first polar body and, from the secondary oocyte, a third second polar body and a larger ootid.

c. Maturation of one egg from the ootid; degeneration of the three second polar bodies.

APPENDIX C

The biologically important amino acids[1]

Alanine (ala)

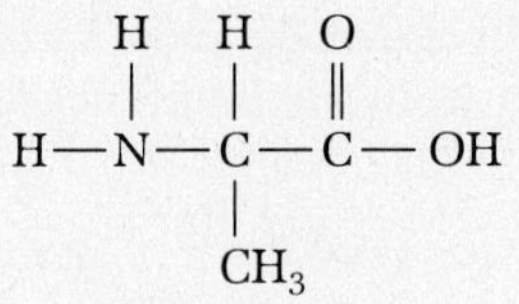

$$\begin{array}{ccccc} & H & & H & & O \\ & | & & | & & \parallel \\ H & - & N & - & C & - & C & - OH \\ & & & & | \\ & & & & CH_3 \end{array}$$

Arginine (arg)

$$\begin{array}{ccccc} & H & & H & & O \\ & | & & | & & \parallel \\ H - N - C - C - OH \\ & & & | \\ & & & CH_2 \\ & & & | \\ & & & CH_2 \\ & & & | \\ & & & CH_2 \\ & & & | \\ & & & NH \\ & & & | \\ & & & C = NH \\ & & & | \\ & & & NH_2 \end{array}$$

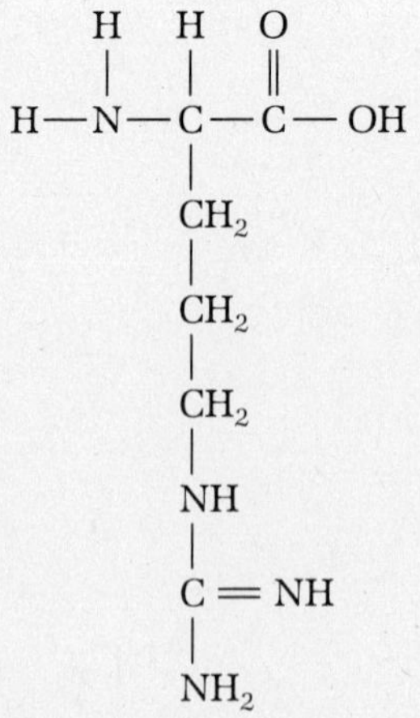

[1]Amino acids marked * are required in the diet of mammals.

Asparagine (asn)

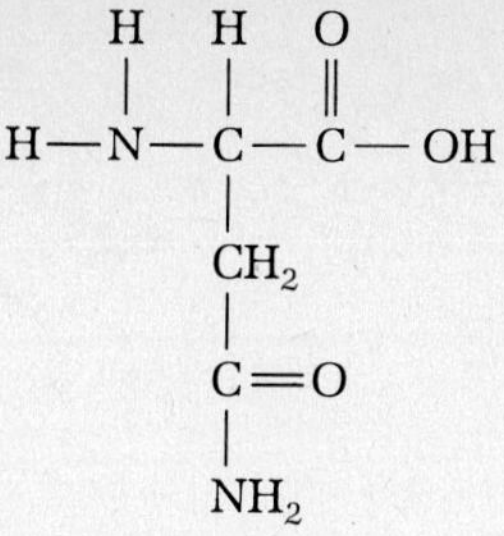

```
        H   H   O
        |   |   ‖
   H — N — C — C — OH
            |
           CH₂
            |
           C=O
            |
           NH₂
```

$$\begin{array}{c}
\quad\ \ \text{H}\quad \text{H}\quad\ \text{O}\\
\quad\ \ |\quad\ \ |\quad\ \ \|\\
\text{H—N—C—C—OH}\\
\quad\ \ \ \ \ \ \ \ \ \ |\\
\quad\ \ \ \ \ \ \ \ \text{CH}_2\\
\quad\ \ \ \ \ \ \ \ \ \ |\\
\quad\ \ \ \ \ \ \ \ \text{C=O}\\
\quad\ \ \ \ \ \ \ \ \ \ |\\
\quad\ \ \ \ \ \ \ \ \text{NH}_2
\end{array}$$

Asparagine (asn)

$$\begin{array}{l}
\text{H—N—C—C—OH}\\
\ \ \ \ \ \ |\ \ |\\
\ \ \ \ \ \text{CH}_2\\
\ \ \ \ \ \text{C=O}\\
\ \ \ \ \ \text{NH}_2
\end{array}$$

Aspartic Acid (asp)

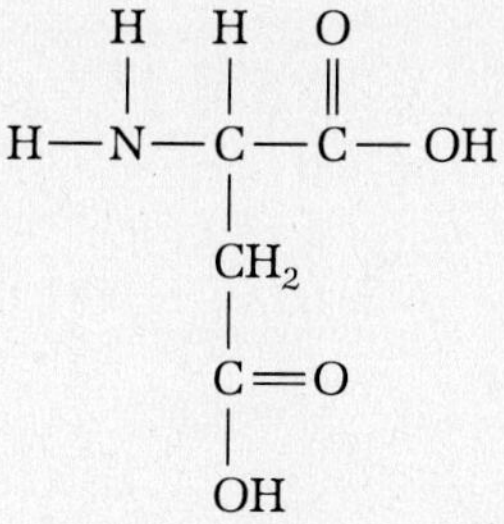

$$\begin{array}{l}
\text{H—N—C—C—OH}\\
\ \ \ \ \ \text{CH}_2\\
\ \ \ \ \ \text{C=O}\\
\ \ \ \ \ \text{OH}
\end{array}$$

Cysteine (cys)

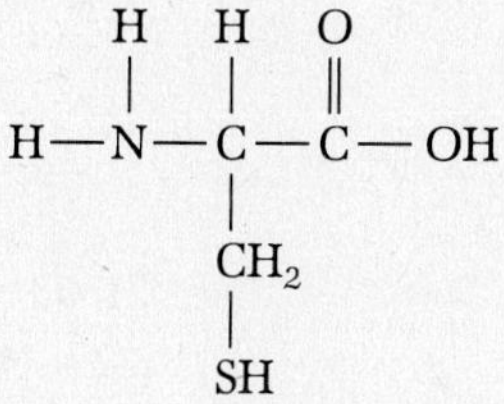

$$\begin{array}{l}
\text{H—N—C—C—OH}\\
\ \ \ \ \ \text{CH}_2\\
\ \ \ \ \ \text{SH}
\end{array}$$

Glutamic Acid (glu)

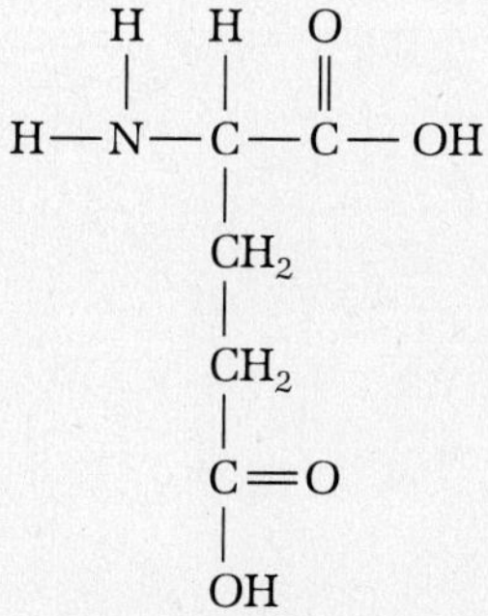

$$\begin{array}{l}
\text{H—N—C—C—OH}\\
\ \ \ \ \ \text{CH}_2\\
\ \ \ \ \ \text{CH}_2\\
\ \ \ \ \ \text{C=O}\\
\ \ \ \ \ \text{OH}
\end{array}$$

Glutamine (gln)

$$\begin{array}{l}
\text{H—N—C—C—OH}\\
\ \ \ \ \ \text{CH}_2\\
\ \ \ \ \ \text{CH}_2\\
\ \ \ \ \ \text{C=O}\\
\ \ \ \ \ \text{NH}_2
\end{array}$$

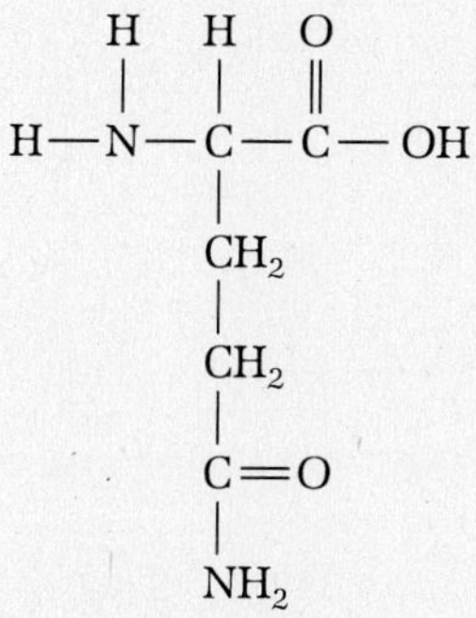

Glycine (gly)

$$\begin{array}{l}
\text{H—N—C—C—OH}\\
\ \ \ \ \ \text{H}
\end{array}$$

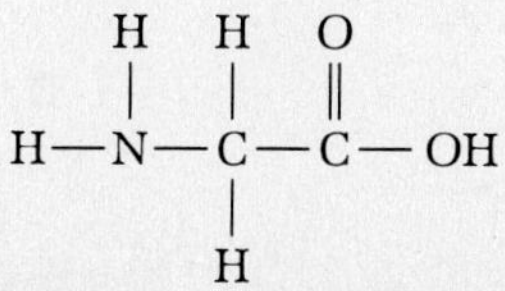

Histidine (his)

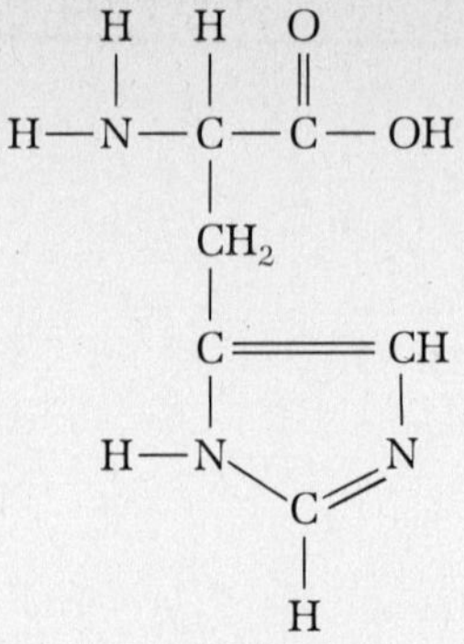

$$
\begin{array}{ccc}
& H & H & O \\
& | & | & \| \\
H- & N- C- C- OH \\
& & | \\
& & CH_2 \\
& & | \\
& C ====== CH \\
& | & | \\
& H- N & N \\
& & C \\
& & | \\
& & H
\end{array}
$$

Isoleucine (ile)

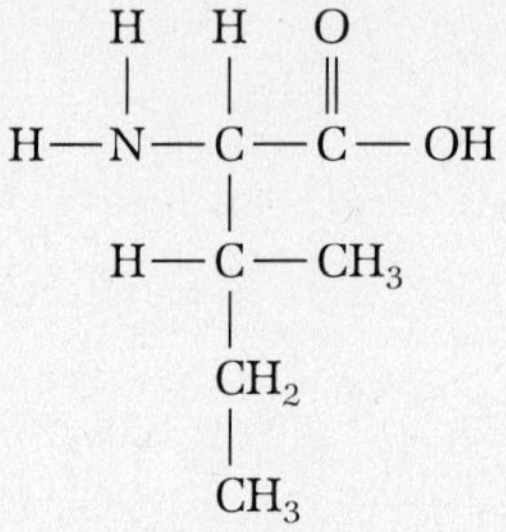

$$
\begin{array}{ccc}
H & H & O \\
| & | & \| \\
H- N- C- C- OH \\
& | \\
& H- C- CH_3 \\
& | \\
& CH_2 \\
& | \\
& CH_3
\end{array}
$$

Leucine (leu)

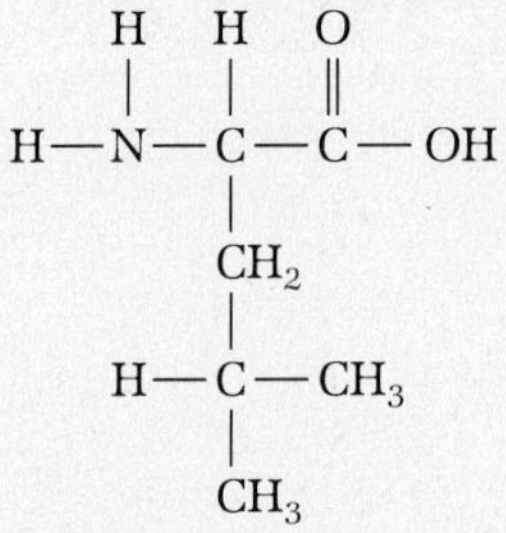

$$
\begin{array}{ccc}
H & H & O \\
| & | & \| \\
H- N- C- C- OH \\
& | \\
& CH_2 \\
& | \\
& H- C- CH_3 \\
& | \\
& CH_3
\end{array}
$$

Lysine (lys)

$$
\begin{array}{ccc}
H & H & O \\
| & | & \| \\
H- N- C- C- OH \\
& | \\
& CH_2 \\
& | \\
& CH_2 \\
& | \\
& CH_2 \\
& | \\
& CH_2 \\
& | \\
& NH_2
\end{array}
$$

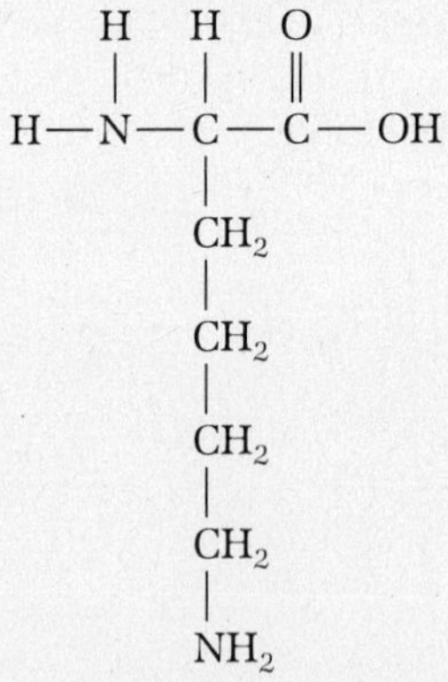

Methionine (met)

$$
\begin{array}{ccc}
H & H & O \\
| & | & \| \\
H- N- C- C- OH \\
& | \\
& CH_2 \\
& | \\
& CH_2 \\
& | \\
& S \\
& | \\
& CH_3
\end{array}
$$

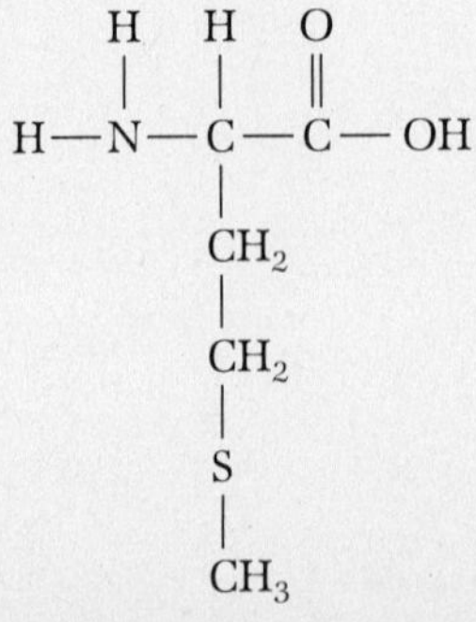

Phenylalanine (phe)

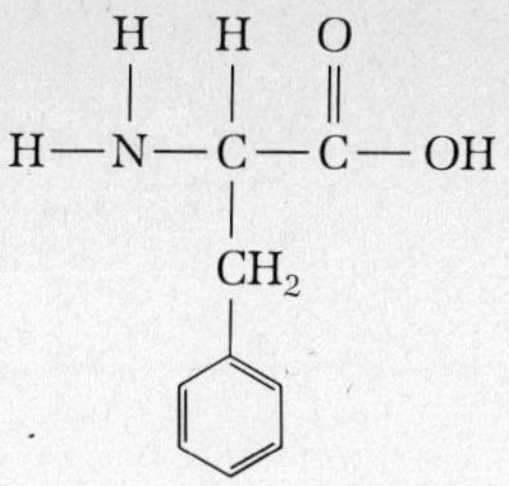

$$
\begin{array}{c}
H \quad\quad H \quad\quad O \\
| \quad\quad\quad | \quad\quad\quad \| \\
H-N-C-C-OH \\
| \\
CH_2 \\
| \\
\bigcirc
\end{array}
$$

Proline (pro)

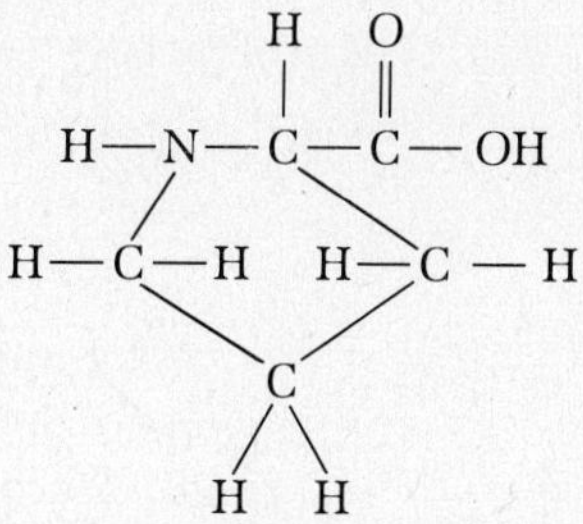

$$
\begin{array}{c}
H \quad\quad O \\
| \quad\quad \| \\
H-N-C-C-OH \\
H-C-H \quad H-C-H \\
C \\
H \quad H
\end{array}
$$

Serine (ser)

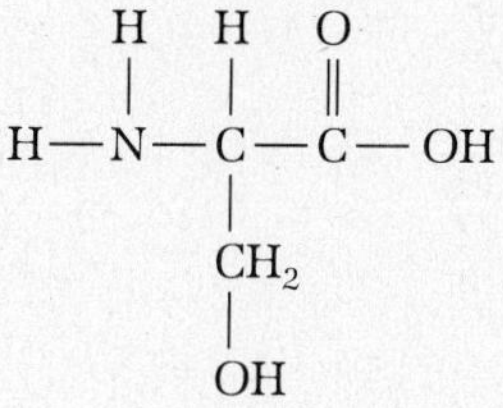

$$
\begin{array}{c}
H \quad\quad H \quad\quad O \\
| \quad\quad\quad | \quad\quad\quad \| \\
H-N-C-C-OH \\
| \\
CH_2 \\
| \\
OH
\end{array}
$$

Threonine (thr)

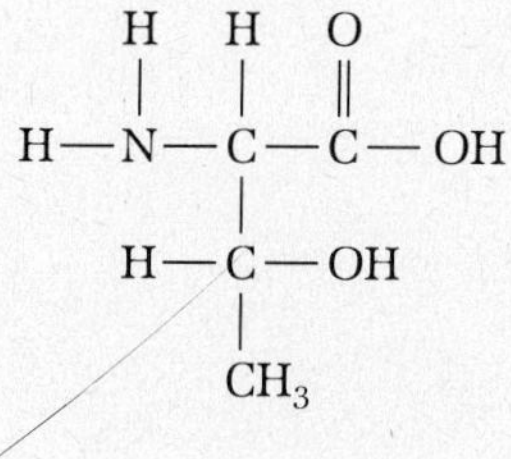

$$
\begin{array}{c}
H \quad\quad H \quad\quad O \\
| \quad\quad\quad | \quad\quad\quad \| \\
H-N-C-C-OH \\
| \\
H-C-OH \\
| \\
CH_3
\end{array}
$$

Tryptophan (trp)

$$
\begin{array}{c}
H \quad\quad H \quad\quad O \\
| \quad\quad\quad | \quad\quad\quad \| \\
H-N-C-C-OH \\
| \\
CH_2 \\
C=C \quad H \\
NH
\end{array}
$$

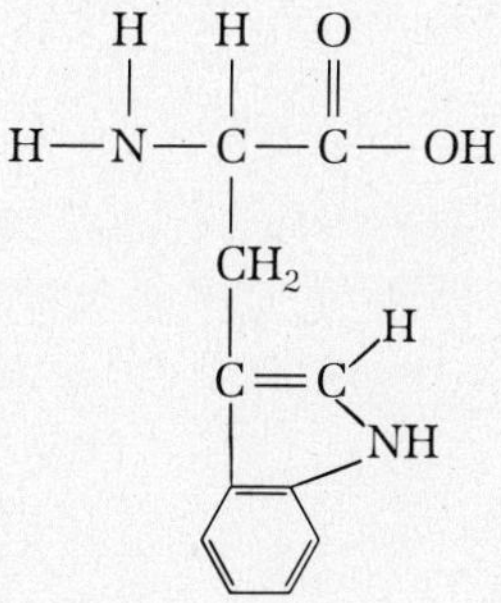

Tyrosine (tyr)

$$
\begin{array}{c}
H \quad\quad H \quad\quad O \\
| \quad\quad\quad | \quad\quad\quad \| \\
H-N-C-C-OH \\
| \\
CH_2 \\
| \\
\bigcirc \\
OH
\end{array}
$$

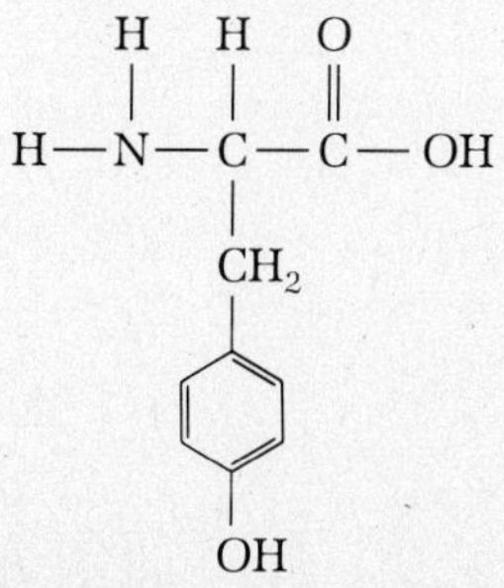

Valine (val)

$$
\begin{array}{c}
\quad\; H \quad\; H \quad\; O \\
\quad\; | \quad\;\; | \quad\;\; \| \\
H\!-\!N\!-\!C\!-\!C\!-\!OH \\
\quad\;\; | \\
\quad\;\; H\!-\!C\!-\!CH_3 \\
\quad\quad\; | \\
\quad\quad CH_3
\end{array}
$$

APPENDIX D

Useful formulas, ratios, and statistics

3:1	Monohybrid phenotypic ratio produced by $Aa \times Aa$.
1:2:1	Monohybrid genotypic ratio produced by $Aa \times Aa$; monohybrid phenotypic and genotypic ratio produced by $a^1a^2 \times a^1a^2$.
1:1	Monohybrid testcross phenotypic and genotypic ratio produced by $Aa \times aa$.
2:1	Monohybrid lethal genotypic and phenotypic ratio produced by $a^1a^2 \times a^1a^2$ where either a^1a^1 or a^2a^2 is lethal; also sex ratio produced by $Aa \times AY$ where a is a sex-linked recessive lethal.
"1:0"	Monohybrid lethal phenotypic ratio produced by $Aa \times Aa$ where aa or $A-$ is lethal; also monohybrid testcross phenotypic and genotypic ratio produced by $AA \times aa$.
9:3:3:1	Dihybrid phenotypic ratio produced by $AaBb \times AaBb$ where phenotypes may be represented as $9\ A-B-$, $3\ A-bb$, $3\ aaB-$, $1\ aabb$; note epistatic possibilities producing "condensations" of this ratio (e.g., 9:7, 9:6:1).
1:1:1:1	Dihybrid testcross genotypic and phenotypic ratio produced by $AaBb \times aabb$ where there is no linkage.

3:6:3:1:2:1

Dihybrid phenotypic ratio produced by $Aab^1b^2 \times Aab^1b^2$. Note that dihybrid and polyhybrid ratios are products of their component monohybrid ratios and may be combined in any way.

2^n

Number of gamete genotypes and progeny phenotypes where n = the number of pairs of heterozygous genes with complete dominance.

3^n

Number of zygote genotypes under the preceding conditions.

4^n

Number of possible zygote combinations under the preceding conditions, which yield 3^n zygote genotypes.

$\frac{n}{2}(n + 1)$

The chance of selecting at random any two items in pairs (e.g., the number of possible genotypes for n multiple alleles in a series).

$(a + b)^{2n}$

Progeny phenotypic ratios in polygene crosses are given by the *coefficients* of the expansion of $(a + b)^{2n}$ where n = the number of *pairs* of polygenes.

$(a + b)^n$

Expansion of the binomial provides probability determinations where n = the number of independent events and the choices, represented by a and b, respectively, are two.

$(p + q)^2 = 1$

Binomial whose expansion permits calculation of the frequency of each member of a pair of alleles.

$(p + q + r)^2 = 1$

Trinomial whose expansion permits calculation of the frequency of each of three multiple alleles.

$(\frac{1}{4})^n$

In polygene cases, the fraction of the F_2 like either P, is given by this expression, where n = the number of pairs of genes in which the parents (P) differ.

$(\frac{1}{2})^n$

In polygene cases, the fraction of the F_2 like either P is given by this expression, where n = the number of effective or contributing alleles.

$\chi^2 = \Sigma \left[\frac{(o - c)^2}{c} \right]$

Consult tables of chi-square for levels of significance (where degrees of freedom equal one less than the number of classes). In general, a value of chi-square $\geq$ than that for P = 0.05 is regarded as significant; i.e., there is significant evidence against the hypothesis. A value of chi-square showing a level of P = 0.05, for example, does *not* mean that a deviation as large or larger will *not* occur by chance alone under the hypothesis adopted; but, it is likely to in only five trials out of 100. This is considered too few; at this level, the chance of rejecting a right hypothesis is only one in 20.

$\bar{x} = \frac{\Sigma fx}{n}$ or $\frac{\Sigma x}{n}$

The sample mean is self-explanatory.

$$s^2 = \frac{\Sigma f(x - \bar{x})^2}{n - 1}$$

The variance (s^2) provides an unbiased estimate of the population variance (σ^2).

$$s = \sqrt{\frac{\Sigma f(x - \bar{x})^2}{n - 1}}$$

The standard deviation measures the variability of the sample; in a normal distribution, 68.26 percent of the sample will lie in the range $\bar{x} \pm s$, and 95.44 percent will fall in the range $\bar{x} \pm 2s$. Used with normal distributions.

$$s_{\bar{x}} = \frac{s}{\sqrt{n}}$$

The standard error of the sample mean indicates the degree of correspondence between $\bar{x}$ and μ; there is 68.26 percent confidence that $\mu = \bar{x} \pm s_{\bar{x}}$ by chance alone, and 95.44 percent confidence that $\mu = \bar{x} \pm 2s_{\bar{x}}$ by chance alone.

$$S_d = \sqrt{(s_{\bar{x}_1})^2 + (s_{\bar{x}_2})^2}$$

The standard error of the difference in means is useful in comparing two samples to determine whether the difference in their means is significant. If $(\bar{x}_1 - \bar{x}_2) > 2S_d$, the difference in sample means is considered significant and the two samples represent two different populations.

$$s = \sqrt{\frac{pq}{n}}$$

The standard deviation (s) of a simple proportionality, such as heads (p) versus tails (q) for n trials.

$$s = \sqrt{\frac{pq}{2N}}$$

The standard deviation of gene frequencies where N represents the number of *diploid* individuals, and p and q represent the frequency of each of a pair of alleles.

$$n = \frac{R^2}{8(s^2_{F_2} - s^2_{F_1})}$$

The number of pairs of polygenes (n) is calculated from this equation where R is the maximum quantitative range between phenotypes, $s^2_{F_2}$ is the variance of the F_2, and $s^2_{F_1}$ is the variance of the F_1.

$$q_n = \frac{q_0}{1 + nq_0}$$

The frequency of a recessive lethal (q_n) after n additional generations equals the initial frequency of the gene (q_0) divided by 1 plus the product of the number of additional generations (n) times the initial frequency of the gene.

$$n = \frac{1}{q_n} - \frac{1}{q_0}$$

The number of additional generations (n) required to reduce the frequency of a gene from its initial value (q_0) to any particular value (q_n) is given by this equation.

$$W = 1 - s, \text{ and } s = 1 - W$$

The adaptive value of a genotype (W) is 1 minus its selection coefficient(s), and the selection coefficient is 1 minus the adaptive value of the genotype.

$$q_1 = \frac{q_0 - s(q_0)^2}{1 - s(q_0)^2}$$

The equation for determining the frequency of a gene after one generation under selection (q_1) when its selection coefficient (s) and its initial frequency (q_0) are known.

$$\Delta q_s = \frac{-sq_0^2 p}{1 - sq_0^2}$$

Change in frequency of a recessive gene under selection in one generation.

$L_{\mu m} = 3.4 \times 10^{-4}P$

Length in micrometers ($L_{\mu m}$) of double-stranded DNA equals 0.00034 times the number of deoxyribonucleotide pairs (P).

$P = \dfrac{M}{650}$

The number of deoxyribonucleotide pairs (P) equals the molecular weight (M) of a DNA molecule divided by 650.

$M = 650P$

The molecular weight of a DNA molecule (M) equals 650 times the number of deoxyribonucleotide pairs (P) of which it consists.

$P = \dfrac{L_{\mu m}}{3.4 \times 10^{-4}}$

The number of deoxyribonucleotide pairs (P) in a DNA molecule equals its length in micrometers ($L_{\mu m}$) divided by 0.00034.

$\Delta q_m = up - vq$

The rate of change in recessive gene frequency under mutation alone equals the rate of forward mutation times the frequency of the dominant allele minus the quantity the rate of back mutation times the frequency of the recessive allele. See page 283.

$\hat{q}_m = \dfrac{u}{u + v}$

The equilibrium frequency of a recessive allele under mutation alone equals the rate of forward mutation divided by the sum of the rates of forward mutation plus back mutation.

$\hat{q} = \sqrt{\dfrac{u}{s}}$

The equilibrium frequency of a recessive allele under the combined effect of mutation and selection equals the square root of the quantity rate of forward mutation divided by the selection coefficient.

APPENDIX E

Useful metric values

Name	Numerical value (m)	Power of 10	Symbol	Synonym
Meter	1.0		m	
Decimeter	0.1	10^{-1}	dm	
Centimeter	0.01	10^{-2}	cm	
Millimeter	0.001	10^{-3}	mm	
Micrometer	0.000 001	10^{-6}	μm	micron (μ)
Nanometer	0.000 000 001	10^{-9}	nm	millimicron (mμ)
	0.000 000 000 1	10^{-10}		Angstrom unit (Å)

Note: 1 Å = 0.0001 or 1×10^{-4} μm; 1 μm = 10,000 or 1×10^{4} Å.

APPENDIX F

Journals and reviews

Advances in Genetics
Advances in Human Genetics
American Journal of Human Genetics
Annals of Human Genetics
Annual Review of Biochemistry
Annual Review of Genetics
Biochemical Genetics
Chromosoma
Cytogenetics
Genetica
Genetical Research
Genetics
Genetics Abstracts
Hereditas
Heredity
Journal of Bacteriology
Journal of Genetics
Journal of Heredity
Journal of Medical Genetics
Journal of Molecular Biology
Journal of Virology
Lancet
Molecular and General Genetics
Nature
New England Journal of Medicine
Proceedings of the National Academy of Science (U.S.)
Science
Scientific American

Glossary

Abortus. An aborted fetus.

Acentric. A chromatid or chromosome that lacks a centromere.

Acquired character. An alteration in function or form resulting from a response to environment; it is not heritable.

Acridine dye. Organic molecules that produce mutations by binding to DNA and causing insertions or deletions of bases.

Acrocentric. A chromosome with a nearly terminal centromere.

Adaptation. Adjustment in some way to the environment.

Adaptive value. The proportion of the progeny of a given genotype surviving to maturity (relative to that of another genotype) is its adaptive value (also called *fitness*). Expressed as a pure number between 0 and 1. Thus if only 60 percent of the progeny of genotype aa survive, whereas 100 percent of those of genotype A − survive, the adaptive value of aa is 0.6.

Adenine. A purine base occurring in DNA and RNA. Pairs normally with thymine in DNA.

Adenosine. The ribonucleoside containing the purine adenine.

Adenylic acid. The ribonucleotide containing the purine adenine.

Agglutinin. An antibody that produces clumping (agglutination) of the antigenic structure.

Albino. An individual characterized by absence of pigment. Ordinarily applied to animals whose skin, hair, and eyes are unpigmented and whose skin and eyes are pinkish because of blood vessels showing through. Also used to describe plants in which a particular pigment (usually the chlorophylls) is absent.

Aleurone layer. Outermost layer of endosperm in some seeds, rich in aleurone (proteinaceous) grains; typically triploid cells.

Allele (also **allelomorph;** adj., **allelic** or **allelomorphic**). One member of a pair or a series of genes that can occur at a particular locus on homologous chromosomes.

Allopolyploid. A polyploid having whole chromosome sets from different species.

Allosteric. Applied to any enzyme (or protein) having two or more nonoverlapping receptor (or attachment) sites.

Alternation of generations.   The regularly occurring alternation of monoploid (gametophyte) and diploid (sporophyte) phases in the life cycle of sexually reproducing plants.

Amber.   The nonsense codon UAG.

Ambiguity.   The coding for more than one amino acid by a given codon. See also *genetic code* and *codon*.

Amino acid.   A class of chemicals containing an amino (NH_2) group and a carboxyl (COOH) group, plus a side chain; the basic constructional unit of proteins.

Amino group.   The $-NH_2$ chemical group.

Amniocentesis.   A method for testing the genotype of an unborn child by withdrawal of a small amount of fluid from the amniotic sac; it contains sloughed fetal cells.

Amphidiploid.   A tetraploid individual, i.e., having two sets of chromosomes from each of two known ancestral species. An allotetraploid in which the species source of the two different genomes is clearly known. See also *allopolyploid*.

Anaphase.   The stage of nuclear division characterized by movement of chromosomes from spindle equator to spindle poles. It begins with separation of the centromeres and closes with the end of poleward movement of chromosomes.

Aneuploidy.   Variation in chromosome number by whole chromosomes, but less than an entire set; e.g., $2n + 1$ (trisomy), $2n - 1$ (monosomy).

Angstrom unit (Å).   A measurement of length or distance, often used in describing intra- or intermolecular dimensions; equal to 1×10^{-10} meter and 1×10^{-1} nanometer (which see).

Anisogamy.   Sexual reproduction involving gametes similar in morphology but differing in size.

Anther.   The microsporangium of flowering plants; distal part of stamen in which microspores, and later pollen grains, are produced.

Antibiotic.   Any chemical substance, elaborated by a living organism (usually a microorganism), that kills or inhibits growth of bacteria.

Antibody.   A substance that acts to neutralize a specific antigen in a living organism.

Anticodon.   The group of three nucleotides in transfer RNA that pairs complementarily with three nucleotides of messenger RNA during protein biosynthesis.

Antigen.   Any substance, usually a protein, that causes antibody production when introduced into a living organism.

Antiparallel.   A term used to refer to the opposite but parallel arrangement of the two sugar-phosphate strands in double-stranded DNA; the 5'-3' orientation of one such strand is aligned along the 3'-5' orientation of the other strand.

Ascospore.   One of the asexual, monoploid (haploid) spores contained in the ascus of ascomycete fungi, such as *Neurospora* and the yeasts.

Ascus (pl., **asci**).   The generally elongate, saclike meiosporangium in which ascospores are produced in ascomycete fungi.

Asexual reproduction.   Any process of reproduction that does not involve fusion of cells. Vegetative reproduction, by portions of the vegetative body, is sometimes distinguished as a separate type.

ATP.   Adenosine triphosphate, an energy-rich compound participating in energy-storing and energy-using reactions in the cell.

Attached-X.   A strain of *Drosophila* in which the two X chromosomes of the female are permanently attached (symbolized $\widehat{XX}$) so that only $\widehat{XX}$ and O eggs are produced.

Autoimmunity.   The production of antibodies against one's own tissues.

Autopolyploid. A polyploid all of whose sets of chromosomes are those of the same species.

Autosome. A chromosome not associated with the sex of the individual and therefore possessed in matching pairs by diploid members of both sexes.

Auxotroph. An individual that, unable to carry on some particular synthesis, requires supplementing of minimal medium by some growth factor.

Backcross. The cross of a progeny individual with one of its parents. See also *testcross.*

Bacteriophage. See *phage.*

Balbiani ring. An RNA-producing puff in the polytene chromosomes (which see) of the dipteran *Chironomus.*

Barr body. The inactive, densely staining, condensed X chromosome, generally found next to the nuclear membrane, in nuclei of somatic cells of XX females. The number of Barr bodies in such nuclei is one less than the total number of X chromosomes.

Base analog. A slightly modified purine or pyrimidine molecule that may substitute for the normal base in nucleic acid molecules.

Bivalent. A pair of synapsed homologous chromosomes.

Carboxyl group. An acidic chemical group,—COOH.

Carotenoid. Any of a group of yellow, orange, or red pigments, in plants associated with chlorophylls, and in animal fat.

Carrier. A heterozygous individual. Term ordinarily used in cases of complete dominance.

Cell culture. A growth of cells in vitro.

Cell-free extract. A fluid extract of soluble materials of cells obtained by rupturing the cells and discarding particulate materials and any intact cells.

Centimeter (cm). 1×10^{-2} meter.

Centimorgan. A unit of relative distance between linked genes; equal to 1 percent crossing-over. See also *Morgan unit.*

Centriole. The central granule in the *centrosome* (which see).

Centromere. A specialized, complex region of the chromosome, consisting of one kinetochore (which see) for each sister chromatid.

Centrosome. A self-propagating cytoplasmic body present in animal cells and some lower plant cells, consisting of a centriole and astral rays at each pole of the spindle during nuclear division.

Chiasma, (pl., **chiasmata**). The visible connection or crossover between two chromatids seen during prophase-I of meiosis.

Chi-square test. A statistical test for determining the probability that a set of experimentally obtained values will be equaled or exceeded by chance alone for a given theoretical expectation.

Chloroplast. Cytoplasmic organelle containing several pigments, particularly the light-absorbing chlorophylls, and also DNA and polysomes (which see).

Chromatid. One of the two identical longitudinal halves of a chromosome, which shares a common centromere with a sister chromatid; this results from the replication of chromosomes during interphase.

Chromatin. Nuclear material comprising the chromosomes; the DNA-protein complex or nucleoprotein.

Chromomere. Small, stainable thickenings arranged linearly along a chromosome.

Chromosome. Nucleoprotein structures, generally more or less rodlike during nuclear division, the physical sites of nuclear genes arranged in linear order. Each species has a characteristic number of chromosomes, although individuals with fewer or more than this characteristic number occur, especially in plants.

Cis arrangement. Linkage of the dominants of two or more pairs of alleles on one chromosome and the recessives on the homologous chromosome.

Cis-trans test. A test to determine whether two mutant sites are in the same or different cistrons. In the cis configuration both mutations are on the same chromosome, and both wild-type sites are on the homolog. In the trans configuration, each homolog has one mutant and one wild type (e.g., $a^1 a^+ / a^+ a^2$). In the cis configuration the wild-type phenotype occurs; in the trans, the mutant phenotype occurs.

Cistron. A segment of DNA specifying one polypeptide chain in protein synthesis. Under the concept of a triplet code, one cistron must contain three times as many nucleotide pairs as amino acids in the chain it specifies.

Clone. A group of cells or organisms, derived from a single ancestral cell or individual and all genetically alike.

Code. See *genetic code.*

Codominance. The condition in heterozygotes where both members of an allelic pair contribute to phenotype, which is then a *mixture* of the phenotypic traits produced in either homozygous condition. In cattle the cross of red × white produces roan offspring whose coat consists of both red hairs and white hairs. Codominance differs from incomplete dominance.

Codon. A set of nucleotides that is specific for a particular amino acid in protein synthesis; generally agreed to consist of three nucleotides, the last of which, in the case of some amino acids, may be any of the four nucleotides.

Coincidence. The observed frequency of double crossovers, divided by their calculated or expected frequency. Expressed as a pure number; a measure of interference. In *positive* interference the coincidence is < 1; in *negative* interference the coincidence is > 1.

Col factors. Plasmids (which see) in bacteria conferring ability to produce colicins (antibiotic lipocarbohydrate-proteins).

Colinearity. Said of a genetic code (which see) in which the sequence of nucleotides corresponds to the sequence of amino acid residues in a polypeptide.

Commaless. Said of a genetic code (which see) in which successive codons (which see) are contiguous and not separated by noncoding bases or groups of bases.

Complementation. The ability of linearly adjacent segments of DNA to supplement each other in phenotypic effect. Complementary genes, when present together, interact to produce a different expression of a trait.

Conjugation. Side-by-side association of two bodies, as of synapsed chromosomes in meiosis or of two organisms during sexual reproduction.

Constitutive enzyme. A continuously produced enzyme.

cpDNA. The DNA located in chloroplasts.

Crossing-over. A process whereby genes are exchanged between nonsister chromatids of homologous chromosomes. *Unequal crossing-over* may occur, with the result that one chromatid receives a given gene twice, whereas the other chromatid lacks that gene entirely. Chiasmata (which see) are visible evidences of crossing-over.

Crossover. Said of a chromatid or gamete resulting from crossing-over.

Crossover unit. A crossover value of 1 percent between linked genes. See also *map unit.*

Cross-reacting material (CRM). A defective protein produced by a mutant gene, enzymatically inactive but antigenically similar to the wild-type protein.

C-terminus. That end of a peptide chain carrying the free alpha carboxyl group of the last amino acid in the sequence; written at the right end of the structural formula.

Cytidine. The ribonucleoside containing the pyrimidine cytosine.

Cytidylic acid. The ribonucleotide containing the pyrimidine cytosine.

Cytogenetics. Study of the cellular structures and mechanisms associated with genetics.

Cytokinesis. The division of the cytoplasm during cell division.

Cytology. The study of the structure and function of cells.

Cytosine. A pyrimidine base that occurs in DNA and RNA. Pairs with guanine in DNA.

Dalton. A unit that equals the mass of the hydrogen atom (1.67×10^{-24} g).

Deficiency. See *deletion*.

Degeneracy. Said of a genetic code (which see) in which a particular amino acid is coded for by more than one codon (which see).

Deletion. The loss of a part of a chromosome, usually involving one or more genes (rarely a portion of one gene).

Deletion loop. The loop formed by the nonsynapsing portion of an unaltered chromosome, caused by a deletion in the homologous chromosome.

Deoxyadenosine. The deoxyribonucleoside containing the purine adenine.

Deoxyadenylic acid. The deoxyribonucleotide containing the purine adenine.

Deoxycytidine. The deoxyribonucleoside containing the pyrimidine cytosine.

Deoxycytidylic acid. The deoxyribonucleotide containing the pyrimidine cytosine.

Deoxyguanosine. The deoxyribonucleoside containing the purine guanine.

Deoxyguanylic acid. The deoxyribonucleotide containing the purine guanine.

Deoxyribonucleic acid (DNA). A usually double-stranded, helically coiled, nucleic acid molecule, composed of deoxyribose-phosphate "backbones" connected by paired bases attached to the deoxyribose sugar; the genetic material of all living organisms and many viruses.

Deoxyribonucleoside. Portion of a DNA molecule composed of one deoxyribose molecule plus either a purine or a pyrimidine.

Deoxyribonucleotide. Portion of a DNA molecule composed of one deoxyribose phosphate bonded to either a purine or a pyrimidine.

Deoxyribose. The 5-carbon sugar of DNA.

Deviation. A departure from the expected or from the norm.

Dicentric. Said of a chromosome or chromatid with two centromeres.

Dihybrid. An individual heterozygous for two pairs of alleles; also said of a cross between individuals differing in two gene pairs.

Dioecious. Individuals producing either sperm or egg, but not both. In dioecious species, the sexes are separate. Compare with *monoecious*.

Diploid. An individual or cell with two complete sets of chromosomes.

Disjunction. The separation of homologous chromosomes during anaphase-I of meiosis.

DNase. Any enzyme that hydrolyzes DNA. (Older spelling *DNAase*.)

DNA ligase. An enzyme catalyzing the covalent union of segments of an interrupted sugar-phosphate strand in double-stranded DNA.

DNA polymerase. Any one of a system of enzymes that catalyze the formation of DNA from deoxyribonucleotides, using one strand of DNA as a template.

Dominance. That situation in which one member of a pair of allelic genes expresses itself in whole (complete dominance) or in part (incomplete dominance) over the other member.

Dominant. Pertaining to that member of a pair of alleles that expresses itself in heterozygotes to the complete exclusion of the other member of the pair. Also, the trait produced by a dominant gene.

Drift. See *genetic drift*.

Duplication. A chromosomal aberration in which a segment of the chromosome bearing specific loci is repeated.

Dysgenic. Any effect or situation that is or tends to be harmful to the genetics of future generations.

Effector (molecule). A substance that combines with regulatory proteins, either activating or inactivating them with respect to their ability to bind to an operator site.

Embryo sac. The female gametophyte of a flowering plant. A large, thin-walled cell within the ovule, containing the egg and several other nuclei, within which the embryo develops after fertilization of the egg.

Endoplasmic reticulum. A double membrane system in the cytoplasm, continuous with the nuclear membrane and bearing numerous ribosomes.

Endosperm. A polyploid (in many species, triploid) food-storage tissue in many angiosperm seeds formed by fusion of two (or more) female cells and a sperm.

Enzyme. Any substance, protein in whole or in part, that regulates the rate of a specific biochemical reaction in living organisms.

Episome. A closed, circular molecule of DNA that may be present in a given (bacterial) cell, either separately in the cytoplasm or integrated into the chromosome. The *F*, or fertility factor, in *Escherichia coli* is an example. See also *plasmid.*

Epistasis. The masking of the phenotypic effect of either or both members of one pair of alleles by a gene of a different pair. The masked gene is said to be hypostatic.

Equatorial plate. The figure formed at the spindle equator in nuclear division.

Equilibrium frequency. In a population, that gene frequency that varies non-directionally about a mean by an amount described by the standard deviation (which see) under conditions of unchanging selection pressure and mutation rate, with no intermixing from other populations.

Euchromatin. Chromatin (which see) that is noncondensed during interphase and condensed during nuclear division, reaching a maximum in metaphase. The banded segments of the polytene chromosomes of *Drosophila* larval salivary glands contain euchromatin.

Eukaryote (eucaryote). Any organism or cell with a structurally discrete nucleus. Contrast with *prokaryote.*

Euploidy. Variation in chromosome number by whole sets or exact multiples of the monoploid (haploid) number, e.g., diploid, triploid. Euploids above the diploid level may be referred to collectively as polyploids.

Exon. DNA sequences of a cistron that *are* transcribed into mRNA. See also *intron.*

F^- cell. A bacterial cell lacking the fertility factor (F); it acts as the receptor cell in conjugation.

F^+ cell. A bacterial cell having the fertility factor (F) as a plasmid in the cytoplasm.

F' cell. A bacterial cell, the fertility factor of which includes some chromosomal genes.

F factor. The fertility factor in the bacterium *Escherichia coli*; it is composed of DNA and must be present for a cell to function as a donor in conjugation. See also *episome.*

F_1. The first filial generation; the first generation resulting from a given cross.

F_2. The second filial generation; the generation resulting from interbreeding or selfing members of the F_1.

Fertility factor. See *F factor.*

Fitness. See *adaptive value.*

Fixation. Attainment of a gene frequency of 1.0.

Frame shift (reading frame shift). The shift in code reading that results from addition or deletion of nucleotides in any number other than 3 or multiples thereof.

Gamete. A protoplast that, in the process of sexual reproduction, fuses with another protoplast.

Gametophyte. In plants, the phase of the life cycle that reproduces sexually by gametes and characterized by having the reduced (usually monoploid or "haploid") chromosome number.

Gene. The particulate determiner of a hereditary trait; a particular segment of a DNA molecule, generally located in the chromosome. See also *cistron*, *muton*, and *recon*.

Gene frequency. The proportion of one allele of a pair or series present in the population or a sample thereof; that is, the number of loci at which a gene occurs, divided by the number of loci at which it could occur, expressed as a pure number between 0 and 1.

Gene interaction. The coordinated effect of two or more genes in producing a given phenotypic trait.

Gene pool. The total of all genes in a population.

Genetic code. The collection of base triplets of DNA and RNA carrying the genetic information by which proteins are synthesized in the cell.

Genetic drift. A change in gene frequency in a population. *Steady drift* is a directed change in frequency toward either greater or lower values; *random drift* is random fluctuation of gene frequencies caused by chance in mating patterns or to sampling errors.

Genetic equilibrium. Constancy of a particular gene frequency through successive generations.

Genetic load. The proportional reduction in average fitness, relative to an optimal genotype; the average number of lethals per individual in a population.

Genetic polymorphism. The continued occurrence in a population of two or more discontinuous genetic variants in frequencies that cannot be accounted for by recurrent mutation.

Genome. A complete set of chromosomes, or of chromosomal genes, inherited as a unit from one parent, or the entire genotype of a cell or individual.

Genotype (adj., **genotypic**). The genetic makeup or constitution of an individual, with reference to the traits under consideration, usually expressed by a symbol, e.g., "$+$," "D," "Dd," "str." Individuals of the same genotype breed alike. See *phenotype*.

Guanine. A purine base occurring in DNA and RNA. Pairs normally with cytosine in DNA.

Gynandromorph; gynander. An individual part of whose body exhibits male sex characters and part female characters.

Haplo-. A prefix before a chromosome number denoting an individual whose somatic cells lack one member of that chromosome pair.

Haploid. An individual or cell with a single complete set of chromosomes. Synonym, *monoploid*.

Hardy-Weinberg equilibrium. Occurrence in a population of three genotypes, two homozygous and one heterozygous (e.g., AA, Aa, and aa), where the frequencies of the three genotypes are p^2 for homozygous dominants, $2pq$ for heterozygotes, and q^2 for homozygous recessives and p = frequency of A and q = frequency of a. At these frequencies there is no change in genotypic frequencies as long as isolation mechanisms, selection mutation frequencies, and drift are constant.

HeLa cells. Cells originally (1952) from a carcinoma of the cervix, maintained as a standard human cell line for tissue and genetic experimentation. "HeLa" is derived from the pseudonym (Helen Lane) of the original patient source (Henrietta Lacks).

Hemizygous. An individual having but one of a given gene; or a gene present only once. Designates either a monoploid organism, a sex-linked gene in XY males (or an XY individual with regard to a particular X-linked gene), or an individual heterozygous for a given chromosomal deletion.

Hemoglobin. An iron-protein pigment of blood functioning in oxygen-carbon dioxide exchange of living cells.

Hemophilia. A metabolic disorder characterized by free bleeding from even slight wounds because of the lack of formation of clotting substances. It is associated with a sex-linked recessive gene.

Hermaphrodite. An individual with both male and female reproductive organs.

Heterochromatin. Condensed chromatin (which see). *Constitutive heterochromatin* occurs in the centromeric region of chromosomes and in various intercalary positions characteristic of a given species; it includes certain important, genetically active regions. *Facultative heterochromatin* is that chromatin that makes up the genetically inactive whole chromosomes (e.g., the X chromosome in human females) or parts of certain other chromosomes.

Heterogametic sex. That sex with either only one or two different sex chromosomes, as XO, XY, or ZW. The heterogametic sex thus produces two kinds of gametes with respect to sex chromosomes, e.g., X and Y sperms in human beings.

Heterogeneous nuclear RNA (hnRNA). Various unprocessed RNA molecules located in the nucleus.

Heterospory. In higher plants the production of two kinds of meiocytes (megasporocytes and microsporocytes), which gives rise to two different kinds of meiospores. Compare with *homospory*.

Heterozygote (adj., **heterozygous**). An individual whose chromosomes bear unlike genes of a given allelic pair or series. Heterozygotes produce more than one kind of gamete with respect to a particular locus.

Histone. Any of several basic, low molecular weight proteins that can complex with DNA.

HLA. Human leukocyte antigens; antigens located on the surfaces of the leukocytes and which affect tissue compatibility in organ transplants and skin grafting. These antigens occur in quite a large number of types and are produced by at least four closely linked loci (in the sequence $D\,B\,C\,A$) on autosome 6. Collectively, they make up the *major histocompatibility complex* (MHC).

Holandric gene. A gene located only on the Y chromosome in XY species.

Homogametic sex. That sex possessing two identical sex chromosomes (XX or ZZ). Gametes produced by the homogametic sex are all alike with respect to sex chromosome constitution.

Homolog. See *homologous chromosomes*.

Homologous chromosomes. Chromosomes that occur in pairs (in diploids), one derived from each of two parents; Normally (except for chromosomes associated with sex), these chromosomes are morphologically alike and bear the same gene loci. Each member of such a pair is the *homolog* of the other.

Homospory. In plants, the production of but one kind of meiocyte, which gives rise to meiospores morphologically indistinguishable from each other.

Homozygote (adj., **homozygous**). An individual whose chromosomes bear identical genes of a given allelic pair or series. Homozygotes produce only one kind of gamete with respect to a particular locus and therefore "breed true."

Hybrid. An individual that results from a cross between two genetically unlike parents.

Hypha (pl. **hyphae**). One of the filaments of a mycelium in fungi.

Imperfect flower. One lacking either stamens or pistil.

Incomplete dominance. The condition in heterozygotes where the phenotype is intermediate between the two homozygotes. In some plants the cross of red × white produces pink-flowered progeny. Incomplete dominance differs from codominance.

Incompletely sex-linked genes. Genes located on homologous portions of the X and Y (or Z and W) chromosomes.

Independent segregation. The random or independent behavior of genes on different pairs of chromosomes.

Inducer. See *effector*.

Inducible enzyme. An enzyme synthesized only in the presence of an effector (which see).

Interference. The increase (*negative interference*) or decrease (*positive interference*) in likelihood of a second crossover closely adjacent to another. In most organisms interference increases with decreased distance between crossovers. See *coincidence*.

Interphase. The stage of cell life during which that cell is not dividing.

Intersex. An individual showing secondary sex characters intermediate between male and female or some of each sex.

Intron (intervening sequence). A sequence of nucleotides in DNA that does not appear in mRNA; probably excised from hnRNA in its processing. Its function is unknown; polypeptide synthesis proceeds unimpeded in vitro without the introns.

Inversion. Reversal of the order of a block of genes in a given chromosome. PQ*UTSR*VWX would represent an inversion of genes *RSTU* if the normal order is alphabetical.

Inversion loop. The loop configuration in a pair of synapsed, homologous chromosomes, caused by an inversion in one of them.

In vitro. Experimentally induced biological processes outside the organism (literally, "in glass").

In vivo. Experimentally induced biological processes within the organism.

Isogametes. Gametes not sexually differentiated. Fusing isogametes are physically similar but in some species are chemically differentiated.

Karyokinesis. The division of the nucleus during cell division.

Karyolymph. A clear fluid material within the nuclear membrane; "nuclear sap."

Karyotype. The somatic chromosome complement of an individual, usually as defined at mitotic metaphase by morphology (including centromere location and often by special staining techniques, as fluorescence banding) and number, arranged in a sequence that is standard for that organism.

Kilobase (abbr., **Kb.**). One thousand nucleotides or nucleotide pairs in sequence. May be used to pertain to either DNA or RNA.

Kinetochore. The attachment region (within the centromere) of the microtubules of the spindle in cells undergoing mitosis or meiosis.

Lampbrush chromosome. A chromosome that has paired loops extending laterally, and occurs in primary oocyte nuclei; they represent sites of active RNA synthesis.

Lethal gene. A gene whose phenotypic effect is sufficiently drastic to kill the bearer. Death from different lethal genes may occur at any time from fertilization of the egg to advanced age. Lethal genes may be dominant, incompletely dominant, or recessive.

Life cycle. The entire series of developmental stages undergone by an individual from zygote to maturity and death.

Linkage. The occurrence of different genes on the same chromosome. They show nonrandom assortment at meiosis. See also *synteny*.

Linkage group. All of the genes located physically on a given chromosome.

Linkage map. A scale representation of a chromosome that shows the relative positions of all its known genes.

Locus (pl., **loci**). The position or place on a chromosome occupied by a particular gene or one of its alleles.

Lysis (n.) (v.i. or v.t., **lyse**). Disintegration or dissolution; usually, the destruction of a bacterial host cell by infecting phage particles.

Lysogenic bacteria. Living bacterial cells that harbor temperate phages (viruses).

Major histocompatibility complex. See *HLA*.

Map unit. A distance on a linkage map represented by 1 percent of cross-overs (recombinants), that is, by a recombination frequency of 1 percent.

Mean. The arithmetic average; the sum of all values for a group, divided by the number of individuals. Symbolized by $\bar{x}$ (sample) or by μ (population).

Median. The middle value of a series of readings arranged serially according to magnitude.

Megasporocyte. In plants, the meiocyte that is destined to produce megaspores by meiosis. Synonym, *megaspore mother cell.*

Meiocyte. Any cell that undergoes meiosis.

Meiosis. Nuclear divisions in which the diploid or somatic chromosome number is reduced by half. In the first of the two meiotic divisions, homologous chromosomes first replicate, then pair (synapse), and finally separate to different daughter nuclei, which thus have half as many chromosomes as the parent nucleus, and one of each kind instead of two. A second division, in which chromosomal replicates separate into daughter nuclei, follows so that meiosis produces four monoploid daughter nuclei from one diploid parent nucleus.

Meiospore. In plants, one of the asexual reproductive cells produced by meiosis from a meiocyte.

Meristem. An undifferentiated cellular region in plants characterized by repeated cell division.

Merozygote. A partially diploid receptor bacterial cell that results from conjugation.

Messenger RNA. Ribonucleic acid conferring amino acid specificity on ribosomes; complementary to a given DNA cistron.

Metacentric. A chromosome with a centrally located centromere.

Metafemale (also **superfemale**). Abnormal females in *Drosophila*, usually sterile and weak, with an overbalance of X chromosomes with respect to autosomes; X/A ratio greater than 1.0.

Metamale (also **supermale**). Abnormal males in *Drosophila* with an overabundance of autosomes to X chromosomes; X/A ratio less than 0.5.

Metaphase. That stage of nuclear division in which the chromosomes are located in the equatorial plane of the spindle prior to centromere separation.

MHC. Major histocompatibility complex. See *HLA*.

Micrometer (μm). A commonly employed unit of measurement in microscopy, it equals 1×10^{-6} meter or 1×10^{-3} millimeter. Synonym, *micron* (μ) in older usage.

Micron. See *micrometer.*

Microspore. In plants, the meiospore that gives rise to the male gametophyte and sperms.

Microsporocyte. In plants, a cell that is destined to produce microspores by meiosis. Synonyms, *microspore mother cell* and *"pollen mother cell."*

Microtubule. A hollow tubular cytoplasmic component that forms the spindle of the mitotic apparatus (which see), and is found especially in motile cells. It has an outside diameter about 15 to 30 nm.

Millimeter (mm). A unit of distance measurement equal to 1×10^{-3} meter.

Missense. A mutation by which a particular codon is changed to incorporate a different amino acid, which often results in an inactive protein. A *missense codon* results when one or more bases of a *sense codon* (which see) are changed so that a different amino acid is coded for.

Mitochondrion (pl., **mitochondria**). Small cytoplasmic organelle where cellular respiration occurs.

Mitosis. Nuclear division in which a replication of chromosomes is followed by separation of the products of replication and their incorporation into two daughter nuclei. Daughter nuclei are normally identical with each other and with the original parent nucleus in both kind and number of chromosomes. Mitosis may or may not involve cytoplasmic division.

Mitotic apparatus. The collection of cytoplasmic structures that are present in mitotic cells and consists of the asters (if present) surrounding each centrosome and the spindle of microtubules.

Mode. The numerically largest class or group in a series of measurements or values.

Monoecius. Individuals producing both sperm and egg.

Monohybrid. The offspring of two homozygous parents that differ in only one gene locus or in which only one such locus is under consideration.

Monohybrid cross. A cross between two parents that differ in only one heritable character or in which only one such character is under consideration.

Monoploid. An individual with a single complete set of chromosomes. Also the fundamental number of chromosomes comprising a single set. Synonym, *haploid.*

Monosomic. An individual lacking one chromosome of a set $(2n - 1)$.

Morgan unit (morgan). A unit of relative distance between genes on a given chromosome; it equals a cross-over value of 100 percent.

Morphology. The study of form and structure in organisms.

Mosaic. An individual part of whose body is composed of tissue genetically different from another part.

mtDNA. DNA that occurs as a normal component within mitochondria.

Multiple alleles. A series of three or more alternative alleles, any one of which may occur at a particular locus on a chromosome.

Multiple genes. See *polygenes.*

Mutagen. Any agent that brings about a mutation.

Mutation. A sudden change in genotype having no relation to the individual's ancestry. Used for changes in a single gene itself ("point mutations") and for chromosomal aberrations.

Muton. The smallest segment of DNA or subunit of a cistron that can be changed and thereby bring about a mutation; can be as small as one nucleotide pair.

Mycelium. The threadlike filamentous vegetative body of many fungi.

Nanometer (nm). A unit of distance equal to 1×10^{-9} meter. Synonym, *millimicron* (mμ).

Negative control. Repression of an operator site (which see) by a regulatory protein that is produced by a regulator site (which see).

Nondisjunction. The failure of homologous chromosomes to separate at anaphase-I of meiosis. *Primary nondisjunction* may occur in an XX female, and lead to production of XX or O eggs (in addition to the normal X eggs); or it may occur (first division) in an XY male, and result in XY and O sperms or (second division) in XX and O or YY and O sperm (in addition to normal Y or X sperms). *Secondary nondisjunction* may occur in an XXY female, giving rise to eggs with XX, XY, X, or Y chromosomal combinations.

Nonsense. A codon that does not specify any amino acid in the genetic code.

Normal curve. A smooth, symmetrical, bell-shaped curve of distribution.

N-terminus. The amino ($—NH_2$) end of a peptide chain, by convention written as the left end of the structural formula.

Nucleolus. Deeply staining body containing rRNA; multiple copies of DNA that code for rRNA, and proteins occur in the nucleus.

Nucleoside. Portion of a DNA or RNA molecule composed of one deoxyribose molecule (in DNA), or ribose (in RNA), plus a purine or a pyrimidine.

Nucleosome. Short, disk-shaped cylinders, composed of nucleoproteins and spaced at roughly 100 Å intervals on chromosomes. Histones (which see) H2A, H2B, H3, and H4 are complexed with DNA in the nucleosomes.

Ocher. The nonsense codon UAA.

Okazaki fragment. Segments of approximately 1,000 to 2,000 nucleotides of the daughter strand of replicating DNA. These are joined in succession by the enzyme DNA ligase to form a normal double-stranded DNA molecule.

Oligonucleotide. A linear sequence of a few (generally not over 10) nucleotides.

Ontogeny. The complete development of the individual from zygote, spore, and so on, to adult form.

Oocyte. The diploid cell that will undergo meiosis to form an egg.

Oogenesis. Egg formation.

Opal. The nonsense codon UGA.

Operator site. A segment of DNA in an operon that affects activity or non-activity of associated cistrons; it may be combined with a repressor and thereby "turn off" the associated cistrons.

Operon. A system of cistrons, operator and promoter sites, by which a given genetically controlled, metabolic activity is regulated.

Organelle. A specialized cytoplasmic structure with a particular function (e.g., chloroplast, mitochondrion).

Ovary. The female gonad in animals or the ovule-containing portion of the pistil of a flower.

Ovule. Structure within the ovary of the pistil of a flower that becomes a seed. It represents a megasporangium, together with some overgrowing tissue (integuments), and ultimately contains the female gametophyte (embryo sac), one of whose nuclei is an egg.

P. The parental generation in a given cross.

Paracentric. Refers to an inversion that does not involve the centromere but lies entirely within one arm of the chromosome.

Parameter. Actual value of some quantitative character for a population. Compare *statistic*.

Parthenogenesis. Development of a new individual from an unfertilized egg.

Pedigree. The ancestral history of an individual; a chart showing such history.

Peptide bond. A chemical bond (CONH) linking amino acid residues together in a protein.

Perfect flower. One with both stamen(s) and pistil(s).

Pericentric. Refers to an inversion that does include the centromere, hence involves both arms of a chromosome.

Petite. A slow-growing strain of yeast (*Saccharomyces*) that lacks certain respiratory enzymes and forms unusually small colonies on agar. *Segregational petites* bear mutant nuclear gene(s), whereas *neutral petites* (sometimes called *vegetative petites*) bear mutant mitochondrial DNA.

Phage. A virus that infects bacteria.

Phenotype (adj., **phenotypic**). The appearance or discernible character of an individual, which is dependent on its genetic makeup, usually expressed in words, e.g., "tall," "dwarf," "wild type," "prolineless." Identical phenotypes may not necessarily breed alike.

Phosphodiester bond. A bond between the sugar and a phosphate group in the sugar-phosphate strand of DNA.

Phylogeny. The evolutionary development of a species or other taxonomic group.

Pilus (pl., **pili**). One of the filamentous projections from the surface of a bacterial cell. One kind of pilus serves as the conjugation tube in bacteria.

Pistil. The entire part of the flower that produces megaspores (which, in turn, produce eggs). Sometimes, and *incorrectly*, referred to as a female floral part.

Plaque. Clear area on culture plate of bacteria where these have been killed by phages.

Plasmagene. A self-replicating, cytoplasmically located gene.

Plasmid. Originally, a closed, circular DNA molecule of restricted size (a few tens of thousands of nucleotide pairs), existing only in the cytoplasm (of bacteria); incapable of integration into the bacterial "chromosome." See also *episome*; *plasmid* is often used to include both episomes and plasmids.

Pleiotropy. The influencing of more than one trait by a single gene.

Polar body. One of the three very small cells produced during meiosis of an oocyte, which contains a monoploid nucleus but little cytoplasm. It is nonfunctional in reproduction.

Pollen. The young male gametophyte of a flowering plant, surrounded by the microspore wall.

Poly-A tail. The initially long sequence of adenine nucleotides at the 3' end of mRNA; added after transcription.

Polygenes. Two or more different pairs of alleles, with a presumed cumulative effect that governs such quantitative traits as size, pigmentation, intelligence, among others. Those contributing to the trait are termed contributing (effective) alleles; those appearing not to do so are referred to as noncontributing or noneffective alleles.

Polymer. A chemical compound composed of two or more units of the same compound.

Polynucleotide. A linear sequence of many nucleotides.

Polypeptide. A compound containing amino acid residues joined by peptide bonds. A protein may consist of one or more specific polypeptide chains.

Polyploid. An individual having more than two complete sets of chromosomes, e.g., triploid ($3n$), tetraploid ($4n$).

Polyribosome. See *polysome*.

Polysome. A group of ribosomes joined by a molecule of messenger RNA.

Polytene chromosome. Many-stranded giant chromosomes produced by repeated replication during synapsis in certain dipteran larval tissues. Synonym, *giant chromosome*.

Population. An infinite group of individuals, measured for some variable, quantitative character, from which a sample is taken.

Position effect. A phenotypic effect dependent on a change in position on the chromosome of a gene or group of genes.

Positive control. Activation of an operator site (which see) by a regulatory protein that is produced by a regulator site (which see).

Probability. The likelihood of occurrence of a given event. Usually expressed as a number between 0 (complete certainty that the event will *not* occur) and 1 (complete certainty that the event *will* occur).

Progeny. Offspring individuals.

Prokaryote. A cell or organism lacking a discrete nuclear body (also *procaryote*).

Promoter site. Site on DNA at which RNA polymerase (which see) attaches in bacterial operons (which see), such as the *lac* operon.

Prophage. Phage nucleoprotein (DNA or RNA, depending on the phage) integrated into the bacterial DNA.

Prophase. The first stage of nuclear division, including all events up to (but not including) arrival of the chromosomes at the equator of the spindle.

Protoplast. A structural unit of protoplasm; all the living (protoplasmic) material of a cell. The two principal parts are nucleus and cytoplasm.

Prototroph. An individual able to carry on a given synthesis; a wild-type individual able to grow on minimal medium.

Pseudoalleles. Nonalleles so closely linked that they are often inherited as one gene, but shown to be separable by crossover studies.

Pseudodominance. The expression (apparent dominance) of a recessive gene at a locus opposite a deletion.

Punnett square. A "checkerboard" grid designed to determine all possible genotypes produced by a given cross. Genotypes of the gametes of one sex are entered across the top, those of the other down one side. Zygote genotypes produced by each possible mating are then entered in the appropriate squares of the grid.

Pure line (pure breeding line). A strain of individuals homozygous for all genes being considered.

Purine. Nitrogenous base occurring in DNA and RNA; these are adenine and guanine.

Pyrimidine. Nitrogenous base occurring in DNA (thymine and cytosine) or RNA (uracil and cytosine).

r. That portion of the R plasmid carrying genes for all drug resistance, with the probable exception of tetracycline resistance. See also *R factors* and *RTF*.

R factors. Bacterial *plasmids* (which see) carrying genes for drug resistance. See also *r* and *RTF*.

Rad. Term meaning radiation-absorbed dose; it is the amount of ionizing radiation that liberates 100 ergs of energy in one gram of matter.

Reading frame. The codon sequence resulting from mRNA "reading" the DNA nucleotides in groups of three.

Recessive. An adjective applied to the member of a pair of genes that fails to express itself in the presence of its dominant allele. The term is also applicable to the trait produced by a recessive gene. Recessive genes express themselves ordinarily only in the homozygous state.

Reciprocal cross. A second cross of the same genotypes in which the sexes of the parental generation are reversed. The cross *AA* (♀) × *aa* (♂) is the reciprocal of the cross *aa* (♀) × *AA* (♂).

Reciprocal translocation. The exchange of segments between two nonhomologous chromosomes.

Recombinant. An individual derived from a crossover gamete.

Recombinant DNA. A DNA molecule (in practice generally a bacterial plasmid) which has been enzymatically cut and DNA of another individual of the same or a different species inserted in the space so produced, then reannealed to the closed, circular form. Many recombinant DNA molecules, therefore, can be so constructed that an individual (e.g., a bacterium) has a genetic capability not previously associated with that organism.

Recombination. The new association of genes in a recombinant individual; this association arises from independent assortment of unlinked genes, from crossing-over between linked genes, or from intracistronic crossing-over.

Recon. The smallest segment of DNA or subunit of a cistron that is capable of recombination; may be as small as one deoxyribonucleotide pair.

Reduction division. See *meiosis*.

Regulator site. The specific segment of DNA responsible for production of a regulatory protein that may serve as a repressor in some cases or as an activator in others. See also *negative control* and *positive control*.

Relational coiling. The loose coiling of chromatids about each other.

Replicate. To form replicas from a model or template; applies to synthesis of new DNA from preexisting DNA as part of nuclear division.

Replicon. A sequentially replicating segment of a nucleic acid, controlled by a subsegment known as a replicator. A single replicator is present in the bacterial "chromosome," whereas the chromosomes of eukaryotes bear large numbers of replicons in series.

Repressor. A protein produced by a regulator gene that can combine with and repress action of an associated operator gene.

Resistance transfer factor. See *RTF.*

Restriction endonuclease. Any of a group of enzymes that break internal bonds of DNA at highly specific points.

Reverse transcriptase. RNA-dependent DNA polymerase (which see).

Ribonucleic acid (RNA). A single-stranded nucleic acid molecule, synthesized principally in the nucleus from deoxyribonucleic acid, composed of a ribose-phosphate backbone with purines (adenine and guanine) and pyrimidines (uracil and cytosine) attached to the sugar ribose. RNA is of several kinds and functions to carry the "genetic message" from nuclear DNA to the ribosomes.

Ribonucleoside. Portion of an RNA molecule composed of one ribose molecule plus either a purine or a pyrimidine.

Ribonucleotide. Portion of an RNA molecule composed of one ribosephosphate unit plus a purine or a pyrimidine.

Ribose. The 5-carbon sugar of ribonucleic acid.

Ribosomal RNA. That ribonucleic acid incorporated into ribosomes; it is nonspecific for amino acids.

Ribosome. Cytoplasmic structure, usually adherent to the endoplasmic reticulum, which is the site of protein synthesis. In bacteria the ribosomes are free in the cytoplasm. See also *polysome.*

RNase. An enzyme that hydrolyzes RNA.

RNA-dependent DNA polymerase. A group of enzymes that catalyze formation of DNA molecules from RNA templates. These occur in some viruses (e.g., those that produce tumors).

RNA polymerase. An enzyme that catalyzes the formation of RNA from ribonucleotide triphosphates, using single-stranded DNA as a template.

Roentgen (R). The amount of ionizing radiation that produces about two ion pairs per cubic micrometer of matter, or about 1.6×10^{12} ion pairs per cubic centimeter.

RTF (resistance transfer factor). That part of the R plasmid (which see) responsible for transfer of the R factor during bacterial conjugation. It appears to include the cistron for tetracycline resistance, whereas other drug resistance factors occur in the r portion (which see) of the R factor.

Sedimentation coefficient. The rate of sedimentation of a solute in an appropriate solvent under centrifugation. An s value of 1×10^{-13} second is one *Svedberg unit.*

Selection coefficient. The measure of reduced fitness of a given genotype; it equals one minus the adaptive value (which see) of that genotype.

Self-fertilization. Functioning of a single individual as both male and female parent. Plants are "selfed" if sperm and egg are supplied by the same individual.

Semiconservative. Replication of DNA in which the two sugar-phosphate "backbones" become separated; each is conserved as one of the two strands of two new DNA molecules.

Sense codon. A codon (which see) specifying a particular amino acid in protein synthesis.

Sex chromosomes. Heteromorphic chromosomes that do not occur in identical pairs in both sexes in diploid organisms; in humans and fruit flies these are designated as X and Y chromosomes, respectively; in fowl, as the Z and W chromosomes.

Sexduction.　Incorporation of bacterial chromosomal genes in the fertility plasmid, with subsequent transfer to a recipient cell in conjugation.

Sex-influenced trait.　One in which dominance of an allele depends on sex of the bearer; e.g., pattern baldness in humans is dominant in males, recessive in females.

Sex-limited trait.　One expressed in only one of the sexes; e.g., cock feathering in fowl is limited to normal males.

Sex-linked gene.　A gene located only on the X chromosome in XY species (or on the Z chromosomes in ZW species).

Siblings (also **sibs**).　Individuals with the same maternal and paternal parents; brother-sister relationship.

Significance.　In statistical treatments, probability values of >0.05 are termed *not significant*, those ≤ 0.05 but >0.01 are *significant*, those ≤ 0.01 but >0.001 are *highly significant*, and those ≤ 0.001 are *very highly significant*. Significant probability values indicate that the results, although possible, deviate too greatly from the expectancy to be acceptable when chance alone is operating.

Soma (adj., **somatic**).　The body, cells of which in mammals and flowering plants normally have two sets of chromosomes, one derived from each parent.

Spermatid.　The monoploid cells, resulting from meiosis of a primary spermatocyte that will mature into sperms.

Spermatocyte.　The cell that undergoes meiosis to produce four spermatids.

Spermatogenesis.　Development of sperms.

Spore.　An asexual reproductive protoplast capable of developing into a new individual. See also *meiospore*.

Sporocyte.　In plants, a meiocyte.

Sporogenesis.　Formation of spores.

Sporophyte.　In plants, the phase of the life cycle that reproduces asexually by meiospores and is characterized by having the double (usually diploid) chromosome number.

Stamen.　That part of a flower that produces microspores (which, in turn, produce sperms). Sometimes, and *incorrectly*, referred to as a male floral part.

Standard deviation.　A measure of the variation in a sample. Symbolized by s.

Standard error of difference in means.　Measure of the significance of the difference in two sample means. Symbolized by S_d.

Standard error of sample mean.　An estimate of the standard deviation of a series of hypothetical sample means that serves as a measure of the closeness with which a given sample mean approximates the population mean. Symbolized by $s_{\bar{x}}$.

Statistic.　Actual value of some quantitative character for a sample from which estimates of parameters may be made.

Stigma.　The pollen-receptive portion of the pistil of a flower.

Structural gene.　See *cistron*.

Svedberg unit.　See *sedimentation coefficient*.

Synapsis (v.i., **synapse**).　The pairing of homologous chromosomes that occur in prophase-I of meiosis.

Syngamy.　The union of the nuclei of sex cells (gametes) in reproduction.

Synteny.　The occurrence of two or more genetic loci on the same chromosome. Depending on intergene distance(s) involved, they may or may not exhibit nonrandom assortment at meiosis.

Tautomer.　An alternate molecular form of a compound, characterized by a different arrangement of its electrons and protons as compared with the common form of the molecule.

Taxon (pl., **taxa**).　A taxonomic group of any rank.

Taxonomy. The study of describing, naming, and classifying living organisms, and the bases on which resultant classification systems rest.

Telocentric. A chromosome with a terminal centromere.

Telophase. The concluding stage of nuclear division, characterized by the reorganization of interphase nuclei.

Temperate phage. A phage (bacterium-infecting virus) that invades and multiplies in but does not ordinarily lyse its host.

Template. A model, mold, or pattern; DNA acts as a template for RNA synthesis.

Testcross. The cross of an individual (generally of dominant phenotype) with one having the recessive phenotype. Generally used to determine whether an individual of dominant phenotype is homozygous or heterozygous, or to determine the degree of linkage.

Tetrad. The four monoploid (haploid) cells arising from meiosis of a megasporocyte or microsporocyte in plants; also, a group of four associated chromatids during synapsis.

Tetraploid. A polyploid cell, tissue, or organism with four sets of chromosomes ($4n$). See also *polyploid*, *allopolyploid*, and *autopolyploid*.

Three-point cross. A trihybrid testcross (e.g., *ABC/abc* $\times$ *abc/abc*, etc.) used primarily in chromosome mapping.

Thymidine. The deoxyribonucleoside that contains the pyrimidine thymine.

Thymidylic acid. The deoxyribonucleotide that contains the pyrimidine thymine.

Thymine. A pyrimidine base occurring in DNA. Pairs normally with adenine.

Totipotency. The property of a cell (or cells) whereby it develops into a complete and differentiated organism.

Trans arrangement. Linkage of the dominant allele of one pair and the recessive of another on the same chromosome.

Transcription. Synthesis of messenger RNA from a DNA template.

Transduction. Recombination in bacteria whereby DNA is transferred by a phage from one cell to another. *Generalized* transduction involves phages that have incorporated a segment of bacterial chromosome during packaging of the phage; *specialized* transduction involves temperate phages that are always inserted into the bacterial chromosome at a site specific for that phage.

Transfer RNA. Amino acid-specific RNA that transfers activated amino acids to mRNA, where polypeptide synthesis takes place.

Transformation. Genetic recombination, particularly in bacteria, whereby naked DNA from one individual becomes incorporated into that of another.

Transgressive variation. Appearance in progeny of a more extreme expression of a trait than occurs in the parents. Assumed to result from cumulative action of polygenes, but careful testing of variation in parental lines is necessary for verification.

Transition. The substitution in DNA or RNA of one purine for another or of one pyrimidine for another.

Translation. The process by which a particular messenger RNA nucleotide sequence is responsible for a specific amino acid residue sequence of a polypeptide chain.

Translocation. The shift of a portion of a chromosome to another part of the same chromosome or to an entirely different chromosome. (See also *reciprocal translocation.*)

Transposon (Transposable element). A segment of DNA, which generally consists of more than 2,000 nucleotides, and is capable of moving into and out of a chromosome or plasmid, and/or from place to place within the chromosome in both prokaryotes and eukaryotes. It is capable of turning genes "on" or "off."

Transversion. The substitution in DNA or RNA of a purine for a pyrimidine or vice versa.

Trihybrid. An individual heterozygous for three pairs of genes.

Triplet. A group of three successive nucleotides in RNA (or DNA) that, in the genetic code (which see), specifies a particular amino acid in the synthesis of polypeptide chains.

Triplo-. Prefix denoting a trisomic individual where the identity of the extra chromosome is known, e.g., triplo-IV *Drosophila*, or triplo-21 human beings.

Triploid. A polyploid cell, tissue, or organism with three sets of chromosomes $(3n)$.

Trisomic. An individual with one extra chromosome of a set $(2n + 1)$.

Universal donor. A person with group O blood, whose erythrocytes therefore bear neither A nor B antigens, and whose blood can be donated to members of groups O, A, B, and AB if necessary.

Universal recipient. A person of blood group AB who can receive blood from members of groups AB, A, B, or O if necessary.

Uracil. A pyrimidine base occurring in RNA.

Uridine. The ribonucleoside that contains the pyrimidine uracil.

Uridylic acid. The ribonucleotide that contains the pyrimidine uracil.

Variance. A statistic providing an unbiased estimate of population variability; it is the square of the standard deviation (which see).

Virulence. The ability to produce disease.

Virulent phage. A phage (virus) that destroys (lyses) its host bacterial cell.

Wild type. The most frequently encountered phenotype in natural breeding populations; the "normal" phenotype.

Wobble hypothesis. The partial or total lack of specificity in the third base of some triplet codons (which see) whereby two, three, or four codons differing only in the third base may code for the same amino acid. See Table 18-1.

X-linkage (sex linkage). Refers to genes located on the X-chromosome.

Y-linkage (holandric genes). Refers to genes located on the Y chromosome.

Zygote. The protoplast resulting from the fusion of two gametes in sexual reproduction; a fertilized egg.

*INDEX**

*f, Figure; n, footnote; t, table.